Statistics for Social Science and Public Policy

Advisors:
S.E. Fienberg   D. Lievesley   J. Rolph

Springer

New York
Berlin
Heidelberg
Barcelona
Hong Kong
London
Milan
Paris
Singapore
Tokyo

Statistics for Social Science and Public Policy

Michael O. Finkelstein
Bruce Levin

Statistics for Lawyers

Second Edition

With 55 Figures

 Springer

Michael O. Finkelstein
299 Park Avenue
New York, NY 10171
USA

Bruce Levin
School of Public Health
The Joseph L. Mailman
   Department of Biostatistics
Columbia University
New York, NY 10027
USA

Advisors:

Stephen E. Fienberg
Department of Statistics
Carnegie Mellon University
Pittsburgh, PA 15213
USA

Denise Lievesley
Institute for Statistics
Room H. 113
UNESCO
7 Place de Fontenoy
75352 Paris 07 SP
France

John Rolph
Department of Information and
   Operations Management
Graduate School of Business
University of Southern California
Los Angeles, CA 90089
USA

Library of Congress Cataloging-in-Publication Data
Finkelstein, Michael O.
     Statistics for lawyers / Michael O. Finkelstein, Bruce Levin. — 2nd ed.
        p.   cm. — (Statistics for social science and public policy)
     Includes bibliographical references and index.
     ISBN 0-387-95007-9 (alk. paper)
     1. Statistics.   2. Law—Statistical methods.   I. Levin, Bruce A.   II. Title.   III. Series.
QA276.12.F56   2001
001.4´22—dc21   00-059479

Printed on acid-free paper.

Production managed by Lesley Poliner; manufacturing supervised by Jeffrey Taub.
Photocomposed copy prepared using LaTeX by the Bartlett Press, Marietta, GA.
Printed and bound by Maple-Vail Book Manufacturing Group, York, PA.
Printed in the United States of America.

9 8 7 6 5 4 3 2 1

ISBN 0-387-95007-9   SPIN 10557619

Springer-Verlag   New York Berlin Heidelberg
A member of BertelsmannSpringer Science+Business Media GmbH

In Memory of
Herbert Robbins
(1915 – 2001)

DISSERTATIO INAUGURALIS
MATHEMATICO JURIDICA

DE

USU ARTIS
CONJECTANDI
IN JURE,

Quam

DIVINA JUVANTE GRATIA

Auctoritate & Juſſu

*Magnifici & Ampliſſimi JCtorum Ordinis
in Academia Patria*

pro

GRADU DOCTORATUS

In Utroque Jure legitime conſequendo

Ad Diem 14 Junii A. C. M DCC IX
L. H. Q. S.

Publice defendet

M. NICOLAUS BERNOULLI,
Baſilienſis.

BASILEÆ,

Typis JOHANNIS CONRADI à MECHEL.

The title page of Nicolaus Bernoulli's dissertation *The Use of the Art of Conjecturing in Law*, dated 1709. Nicolaus's thesis appears to be the first extended work applying statistical methods to legal questions.

Preface to the Second Edition

In the decade that has passed since the first edition of this book appeared, the crest of the wave of interest in statistical evidence has broadened and moved beyond its origins in civil rights law. Significant new developments, reflected in this edition, include, for example, DNA evidence (Sections 2.1.1, 3.1.2, and 3.2.2), epidemiologic studies in toxic substance litigation (Chapter 10), statistical models for adjusting census counts (Section 9.2.1), and vote-dilution cases (Section 13.2.3). It is emblematic of the importance of statistics in the pantheon of scientific evidence that the leading Supreme Court precedent on such evidence—the *Daubert*[1] case—involved toxic substance claims in which epidemiologic studies played a key role. In *Daubert*, the Court rejected the old *Frye* test of general acceptance in the scientific community as the basis for admissibility, and explicitly imposed on federal trial judges a gatekeeping function: they must now assess whether the proffered evidence is both relevant and *reliable*. The new formulation puts judges in the awkward position not only of counting scientific noses, but also of understanding and appraising the scientific basis of what an expert proposes to say, or calculate, on the witness stand. Fortuitously, about a year after *Daubert*, in 1994, the Federal Judicial Center issued and distributed to all federal judges a *Reference Manual on Scientific Evidence*, which is largely a primer on the applications of statistical methods. A new edition of the *Manual*, which unexpectedly turned out to be a best seller, is due to appear this year. Those who find this book heavy going may wish to consult the *Manual* as a useful introduction to at least some subjects.

But new, case-driven applications of statistics are only part of the development. Perhaps even more important, in the long run, is the continuing flow

---

[1] *Daubert v. Merrill Dow Pharmaceuticals*, 509 U.S. 579 (1993).

of statistical studies of the legal system itself. Studies of this sort can offer insights that sometimes challenge commonly held views of venerable legal institutions. Section 5.6.3 gives an example of such a study, involving peremptory challenges of prospective jurors, in which the authors analyze data and find that most peremptory challenges are "guesses." For another example, as this is being written, the media are prominently reporting a large-scale statistical study of the death penalty, undertaken at Columbia Law School, which paints a startling picture of the high rate of serious errors in criminal trials leading to death sentences. The study will almost certainly influence pending legislation and promises to provide important data in the debate over capital punishment itself. One must note that in both these studies it is the statistical pattern, emerging from the details of individual cases, that tells the most compelling story.

As in the first edition, much of the material for the new portions of this second edition was collected from lawyers, statisticians, or economists who were involved in the cases. We thank them all for their generosity in assisting us. In this connection, we would like to acknowledge in particular Orley Ashenfelter, David Baldus, William Fairley, David Freedman, and Sol Schreiber for their help in furnishing us with their materials and consulting with us on matters of interpretation.

Michael O. Finkelstein
Bruce Levin
New York, New York
June 2000

Preface to the First Edition

For the rational study of the law the black letter man may be the man of the present, but the man of the future is the man of statistics and the master of economics.

—Oliver Wendell Holmes
The Path of the Law (1897)

The aim of this book is to introduce lawyers and prospective lawyers to methods of statistical analysis used in legal disputes. The vehicle of this entertainment is a series of case studies interlaced with sections of mathematical exposition. The studies consist of summaries drawn primarily (but not exclusively) from actual cases, which are cast in the form of problems by questions posed to focus discussion. They are designed to illustrate areas of the law in which statistics has played a role (or at least has promised to do so), and to illustrate a variety of ways to reason quantitatively. Also included are some statistical studies of the legal system, and of the impact of proposed legislation or regulation. Wherever possible, excerpts of data are given to expose the reader to the sobering, hands-on experience of calculation statistics and drawing inferences. Judicial opinions are not given because they generally do not elucidate the statistical issues that are our primary concern. On the other hand, some judicial missteps are included so that the reader may exercise critical faculties and enhance self-esteem as a newly minted expert by correcting the bench.

Knowledge of probability or statistics is not required to calculate most of the answers called for by the snippets of data in the case studies. For the uninitiated, the statistical notes supply the technical tools. Some of these notes deal (in rather condensed fashion) with material that is covered in elementary texts; others go beyond that. For a more leisurely, detailed, or expansive discussion of the material, the reader may wish to consult a statistics text; some references are given in the text sections and in the bibliography. Our calculations for the mathematical questions in the case studies are given in Appendix I. The legal issues and the statistical issues not involving calculation are for the most part left to the reader.

Apart from the riveting intellectual interest of the subject, the lawyer or prospective lawyer may fairly question whether one needs to know quite as much about statistics as this book would teach. Of course, not all will. But for increasing numbers of legal scholars, lawyers, judges, and even legislators, an acquaintance with statistical ideas, to paraphrase Justice Holmes, is not a duty, it is only a necessity. In diverse fields of learning, our knowledge is expressed in data that are appraised statistically. What is true of the general world has filtered into the courtroom. Economists of all stripes, social scientists, geneticists, epidemiologists, and others, testifying in their fields of expertise, make use of statistical tools for description and inference. In economics in particular, and in problems translated into economic terms, the ubiquity of data and computers, and the current fashion, have encouraged the creation of elaborate econometric models that are sufficiently plausible to be accepted in learned journals. But even models with impressive and intimidating technical apparatus may rest on shaky assumptions that, when exposed, undermine their credibility.

Frequently, statistical presentations in litigation are made not by statisticians but by experts from other disciplines, by lawyers who know a little, or by the court itself. This free-wheeling approach distinguishes statistical learning from most other expertise received by the courts and undoubtedly has increased the incidence of models with inappropriate assumptions, or just plain statistical error. The knowledgeable lawyer will be far more effective in proposing useful studies, exposing serious failings in complex models, and making the issues intelligible to a lay decisionmaker than one who is wholly dependent on a consultant for the next question. And although the lawyer usually will not need to make calculations, an appreciation of the principles—which is needed—is best gained from some modest grappling with the data.

Do statistics really matter? This is a question that sometimes vexes statisticians. In the legal setting, the questions are whether statistical models are fairly evaluated in the adversary process and whether statistical findings are given their due in the decisions. Unfortunately, the record here is spotty, even perverse. In some cases the courts have appraised statistical evidence well, but in some important public-issue litigation very good statistical models have been summarily rejected (and very bad ones uncritically embraced) by judges and justices in pursuit of their own agendas. The lawyer of the future predicted by Holmes ninety years ago has not yet come into his or her own.

Despite the trampling of statistical evidence that has occurred in some notable cases, it seems inevitable that studies based on data will continue to be pursued by the scholarly community and presented with increasing frequency in litigated matters involving public issues. A fuller appreciation of the standards for analyzing data and making inferences should at least lead to more accurately focussed studies and more discerning treatment of such evidence by the courts. Beyond that, one may hope that the realities exposed by statistical work will in time influence perceptions of justice, even in the courtroom. A knowledgeable lawyer may not dispatch questions of legal policy with statistics, but by knowing more of the subject may hope to contribute to the store of rational and civilized discourse by which insights are gained and new accommodations reached. That, in any event, is the larger purpose of this book.

*   *   *

Much of the material in this book was collected from lawyers and statisticians who were involved in the cases. We thank them all for their generosity in furnishing us with their papers. They are too numerous to list, but we would like to mention in particular David Baldus, Jack Boger, Will Fairley, David Freedman, Elizabeth Holtzman, Jay Kadane, and Jack Weinstein. We would also like to acknowledge Joseph Fleiss, Mervyn Susser, and Zena Stein, and their respective institutions, the Division of Biostatistics, the Sergievsky Center of the Faculty of Medicine, Columbia University, and the New York State Psychiatric Institute, for their liberal support of this project.

We would like especially to thank Margaret Murphy, who steadfastly typed and retyped the manuscript until the error rate was vanishingly small; Marcia Schoen, who typed the calculation notes; Stephen Sullivan, Lynn Cushman, and Matthew Herenstein for checking the citations; and Ann Kinney for editing the case studies and checking the calculations.

We own a debt of gratitude to our families—Claire, Katie, Matthew, Betty, Joby, Laura, and also Julie—for their patience, encouragement, and support during the long evolution of this book.

Michael O. Finkelstein
Bruce Levin
New York, New York
August 1989

Acknowledgments

Figure 1.2.2. Reprinted by permission of the *Journal of Business and Economics Statistics*. Copyright 1985 by American Statistical Association, Alexandria, VA.

Figure 2.1.1. Reprinted by permission from the National Research Council, *The Evaluation of Forensic DNA Evidence*. Copyright 1996 by the National Academy of Sciences, Washington, D.C.

Figure 9.1.1. Reprinted by permission of *Science*. Copyright 1971 by The American Association for the Advancement of Science, Washington, D.C.

Figure 9.2.1. Reprinted by permission from *Jurimetrics*. Copyright 1999 by the American Bar Association, Chicago, IL.

Figure 10.2.1. Reprinted by permission from Howard Ball, *Justice Downwind: America's Atomic Testing Program in the 1950's*. Copyright 1986 by Oxford University Press, Inc., New York, NY.

Figure 10.3.4. Reprinted by permission from the *Evaluation Review*. Copyright 1999 by Sage Publications, Inc., Newbury Park, CA.

Figure 13.2.3. Reprinted by permission from the *Evaluation Review*. Copyright 1991 by Sage Publications, Inc., Newbury Park, CA.

Figure 13.3c. Reprinted by permission from F.J. Anscombe, *Graphs in Statistical Analysis*, in 27 *The American Statistican* 17 (1973). Copyright 1973 by the American Statistical Association, Alexandria, VA.

Figure 14.1. Reprinted by permission from R.J. Wonnacott & TLL Wonnacott, *Econometries*. Copyright 1970 by John Wiley & Sons, New York, NY.

Table A1. Reprinted by permission from *The Kelly Statistical Tables* by T.I. Kelley. Copyright 1948 by Harvard University Press, Cambridge, MA.

Tables C, E. and F and Figures 5.3c and 5.3d. Reprinted by permission from *Biometrika Tables for Statisticians, Volume I* by *E.S. Pearson and H.O.*

Hartley. Copyright 1954 by Cambridge University Press, Cambridge, England.

Tables G1, G2, H1 and H2. Reprinted by permission from *Handbook Tables for Probability & Statistics 2nd Ed*. Copyright 1968 by CRC Press, Inc., Boca Raton, Fla.

Contents

List of Figures

List of Tables

CHAPTER 1

Descriptive Statistics

1.1 Introduction to descriptive statistics

Population parameters

Statistics is the science and art of describing data and drawing inferences from them. In summarizing data for statistical purposes, our focus is usually on some characteristic that varies in a population. We refer to such characteristics as *variables*; height and weight are variables in a human population. Certain *parameters* are used to summarize such variables. There are *measures of central location*—principally the mean, median, and mode—that describe in various ways the center of the data; these are discussed in Section 1.2. There are *measures of variability* or *dispersion*—most notably the variance and standard deviation—that describe how widely data vary around their central value; these are described in Section 1.3. Finally, there are *measures of correlation*—particularly the Pearson product-moment correlation coefficient—that describe the extent to which pairs of characteristics (such as height and weight) are related in members of the population; this coefficient is described for measured data in Section 1.4. Methods for expressing association in binary data (i.e., data that can take only two values, such as 1 or 0) are described in Section 1.5.

Typically, population values for such parameters are unknown and must be estimated from samples drawn at random from the population. The measures of central location, variability, and correlation have both a theoretical or population version and a sample version, calculated for observed data, and used to estimate the population version when it is unknown. In conventional notation, population parameters are represented by Greek letters: μ for the mean; σ^2 and σ for the variance and standard deviation, respectively; and ρ for the correlation coefficient. The corresponding sample versions are: $\bar{x}$ for the mean of the sample data; s^2 and s or sd for the variance and standard deviation, respectively; and r for the correlation coefficient.

Random variables

Sample estimates of population parameters will vary from sample to sample. The extent of such variation can be described probabilistically using idealized entities called *random variables*. Suppose, for example, that a single member of a human population is selected "at random," meaning that all members have an equal likelihood of selection, and his or her height measured. The possible values of that measurement define a random variable. If n members of a human population are selected at random, their heights measured and the average taken, the values of the sample mean over all possible samples of size n from the population also define a random variable.

Contrary to what the name might suggest, random variables, like individual height and sample mean height, are not totally wayward and unpredictable. In general, the probability of observing any value or set of values of the variables can be known, at least in theory. Thus, while we cannot predict an individual's exact height, we can say that the individual height variable will lie between $5'6''$ and $5'9''$ with probability P, and that the sample mean height variable will lie between those limits with probability P'. The definition thus subsumes a considerable degree of order.

Sample realizations

Sample data can be considered as particular realizations of random variables, just as mathematical functions have realized sets of values. Thus, we may speak of the random variable "individual income in the U.S." contrasted with its observed value for any person sampled from the population. To help distinguish between random variables and their realizations, one uses capitals for the variables and lower case letters for their values, as when we write $P[X = x]$ for the probability that the random variable X takes on the value x. We distinguish the sample mean random variable from any particular realization of it by writing $\bar{X}$ for the former and, as noted, $\bar{x}$ for the latter.

In this chapter we do not make probability statements about random variables. Our purpose here is limited to describing what was actually observed in a sample. For the measures of central location, variability, and correlation referred to above, we give the theoretical population formulas and then the versions for sample data, which in some cases are slightly modified. We have introduced the notion of random variables here because it provides a compact and convenient notation for the formulas and is the foundation for the probability statements made in Chapter 4, which carry the subject to a deeper level. We return to the general discussion of random variables in Section 4.1.

1.2   Measures of central location

Mean

By far the most common measure of location for sample data is the familiar "average," or arithmetic mean. The arithmetic mean, μ, of a characteristic, X, in a population of size N is calculated by summing the values of the characteristic over the population and dividing the sum by the population size. In symbols,

$$\mu = \frac{1}{N} \sum_{i=1}^{N} X_i,$$

where $\sum$ denotes the operation of summation.

We note the following characteristics of the mean:

- The mean is the unique number for which the algebraic sum of the differences of the data from that number is zero.

- Consequently, if each value in the population is replaced by the mean, the sum is unchanged.

- The mean is the number that minimizes the average squared deviation of data points from that number.

The population mean is a central value as described above, but it is not necessarily a "representative" or "typical" value. It is a result of arithmetic that may be possessed by none of the population. No one has the average 2.4 children. Nor is the mean necessarily the most useful number to summarize data when variability is important. One could drown in a river that has an average depth of six inches or have been very warm in July, 1998, when the Earth's average temperature was 61.7° Fahrenheit. A woman's life expectancy exceeds a man's, but charging women more for pensions is discriminatory, the Supreme Court has held, because not all women will outlive the male expectancy. *City of Los Angeles Dep't of Water and Power v. Manhart*, 435 U.S. 702, 708 (1978) ("Even a true generalization about a class is an insufficient reason for disqualifying an individual to whom the generalization does not apply.").

The mean in a sample is computed in the same way as the population mean, i.e., it is the sum of the sample values divided by the sample size. In a random sample, the sample mean is a useful estimator of the population mean because it is both *unbiased* and *consistent*. An estimator is said to be unbiased if its average over all possible random samples equals the population parameter, no matter what that value may be. An estimator is said to be consistent if, as the sample size increases, the probability that the sample mean will differ from the population mean by any given amount approaches zero.[1] The sample mean is

---

[1]This is a statement of the famous *law of large numbers*. For a derivation see Section 1.3 at p. 21.

also desirable because in many situations it will have greater precision—i.e., it will vary less from sample to sample—than other commonly used estimators.

Median and mode

The *median* of a population is any value such that at least half of the values lie at or above and at least half lie at or below it.[2] When there are an odd number n of points in a sample, the sample median is the middle observation. When n is even, it is customary to take the sample median as midway between the two middle observations. Income distributions, for example, are often summarized by their median because that figure is not unduly influenced by a few billionaires. The sample median is a central value that minimizes the average absolute deviation of the data points from that value.

The *mode* of a population is that value with highest relative frequency, and is often chosen as expressing the most likely value. For example the modal number of children in a household reflects the most frequently seen family size and may be a more useful statistic than the mean.

In symmetrical distributions, such as illustrated in Figure 1.2a, there is no ambiguity in the notion of center of location: the mean, median, and (in unimodal distributions) the mode coincide. In skewed distributions, however, such as in Figure 1.2b, they (and a variety of other measures) separate, each reflecting a different aspect of the underlying distribution. In right-skewed distributions, such as an income distribution, it is common to find that the mean exceeds the median, which in turn exceeds the mode.

Variants of the mean

It is sometimes appropriate to use a *weighted mean*, as for example when (i) an overall figure is to be derived from figures for strata of different sizes, in which case weights are proportional to the sizes of the strata; or (ii) the degree of variability in the data changes from one observation to the next, in which case weights inversely proportional to variability provide an estimate of the population mean that will vary less from sample to sample than any other weighted or unweighted average.

A major disadvantage of the sample mean is that it is sensitive to deviant or outlying data points arising from a distribution with "heavy tails." In these circumstances, the sample mean loses some of its useful properties described above and may not be seen as reliably "representing" the center of the data. Various measures have been proposed to deal with this sensitivity. The prototype of these is the median, which is not affected even by several highly deviant

---

[2] "At least" appears in the definition to cover the case in which there are an odd number of discrete data points. For example, if there are five points, the third point is the median: at least half the points are at or above it and at least half are at or below it.

data points. Another is the *trimmed mean*, which is the mean computed after discarding the top and bottom $P\%$ of the data.

The *geometric mean* of n positive numbers is the nth root of their product. Thus, the geometric mean of the numbers 1 and 100 is $(100 \times 1)^{1/2} = 10$. The arithmetic mean is $(1 + 100)/2 = 50.5$. By de-emphasizing differences at the large end of a scale compared with equal differences at the small end, the geometric mean in effect gives less weight to the large numbers and is thus always less than the arithmetic mean when all numbers are positive. The arithmetic mean of the logarithms of numbers is equal to the logarithm of their geometric mean. Hence, the geometric mean arises, in particular, when data are expressed in terms of their logarithms.

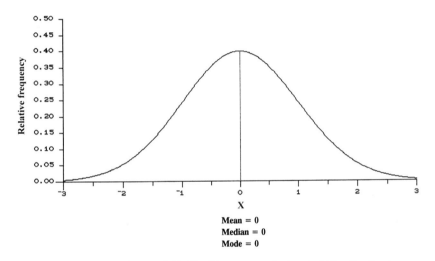

FIGURE 1.2a. Random variable X with a symmetrical probability distribution

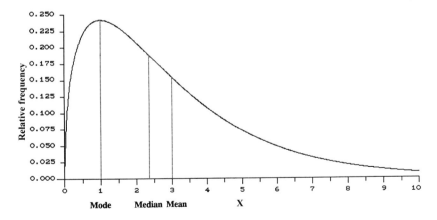

FIGURE 1.2b. Random variable X with a right-skewed probability distribution

The geometric mean is encountered in regulations incorporating some average when the data are bounded by zero on one side and unbounded on the other, so that a skewed distribution is likely to result. For example, the geometric mean length of hospital stay is used for purposes of Medicare in-hospital reimbursement, 62 Fed. Reg. 45966 at 45983 (1997); in an Idaho water quality standard, fecal coliform bacteria may not exceed a geometric mean of 50/100 milliliters based on a minimum of five samples taken over a 30-day period, 40 C.F.R. Part 131 (1997); in an air quality standard, sulfur dioxide emissions are limited to 29 parts per million, by volume, based on a 24-hour daily geometric mean, 40 C.F.R. Part 60 (1997).

The *harmonic* mean of N positive numbers is defined as the reciprocal of the arithmetic mean of the reciprocals of the N numbers. This type of average de-emphasizes differences among large numbers relative to equal differences among small numbers to an even greater extent than does the geometric mean. The harmonic mean is often used for waiting times to events, when differences in relatively short waiting times may be much more important than differences in very long waiting times.

To illustrate the three types of averages, consider the numbers 10 and 1000. Their arithmetic mean is $(10 + 1000)/2 = 505$. The geometric mean is $(10 \times 1000)^{1/2} = 100$; and their harmonic mean is $1/[(1/2)(1/10 + 1/1000)] = 19.8$.

Mathematical expectation

The arithmetic mean is a special case of the more general notion of mathematical expectation. The expected value of a characteristic X in a population, denoted EX, is the weighted sum of the Xs, with the weights given by their probabilities. In symbols,

$$EX = \sum x \cdot P[X = x].$$

If all outcomes are equally likely, the expected value is the familiar average or mean value.

Example 1.1. The expected value of the roll of two dice X and Y is equal to the sum of the values of all possible outcomes multiplied by their probabilities. The expected value is:

$$E(X + Y) = 2\left(\frac{1}{36}\right) + 3\left(\frac{2}{36}\right) + 4\left(\frac{3}{36}\right) + 5\left(\frac{4}{36}\right) + 6\left(\frac{5}{36}\right) + 7\left(\frac{6}{36}\right)$$
$$+ 8\left(\frac{5}{36}\right) + 9\left(\frac{4}{36}\right) + 10\left(\frac{3}{36}\right) + 11\left(\frac{2}{36}\right) + 12\left(\frac{1}{36}\right) = 7.$$

We note two properties of mathematical expectation:

- The expected value of a constant multiple of X is that constant times the expected value of X. Changing units of height from inches to centimeters multiplies the expected height by centimeters per inch (about 2.54).

- The expected value of the sum of two random variables is the sum of their expectations, irrespective of the correlation between them. In the dice example, the expected value for each die is 3.5 and the expected value of their sum is simply 3.5 + 3.5 = 7. The average income of a two-income (heterosexual) household is the sum of the average incomes of male and female wage earners in the population, irrespective of the fact that the paired incomes generally are correlated.

Measures of central location are frequently used to "smooth" away random variation in data to disclose underlying patterns, as the following problem shows.

1.2.1   Parking meter heist

In the late 1970s, the City of New York owned and operated approximately 70,000 parking meters. Daily collections averaged nearly $50,000. Beginning in May 1978, Brink's Inc. was awarded a contract to collect coins from these meters and deliver them to the City Department of Finance. The predecessor collection company was Wells Fargo.

The process of collection involved 10 three-person teams provided by Brink's, with one individual going to each meter on a city-prescribed schedule. The collector inserted a metal key, opened the bottom of the meter head, and removed a sealed coin box. The coin box was placed upside down onto a gooseneck protruding from a large metal canister on wheels, which the collector rolled from meter to meter. By twisting the coin box, the collector unlocked the canister and allowed the coins from the box to drop into the canister without the collector's having access to the coins. The empty coin box was then replaced and the meter relocked. At the end of the route the canister was placed in a collection van that was driven to a central collection location where the canisters were turned over to City personnel. Brink's was responsible for the supervision of its collection personnel.

Acting on an anonymous tip, the City began an investigation of parking meter collections. Surveillance of Brink's collectors revealed suspicious activity. Investigators then "salted" selected meters, inserting coins treated with a fluorescent substance. Most (but not all) of the "salted" meters were checked to ensure that they had been completely emptied by Brink's employees, and collections from the meters were reviewed to see if any treated coins were missing. The salting process indicated that some of the coins collected by Brink's employees were not returned to the City. Surveillance at one location revealed that on several occasions Brink's employees transferred heavy bags from the collection vans to their personal automobiles, and later from the automobiles to their residences.

On April 9, 1980, five Brink's collectors were arrested and charged with grand larceny and criminal possession of stolen property—$4,500 in coins allegedly taken from that day's parking meter collections. They were convicted and sentenced to varying jail terms.

The City terminated the Brink's contract as of April 9, hired a replacement firm (CDC), and took draconian steps to ensure that there was no theft by employees of the new firm.

There was a gasoline shortage in New York City from May to December 1979, and gasoline rationing was in effect from June to September 1979. There was a suburban commuter rail strike from June to August 1980.

The City sued Brink's for negligent supervision of its employees and sought to recover the amount stolen. No one knew how much was taken, but the City proposed to estimate the amount from the collection data shown in Table 1.2.1. In these data "1-A" stands for a small section (47 meters) near City Hall for which City employees did the collections at all times, "# Cols" stands for the number of collection days. There was no indication that CDC or City employees were dishonest.

The judge charged the jury that, since the fact of theft had been established as well as damage, the amount of liability need not be determined precisely, but only as "a just and reasonable inference." However, he cautioned the jurors that they might not "guess or speculate."

Questions

1. As attorney for the City, what damage calculation would you make with respect to these data, assuming that the City seeks damages only for theft occurring during the last ten months of the Brink's contract preceding April 1980? What objections do you anticipate and how would you resolve them?

2. As attorney for Brink's, what objections would you make to the City's method of estimating damages? Use the data for 1-A to make an alternative calculation.

3. Figure 1.2.1a shows the data as presented by the City. Figure 1.2.1b shows the data from Brink's point of view. Using these Figures what would you argue for each party?

Source

Brink's Inc. v. *City of New York*, 717 F.2d 700 (2d Cir. 1983). For a debate on the validity of the statistical inference, see W. Fairley and J. Glen, "A Question of Theft," in *Statistics and the Law* 221 (M. DeGroot, S. Fienberg, and J. Kadane, eds., 1986) (with comment by B. Levin and rejoinder by Fairley and Glen).

Notes

Theft from parking meter collections has been a persistent problem for New York City. Perhaps the nadir was reached when members of a special city unit

TABLE 1.2.1. Parking meter collection data (revenue in dollars)

			#COLS INCL 1-A	TOTAL INCL 1-A	AVG/Col INCL 1-A	Total 1-A	Avg Col 1-A	#Cols 1-A
W	1	5/77	21	2,231,006	106,238	6,729	747	9
E	2	6/77	22	1,898,423	86,291	5,751	821	7
L	3	7/77	20	1,474,785	73,739	6,711	745	9
L	4	8/77	23	1,626,035	70,697	7,069	783	9
S	5	9/77	21	1,516,329	72,206	7,134	792	9
	6	10/77	20	1,516,968	75,848	5,954	744	8
F	7	11/77	19	1,512,424	79,601	5,447	680	8
A	8	12/77	21	1,527,011	72,714	6,558	728	9
R	9	1/78	20	1,076,158	53,807	5,222	746	7
G	10	2/78	14	798,341	57,024	4,150	691	6
O	11	3/78	23	1,609,606	69,982	6,765	731	9
	12	4/78	24	1,253,164	52,215	6,681	835	8
	Totals		248	18,040,250	72,743	74,171	757	98
	13	5/78	22	1,337,159	60,779	7,016	779	9
	14	6/78	22	1,532,310	69,673	7,440	826	9
	15	7/78	20	1,318,521	65,926	6,264	783	8
	16	8/78	23	1,502,054	65,306	7,337	815	9
	17	9/78	20	1,393,093	69,654	7,271	807	9
	18	10/78	21	1,564,212	74,486	6,694	743	9
	19	11/78	19	1,474,861	77,624	5,795	724	8
	20	12/78	20	1,554,116	77,705	7,105	789	9
	21	1/79	22	1,572,284	71,467	6,613	734	9
	22	2/79	16	1,129,834	70,614	5,258	657	8
B	23	3/79	22	1,781,470	80,975	7,664	851	9
R	24	4/79	21	1,639,206	79,009	6,716	839	8
I	25	5/79	22	1,732,172	79,644	7,614	846	9
N	26	6/79	21	1,685,938	80,282	7,652	850	9
K	27	7/79	21	1,644,110	78,290	7,513	834	9
S	28	8/79	23	1,746,709	75,943	7,862	873	9
	29	9/79	19	1,582,926	83,311	6,543	817	8
	30	10/79	22	1,853,363	84,243	6,855	761	9
	31	11/79	19	1,754,081	92,320	7,182	798	9
	32	12/79	20	1,692,441	84,622	6,830	758	9
	33	1/80	22	1,801,019	81,864	6,552	819	8
	34	2/80	19	1,702,335	89,596	7,318	813	9
	35	3/80	21	1,678,305	79,919	6,679	834	8
	36	4/80	22	1,527,744	69,442	6,637	737	9
	Totals		499	38,240,763	76,635	166,410	796	209
	37	5/80	21	1,980,876	94,327	7,912	879	9
	38	6/80	21	1,941,688	92,461	7,314	914	8
	39	7/80	22	1,889,106	85,868	7,803	867	9
	40	8/80	21	1,741,465	82,926	8,126	902	9
C	41	9/80	21	1,832,510	87,262	7,489	832	9
D	42	10/80	22	1,926,233	87,556	7,986	887	9
C	43	11/80	17	1,670,136	98,243	6,020	860	7
	44	12/80	22	1,948,290	88,558	6,442	920	7
	45	1/81	21	1,627,594	77,504	7,937	881	9
	46	2/81	18	1,655,290	91,960	6,685	835	8
	47	3/81	22	1,844,604	83,845	7,470	830	9
	Totals		228	20,057,792	87,973	81,199	873	93

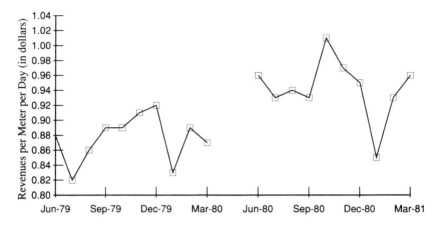

FIGURE 1.2.1a. Revenue per meter per day for two ten-month periods

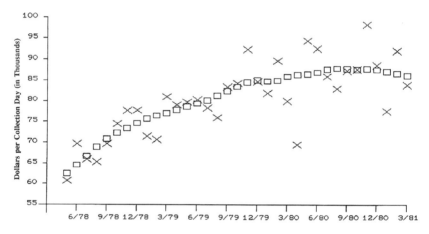

FIGURE 1.2.1b. Revenue per collection day, May 1978–March 1981. Boxes represent the trend obtained by locally weighted scatterplot smoothing (see Section 14.4)

created to prevent meter revenue theft were themselves charged with theft. See Selwyn Raab, *20 In Meter Anti-Theft Unit Accused of Stealing*, N.Y. Times December 10, 1993, Sec. A, at 2. *Sed quis custodiet ipsos Custodes?*

1.2.2   Taxing railroad property

Property taxes are levied on the basis of assessed value, but the relation between assessed value and market value varies for different classes of property, and for different taxpayers. Railroads believed that state and local authorities discriminated against them by overassessing their property relative to other taxpayers' property. To protect the railroads, Congress passed Section 306 of the Railway Revitalization and Regulatory Reform Act (4-R Act), which was

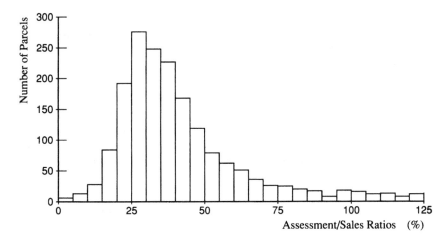

FIGURE 1.2.2. Histogram for distribution of assessment/sales ratios. Expressed as percentages: Los Angeles County, 1981–1982 sales, 1975 base year

intended to stop certain forms of property tax discrimination against railroads. This section (49 U.S.C. §11501 (1994)) provides that a state may not assess rail transportation at a value that has a higher ratio to the true market value of the rail transportation property than the ratio that the assessed value of other commercial and industrial property in the same assessment jurisdiction has to the true market value of the other commercial and industrial property.

Under the Act, railroads are entitled to relief if the ratio applicable to them exceeds the ratio applicable to other property by 5% or more.

In applying the Act, assessment is determined from tax rolls and true market value is determined on a sample basis from recent sales.

A sample distribution of the ratios of assessments to sales for Los Angeles County is shown in the histogram in Figure 1.2.2.

The legislative history indicates that the ratio for railroads was to be compared with the ratio for the "average" or "weighted average" commercial and industrial taxpayer in the taxing district. The word "average" was not used in the statute because, according to a railroad spokesman at the hearings on the bill, it "has a precise mathematical connotation which makes it unsuitable in this context."

In litigation under this Act, the defendant local and state taxing authorities argue that the correct measure of the average is the weighted mean.

When sampling from recent sales by each type (stratum) of property (commercial, industrial, or utility), the weighted mean is obtained as follows: (a) in each stratum, take the ratio of total sales value for the parcels sampled to the total assessed value for these parcels; this gives the dollars of sales per dollar of assessment in the sample; (b) multiply that sales value per dollar of assessed value figure by the total assessed value of all property in the entire stratum to estimate the total sales (market) value of all property in the stratum; (c) take the sum of assessments for all strata and divide it by the total sales value for all property in all strata as obtained in (b); this is the weighted mean assessment to sales ratio for all property in the population.

Railroads argue that the parcel-weighted median should be used to express the value of other industrial and commercial property.

The parcel-weighted median is obtained as follows: (a) for each parcel in the sample, compute the ratio of its assessed value to its sales price; (b) array these parcels so that the ratios increase from left to right; (c) attach a weight to each parcel, representing the number of parcels in the corresponding stratum of the population divided by the number of sample parcels in that stratum; (d) find the ratio in the array that has half of the total weight to the left and half to the right.

Generally, railway and utility properties are assessed centrally by state authorities, while other commercial and industrial properties are assessed by counties. Utility property is generally taxed at much higher ratios of assessed value to market value ratios than locally assessed property. In California, for example, a few hundred utilities in the state account for about 14% of the assessed value of "other commercial and industrial property." It was assumed in the litigation that utilities were assessed at 100% of market value.

Questions

1. Which measure–the mean, the weighted mean, or the weighted median–is most appropriate, given the language and history of the statute?

2. Which is the more appropriate measure of discrimination against the railroads: the sum of the absolute or algebraic differences between the assessment/market value ratio for the railroads and other properties? Which measure does your choice suggest?

3. Which measure, if applied to all properties, would produce no change in total revenue?

4. Do you agree with the rationales for picking the median given by the Fourth Circuit in the *Clinchfield* and *CSX Transp. Inc.* cases (described in the Notes, below).

5. The weighted mean assessment/sales ratio is a type of weighted harmonic mean of the sample assessment/sales ratios within each stratum. Do you see why?

Source

David Freedman, *The Mean Versus the Median: A Case Study in 4-R Act Litigation*, 3 J. Bus. & Econ. Stat. 1 (1985).

Notes

In litigation under the 4-R Act, some courts have favored the mean, others the median. *See, e.g., Southern Pacific v. California State Board of Equalization,* (unpublished; discussed in Freedman, *supra*) (mean); *ACF Industries, Inc. v. Arizona,* 714 F.2d 93 (9th Cir. 1983)(median in *de facto* discrimination cases; mean in *de jure* cases); *Clinchfield R.R. Company v. Lynch,* 527 F. Supp. 784 (E.D. N.C. 1981), *aff'd,* 700 F.2d 126 (4th Cir. 1983)(median); *CSX Transp. Inc. v. Bd. of Pub. Works of West Virginia,* 95 F.3d 318 (4th Cir. 1996) (median). Generally, the choice has not evoked much discussion. However, in *Clinchfield R.R. Co. v. Lynch, supra,* the court endorsed use of the median, arguing that the weighted mean would be skewed upward due to the large amounts of property owned by public service companies that were centrally assessed at higher values in much the same way as the railroads were assessed. The court argued that railroads should not be compared primarily with utilities, but on a broader basis with other industrial and commercial property. In *CSX Transp. Inc.* the court chose the median because the mean was too highly affected by the inclusion or exclusion of a few large properties.

1.2.3   Capital infusions for ailing thrifts

In 1986, the Federal Home Loan Bank Board adopted a program to increase regulatory capital of insured thrift institutions. The goal was to increase capital from 3% to 6% of assets. The required annual increase for each institution was computed as a percentage of the average return on assets (ROA) for the industry as a whole. The percentage depended on whether the institution was in the standard group, with base ratios of at least 3%, or in the lower group, with base ratios under 3%. The theory was that institutions would be able to retain these percentages of earnings to bolster capital; those that lacked sufficient earnings would have to raise outside capital. However, a segment of the industry in deep financial trouble had a negative ROA; as a result, the mean return on assets for the industry was only 0.09%. This meant that it would take about 40 years to raise capital from an average of 3% to 6% for the industry as a whole. To speed up the process, the Board announced that it would shift to the median ROA, which was about 0.33%. This would raise sufficient capital in about 12 years. To determine the median, each institution was treated as a single data point. The Board explained:

> The fundamental reason for this decision is that the mean is too sensi-
> tive to extremely high or low ROA's. This sensitivity becomes especially
> critical in an environment in which the unhealthy segment of the indus-
> try operates at a severe loss .... [The median] ROA should be used
> because it focuses on the probability of the fiftieth percentile institution
> and on the ranking that generates the fiftieth percentile institution ....
> [The median] more accurately reflects the ability of a large majority of

insured institutions to advance toward higher minimum required capital levels.

Questions

1. What asymmetry in the distribution of ROAs produces the difference between the mean and median ROA?

2. Is the median ROA likely to be superior to the mean as a measure of general industry capacity to raise levels of capital from retained earnings?

3. What other measures of central location could the Board have used?

Source

Federal Home Loan Bank Board, *Regulatory Capital of Insured Institutions*, 53 Fed. Reg. 11243 (March 30, 1988).

1.2.4   Hydroelectric fish kill

The Constantine Project is a 94-year-old hydroelectric generating facility located on the St. Joseph River in Constantine, Michigan; it is owned by the Indiana Michigan Power Company. In licensing proceedings before the Federal Energy Regulatory Commission (FERC), the Michigan Department of Natural Resources sought to have the FERC impose certain conditions designed to reduce the number of fish entrained in the project's turbines. The FERC refused, but did agree that the Company should compensate the state for the fish killed. The annual number of fish killed was estimated from samples taken on random days. The Company advocated the use of a geometric mean to derive an annual kill rate because the daily samples fluctuated, and on some days appeared to be unusually high. Using the geometric mean of the sample data, the Company estimated an annual kill rate of 7,750 fish. Michigan objected that there was too little data to determine skewness, and proposed the arithmetic mean instead. Using the arithmetic mean applied to essentially the same sample data, Michigan estimated an annual kill rate of 14,866 fish.

Question

1. Which is the better measure for estimating annual fish kill, the geometric or the arithmetic mean?

Source

Kelley v. Federal Energy Regulatory Com'n, 96 F.3d 1482 (D.C. Cir. 1996).

Notes

Michigan's recommendations were made to the FERC pursuant to a 1986 Congressional Act, which required the FERC to heed conservation recommendations of federal and state fish and wildlife agencies when renewing dam licenses. 16 U.S.C.A. §803(j)(West Supp. 1998). Nevertheless, the agency did not invoke that mandate until 1997, when it refused to relicense a small, aging hydroelectric dam on Maine's Kennebec River and ordered that it be removed. The dam's removal permitted nine species of Atlantic fish, including salmon and sturgeon, to regain their traditional upriver spawning grounds. Editorial, *Weighing the Needs of Nature*, N.Y. Times, November 27, 1997 at A38.

1.2.5   Pricey lettuce

Assume that romaine lettuce and iceberg lettuce each costs $1/lb, and that the consumer buys 1 lb of each. If the price of romaine lettuce rises to $1.50/lb, the Bureau of Labor Statistics had calculated its consumer price indices (collectively, the CPI) by assuming that the proportionate *quantities* of the two kinds of lettuce remained unchanged and used the arithmetic mean to calculate the percentage increase in the cost of lettuce. Under this assumption, the arithmetic mean percentage increase is known as the Laspeyres Index[3] The Laspeyres Index is regarded as an upper bound to the increase in the cost of living because it takes no account of the substitution phenomenon: if the price of one commodity rises, consumers tend to switch to similar commodities that are lower in price. In our example, consumer satisfaction is assumed to be undiminished if the loss of romaine is offset by purchase of more iceberg lettuce. This may seem improbable, but that is the theory. The amount of extra iceberg needed is affected by another phenomenon, namely, the law of diminishing returns: each additional dollar spent for a more expensive commodity buys a declining quantum of satisfaction compared with the satisfaction purchased with a dollar spent for the less expensive commodity.

Although the CPI does not purport to be a cost-of-living index, it is frequently used in private contracts and in legislation (most notably in social security benefits) to adjust payments for increases in cost of living. In April 1998, responding to criticisms that its index overstated increases in cost of living, the Bureau stated that, beginning in 1999, for certain types of commodities at the lowest level of aggregation, it would take substitution into account by: (i) assuming that the *dollar share* for each commodity (e.g., in our example 50% for each kind of lettuce) would remain the same after the price increase; and (ii) using the geometric mean for averaging. The Bureau estimated that the effect of these changes would be to reduce annual increases in the CPI by about 0.2 of a percentage point.

---

[3]If the proportionate quantities purchased after the price change are used, the index is called the Paasche Index.

Questions

1. Compute the weighted arithmetic mean ratio of the new to the old lettuce prices, using constant dollar shares as weights. What quantities of each kind of lettuce are implied by this average? Looking at the reduction in quantity for romaine lettuce and the increase in quantity for iceberg lettuce, what is the assumed relation between quantum of satisfaction, and price and quantity of lettuce? Is this assumed relation reasonable?

2. Answer the same questions as above, this time using the weighted geometric mean ratio of new to old lettuce prices. [Hint: The weighted geometric mean ratio is the anti-log of the weighted arithmetic mean of log ratios.]

3. Now assume that the price of romaine lettuce falls back $.50 to its original $1/lb. Compare the changes in the ratios when the price rises and falls, using the arithmetic mean and the geometric mean. Which result is superior?

Source

Bureau of Labor Statistics, *Planned Changes in the Consumer Price Index Formula* (April 16, 1998); *id., The Experimental CPI Using Geometric Means*; avail. at <www.stats.bls.gov/cpigmrp.htm>.

1.2.6   Super-drowsy drug

The Able Company claimed in TV advertisements that taking its product Super-Drowsy (the active ingredient of which is an antihistamine) reduced the time it took to fall asleep by 46% over the time necessary without pills. Able based this claim on a sleep study. Thereafter Able was acquired by Baker, which conducted a new study.

The Federal Trade Commission (FTC) began an administrative proceeding against Baker, claiming that the advertisements were not substantiated by the studies.

In the sleep studies, people were asked to record how long they took to fall asleep ("sleep latency"), and their average for a week was calculated. In Able's study, the statistician excluded those nights on which a person took less than 30 minutes to fall asleep on the ground that the subject was not insomniac on that night. In the Baker study, only those who took more than 30 minutes on at least 4 nights out of 7 in the study week were included, but all seven nights were tallied in the average time. In the next week, these people received Super-Drowsy and recorded their sleep times (they took a pill only when they wished to). Baker's expert computed the average time to fall asleep for the selected subjects in the first week and compared that with their average time in the second week. (The measured sleep latency from which the averages were obtained was computed using the mid-points of 15 minute intervals.) The

TABLE 1.2.6. Sleep study data

	(Average sleep latency in minutes)							
	Week			Week			Week	
#	1	2	#	1	2	#	1	2
1	61.07	20.36	26	16.07	7.50	51	41.79	11.79
2	46.07	37.50	27	33.21	26.79	52	22.50	22.50
3	84.64	61.07	28	51.43	20.36	53	39.64	18.21
4	31.07	9.64	29	28.93	18.21	54	78.21	26.79
5	54.64	28.93	30	139.29	37.50	55	46.07	16.07
6	26.79	11.79	31	78.21	39.64	56	46.07	28.93
7	58.93	31.07	32	43.93	35.36	57	61.07	33.21
8	13.93	9.64	33	111.43	7.50	58	39.64	28.93
9	71.79	35.36	34	56.79	31.07	59	56.79	18.21
10	82.50	33.21	35	106.07	43.93	60	34.29	32.14
11	37.50	24.64	36	98.57	77.14	61	43.93	24.64
12	35.36	46.07	37	43.93	70.00	62	31.07	24.64
13	37.50	9.64	38	114.64	26.79	63	46.07	24.64
14	114.64	16.07	39	39.64	33.21	64	22.50	9.64
15	35.36	9.64	40	91.07	31.07	65	50.36	41.79
16	73.93	35.36	41	18.21	20.36	66	41.79	18.21
17	17.50	11.79	42	63.21	18.21	67	35.36	18.21
18	94.29	20.36	43	37.50	20.36	68	65.36	78.21
19	22.50	9.64	44	41.79	22.50	69	37.50	31.50
20	58.93	30.00	45	40.00	43.93	70	48.21	9.64
21	46.07	100.71	46	50.36	41.79	71	102.86	43.93
22	31.07	9.64	47	48.21	60.00	72	86.79	43.93
23	62.14	20.36	48	63.21	50.36	73	31.07	9.64
24	20.36	13.93	49	54.64	20.36			
25	56.79	32.14	50	73.93	13.93			

average latency for week 1 was 53.8 and for week 2 was 28.8; the percentage reduction was 46.5%.

The FTC's expert computed a percentage reduction for each person and took the average of those percentages, which was 40.9%.

The data are shown in Table 1.2.6; a scatterplot of the data is shown in Figure 1.2.6.

Questions

1. What is the reason for the difference between the average computed by Baker's expert and the average computed by FTC's expert?

2. Assuming that a single percentage-reduction figure would be appropriate, which calculation is the more appropriate average?

3. What other types of disclosure might be used to resolve the dispute between Baker and the FTC?

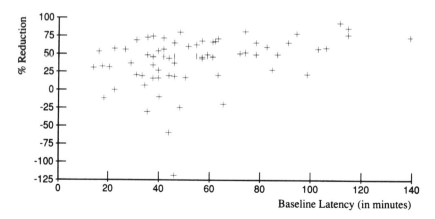

FIGURE 1.2.6. Percentage reduction in sleep latency vs. baseline latency

1.3 Measures of dispersion

In addition to measures of central location, summary measures of dispersion (variability or scatter) are important descriptive devices because they indicate the degree of likely deviation in the data from the central values. Like measures of central location, dispersion measures come in various forms that emphasize differing aspects of the underlying data.

Variance and standard deviation

By far the most important measure of dispersion is the *variance*, denoted by σ^2 or Var, and its positive square root, the *standard deviation*, denoted by σ. The variance is the mean squared deviation of the data from their mean value, and the standard deviation is the root mean squared deviation. Thus, if $X_1, \ldots, X_N$ denote the values of some characteristic in a population of size N with μ as their mean value, then

$$\sigma^2 = Var\, X = \frac{1}{N} \sum_{i=1}^{N} (X_i - \mu)^2.$$

More generally, if X is a random variable, $Var\, X = E(X - \mu)^2$. An alternative expression for the variance that is useful for hand-held calculator computation is:

$$\sigma^2 = Var\, X = E X^2 - \mu^2,$$

where $E X^2$ is the expected or average value of the squared X's and μ is the average or expected value of X. The variance and standard deviation measure dispersion in the data because the farther the values of X lie from the mean, the larger the squared deviations become, with corresponding increases in the mean squared deviation and root mean squared deviation. The following properties follow immediately from the definitions:

- Variance is never negative, and is zero only when the data have no variation.

- Adding a constant to the data does not change the variance.

- Multiplying the data by a constant, c, multiplies the variance by c^2, and multiplies the standard deviation by the absolute value of c. A change of scale thus changes the σ measure of spread by the same scale factor.

- The standard deviation is in the same units as the data.

When sampling from a population with unknown mean and variance, we usually estimate the mean with the sample mean $\bar{X}$, based on a sample of size n, and estimate the variance with the sample variance

$$s^2 = [(X_1 - \bar{X})^2 + \cdots + (X_n - \bar{X})^2]/(n - 1).$$

The sample standard deviation is the positive square root of s^2, usually denoted by s or sd. The denominator $n - 1$ is used here rather than n to make the estimator s^2 unbiased. The reason for bias is that in computing squared deviations, the sample mean minimizes the sum of squares, which makes the expected value of the sum slightly less than if the population mean were known and used. Dividing by $n - 1$ compensates for this and produces an estimator, s^2, whose expected value is exactly equal to σ^2. Note that when a weighted mean is used as a central value, it is appropriate to use weighted sums of squared deviations from that value to estimate variance.

Variance of sample sums and the sample mean

The variance of the sample mean is of particular importance in statistics. To illustrate its derivation, consider a large population of men whose characteristic of interest is their height.

- If X_1 and X_2 are the heights of two different men selected independently from the population, the variance of the sum $X_1 + X_2$ is equal to the sum of the component variances. Thus, if height has a population variance σ^2, the variance of the sum of the heights of two men selected independently from that population is $2\sigma^2$. This principle generalizes: if n men are selected independently from a large population, the variance of the sum of their heights is $n\sigma^2$.

- The variance of the *difference* between the heights of two men, $X_1 - X_2$, is still equal to the sum of the component variances. This may seem counter-intuitive, but follows immediately from the fact that putting a minus sign in front of the height for X_2 doesn't change the variance for that data. Then the variance $X_1 - X_2$ is equal to the variance of $X_1 + (-X_2)$, which is the sum of the component variances.

- If X and Y are not independent, but are positively correlated (e.g., a man's height and his adult son's height), then the variance of their sum $X + Y$ will be greater than the sum of the component variances and the variance of $X - Y$ will be smaller than the sum of the component variances. Do you see why? If X and Y are negatively correlated, then the variance of $X + Y$ will be smaller than the sum of their variances and the variance of $X - Y$ will be greater than the sum of the component variances.

These simple results lead to formulas for the variance and standard deviation of the sample mean. Recall that the mean height of a sample of n men drawn from the population is equal to the sum of the sample heights divided by the sample size. If the sample selections are made randomly and independently,[4] the variance of the sum of the n heights, each with variance σ^2, is $n\sigma^2$. Since the sum of the heights is multiplied by $1/n$, the variance of the sum is multiplied by $1/n^2$. The variance of the sample mean is thus $(1/n^2) \cdot n\sigma^2 = \sigma^2/n$ and the standard deviation is $\sigma/\sqrt{n}$. The standard deviation of a sample mean is often called the *standard error* of the mean to distinguish its variability from that of the underlying population. Since σ is frequently unknown, the standard error of the mean must itself be estimated from the sample. If a sample of size n is drawn from a population with unknown variance σ^2, the standard error of the mean is taken as $s/\sqrt{n}$, where s^2 estimates σ^2.

The fundamental result for the standard error of the mean shows that the mean of a sample is more stable, i.e., it varies less than its components. The stability of the sample mean is the basis for the statistical regularity of mass phenomena. The insurance industry lives on the fact that, although individuals experience unpredictable losses, in the aggregate claim amounts are relatively stable. Therein lies the economic advantage of pooling insured risks. The formula also shows that, although the precision of a sample estimate of the mean increases with sample size, it does so more slowly than one might think because the increase is a function of the square root of the sample size. A doubling of precision requires a quadrupling of the sample.

Chebyshev's inequality

The standard deviation is particularly useful as a measure of variability because many statements about the distribution of data or random variables can be made in terms of the standard deviation. Roughly, the greater the number of standard deviations a given value is from the mean, the less likely it is that a data point or

---

[4]Unless the sampling is done with replacement of the sampled item after each selection, which is not usually the case, the requirement of independence is not perfectly satisfied because prior selections affect the population for subsequent selections. However, if the sample is small relative to the population (e.g., less than 10%), the departure from perfect independence is not material and the imperfection is usually ignored. When the lack of independence is material, the formulas must be adjusted. See Section 4.5.

a random variable will exceed that value. A more precise statement is given by a famous theorem of the Russian mathematician Pafnuty Lvovich Chebyshev (1824–1894).[5] The theorem, "Chebyshev's inequality," can be expressed as follows:

> For any random variable or any set of data (population or sample) the probability that the variable or a randomly selected data point would lie more than k standard deviations on either side of the mean is less than $1/k^2$.

Chebyshev's inequality thus tells us that less than 1/4 of the data from *any* distribution will lie more than two standard deviations from the mean.[6] When the nature of the distribution is known more precisely, stronger statements usually can be made. For example, in normal distributions, approximately 5% of the data lie beyond 1.96 standard deviations from the mean (see Section 4.3).

Standardized random variables and data

An important use of the standard deviation is in the production of standardized random variables and data. To standardize a random variable, one subtracts the mean or expected value and divides by the standard deviation. The first step "centers" the variable or the data by giving them a zero mean, and the second step expresses the variable or data in terms of numbers of standard deviations from the mean. The variable or data then have zero mean and unit standard deviation, by construction. This standardization is important because probabilities often can be calculated from standardized quantities. Chebyshev's theorem is an example of such a calculation. In symbols, the transformation is expressed as

$$Y^* = \frac{Y - \mu}{\sigma},$$

where Y is the original random variable or data point and Y^* is the standardized version.

The law of large numbers

Chebyshev's inequality provides a proof of the weak law of large numbers. Recall the law, which states that, as sample size increases, the probability that the sample mean will differ from the true mean by any given non-zero amount

---

[5]Philip Davis in *The Thread: A Mathematical Yarn* (1983) gives a charmingly whimsical account of world-wide negative reaction to his rendering of Chebyshev's name as "Tschebyscheff" in a scholarly book.

[6]The proof of this theorem is not difficult. See J. Freund and E. Walpole, *Mathematical Statistics* 96 (4th ed. 1987).

approaches zero. What is the probability that the sample mean differs from the true mean by a given amount, say d, or more? Express d in terms of the number of standard deviations, say, $k = d/(\sigma/\sqrt{n})$ standard deviations. Then Chebyshev's inequality states that the probability that d would be exceeded is less than $1/k^2 = \sigma^2/(nd^2)$, a quantity that goes to zero as n increases.

Coefficient of variation

Since the size of the standard error depends upon the units in which the data are expressed, the standard errors of two estimators cannot be compared without adjusting for differences in units. When a quantity that has only positive values is involved, it is useful to compare the size of the standard error with the mean. A frequently used measure is the *coefficient of variation*, which is the standard error expressed as a percentage of the mean. This dimensionless statistic can be used to compare the relative precision of two estimators.

Other measures

Although the variance and standard deviation are used almost reflexively by statisticians, other measures of variability may be more appropriate in certain contexts.

As in the case of the mean, problems may arise when there are significant outliers (data points far from the central body). Because the variance uses squared differences, it gives greater weight to large deviations and this may seriously distort the measure of dispersion. Moreover, quite apart from the distorting effect of outliers, the problem at hand may be one for which squared deviations are simply inappropriate as a description.

The simplest class of measures involves the difference between upper and lower percentiles. For example, the *range* of a set of numbers is the difference between the largest and smallest number. The range is highly sensitive to outliers. Less sensitive is the *interquartile range*, which is the difference between the 75th and 25th percentiles and thus gives an interval containing the central 50% of the data.[7] Another measure of dispersion, the *mean absolute deviation*, is the average absolute magnitude of departures from the mean value. Since the deviations are not squared, the effect of outliers is deemphasized.

1.3.1   Texas reapportionment

The one-person, one-vote doctrine of the U.S. Supreme Court, based on Article I, Section 2 of the U.S. Constitution, permits in congressional districts "only the

---

[7]To compute the interquartile range proceed as follows: In a sample of n observations, first take $n/4$ and round it up to the next integer m if $n/4$ is fractional. Find the mth largest and m^{th} smallest observation and take the difference. If $n/4 = m$ is an integer, average the mth and $(m + 1)^{st}$ value from either end of the data, and take the difference of the averages.

limited population variances which are unavoidable despite a good-faith effort to achieve absolute equality, or for which justification is shown." *Kirkpatrick v. Preisler*, 394 U.S. 526, 531 (1969). The constitutionality of state legislative reapportionments is governed by the less strict standard of the Fourteenth Amendment as applied in *Reynolds v. Sims*, 377 U.S. 533 (1964). Under that case, small divergences from a strict population equality are permitted, but only if "based on legitimate considerations incident to the effectuation of a rational state policy." *Id.* at 579.

A reappointment plan for the Texas House of Representatives provided for 150 representatives from 90 single-member districts and 11 multimember districts. The district data are shown on Table 1.3.1.

For these data, the average district size is 74,645 (74,644.78 exactly) and the average squared district population is 5,573,945,867.

TABLE 1.3.1. Redistricting plan for the Texas House of Representatives

Dist. #	# Mem.	Total or average pop. per mem.	(Under) Over	%	Dist. #	# Mem.	Total or average pop. per mem.	(Under) Over	%
3		78,943	4,298	5.8	42		74,706	61	0.1
38		78,897	4,252	5.7	21		74,651	6	0.0
45		78,090	3,445	4.6	36		74,633	(12)	(0.0)
70		77,827	3,182	4.3	64		74,546	(99)	(0.1)
27		77,788	3,143	4.2	53		74,499	(146)	(0.2)
77		77,704	3,059	4.1	68		74,524	(121)	(0.2)
54		77,505	2,860	3.8	72	(4)	74,442	(203)	(0.3)
39		77,363	2,718	3.6	90		74,377	(268)	(0.4)
18		77,159	2,514	3.4	8		74,303	(342)	(0.5)
57		77,211	2,566	3.4	50		74,268	(377)	(0.5)
2		77,102	2,457	3.3	73		74,309	(336)	(0.5)
30		77,008	2,363	3.2	16		74,218	(427)	(0.6)
55		76,947	2,302	3.1	43		74,160	(485)	(0.6)
9		76,813	2,168	2.9	89		74,206	(439)	(0.6)
15		76,701	2,056	2.8	97		74,202	(443)	(0.6)
14		76,597	1,952	2.6	99		74,123	(522)	(0.7)
52		76,601	1,956	2.6	56		74,070	(575)	(0.8)
29		76,505	1,860	2.5	24		73,966	(679)	(0.9)
1		76,285	1,640	2.2	37	(4)	73,879	(766)	(1.0)
47		76,319	1,674	2.2	75	(2)	73,861	(784)	(1.1)
49		76,254	1,609	2.2	95		73,825	(820)	(1.1)
6		76,051	1,406	1.9	7	(3)	73,771	(874)	(1.2)
34		76,071	1,426	1.9	26	(18)	73,740	(905)	(1.2)
76		76,083	1,438	1.9	35	(2)	73,776	(868)	(1.2)
82		76,006	1,361	1.8	74		73,743	(902)	(1.2)
13		75,929	1,284	1.7	41		73,678	(967)	(1.3)
23		75,777	1,132	1.5	71		73,711	(934)	(1.3)
51		75,800	1,155	1.5	48	(3)	73,352	(1,293)	(1.7)
83		75,752	1,107	1.5	6		73,356	(1,289)	(1.7)
65		75,720	1,075	1.4	91		73,381	(1,264)	(1.7)
81		75,674	1,029	1.4	22		73,311	(1,334)	(1.8)
100		75,682	1,037	1.4	94		73,328	(1,317)	(1.8)
20		75,592	947	1.3	11		73,136	(1,509)	(2.0)
25		75,633	988	1.3	86		73,157	(1,488)	(2.0)
84		75,634	989	1.3	33		73,071	(1,574)	(2.1)
44		75,278	633	0.8	87		73,045	(1,600)	(2.1)
46	(11)	75,154	509	0.7	17		72,941	(1,704)	(2.3)
63		75,191	546	0.7	93		72,761	(1,884)	(2.5)
79		75,164	519	0.7	59	(2)	72,498	(2,148)	(2.9)
101		75,204	559	0.7	96		72,505	(2,140)	(2.9)
19	(2)	75,104	459	0.6	10		72,410	(2,235)	(3.0)
58		75,120	475	0.6	98		72,380	(2,265)	(3.0)
80		75,111	466	0.6	28		72,367	(2,278)	(3.1)
88		75,076	431	0.6	66		72,310	(2,335)	(3.1)
5		75,014	369	0.5	62		72,240	(2,405)	(3.2)
31		75,025	380	0.5	4		71,928	(2,717)	(3.6)
32	(9)	75,055	410	0.5	78		71,900	(2,745)	(3.7)
60		75,054	409	0.5	92		71,908	(2,737)	(3.7)
67		75,034	389	0.5	40		71,597	(3,048)	(4.1)
69		74,765	120	0.2	85		71,564	(3,081)	(4.1)
12		74,704	59	0.1					

Questions

1. Treating the multimember districts as if they were separate equal districts, each with one member, what summary statistics would be available to describe the variation in district size?

2. Which measures are most appropriate from a legal point of view?

Source

White v. Regester, 412 U.S. 755 (1973).

1.3.2 Damages for pain and suffering

Patricia Geressy worked as a secretary for five years in the 1960s, and from 1984 to at least 1997. She used a keyboard manufactured by Digital Equipment Corporation. She suffered numbness, tingling in her hands, and burning in her wrists, ultimately leading to multiple (unsuccessful) operations and substantial loss of the use of both hands. In a suit against Digital, her experts attributed her injuries to repetitive stress injury (RSI) caused by defendant's keyboard. Defendant's experts disputed this conclusion. The jury returned a verdict of $1,855,000 for quantifiable damages and $3,490,000 for pain and suffering.

After the verdict, the district court applied the New York rule that the court must set aside an award and order a new trial unless a stipulation is entered to a different award, if the verdict "deviates materially from what would be reasonable compensation." This standard replaced the old "shock the conscience" standard in a legislative effort to curb runaway awards for pain and suffering in personal injury cases.

In considering the reasonableness of the pain and suffering award, the court followed a two-step process. First, it identified a "normative" group of cases in which awards had been approved. These were cases of "reflex sympathetic dystrophy," a syndrome involving continuous pain in a portion of an extremity after sustaining trauma. Twenty-seven cases were identified as the normative group. The pain and suffering awards in this group ranged from $37,000 for a work-related hand and wrist injury necessitating surgery, to $2,000,000 for a car accident at work, which caused herniated discs requiring spinal surgery and three knee operations. As for the permitted range of variation, the court considered both one- and two-standard deviation rules and opted for allowing two standard deviations: "For present purposes, it is assumed that there are no extraordinary factors that distinguish the case at bar from other cases in the normative group and that the pain and suffering award should fall within two standard deviations." The data for pain and suffering awards in the 27 cases found to be comparable are shown in Table 1.3.2.

In these data, the average squared award is $913,241 and average award squared is $558,562 ($ in 000s).

TABLE 1.3.2. Normative case awards ($ in 000s)

Case #	Award	Case #	Award	Case #	Award
1	37	10	290	19	1,139
2	60	11	340	20	1,150
3	75	12	410	21	1,200
4	115	13	600	22	1,200
5	135	14	750	23	1,250
6	140	15	750	24	1,576
7	149	16	750	25	1,700
8	150	17	1,050	26	1,825
9	238	18	1,100	27	2,000

Questions

1. What is the maximum reasonable award for pain and suffering in this case under the court's two-standard deviation rule?

2. Is the court's two-standard deviation interval based on the normative cases an appropriate way to determine whether the compensation for pain and suffering allowed in *Geressy* exceeded reasonable limits?

Source

Geressy v. Digital Equipment Corp., 950 F. Supp. 519 (E.D.N.Y. 1997) (Weinstein, J.).

1.3.3   Ancient trial of the Pyx

The trial of the Pyx, an ancient ceremony of the Royal Mint of Great Britain, was the final stage of a sampling inspection scheme for gold and silver coinage. Over a period of time one coin was taken from each day's production and placed in a thrice-locked box called the Pyx (from the Greek word for box; in early ecclesiastical literature, the Pyx was the vessel in which the bread of the sacrament was reserved). At irregular times, usually separated by three or four years, a trial of the Pyx was declared, a jury of Governors of the Goldsmiths assembled, and a public trial held before the officers of the Mint. At the trial, the Pyx was opened and the contents counted, weighed, and assayed, and the results compared with the standard set in the indenture (contract) between the Crown and the Mint. The ceremony was well established by 1279 when Edward I issued a proclamation describing the procedure to be followed.

From the earliest times, the Master of the Mint was allowed a tolerance, or "remedy," with respect to the weight and fineness of the coins. If the remedy was exceeded he was charged. At the end of the eighteenth century, the remedy was 1/6 of a carat, or 40 grains (gr) per lb. of gold; by the middle of the nineteenth century, improvements in mint technology had reduced the remedy to 1/20 of a carat, or 12 gr per lb.[8] Because a single sovereign had a standard weight of 123 gr and because there are 5,760 gr in a lb, the remedy for an individual sovereign would have been 0.85417 gr. (Studies of later coins indicated that the remedy for an individual coin was about twice the standard deviation for such coins.)

The trial was conducted by weighing the coins in the Pyx and comparing their aggregate weight with the indenture standard for that number and denomination of coins. For example, in a trial conducted in 1799, when the Pyx was opened after a four-year period, there were 10,748 gold coins in three denominations with a weight of 190 lb, 9 oz, 8 dwt. According to the standard in force at the time, they should have weighed 190 lb, 9 oz, 9 dwt, 15 gr; thus, they were in deficit 1 dwt, 15 gr, or 39 gr. However, since the remedy was 40 gr/lb, the aggregate remedy for this weight was 1 lb, 3 oz, 18 dwt, so that the deficit was well within this figure. In fact, over many centuries, the remedy was almost never exceeded.

The most famous Master of the Mint was Sir Isaac Newton, who held the post for many years, starting in 1696. He survived without fine the "repeated ordeal" of the trial of the Pyx, and successfully fought to prevent the Goldsmiths Company from conducting the trial without the officials of the Mint present to protect their interests.

Questions

1. Assume that the standard deviation of weights of sovereigns is about $0.85417/2 = 0.42708$ gr, and that there were 8,935 sovereigns (and only these coins) in the Pyx. Use Chebyshev's inequality to determine the probability that the remedy would have been exceeded in the 1799 trial described in the text, assuming the variations in weight were caused by random and independent factors operating with respect to the manufacture of each coin.

2. Given the results of (1), would it be appropriate to limit the remedy to, e.g., $3\sigma\sqrt{n}$, where σ is the standard deviation of a single coin and n is the number of such coins?

3. Would Newton have cared how many coins were in the Pyx at the time his trials were conducted?

4. Why not conduct sampling inspection by weighing individual coins?

Source

Stigler, *Eight Centuries of Sampling Inspection: The Trial of the Pyx*, 72 J. Am. Stat. Assoc. 493 (1977); R. Westfall, *Never at Rest: A Biography of Isaac Newton* 607–10 (1980).

---

[8]In troy weights, 24 gr = 1 dwt, 20 dwt = 1 oz, and 12 oz =1 lb; thus 5,760 gr = 1 lb.

1.4   A measure of correlation

Suppose that X is a man's height and Y is his weight. To what extent are the two associated in a population of men? A useful measure of association is the *covariance* between X and Y, defined as the average value of the product of the deviation of each variable from its mean. In symbols,

$$cov(X, Y) = E[(X - EX)(Y - EY)].$$

Since above-average values of weight tend to occur with taller men, and below-average weights with shorter men, the average product of deviations will be positive. On the other hand, if short men tended to be obese and tall men to be skinny, the average product would be negative. If there were no association between height and weight, the covariance would be zero.

The covariance does not change when constants are added to X and Y, i.e., covariance is invariant under changes in the means of the variables. If a scale agreeably short-weights everyone by five lbs, the covariance of weight with height is unchanged. However, $cov(X, Y)$ does depend on the unit scale of the variables, so that multiplying X by a constant multiplies the covariance by that constant, and similarly for Y. Shifting from inches to centimeters in measuring height will change the covariance with weight. To achieve a dimensionless measure that is invariant under changes in location or scale, we apply the covariance to the standardized versions of X and Y, say $x^* = (X - EX)/sd(X)$, and similarly for Y^*, thereby arriving at Pearson's *product-moment correlation coefficient*, denoted by ρ. In symbols,

$$\rho = cov(X^*, Y^*) = E[X^* \cdot Y^*] = \frac{cov(X, Y)}{sd(X) \cdot sd(Y)}.$$

When the population correlation coefficient, ρ, is being estimated from a sample, the sample coefficient, usually denoted by r, is calculated in the same way, using the average in the data of the products of deviations in the numerator and sample standard deviations in the denominator. Note that the adjustment required for the sample standard deviation no longer pertains because the same adjustment should be used in both the numerator and denominator of r, and thus cancels out.

The correlation coefficient takes on a maximum value of plus or minus one when Y is a linear function of X, the case of perfect positive or negative correlation. A large positive or negative value of r signifies a strong *linear* dependence between X and Y. When r is large, prediction of Y from X is feasible; when $r = 0$, (linear) prediction of Y based on X is no more accurate than prediction of Y when X is unknown. Figures 1.4a–1.4c illustrate three sample data sets from populations with varying degrees of positive correlation.

The correlation coefficient is by far the most common measure of dependence in data. While there are no universal rules defining strong vs. weak associations, it is often the case in the social sciences that correlation coef-

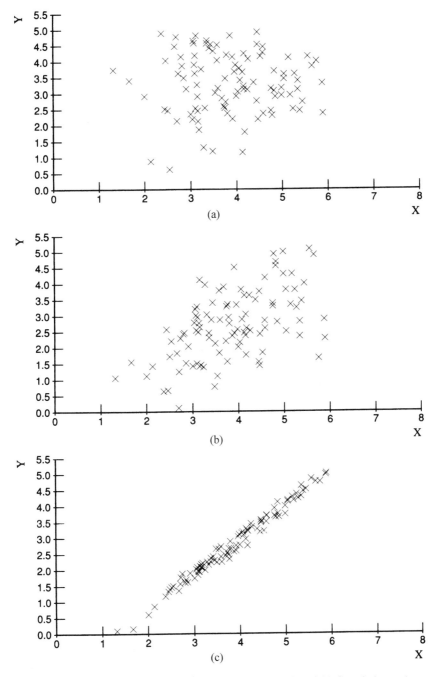

FIGURE 1.4. (a) Correlation $\approx$ 0; (b) Correlation $\approx$ 0.5; and (c) Correlation $\approx$ 1

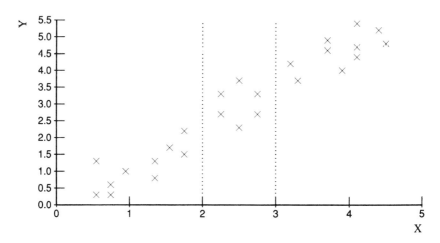

FIGURE 1.4d. Correlation depends on the range of the variables

ficients of 0.50 or more are regarded as signifying strong relationships. For example, LSAT scores are regarded as sufficiently good predictors of law school grades that law school admissions are based on them. Yet the correlation between first-year law school grade point averages and the LSAT test scores is usually less (and sometimes much less) than 0.60.

An important property of r (or its population version, ρ) that helps to interpret the coefficient is that r^2 measures the proportion of total variation of Y that is "explained" by variation in X, while $1 - r^2$ measures the proportion of total variation in Y that is unexplained by variation in X. By "explained" we do not mean to imply that Y is necessarily caused by X, but only that because of the association between them some of the variation in Y may be predicted from X. This notion is made more precise in Section 13.3.

Care must be exercised when interpreting association based on the correlation coefficient for the following reasons:

- The coefficient may show a high value due to correlation of extreme values (outliers) even though there is a weak or non-existent linear relation in the center of the data. See Figure 13.3c at p. 371 (top panel) for an example.

- Conversely, a strong linear correlation in a body of data may be diminished by outliers.

- Even if there is a strong (although not perfect) linear relation between X and Y, the proportion of Y's variability that is explained by X's variability will be diminished as the observed range of X is decreased. See Figure 1.4d, where over the entire range of X there is a correlation between X and Y and, relative to the total variation in Y, the error in predicting Y from X is small. In the narrow range indicated between the dotted lines, however, the observed correlation is weak, and the error in predicting Y from X will be large compared with the reduced variability in Y.

- ρ measures the strength of linear dependence between X and Y, but may fail to be a good measure of even strong non-linear relations. For example, if X is symmetrically distributed around zero, and if $Y = X^2$, then Y is totally (albeit non-linearly) dependent on X, and may be perfectly predicted given X, yet it can be shown that $\rho = 0$. While X and $Y = X^2$ does not often occur naturally as a variable pair, statisticians often add a squared variable to their statistical models to check for the presence of non-linear relations. The example demonstrates that, while in general if X and Y are independent then $\rho = 0$, the converse is not necessarily true. See Figure 13.3c (bottom panel) for another example.

- Correlation does not necessarily imply causation. See Section 10.3 at p. 291 for a discussion of the distinction.

Correlation measures for binary variables are discussed in Section 1.5 and some other correlation measures for discrete variables are discussed in Section 6.3.

1.4.1   Dangerous eggs

A national trade association of egg manufacturers sponsored advertisements stating that there was no competent and reliable scientific evidence that eating eggs, even in quantity, increased the risk of heart attacks. The Federal Trade Commission brought a proceeding to enjoin such advertisements on the ground that they were false and deceptive. In the trial, the Commission staff introduced data prepared by the World Health Organization reporting cholesterol consumption and rates of ischemic heart disease (IHD) mortality in 40 countries. Experts who testified for the FTC have since collected data for egg consumption and IHD mortality in 40 countries; these are shown in Table 1.4.1 and Figure 1.4.1. For purposes of reference, a single medium size egg weighs approximately 60 grams.

In these data, the sum of the IHD mortality rates is 15,807 and the sum of the squares of the rates is 8,355,365. The sum of the egg consumption rates is 1,130.5 and the sum of the squares of the rates is 37,011.87. The sum of the products of the IHD mortality and egg consumption rates is 490,772.4.

Questions

1. Estimate Pearson's correlation coefficient for the data by comparing Figure 1.4.1 with Figures 1.4a–1.4c. Check your estimate by calculating the correlation. Interpret your result. [Hint: $\mathrm{Cov}(X, Y)$ can be estimated by $(\sum X_i Y_i - n \bar{X} \cdot \bar{Y})/(n - 1)$.]

2. What inference would you, as counsel to the FTC, draw from this result with respect to U.S. egg consumption and IHD? As counsel for the egg trade association, what points would you make in reply?

TABLE 1.4.1. Ischemic heart disease mortality and egg consumption among men aged 55–64 in 40 countries

	Countries	(1) Men aged 55–64 ischemic heart disease mortality per 100,000 population	(2) Egg consumption in grams per capita per day	(3) Calories per day	(4) Rank of column (1)	(5) Rank of column (2)
1.	Argentina	411.5	20.2	3,188	23	10
2.	Australia	730.9	27.3	3,121	37	17
3.	Austria	442.0	37.3	3,532	26	33
4.	Belgium	435.7	30.8	3,617	25	23
5.	Bulgaria	375.5	22.8	3,658	22	14
6.	Canada	689.7	35.2	3,245	34	30
7.	Chile	202.0	13.7	2,589	10	5
8.	Costa Rica	219.8	20.6	2,567	12	11
9.	Cuba	371.3	22.8	2,527	21	13
10.	Denmark	578.2	28.3	3,354	33	19
11.	Dominican Republic	129.3	10.2	2,429	4	3
12.	Ecuador	91.4	8.6	2,206	2	2
13.	Egypt	134.7	4.8	3,087	5	1
14.	Finland	1,030.8	28.4	3,098	40	20
15.	France	198.3	35.5	3,626	9	31
16.	Greece	290.2	30.9	3,783	16	24
17.	Hong Kong	151.0	32.3	2,464	7	27
18.	Hungary	508.1	43.5	3,168	28	38
19.	Ireland	763.6	33.8	3,430	38	29
20.	Israel	513.0	57.9	3,386	29	40
21.	Italy	350.8	29.7	3,648	20	22
22.	Japan	94.8	42.5	3,091	3	36
23.	Mexico	140.9	17.6	2,724	6	8
24.	Netherlands	519.8	30.9	3,302	30	25
25.	New Zealand	801.9	41.9	3,120	39	35
26.	Nicaragua	44.5	28.9	2,568	1	21
27.	Norway	570.5	25.0	3,052	32	15
28.	Paraguay	163.0	16.8	3,084	8	7
29.	Poland	346.7	32.5	3,618	19	28
30.	Portugal	239.1	12.6	2,740	13	4
31.	Rumania	246.8	28.2	3,422	14	18
32.	Spain	216.5	41.6	3,356	11	34
33.	Sweden	563.6	31.2	3,087	31	26
34.	Switzerland	321.2	25.9	3,225	17	16
35.	U.K.	710.8	35.6	3,182	35	32
36.	U.S.	721.5	43.4	3,414	36	37
37.	Uruguay	433.9	13.9	2,714	24	6
38.	Venezuela	339.2	18.9	2,435	18	9
39.	West Germany	462.5	46.3	3,318	27	39
40.	Yugoslavia	251.5	22.2	3,612	15	12

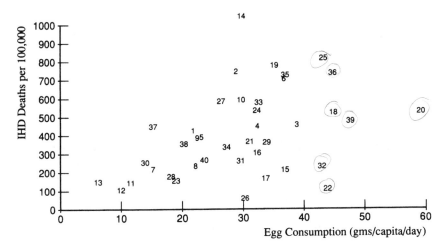

FIGURE 1.4.1. Ischemic heart disease mortality per 100,000 vs. egg consumption

3. What effect would errors in the data have on the correlation coefficient?

Source

In re National Comm'n on Egg Nutrition, 88 F.T.C. 89 (1976), *modified*, 570 F.2d 157 (7th Cir. 1977), *cert. denied*, 439 U.S. 821 (1978).

Notes

A large international study has suggested that variation in coronary artery disease among men may also be explained by variation in levels of hepatic iron. R.B. Laufer, *Iron Stores and the International Variation in Mortality from Coronary Artery Disease*, 35 Medical Hypotheses 96 (1990). For example, in 19 countries for which data were available, CAD mortality rates were correlated with iron ($r = 0.55$) and quite highly correlated with a variable consisting of the product of iron and cholesterol ($r = 0.74$). *Id.*, Table 2, at 99. Recent studies suggest that, although egg yolks contain a large amount of cholesterol (e.g., 213 mg.), two eggs a day when added to a low-fat diet have little effect on plasma cholesterol levels. See, e.g., Henry Ginsberg, et al., *A dose response study of the effects of dietary cholesterol on fasting and postprandial lipid and lipoprotein metabolism in healthy young men*, 1.4 Arteroscler. Thrombosis 576–586 (1994).

1.4.2   Public school finance in Texas

In many states, interdistrict disparities in school expenditures (due primarily to differences in amounts raised by local property taxation) led to widespread attacks on local financing of education in the 1970s. The seminal case was *Serrano v. Priest*, 5 Cal.3d 584, 487 P.2d 1241 (1941), which held that reliance on local property taxes to fund public education "invidiously discriminates against the poor because it makes the quality of a child's education a function of the wealth of his parents and neighbors." 5 Cal.3d at 589, 487 P.2d at 1244.

In the *Rodriguez* case, brought after *Serrano*, plaintiffs attacked the Texas system of dual financing, which depended in part on local property taxes. Plaintiffs sought to prove that in Texas "a direct correlation exists between the wealth of families within each district and the expenditures therein for education." As a result, plaintiffs argued, the financing system was designed to operate to the detriment of the comparatively poor.

On appeal from a judgment for the plaintiffs, the Supreme Court reversed. The Court observed that the principal evidence adduced in support of the comparative-discrimination claim was a study by a Professor Berke, who asserted that there was a positive correlation between expenditures on education per pupil and the wealth of school districts, measured in terms of assessable property per pupil, and a positive correlation between district wealth and the personal wealth of its residents, measured in terms of median family income. See Table 1.4.2. The method of sample selection was not given. The Court observed that if these correlations could be "sustained" then it might be argued that expenditures on education were dependent on personal wealth. However, the Court found that the evidence did not support the district court's conclusion to that effect:

TABLE 1.4.2. Texas school finance sample data

| Market value of taxable property per pupil | 110 districts out of 1200 | | |
	Median family income from 1960	Percent minority pupils	State and local revenues per pupil
Above $100,000 (10 districts)	$5,900	8%	$815
$100,000–50,000 (26 districts)	4,425	32	544
$50,000–30,000 (30 districts)	4,900	23	483
$30,000–10,000 (40 districts)	5,050	31	462
Below $10,000 (4 districts)	3,325	79	305

Professor Berke's affidavit is based on a survey of approximately 10% of the school districts in Texas. His findings previously set out in the margin [Table 1.4.2] show only that the wealthiest few districts in the sample have the highest median family income and spend the most on education, and that the several poorest districts have the lowest family incomes and devote the least amount of money to education. For the remainder of the districts—96 districts composing almost 90% of the sample—the correlation is inverted, i.e., the districts that spend next to the most money on education are populated by families having next to the lowest median family incomes while the districts spending the least have the highest median family incomes. It is evident that, even if the conceptual questions were answered favorably to appellees no factual basis exists upon which to found a claim of comparative wealth discrimination.

Id. at 26–27.

Questions

The Court's conclusion is susceptible to two interpretations: (i) there is no positive overall correlation between district expenditures per pupil and median family income because of the inverse relation in the central body of data, or (ii) there is such a correlation, but it depends on a few extreme districts and that is insufficient to brand the entire system.

1. Test proposition (i) by computing and interpreting a correlation coefficient for the data in Table 1.4.2. Does the size of the coefficient take account of the inverse relation in the center of the data?

2. What objections do you have to proposition (ii)?

Source

San Antonio Independent School District v. Rodriguez, 411 U.S. 1 (1973).

Notes

For more recent litigation over the validity under federal and state constitutions of using local property taxes to finance public schools, see Jonathan M. Purver, *Annotation: Validity of Basing Public School Financing System on Local Property Taxes*, 41 A.L.R. 3d 1220 (1996). Educational production models estimated from statistical data have been used in a number of these cases. See Sections 13.6.2 and 13.6.3 for two examples.

1.5   Measuring the disparity between two proportions

It is remarkable how many important legal issues involve a comparison of two proportions. To make the discussion specific, consider the example of a test taken by men and women on which men are observed to have a higher pass rate. Here are three candidates (there are more to come) to describe the disparity in pass (or fail) rates:

The difference between pass (or fail) rates

This measure has the desirable properties that (i) the difference between pass rates is the same as the difference between fail rates (apart from a change in sign), and (ii) it allows calculation of the number of women adversely affected by their lower pass rate. To illustrate the first property, if 72% of the men pass but only 65% of the women, the difference is 7 percentage points, the same absolute difference that arises from using the 28% failure rate for men and the 35% failure rate for women. As for the second property, the 7 percentage-point shortfall for women multiplied by the number of women who have taken or will take the test yields the number of women adversely affected by the disparity. However, the difference may be interpreted as more or less substantial depending on the level of rates. As one court observed, "[a] 7% difference between 97% and 90% ought not to be treated the same as a 7% difference between, e.g., 14% and 7%, since the latter figures evince a much larger degree of disparity."[9]

The ratio of the pass (or fail) rates

This measure, sometimes referred to as relative risk, takes account of the level of rates. The relative risk is 1.08 in the 97%–90% case and 2.0 in the 14%–7% case. However, relative risk is generally different for pass and fail rates. If fail rates are used, the 97%–90% case shows a *larger* disparity (3.33 against women) than the 14%–7% (1.08 against women), reversing the order of disparity based on pass rates. When pass rates are high (i.e., above 0.5) there is a "ceiling effect" that constrains rate ratios to at most 2.0. At the same time, small differences produce highly discrepant ratios between fail rates. This inconsistency is most troublesome when there is no clear rationale for preferring one rate to the other.

   The relative risk is frequently used in biostatistics and epidemiology (see Chapter 10) to describe hazards to human health. We use this terminology for both pass and fail rates even though it is a little inappropriate to describe passing a test as a risk.

---

[9]*Davis v. City of Dallas*, 487 F. Supp. 389, 393 (N.D. Tex. 1980).

The odds ratio

The odds in favor of an event are defined as the probability that the event will occur divided by the probability that it will not. That is, if the probability of an event is p, the odds on the event are $p/(1 - p)$.

The ratio of two odds, unsurprisingly called the odds ratio, is another statistic frequently encountered in statistical and epidemiological work. For the test taken by men and women, the odds ratio is simply the odds on passing for men divided by the odds on passing for women. In the 97%–90% case, the odds on passing for men are $0.97/0.03 = 32.3$; the odds on passing for women are $0.90/.10 = 9$. The odds ratio is then $32.3/9 = 3.6$ in favor of men. In contrast, the odds ratio in the 14%–7% case is $(0.14/0.86)/(0.07/0.93) = 2.16$ in favor of men. Thus, the odds ratio ranks the disparity in the first case as greater than that in the second, contrary to the relative risk per ratio (1.08 v. 2.0).

An important property of the odds ratio is invariance: it remains unchanged if both the outcome and the antecedent factor are inverted. For example, the ratio of the odds on passing for men to the odds on passing for women is equal to the ratio of the odds on failing for women to the odds on failing for men. In the 97%–90% case, the odds on failing for women are 1/9 while the odds on failing for men are 3/97, yielding the same odds ratio as before, 3.6. If only one variable is inverted, the resultant odds ratio is equal to the reciprocal of the original ratio. When fail rates are substituted for pass rates, the odds ratio for men vs. women is reciprocated (in the 97%–90% case, $1/3.6 = 0.278$).

In general, whenever the relative risk is greater than 1, the odds ratio is greater than the relative risk. Whenever the relative risk is less than 1, the odds ratio is smaller. Most important, when the outcome rate is low, the odds ratio is approximately equal to the relative risk. Compare, for example, the relative risk of failing with the odds ratio on failing for women vs. men in the 97%–90% case. The relative risk is 3.33; the odds ratio is 3.6.

An additional and important virtue of the odds ratio is that it is the same regardless of which factor is treated as antecedent and which as outcome. Thus, the ratio of the odds on passing for men to the odds on passing for women is equal to the ratio of the odds on being male among those who passed to the odds on being male among those who failed. This invertibility is important when "prospective" information is required to show how the probability of a certain outcome depends on an antecedent factor. So-called cohort studies—in which a selected group is divided on the basis of an antecedent factor and then followed for some time to observe the outcome—provide this type of information directly. But it may be difficult or impossible to use this study design. An outcome may be so rare, e.g., the occurence of cancer due to an environmental toxin, that identification of a reasonable number of outcomes would require too large a sample and too long a follow-up period. The alternative is a "retrospective" study in which the investigator selects two populations based on the outcome factor and then compares the occurrence rates of the antecedent factor in the two groups. So-called case-control studies are of this type.

A retrospective study in which one chooses equal numbers in both outcome groups can be statistically more powerful than a prospective study, especially for rare outcomes. A drawback is that the individual rate estimates obtained from retrospective studies are seldom of direct interest. If, for example, we take samples of people who pass a test and people who do not and determine the respective sex ratios, we can estimate the probability that a person passing the test was a man, but this is not the same as the probability that a man would pass the test. Do you see why not?

Because the odds ratio is invertible, however, the odds ratio computed from retrospective data can be interpreted in the more relevant prospective sense. If a retrospective study shows that the odds on being male among persons passing a test are three times the odds on being male among persons failing the test, it is also true that the odds on passing for men are three times the odds on passing for women. Given the approximate equality of the odds ratio and the relative risk when an outcome rate is low, the odds ratio is a useful surrogate for the relative risk, which could not otherwise be estimated from retrospective data.[10]

A drawback to the odds ratio (and the relative risk) is that, because they deal with ratios of rates, no account is taken of the absolute numbers involved. Suppose we compare a 97%–90% pass rate disparity with a 14%–5% disparity. The odds ratio for the first case (3.6) shows a greater disparity than the odds ratio for the second (3.1). If, however, 1,000 men and 1,000 women take each test, the excess number of women failing the first test is 70 and the excess number failing the second is 90. When the number of persons adversely affected is arguably the best measure of practical impact, the difference in rates, or attributable risk (see Section 10.2), is a more direct measure than the odds ratio. For example, in the United States, the odds ratio relating smoking to lung cancer is about five times larger than the odds ratio relating smoking to heart disease. However, because the incidence of heart disease is much greater than the incidence of lung cancer, a smoking-cessation policy would save many more people from heart disease than from lung cancer. The odds ratio, perhaps the best overall measure for studies of causality, may not be as useful as attributable risk in allocating resources for public policy purposes.

See Chapter 10 for further discussion of this subject.

---

[10]While case-control (retrospective) studies have efficiency advantages over cohort (prospective) studies, they are not easily constructed to be free of bias. In particular, ensuring that controls are similar to cases with respect to potentially confounding factors is notoriously difficult, and many epidemiologic studies are fatally flawed because of failure to adjust adequately for such factors, either by design or in the analysis. For this reason experimental studies, unlike observational studies, incorporate randomization whenever possible to diminish the likelihood that confounding factors are substantial sources of bias.

TABLE 1.5.1. Department store proficiency test:
Test performance by race

	Pass	Fail	Totals
Blacks	448	322	770
Whites	240	101	341
Totals	688	423	1,111

1.5.1 Proficiency test with a disparate impact

An African-American woman was hired as a "fitter" in a department store during the Christmas season. A condition of employment was that all employees were required to take a proficiency test, which included an interview, before being hired. Due to the rush of the season the woman was hired and took the test two weeks later. She failed and was discharged. She sued, contending that a disparate number of African-Americans failed the test. The data are in Table 1.5.1.

Under the four-fifths rule of the U.S. Equal Employment Opportunity Commission (29 C.F.R. 1607.4(D) (1998)), the EEOC tests for adverse impact on a minority group by calculating the rate of selection for each race, sex, or ethnic group and determining whether the rate for that group is less than 80% of the rate for the group with the highest rate. If so, adverse impact is presumed, and the procedure must be validated as job-related. The rule provides as follows:

A selection rate for any race, sex, or ethnic group which is less than four-fifths (4/5) (or 80%) of the rate for the group with the highest rate will generally be regarded by the Federal enforcement agencies as evidence of adverse impact, while a greater than four-fifths rate will generally not be regarded by Federal enforcement agencies as evidence of adverse impact. Smaller differences in selection rate may nevertheless constitute adverse impact, where they are significant in both statistical and practical terms or where a user's actions have discouraged applicants disproportionately on grounds of race, sex, or ethnic group. Greater differences in selection rate may not constitute adverse impact where the differences are based on small numbers and are not statistically significant, or where special recruiting or other programs cause the pool of minority or female candidates to be atypical of the normal pool of applicants for that group. Where the user's evidence concerning the impact of a selection procedure indicates adverse impact but is based upon numbers which are too small to be reliable, evidence concerning the impact of the procedure over a longer period of time, and/or evidence concerning the impact which the selection procedure had when used in the same manner in similar circumstances elsewhere, may be considered in determining adverse impact.

Questions

1. Compute the difference in rates, the ratios of rates, and the odds ratio for these data. Do this first with passing as the primary outcome of interest and then with failing as the primary outcome.

2. Does the test show a disparate impact on African-Americans under the express terms of the four-fifths rule or under the policies reflected in the rule?

3. Suppose that the test is only weakly predictive of proficiency for those who pass it, but is strongly predictive of a lack of proficiency for those who fail it (i.e., it is akin to a minimum qualifications test). Does that affect your answers to question 2?

Source

Adapted from J. Van Ryzin, *Statistical Report in the Case of Johnson v. Alexander's* (1987). For other cases on the subject of the four-fifths rule see Sections 5.5.1 and 6.2.1.

1.5.2 Bail and bench warrants

Elizabeth Holtzman, the Brooklyn district attorney, asked an investigator to determine the extent to which factors could be identified that would indicate whether an accused would not appear for trial if granted bail. One factor the investigator considered was prior bail violations for which a bench warrant had been issued. To study the relation between these factors, the investigator selected a sample of 293 defendents on the basis of outcome: in approximately half the cases (147) no bench warrant had been currently issued and in the other half (146) a bench warrant had been currently issued. The investigator then looked to see what numbers of warrants had previously been issued for each group, with the results shown in Table 1.5.2.

TABLE 1.5.2. Issuance of bench warrants: current issuance by previous issuance.

	Bench warrant previously issued	No bench warrant previously issued	Total
Bench warrant currently issued	A 54	C 92	146
No bench warrant currently issued	B 23	D 124	147
Total	77	216	293

Questions

1. What are the rates of bench warrants previously issued among those currently issued and not currently issued? Are these rates dependent upon the sample sizes chosen?

2. What are the rates of bench warrants currently issued among those previously issued and not previously issued? Are these rates dependent upon the sample sizes chosen?

3. How would you express the value of using prior issuance of a bench warrant as a predictor of the current need for a warrant?

4. Can you tell from these data how many non-appearances would be avoided if prior issuance of a warrant invariably resulted in denial of bail the second time?

Notes

Some important statistical studies have been made of the factors that are likely to indicate whether an arrested person, if released, would return for trial. Perhaps the most notable was the Manhattan Bail Project, conducted by the Vera Foundation in the early 1960s. This was an experiment to see whether arrested persons with close ties to the community would return for trial even if not required to post bail. The result was an overwhelming success: only about 1% (3/250) of those recommended by Vera for parole on this basis and who were paroled failed to show up for trial. Ares, Rankin and Sturz, *The Manhattan Bail Project: An Interim Report on the Use of Pre-Trial Parole*, 38 N.Y.U.L. Rev. 67 (1963). The success of the experiment caused it to be institutionalized and expanded across the country. See B. Botein, *The Manhattan Bail Project: Its Impact on Criminology and the Criminal Law Processes*, 43 Texas L. Rev. 319 (1964–65).

A significant and unanticipated finding of the Manhattan Bail Project study was that a disproportionately large number of those recommended for release (60%) were acquitted or had their cases dismissed, compared with 25% in the control group. A multiple regression statistical study (see Chapter 13) by the Legal Aid Society suggested that, after accounting for other relevant factors, pretrial incarceration alone increased the probability of conviction. However, a constitutional challenge to bail based on this study was rejected by the New York courts. As the Appellate Division held, evidently not understanding the import of the statistical findings, "It is not because bail is required that the defendant is later convicted. It is because he is likely to be convicted that bail is required." *Bellamy v. The Judges*, 41 A.D.2d 196, 202, 342 N.Y.S.2d 137, 144 (1st Dept.), *aff'd without opinion*, 32 N.Y.2d 886, 346 N.Y.S.2d 812 (1973).

For a case on a related subject see Section 6.3.1.

1.5.3   Non-intoxicating beer

Oklahoma passed a statute in 1958 that prohibited sale of "non-intoxicating" 3.2% beer to males under the age of 21 and to females under the age of 18. In a challenge to the statute, Justice Brennan, writing for the Court in *Craig v. Boren*, 429 U.S. 190 (1976), held that the gender-based differential constituted a denial to males 18–20 years of age of equal protection of the laws. The key passage in Brennan's opinion appears to be this:

> The most focused and relevant of the statistical surveys, arrests of 18–20 year-olds for alcohol-related driving offenses, exemplifies the ultimate unpersuasiveness of this evidentiary record. Viewed in terms of the correlation between sex and the actual activity that Oklahoma seeks to regulate—driving while under the influence of alcohol—the statistics broadly establish that 0.18% of females and 2% of males in that age group were arrested for that offense. While such a disparity is not trivial in a statistical sense, it hardly can form the basis for employment of a gender line as a classifying device. Certainly if maleness is to serve as a proxy for drinking and driving, a correlation of 2% must be considered an unduly tenuous "fit."

Id. at 201–202.

   This statement was based on a study showing that, in a four-month period in 1973 for 18–20 year-olds, there were 427 males and 24 females arrested in Oklahoma for driving under the influence of alcohol. The Court apparently (and inexplicably) derived its percentages by adding to the above figures arrests for drunkenness in the same period (966 males and 102 females) and dividing by figures for the Oklahoma total population in each category that were obtained from the census (69,688 males and 68,507 females).

Questions

1. In defense of the statute, using Brennan's percentages, compute (i) the relative risk of DWI arrest by gender, and (ii) the fraction of all DWI arrests that would be avoided if the male arrest rate could be reduced to the female rate. (This is the attributable risk due to maleness—see Section 10.2.)

2. Is Brennan's position reasonable, given the defects in the statistics he used, as a measure of the extent of the drinking and driving problem arising from 3.2% beer among the young?

CHAPTER 2

How to Count

2.1 Permutations and combinations

The key to elementary probability calculations is an ability to count outcomes of interest among a given set of possibilities called the sample space. Exhaustive enumeration of cases often is not feasible. There are, fortunately, systematic methods of counting that do not require actual enumeration. In this chapter we introduce these methods, giving some applications to fairly challenging probability problems, but defer to the next chapter the formal theory of probability.

Basic formula

A basic counting principle is this: if A can occur in m different ways and B in n different ways, then, absent any restrictions on either A or B, both can occur in m times n different ways. For example, if in a given political party there are three potential candidates for president, and four for vice-president, then, barring any prohibited marriages, there are twelve possible tickets for the two executive positions.

Permutations

A permutation is an ordered arrangement of distinct items into distinct positions or slots. If there are n items available to fill slot 1, there will be $n - 1$ items remaining to fill slot 2, resulting in $n(n - 1)$ ways to fill two slots. Continuing in this manner, there are $n \cdot (n - 1) \cdots (n - r + 1)$ ways to fill $r \leq n$ slots. This permutation number is given the symbol $_nP_r$. In particular, there are $_nP_n = n \cdot (n - 1) \cdots 2 \cdot 1$ ways to arrange all n items, and this number is given the special symbol $n!$, read "n factorial." Factorial notation is well defined for positive integers. For $n = 0$, because there is precisely one way to arrange an empty set of items into zero slots (the arrangement is vacuous, but it is an arrangement), the definition $0! = 1$ is adopted as consistent and useful. Note that we can write $_nP_r$ with factorial notation as $n!/(n-r)!$

Combinations

Suppose now that in selecting r items out of n our only interest is in the final *collection* of items as opposed to the number of *arrangements*. Then among the $_nP_r$ possible arrangements of these r items, each collection or *combination* is repeated $r!$ times (once for each rearrangement of the same r items), so that the number of combinations of r items out of n is $_nP_r/r!$ This combination number, sometimes given the symbol $_nC_r$, is most often denoted by the *binomial coefficient*, $\binom{n}{r}$, read "n choose r." Thus

$$\binom{n}{r} = {}_nC_r = \frac{{}_nP_r}{r!} = \frac{n!}{r! \cdot (n-r)!} = \frac{n \cdot (n-1) \cdots (n-r+1)}{r \cdot (r-1) \cdots 2 \cdot 1}.$$

Note that in the final expression both numerator and denominator contain r factors. Also note that $\binom{n}{r} = \binom{n}{n-r}$. As an example, there are $\binom{5}{2} = 5!/(2! \cdot 3!) = (5 \cdot 4)/(2 \cdot 1) = 10$ committees of two that can be formed from a pool of five people. There are also $\binom{5}{3} = 10$ committees of three that can be formed (e.g., those left in the pool after selecting committees of two).

The binomial coefficient also counts the number of different patterns of n items that are of two types, say r of type 1 and $n-r$ of type 0, when items of the same type are *indistinguishable*. Among the $n!$ potential arrangements, there are $r!$ repetitions due to indistinguishable rearrangements of type 1 items, and there are $(n-r)!$ repetitions due to rearrangements of type 0 items. Thus the $r! \cdot (n-r)!$ repetitions leave a total of only $n!/[r! \cdot (n-r)!]$ distinguishable patterns. For example, if a coin is tossed n times, there are $\binom{n}{r}$ distinct arrangements of outcomes with r heads and $n-r$ tails. More generally, if there are k types among n items—say, r_1 of type 1 ... and r_k of type k—with objects of the same type indistinguishable, then the number of distinct patterns is $n!/(r_1! \cdots r_k!)$; this is known as a *multinomial coefficient*. For example, there are 2,520 distinct gene sequences of length eight that can be formed from the genome $\{a, a, b, b, c, c, d, d\}$. Do you see why?

Stirling's formula

A beautiful approximation known as *Stirling's formula* may be used to calculate $n!$ when n is large:

$$n! \approx \sqrt{2\pi n} \cdot n^n \cdot e^{-n} = \sqrt{2\pi} \cdot n^{n+1/2} \cdot e^{-n},$$

where e is Euler's constant ≈ 2.718. A slightly more accurate version of Stirling's formula is

$$n! \approx \sqrt{2\pi} \cdot \left(n + \frac{1}{2}\right)^{n+1/2} \cdot e^{-(n+1/2)}.$$

For example, when $n = 10$, $n! = 10 \cdot 9 \cdot 8 \cdots 3 \cdot 2 \cdot 1 = 3{,}628{,}800$. Stirling's formula gives 3,598,696 (99.2% accuracy), while the second approximation gives 3,643,221 (99.6% accuracy). Stirling's formula may be applied to the factorials in a binomial coefficient to yield, for large n and r,

$$\binom{n}{r} \approx \frac{1}{\sqrt{2\pi}} \cdot \frac{n^{n+1/2}}{r^{r+1/2} \cdot (n-r)^{n-r+1/2}}.$$

For example, when $n = 2r$, the binomial coefficient $\binom{n}{r}$ is approximately $2^n/\sqrt{\pi r}$. This result is used in Sections 2.1.2 and 2.2.

Occupancy problems

Some probability problems correspond to placing balls at random into cells. The number of balls in each cell is known as the occupancy number for that cell, or the cell frequency. To compute the number of ways of obtaining a *set* of occupancy numbers (without regard to the order of the cells or the order of the balls), multiply (i) the number of ways of distributing the balls to arrive at a specific sequence of cell frequencies by (ii) the number of different sequences of cell frequencies with the given set of occupancy numbers. To calculate the probability of obtaining a particular set of occupancy numbers, assuming equally likely outcomes, divide the total number of distributions that produce that set of occupancy numbers by the total number of ways of putting the balls into the cells. If there are n balls distributed into k cells, the total number of possible distributions is k^n.

The birthday problem is an example of an occupancy problem. Given a room with N people, what is the probability that two or more people have the same birthday? How large must N be for this chance to be at least 1/2? In this problem there are $k = 365$ cells and N balls. No coincidences implies a set of occupancy numbers all zero or one. Assuming that all birth dates are equally likely, the probability of the sequence of cell frequencies $(1, 1, \ldots, 1, 0, 0, \ldots, 0)$ is $[N!/(1! \ldots 0!)]/k^N = N!/k^N$, and there are $\binom{k}{N} = k!/[N!(k-N)!]$ such sequences. Thus the probability of obtaining a set of occupancy numbers all zero or one is $(N!/k^N)k!/[N!(k-N)!] = k(k-1)\ldots(k-N+1)/k^N$, or $\left(\frac{365}{365}\right)\left(\frac{364}{365}\right)\ldots\left(\frac{365-N+1}{365}\right)$. For $N = 24$, this works out to just under 0.50 so that the chance of at least one coincidence is just over 0.50.

2.1.1   DNA profiling

DNA profiling has become standard in criminal proceedings in which identification of a defendant is an issue. We describe briefly the genetic background and introduce the statistical aspects raised by the method. Further issues are discussed at Sections 3.1.2 and 3.2.2.

Human beings have 23 pairs of chromosomes in every cell except egg and sperm cells, which have 23 single chromosomes. A chromosome is a very thin thread of DNA (deoxyribonucleic acid). The thread consists of two long strings of four chemical bases twisted to form a double helix. The four bases are abbreviated A,T,G, and C (which stand for adenine, thymine, guanine, and cytosine). In double-stranded DNA, the bases line up in pairs, an A opposite a T and a G opposite a C, so that if a sequence on one strand is known the other is determined. Before cell division, the two strands separate into single strands. Each strand then picks up free-floating bases from the cell in accordance with the A-T and G-C pairing, thus creating two identical double-stranded DNA helixes, one for each cell. The process is completed when the replicated chromosome pairs separate into daughter cells. This process assures uniformity of DNA throughout the cells of the body.

A *gene* is a stretch of DNA, ranging from a few thousand to tens of thousands of base pairs, that produces a specific product, usually a protein. The position that a gene occupies on the DNA thread is its *locus*. Genes are interspersed along the length of the DNA and actually compose only a small fraction of the total molecule. Most of the rest of the DNA has no known function.

Alternative forms of genes at the same locus, like those producing both normal blood and sickle-cell anemic blood, are called *alleles*. A person has two genes at each locus, one from the maternal chromosome and the other from the paternal chromosome: the two genes together are referred as the person's *genotype* at the locus. If the same allele is present in both chromosomes of a pair, the genotype is said to be *homozygous*. If the two are different, the genotype is said to be *heterozygous*. A heterozygous genotype with allele A from the maternal chromosome and allele B from the paternal chromosome cannot be distinguished from one in which allele A is from the paternal chromosome and allele B is from the maternal chromosome. However, genes on the Y chromosome can only have come from the father, which permits some spectacular lineage tracing– such as the evidence that Thomas Jefferson was indeed the father of Eston Hemings Jefferson, the younger son of his slave Sally Hemings. *DNA Test Finds Evidence of Jefferson Child by Slave*, N.Y. Times, November 1, 1998, at A1, col 5. Matrilineal descent can be traced using genes from mitochondrial DNA. Mitochondria are microscopic organelles responsible for energy storage and release found in the cell, but outside the nucleus, so they are not associated with the chromosomes. The transmission of mitochondria is from mother to child because the sperm has very little material other than chromosomes. All children of one woman will have identical mitochondrial DNA and this will be passed down through the female line to successive generations.

VNTR analysis

One group of DNA loci that were used extensively in forensic analysis are those containing Variable Numbers of Tandem Repeats (VNTRs). Technically, these are not genes because they have no known effect on the person. This is an important attribute for forensic work because it makes it less likely that the VNTRs would be influenced by natural selection or selection of mates, which could lead to different frequencies in different populations.

A typical VNTR region consists of 500 to 10,000 base pairs, comprising many tandemly repeated units, each some 15 to 35 base pairs in length. The number of repeats, and hence the length of the VNTR region, varies from one region to another, and different regions can be distinguished by their lengths. The variation in length of VNTR regions is a form of *length polymorphism. The term restriction fragment length polymorphism* (RFLP) refers to length polymorphism found in fragments of DNA snipped out by a biochemical method (using restriction enzymes) which isolates the same regions of DNA on the two chromosomes. The term allele is usually applied to alternative forms of a gene; here we extend the term to include nongenic regions of DNA, such as VNTRs. Each VNTR allele is distinguished by a characteristic band on an autoradiograph in which the molecular weight of the band reflects the number of repeats in the VNTR and determines its location. If the genotype is homozygous, only a single band will appear; if heterozygous, two bands will appear. To allow for measurement error that is roughly proportional to the fragment size, preset "match windows" (e.g. $\pm 2.5\%$) around each autorad band are used and two bands are declared to match only if their match windows overlap.

Typically, there is a large number of alleles at VNTR regions (usually 15 to 25 can be distinguished), no one of which is common. The number of genotypes (pairs of alleles) is far larger; and when the possibilities for different loci are combined, the number of allelic combinations quickly becomes astronomical. An example of an autorad in an actual case is shown in Figure 2.1.1.

PCR-based methods

The *polymerase chain reaction* (PCR) is a laboratory process for copying a chosen short segment of DNA millions of times. The process is similar to the mechanism by which DNA duplicates itself normally, except that by the use of enzymes only a segment of the DNA is reproduced. At present, the method is used to reproduce relatively short segments of DNA, up to 1,000 nucleotides in length, which is much shorter than most VNTRs.

There are significant advantages to this process over the VNTR process. First, it is possible to work with much smaller amounts of DNA, which is

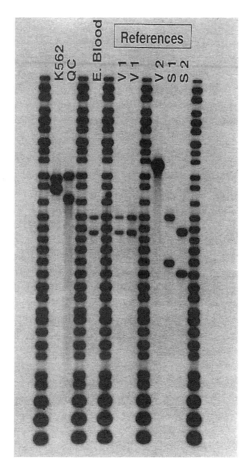

FIGURE 2.1.1. An autoradiograph from an actual case illustrating RFLP at the D1S7 locus. In the case, the suspects (S-1 and S-2) were charged with having beaten to death the two victims (V-1 and V-2). Blood stains were found on the clothing of one of the suspects. The lane marked E Blood is the DNA from those stains. The lanes marked V-1 and V-2 are DNA from the first and second victims, respectively; the lanes marked S-1 and S-2 are DNA from the first and second suspects, respectively. The other lanes are for molecular sizing and quality control purposes. Note that E blood matches Victim 1's blood, a result confirmed in the case by matching on 10 loci.

significant because forensic traces may involve minute amounts of DNA. Second, amplification of samples of degraded DNA is possible, which permits analysis of old and decayed samples. Third, it is usually possible to make an exact identification of each allele copied so that the measurement uncertainties associated with the identification of VNTRs by weight are largely eliminated.

There are also some disadvantages. The amplification process is so efficient that even a few molecules of contaminating DNA can be amplified with the intended DNA. Second, most markers used in PCR-based typing have fewer alleles than VNTRs and the distribution of allele frequencies is not as flat. Hence, more loci are required to produce the same amount of information about the probability that two persons share a profile. Furthermore, some of these loci are functional (they are genes, not just markers). Consequently, they are more likely to be subject to mating and natural selection and may not conform to the population-genetics assumptions used in evaluating the significance of a match. (For a discussion of such assumptions, see Section 3.1.2.)

On balance, PCR methods are definitely superior to VNTR methods and have largely replaced them in forensic analyses.

Questions

1. If there are 20 possible alleles at a locus, how many distinguishable homozygous and heterozygous genotypes are possible at that locus?

2. If four similar loci are considered, how many possible distinguishable genotypes are there?

Source

National Research Council, *The Evaluation of Forensic DNA Evidence* (1996).

Notes

Works on DNA profiling abound. Besides the NRC Report cited above, see, e.g., David H. Kaye & George F. Sensabaugh, Jr., *Reference Guide on DNA Evidence*, in Federal Judicial Center, *Reference Manual on Scientific Evidence*, (2d ed. 2000).

2.1.2   Weighted voting

A nine-person board of supervisors is composed of one representative from each of nine towns. Eight of the towns have approximately the same population, but one larger town has three times the population of each of the others. To comply with the one-man, one-vote doctrine of the federal constitution, the larger town's supervisor has been given 3 votes, so that there is a total of 11 votes.

In *Iannucci v. Board of Supervisors*, 20 N.Y.2d 244, 282 N.Y.S.2d 502 (1967), the New York Court of Appeals held that, "The principle of one man one vote is violated, however, when the power of a representative to affect the passage of legislation by his vote, rather than by influencing his colleagues, does not roughly correspond to the proportion of the population in his constituency ... Ideally, in any weighted voting plan, it should be mathematically possible for every member of the legislative body to cast the decisive vote on legislation in the same ratio which the population of his constituency bears to the total population." *Id.*, 20 N.Y.2d at 252, 282 N.Y.S.2d at 508.

Test whether the weighted-voting scheme complies with the *Iannucci* standard by computing the number of ways the supervisors can vote so as to permit the larger town, on the one hand, or a smaller town, on the other, to cast a decisive vote. A decisive vote may be defined as a vote which, when added to the tally, could change the result. Measures are carried by a majority; a tie defeats a measure. Note that the larger town can cast a decisive vote if the other eight towns are evenly split (4-4) or if they are divided 5-3, either for or against a measure. A smaller town can cast a decisive vote if the larger town is joined by two smaller towns, again either for or against a measure.

Questions

1. In how many ways can the eight small-town supervisors vote on a measure so that the larger-town supervisor has a deciding vote?

2. In how many ways can the larger-town supervisor and seven small-town supervisors vote on a measure so that the eighth small-town supervisor has a deciding vote?

3. Does the ratio of the two results indicate that by the *Iannucci* standard the larger town has been correctly compensated for its larger size?

4. Using Banzhaf's theory and Stirling's formula, as applied at the voter level, show that a voter in the larger town has $1/\sqrt{3}$ the voting power of a voter in one of the smaller towns.

Notes

The measure of voting power referred to in the problem was first suggested in John F. Banzhaf III, *Weighted Voting Doesn't Work: A Mathematical Analysis*, 19 Rutgers L. Rev. (1965). The *Iannucci* case was the first to adopt Banzhaf's theory.

At the voter level, the Banzhaf theory is in effect a square root rule: the probability of breaking a tie in a district with N votes is approximately $[2/(\pi N)]^{1/2}$ by Stirling's formula (see Section 2.1 at p. 44). At this level, the probability of breaking a tie seems too remote to influence a voter. The U.S. Supreme Court has so held. See *Whitcomb v. Chavis*, 403 U.S. 124 (1971); *Board of Estimate v. Morris*, 489 U.S. 688 (1989).

However, voters are probably influenced by the size of the anticipated plurality and may feel that individual votes are more important if the plurality will be smaller rather than larger. Under quite general conditions—particularly in close elections—the plurality will be approximately normally distributed, with an ascertainable probability P of exceeding a given value d in a district of a given size (in which everyone votes). If a second district is k times larger than the first, it follows from the assumption of normality that the plurality event in the larger district with the same probability P is not dk, but $d\sqrt{k}$. Thus, the square root relation derived from the tie-breaking situation has a more general justification in terms of pluralities than Banzhaf had given to it.

2.1.3   Was the bidding rigged?

Every six months the U.S. Maritime Administration issued requests for sealed bids for sale of its obsolete ships. Seven firms in the ship dismantling business ostensibly compete in bidding for these ships. In the last nine requests for bids, five firms submitted lowest bids once each, and two submitted lowest bids twice each. The firms deny collusion, arguing that the work is standard and the bidders have the same cost structure.

Questions

1. Assuming that each firm has the same probability of success on a bid and that success on one bid is independent of success on another, use a simple probability model to argue that the distribution of successes suggests collusive allocation.

2. Is the observed distribution of successes the most probable?

3. Are the assumptions of the model reasonable?

2.1.4   A cluster of leukemia

Between 1969 and 1979 there were 12 cases of childhood leukemia in Woburn, Massachusetts, when only 5.3 were expected on the basis of national rates. There are six approximately equal census tracts in Woburn. Six cases were clustered in census tract 3334. Lawsuits were brought on the theory that the leukemia had been caused by contaminated well water, although census tract 3334 did not receive the largest amount of this water. See Figure 2.1.4. For additional facts, see Section 11.2.2.

Question

In principle, how would you calculate the probability that a cluster of six or more leukemia cases would occur as a matter of chance in one district, assuming that the probability of leukemia is the same in all tracts and that 12 cases occurred?

Source

See Section 11.2.2.

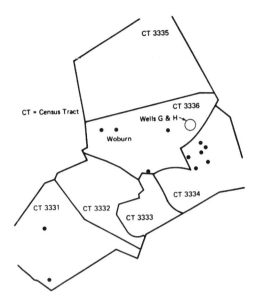

FIGURE 2.1.4. Residences of childhood leukemia patients at time of diagnosis, Woburn, Massachusetts, 1969-1979

2.1.5 *Measuring market concentration*

The U.S. Department of Justice guidelines to its enforcement policy for horizontal mergers relate enforcement to a measure of market concentration known as the Herfindahl Index (HHI). The HHI is an index of concentration calculated by squaring the percentage market share of each firm in the market and then adding those squares. To interpret the HHI, note that if all firms have equal market shares, the index divided into 100^2, or 10,000, is the number of such firms. Thus, when firms have unequal market shares, dividing the HHI into 10,000 gives the number of equal firms in a market of equivalent concentration. The index divided by 10,000 has another interpretation: it gives probability that two customers selected at random with replacement would be customers of the same firm. This probability is also called the "repeat rate."

Use of the HHI depends on the degree of concentration in the market and the effect of the merger. Here are two examples from the guidelines.

- Where the post-merger market is "unconcentrated", which is defined as a market with an HHI that is below 1,000. In such an "unconcentrated" market, the Department would be unlikely to challenge any merger. An index of 1,000 indicates the level of concentration that exists, for instance, in a market shared equally by 10 firms.

- Where the post-merger market is "moderately concentrated," with an HHI between 1,000 and 1,800. A challenge would still be unlikely, provided the merger increases the HHI by less than 100 points. If the merger increased the index by more than 100 points, it would "potentially raise significant competitive concerns," depending on the presence or absence of other relevant factors specified in the guidelines.

Questions

1. If there are N equal customers in a market, and the ith firm has a_i customers, write an expression for the number of different ways the customers may be distributed among firms without changing market shares.

2. Write the above expression in terms of market shares, with the ith firm having $1/n_i$ share of market, so that $1/n_i = a_i/N$.

3. Use the approximation[1] $n! \approx n^n/e^n$ to eliminate the number of customers, leaving only market shares, and take the Nth root to make it equal the number of firms in the market when each firm has the same percentage share of market. Is this "entropy" index a plausible measure of concentration?

4. Compare the "entropy" measure with the HHI for the situation in which four large firms share equally 80% of the market, with the remaining 20% shared first by 10, and then by 20, firms. Which measure shows the more concentrated market in terms of an equivalent number of equal firms as the number of small firms increases?

Source

Merger Guidelines of Department of Justice, Trade Reg. Rep. (CCH) ¶13,104 (1997); Michael O. Finkelstein and Richard Friedberg, *The Application of an Entropy Theory of Concentration to the Clayton Act*, 76 Yale L.J. 677, 696-97 (1967), reprinted in Michael O. Finkelstein, *Quantitative Methods in Law*, ch. 5 (1978) (hereinafter *Quantitative Methods in Law*).

2.2   Fluctuation theory

A sequence of equal deposits and withdrawals to and from an account can be likened to a series of coin tosses, with a deposit corresponding to heads and a withdrawal corresponding to tails. The amount on deposit at any time is represented by the excess of heads over tails. This process can be represented as a polygonal line that starts at the origin, with vertices at abscissas $0, 1, 2\ldots$ representing the aggregate number of tosses and ordinates at each vertex equal to the net lead of heads over tails (or tails over heads). The behavior of these paths is the subject of fluctuation theory, which deals with problems such as

---

[1]This is a crude form of Stirling's formula (see Section 2.1 at p. 44.)

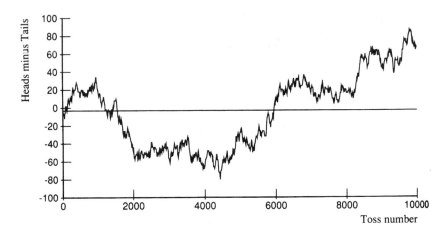

FIGURE 2.2a. Record of 10,000 tosses of an ideal coin

leads in coin tossings and random walks. The results of this theory are frequently counterintuitive because people tend to believe that a random process
that is balanced in probability will remain closely balanced in result throughout a sequence. The more correct view, reflected in various theorems, is that
leads in one direction, even over long stretches, are surprisingly likely to occur.
This behavior is illustrated by a result of 10,000 tosses of a fair coin, shown in
Figure 2.2a. The result is that, in a coin-tossing game, large leads (on the order
of the square root of the number of tosses) in favor of one party or the other, and
the persistence of a lead in favor of one party or the other, are more probable
than one might imagine. W. Feller gives the following example: If a fair coin
is tossed once a second for 365 days a year, "in one out of twenty cases the
more fortunate player will be in the lead for more than 364 days and 10 hours.
Few people believe that a perfect coin will produce preposterous sequences
in which no change in lead occurs for millions of trials in succession, and yet
this is what a good coin will do rather regularly." W. Feller, *An Introduction to
Probability Theory and Its Applications* 81 (3d ed. 1968).

Two basic points about paths representing fluctuations are noted here. First,
the number of possible paths representing n coin tossings in which there are a
heads and b tails is $\binom{n}{a} = \binom{n}{b}$. Second, the number of paths that start at ordinate
A, touch or cross the line corresponding to ordinate B, and end at ordinate C
(B lying below A and C) is equal to the number of paths that start at A and end
at a point C' that is as far below B as B is below C. This is the "reflection"
principle of D. André. See Figure 2.2b.

Assume that n is the total number of deposits and withdrawals, not necessarily equal in number, but each deposit or withdrawal equal in amount;
w is the size of each single deposit or withdrawal; and aw is the amount by
which the closing balance is less than the opening balance (a is a negative
integer if the closing balance exceeds the opening balance). The probability,

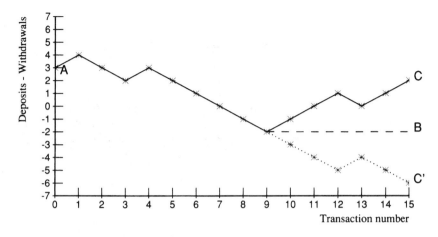

FIGURE 2.2b. The André Reflection Principle

P, that at some point the opening balance was decreased by kw or more is: $P = exp[-2k(k - a)]$, approximately, for $k \geq a$. (The term exp $[x]$ is alternate notation for the exponential function e^x). In the special case in which the opening and closing amounts are equal and $w = 1$, the expected maximum dissipation is $0.627\sqrt{n}$ and the median maximum dissipation is $0.589\sqrt{n}$.

If deposits and withdrawals are not equal in amount, no closed form approximation is known for the probability of any given dissipation of the account. In such cases, computer simulation must be used instead. See Finkelstein and Robbins, *infra*, for an example.

Further Reading

W. Feller, *An Introduction to Probability Theory and Its Applications*, ch. 14 (3d ed. 1968).

2.2.1   Tracing funds for constructive trusts

A person who defrauds another of money may be declared a constructive trustee of the funds for the defrauded party—if that party can trace the funds. There are various rules for tracing. If the funds are deposited in a bank account, withdrawals are deemed made first from non-trust funds (if they cannot be traced into new assets), but to the extent the account is reduced at any point a subsequent restoration by fresh deposits will not restore the trust. The sequence of deposits and withdrawals may therefore be critical. Suppose that there are $10 in trust funds in an account, and that during a day there are 50 one-dollar deposits and 50 one-dollar withdrawals. As in the case of most banks, the order of deposits and withdrawals within the day cannot be determined.

Questions

1. What is the probability that at some point during the day the trust funds were reduced to 0?

2. What is the expected maximum reduction in the trust?

3. If this situation persists for 10 days (i.e., 50 repeated deposits and withdrawals leaving the account even at the end of the day), is it reasonable to presume that the fund has been reduced to 0 at some point during the period?

Further Reading

Michael O. Finkelstein and Herbert Robbins, *A Probabilistic Approach to Tracing Presumptions in the Law of Restitution*, 24 Jurimetrics 65 (1983).

CHAPTER 3

Elements of Probability

3.1 Some fundamentals of probability calculation

Many probability puzzles can be made transparent with knowledge of a few basic rules and methods of calculation. We summarize some of the most useful ideas in this section.

Probabilities

In the classic formulation, probabilities are numbers assigned to elements of a *sample space*. The sample space consists of all possible outcomes of some conceptual experiment, such as flipping a coin or rolling dice. In making calculations, probability numbers are assigned to the "simple" (indecomposable) elements, or *events*, of the sample space. A subset of simple events is called a "compound" event, and its probability is defined as the sum of the probabilities of the simple events it contains. For example, if a coin is tossed four times, the simple events comprising the sample space are the possible sequences of heads and tails in four tosses; there are 16 such sequences, of which heads-tails-heads-tails (HTHT) is an example. A compound event is some subset of the 16 sequences, such as tossing two heads in four tails, which consists of the simple events HHTT, HTHT, HTTH, HHTT, THHT, and TTHH. The rules of the calculus of probabilities discussed in this section give some short cuts for calculating probabilities of compound events. Some probability chestnuts turn on the proper definition of the sample space, as the example below shows.

Example. Assume that boys and girls are born with equal frequency. Mr. Able says, "I have two children, and at least one of them is a boy." What is the probability that the other child is a boy? Mr. Baker says, "I went to the house of a two-child family, and a boy answered the door." What is the probability that the other child is a boy?

The answer to the first question is 1/3, but the answer to the second question is 1/2. The reason for this small paradox is that the sample space defined by Mr. Able consists of families with three birth sequences of children: boy-girl, girl-boy, and boy-boy. Since each has the same probability, and since only in the boy-boy case is the "other" child a boy, the probability of that event is 1/3. In Mr. Baker's case, by designating a particular boy (the one who answered the door) the sample space of family types is reduced to two: the door-answering child is a boy, and the other is a girl, or the door-answering child is a boy, and the other is also a boy. Since there are only two possible family types, the probability of boy-boy is 1/2, i.e., the probability of the other child being a boy is 1/2.

The probability assigned to a simple event can be any number between 0 and 1. For purposes of applications, useful assignments represent the long-range frequency of the events, such as 1/2 for the probability of tossing heads with a coin. However, nothing in mathematical theory compels any particular assignment of probabilities to simple events, except that an event certain *not* to occur has probability 0,[1] and the probabilities of the simple events constituting a sample space must sum to 1.

The concept of probability also applies to the degree of belief in unknown events, past or future. It has been shown that the calculus of probabilities introduced in this section can consistently be applied to both interpretations so that what probability "really is" need not be resolved to make use of the mathematical theory. See Section 3.6.

Complementary events

The probability of the negation of a given event is one minus the probability of the event. In symbols,

$$P[\bar{B}] = 1 - P[B].$$

Examples.

- If there is a 1 in 5 chance of selecting a black juror from a venire, there is a 4 in 5 chance of selecting a non-black juror.
- A game-show host (Monty Hall) presents a contestant with three closed doors. Behind one of them is a prize, with equal likelihood for doors A, B, or C. The contestant is asked to select a door; say he picks A. Before opening that door, Monty opens one of the other doors, which both he and the

---

[1]In infinite sample spaces, an event with probability 0 may still occur if the infinitude of possibilities is non-denumerable. For example, when a dart is thrown at a target, the probability that any specified point will be hit is zero because points have zero width, but still the dart hits the target at some point.

contestant know will be empty; say B. Then Monty presents the contestant with a choice: he may either stick with A or switch to C. Question: Is it better to stick, or to switch, or does it not matter?

Answer: The contestant has a 1/3 chance that the prize is behind A and a $1 - 1/3 = 2/3$ chance that the prize is behind B or C. Monty's opening of B doesn't change the probability that the prize is behind A, because it gives no information about the correctness of that choice. But by eliminating B, the probability of the prize being at C is now 2/3. In short, the best strategy is to switch.

Disjoint unions

- The union of two events A and B is the occurrence of A or B (or both), i.e., their *inclusive disjunction*. Thus, the union of two events encompasses both the occurrence of either event without the other (their *exclusive disjunction*) or the occurrence of both together (their *exclusive conjunction*).

- A and B are *disjoint* if they are mutually exclusive (i.e., their joint occurrence, or logical conjunction, is impossible).

- The probability of a union of disjoint events A and B is the sum of their individual probabilities. In symbols,

$$P[A \text{ or } B] = P[A] + P[B].$$

Example. Suppose:

$A = $ [Mr. F. is the girl's father and has her blood type], with $P[A] = 0.5$, and
$B = $ [Mr. F. is *not* the girl's father, but has her blood type], with $P[B] = 0.2$.

Then:

$$P[A \text{ or } B] = P[\text{Mr. F has the girl's blood type}] = 0.7.$$

Unions in general

For events A and B, not necessarily disjoint, the probability of A or B equals the sum of the individual probabilities less the probability of their conjunction, A and B. In symbols,

$$P[A \text{ or } B] = P[A] + P[B] - P[A \text{ and } B].$$

The joint probability $P[A \text{ and } B]$ is subtracted because the sum of the probability of A and the probability of B generally overestimates $P[A \text{ and } B]$ because it counts the probability of the joint occurrence of A and B twice: once in the probability of A alone and again in the probability of B alone. This double counting is shown by the top Venn diagram in Figure 3.1, where the areas within the circles are proportional to the probabilities of events A and B, respectively. The union of the two events is represented by the total area covered by A and B. The region labelled AB represents their intersection (conjunction).

It follows that the probability of A or B is no greater than the sum of their individual probabilities. In symbols,

$$P[A \text{ or } B] \leq P[A] + P[B].$$

The same reasoning leads to the more general inequality

$$P(A \text{ or } B \text{ or } C \ldots) \leq P(A) + P(B) + P(C) + \ldots$$

Although there are other inequalities due to Bonferroni, this is known as the Bonferroni inequality because it is the one most frequently encountered in practice.[2] See Section 6.2 for a discussion of the use of this inequality in setting levels of statistical significance for tests involving multiple comparisons.

Example. Referring to Section 2.1.4, assume that the probability of observing six or more cases of leukemia in any single census tract is 0.007925. Let $P[A_i]$ be the probability of observing such an event in census tract i. Then the probability of observing six or more cases of leukemia in at least one of six census tracts is no greater than

$$P[A_i] + P[A_2] + \ldots + P[A_6] = 6 \times 0.007925 = 0.04755.$$

In Section 2.1.4, where there were only 12 cases in total, the joint probability of finding six cases in two tracts is the minuscule 0.000006, and the probability of this occurrence in three or more tracts is zero. Bonferroni's first approximation is extremely accurate in this case.

Example. Assume that a person is infected with a virus for which there are two diagnostic tests. Let A = [diagnostic test 1 is positive], with $P[A] = 0.95$. Let B = [diagnostic test 2 is positive], with $P[B] = 0.99$

What is the probability that the infected person would test positive on at least one of the tests? By Bonferroni's inequality $P[A \text{ or } B] \leq 0.95 + 0.99 = 1.94$. Clearly this is not helpful. If the probability that both tests would be positive were $0.95 \times 0.99 = 0.9405$ (independent tests), then the probability of the union would be $0.95 + 0.99 - 0.9405 = 0.9995$. In this case the Bonferroni approximation is not useful because the probability of the joint event

---

[2]This is Bonferroni's first inequality. More generally, if $A_1, \cdots, A_n$ are n events, the probability of the union of $A_1, \cdots, A_n$ is obtained as follows: first, take the sum of each event separately, $\sum_i P[A_i]$; second, subtract the sum of all pairwise joint probabilities $\sum_{i \neq j} P[A_i \cap A_j]$ since the previous sum overestimates the union's probability; next, add back the sum of all joint probabilities in triples, $\sum_{i \neq j \neq k} P[A_i \cap A_j \cap A_k]$, since the previous step overcorrected slightly. Continue in this way by alternately subtracting and adding sums of joint probabilities until one adds or subtracts the final term $P[A_1 \cap \cdots \cap A_n]$. At any stage, the probability of the union may be approximated by the terms included up to that point, incurring an error no larger than the magnitude of the first omitted term.

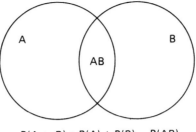

P(A or B) = P(A) + P(B) − P(AB)

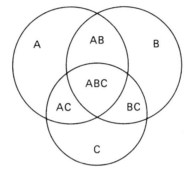

P(A or B or C) = P(A) + P(B) + P(C) − P(AB) − P(AC) − P(BC) + P(ABC)

FIGURE 3.1. Venn diagrams illustrating conjunction ("and") and inclusive disjunction ("and/or")

is not negligible. In fact, Bonferroni's inequality shows that $P[A\,\text{and}\,B]$ must be substantial, as follows. The probability that one or the other test shows a false negative reading, $P[\bar{A}\text{ or }\bar{B}]$, is, by Bonferroni's inequality, no greater than $P[\bar{A}]+P[\bar{B}] = 0.05+0.01 = 0.06$, which is quite accurate because the probability of two false negatives is very small ($0.01 \times 0.05 = 0.0005$, for independent tests). Taking complements implies that $P[A \text{ and } B] = P[\text{ not }(\bar{A} \text{ or } \bar{B})] = 1 - P[\bar{A} \text{ or } \bar{B}] \geq 1 - 0.06 = 0.94$.

Intersection

- The intersection of two events, A and B, is their conjunction, A and B.

- The probability of A and B is no greater than the probability of either event alone. This follows from the rule for mutually exclusive events, since

$$P[A] = P[A \text{ and } B] + P[A \text{ and } \bar{B}] \geq P[A \text{ and } B].$$

- If event B implies event A then $P[B] = P[A \text{ and } B]$, whence $P[B] \leq P[A]$. In words, an event is at least as likely to occur as any other event which implies it, and an event is no more likely to occur than any other event which it implies.

Conditional probability

The conditional probability of an event A given an event B with $P[B] > 0$ is defined as $P[A|B] = P[A \text{ and } B]/P[B]$. The conditional probability of A given B is the *relative* likelihood of occurrence of A, among all times when B occurs.

Example. Referring to the boy-girl family example at p. 57, the unconditional probability of a boy-boy family (A) is P[A]= 1/4. The probability of a family with at least one boy (B) is P[B]= 3/4. The probability of a boy-boy family conditional on there being at least one boy (B) is P[A and B]/P[B]= 1/4÷3/4 = 1/3. In Mr. Baker's case, the probability of the door-answering child being a boy (C) is P[C]= 1/2. Hence, the probability of a boy-boy family conditional on a boy answering the door is P[A and C]/P[C]=1/4 ÷ 1/2 = 1/2.

For non-vacuous events, one can always write the joint probability of events A and B as

$$P[A \text{ and } B] = P[A|B]P[B] = P[B|A]P[A],$$

although note that in general $P[A|B] \neq P[B|A]$. In the preceding example, P[A|B] = 1/3, but P[B|A] = 1.

Independent events

Events A and B are said to be *independent* if

$$P[A|B] = P[A] = P[A|\bar{B}].$$

In words, A and B are independent if the occurrence (or non-occurrence) of B does not affect the likelihood of occurrence of A. Thus, for independent events we have the multiplication rule

$$P[A \text{ and } B] = P[A|B] \cdot P[B] = P[A] \cdot P[B].$$

Averaging conditional probabilities

An overall probability equals a weighted average of conditional probabilities. In symbols,

$$P[A] = P[A|B] \cdot P[B] + P[A|\bar{B}] \cdot P[\bar{B}].$$

The weights are given by $P[B]$ and its complement. This follows by writing

$$P[A] = P[(A \text{ and } B) \text{ or } (A \text{ and } \bar{B})]$$
$$= P[A \text{ and } B] + P[A \text{ and } \bar{B}].$$

The conditional probabilities are often interpreted as specific rates.

Example. The overall promotion rate $P[A]$ in a company may be obtained from the specific rates for black and non-black employees (respectively $P[A|B]$ and $P[A|\bar{B}]$) by weighting the rates by the proportion of blacks and non-blacks, $P[B]$ and $P[\bar{B}]$.

3.1.1   Interracial couple in yellow car

In *People v. Collins*, 68 Cal. 2d 319, 438 P.2d 33 (1968)(*en banc*) an elderly woman, while walking through an alley in the San Pedro area of Los Angeles, was assaulted from behind and robbed. The victim said that she managed to see a young woman with blond hair run from the scene. Another witness said that a Caucasian woman with dark-blond hair and a ponytail ran out of the alley and entered a yellow automobile driven by a black man with a mustache and a beard.

A few days later, officers investigating the robbery arrested a couple on the strength of these descriptions and charged them with the crime.[3] It is not clear what led police to the Collins couple. A police officer investigating the robbery went to their house and took them to the police station, where they were questioned, photographed, and then released. When police officers came a second time to their house, apparently to arrest them, Malcolm was observed running out the back and was found hiding in a closet in a neighboring house. The officer found two receipts in Malcolm's pocket, indicating that he had just paid two traffic fines in a total amount equal to the amount of money stolen. Questioned as to the source of the money, Malcolm and Janet gave conflicting accounts. Finally, first Janet alone, and then the two together, engaged in a bargaining session with police in an effort to have charges dismissed against Malcolm, in particular, because he had a criminal record. Although no admissions were made, the tone of the conversion, according to the appellate court, "evidenced a strong consciousness of guilt on the part of both defendants who appeared to be seeking the most advantageous way out."

At their trial, the prosecution called an instructor of mathematics to establish that, assuming the robbery was committed by a Caucasian blonde with a ponytail who left the scene in a yellow car driven by a black man with a beard and mustache, the probability was overwhelming that the accused were guilty because they answered to this unusual description. The mathematician testified to the "product rule" of elementary probability theory, which states that the probability of the joint occurrence of a number of mutually independent events equals the product of their individual probabilities. The prosecutor then had the witness assume the following individual probabilities for the relevant characteristics:

---

[3]When defendants were arrested, the woman's hair was light, not dark, blond and the man did not have a beard. There was some evidence that the man had altered his appearance after the date on which the offense had been committed. The car was only partly yellow.

(a)	Yellow automobile	1/10
(b)	Man with mustache	1/4
(c)	Girl with ponytail	1/10
(d)	Girl with blond hair	1/3
(e)	Black man with beard	1/10
(f)	Interracial couple in car	1/1000

Applying the product rule to the assumed values, the prosecutor concluded that there was but one chance in twelve million that a couple selected at random would possess all these incriminating characteristics. The prosecutor gratuitously added his estimation that the "chances of anyone else besides these defendants being there ... having every similarity ... is somewhat like one in a billion." The jury convicted the defendants. On appeal, the Supreme Court of California reversed, holding that the trial court should not have admitted the evidence pertaining to probability.

In an appendix to the opinion, the court proposed a mathematical model to prove that, even on the prosecution's assumption that the probability that a random couple would answer to the description of the Collins couple (a "C-couple") was 1 in 12,000,000, there was a 41% probability of there being at least a second C-couple, if 12,000,000 selections were made. One way of describing the court's model involves imagining that there is a very large population of couples in cars, with the rate of C-couples in that population being 1 in 12,000,000. Out of this population one picks a couple at random, checks whether it is a C-couple, returns it to the population, and picks again. Assuming that one makes 12,000,000 selections to simulate the creation of the population, count the number of C-couples picked. Repeat this process many times and one finds that, in about 41% of these hypothetical populations in which there appeared at least one C-couple, there was more than one. The appendix concludes:

> Hence, even if we should accept the prosecution's figures without question, we would derive a probability of over 40 percent that the couple observed by the witnesses could be "duplicated" by at least one other equally distinctive interracial couple in the area, including a Negro with a beard and mustache, driving a partly yellow car in the company of a blonde with a ponytail. Thus, the prosecution's computations, far from establishing beyond a reasonable doubt that the Collinses were the couple described by the prosecution's witnesses, imply a very substantial likelihood that the area contained more than one such couple, and that a couple other than the Collinses was the one observed at the scene of the robbery.

Id. 438 P. 2d at 42.

Questions

1. Are the identifying factors listed likely to be statistically independent?

2. Suppose the frequency of identifying factors had been determined as follows: A survey was made of couples in cars and one in a thousand of them was interracial. In one of ten of those interracial couples there was a black man with a beard. In one of three interracial couples in which the black man had a beard the woman had blond hair. And so forth for the rest of the factors. In those circumstances would multiplication together of the frequencies have been correct?

3. Assuming the 1 in 12,000,000 figure were correct as the probability of selecting at random a couple with the specified characteristics, what objections do you have to the prosecutor's argument?

4. Assume, as in the appendix to the court's opinion, that the couples who might conceivably have been at the scene of the crime were drawn from some larger population in which the rate of C-couples was 1 in 12,000,000. Does it follow that the 41% probability computed in the appendix is relevant to the identification issue?

5. Does the conclusion of the court's appendix that the prosecutor's computation implies a very substantial likelihood that a couple other than the Collinses was the one observed at the scene of the robbery follow from the preceding mathematical demonstration?

Source

Michael O. Finkelstein and William Fairley, *A Bayesian Approach to Identification Evidence*, 83 Harv. L. Rev. 489 (1970), reprinted in *Quantitative Methods In Law*, ch. 3 (1978); see also, William Fairley and Frederick Mosteller, *A Conversation About Collins*, 41 U. Chi. L. Rev. 242 (1974).

Notes

The product rule has been involved in a number of cases both before and after *Collins*.

In *People v. Risley*, 214 N.Y. 75, 108 N.E. 200 (1915), the issue was whether defendant had altered a court document by typing in the words "the same." Defendant was a lawyer and the alteration helped his case. Eleven defects in the typewritten letters on the court document were similar to those produced by the defendant's machine. The prosecution called a professor of mathematics to testify to the chances of a random typewriter producing each of the defects found in the words. The witness multiplied these component probabilities together to conclude that the joint probability of all the defects was one in four billion. On appeal the court reversed, objecting that the testimony was "not based upon actual observed data, but was simply speculative, and an attempt to make inferences deduced from a general theory in no way connected with the matter under consideration supply [sic] the usual method of proof." *Id.* at 85, 108 N.E. at 203.

In *Miller v. State,* 240 Ark. 340, 399 S.W.2d 268 (1966), the expert testified that dirt found on defendant's clothing matched dirt at the burglary site as to color, texture, and density and that the probability of a color match was 1/10, a texture match was 1/100, and a density match was 1/1000. Multiplying these together, the expert concluded that the probability of an overall match was 1/1,000,000. On appeal, the conviction was reversed. The expert's testimony as to probability was inadmissible because he had neither performed any tests, nor relied on the tests of others, in formulating his probability estimates.

On the other hand, in *Coolidge v. State,* 109 N.H. 403, 260 A.2d 547 (1969), the New Hamphire Supreme Court cited *Collins,* but came to a different conclusion. The expert in that case obtained particles by vacuuming the victim's clothes and the defendant's clothes and automobile (where the crime was believed to have taken place). Forty sets of particles (one from the victim and the other from the defendant) were selected for further testing on the basis of visual similarity under a microscope. In these further tests the particles in 27 of the 40 sets could not be distinguished. Previous studies made by the expert indicated that "the probability of finding similar particles in sweepings from a series of automobiles was one in ten." The expert concluded that the probability of finding 27 similar particles in sweepings from independent sources would be only one in ten to the 27th power. On cross-examination, the expert conceded that all 27 sets may not have been independent of one another, but the court found that this went to weight rather than admissibility and affirmed the conviction.

3.1.2   Independence assumption in DNA profiles

We continue the discussion of DNA profiles that was begun in Section 2.1.1.

In forensic science applications, a sample of DNA that is connected with the crime is compared with the suspect's DNA. For example, in a rape case, DNA in semen found in the victim may match the suspect's DNA (which would be incriminating) or may not (which would be exculpatory). If there is a match, its forensic power depends on the probability that such a profile would have been left by a randomly selected person, if the suspect was not responsible. This in turn depends on the frequency of matching profiles in the population of persons who could have left the trace.

To estimate such frequencies, DNA laboratories begin with the frequency of the observed allele at each locus included in the profile. Each heterozygous genotype frequency is determined by multiplying together the frequencies of the maternal and paternal alleles that constitute the genotype, and multiplying that product by two. For homozygous genotypes, the frequency of the observed allele should be squared. The independence assumption that underlies the multiplications is justified on the belief that people mate at random, at least with respect to VNTR alleles, which are not known to correspond to any observable trait. However, the same assumption is made for PCR alleles, which may involve observable traits. A population whose genotypes are in these proportions is said to be in Hardy-Weinberg (HW) equilibrium.

Genes that are on the same chromosome are *linked*, that is, they tend to be inherited together. However, during the formation of a sperm or egg, the two members of a chromosomal pair lined up side by side can randomly exchange parts, a process called *crossing over* or *recombination*. Genes that are very close together on the same chromosome may remain associated for many generations while genes that are far apart on the same chromosome or on different chromosomes become randomized more rapidly.

To arrive at an overall frequency for a multilocus genotype, it is usual to take the product of the frequencies of the genotypes at the separate loci. This is justified on the assumption that genotypes at different loci are independent. A population in such a state is said to be in *linkage equilibrium* (LE). The state of LE, like HW, is the result of random mating, but a population only arrives at LE after several generations, whereas HW is arrived at in one generation. Because of recombination, loci that are close together on the same chromosomal pair approach LE more slowly than those far apart on the same pair or on different pairs. Departure from LE is called *linkage disequilibrium*, and is an important tool for locating marker genes close to true disease genes.

It has been objected that frequencies of apparent homozygotes will be greater than expected under HW if either (i) there are subgroups in the population that tend to in-breed and have higher rates of the particular alleles observed as homozygotes, or (ii) only a single band is found at a locus because the autorad band for the other allele erroneously has been missed. To protect against these possibilities, some laboratories conservatively estimate the frequency of a homozygote as twice the frequency of the observed allele, instead of the square of its frequency. This change will generally favor defendants by increasing the estimated frequency of matching homozygous genotypes in the population.

To test the HW assumption, Table 3.1.2 shows allele frequencies for 3 out of 28 different alleles found at locus D2A44 in samples of varying sizes from four

TABLE 3.1.2. Numbers of three alleles at locus D2S44 in samples from four populations

Allele type (i)	Canadian	Swiss	French	Spanish	Total alleles by type
⋮					
9	130	100	68	52	350
10	78	73	67	43	261
11	72	67	35	48	222
⋮					
Total alleles/sample	916	804	616	508	2844

white populations–Canadian, Swiss, French, and Spanish. Presumably there is not extensive mating at random across these populations so that hypothetical combined populations (e.g., the U.S. white population) would be vulnerable to departures from HW.

Questions

1. To arrive at an average figure for a total population consisting of the four subpopulations combined, compute the weighted average frequency of the homozygous genotype consisting of allele 9 (without adjustment for the risk of missing a band) and the heterozygous genotypes consisting of alleles 9 and 10 across the four subpopulations, with weights proportional to the sizes of the subpopulation samples. (These calculations assume that there is mating at random *within* each subpopulation, but not necessarily across subpopulations.)

2. Compute the frequency of the same alleles using the total population figures. (These calculations assume that there is mating at random within and across the subpopulations.)

3. Compare the results. Is HW justified?

4. Since there is not, in fact, mating at random across these subpopulations, what is a sufficient condition for HW in the total population, given HW in each of the subpopulations?

5. Consider a hypothetical population comprised of Canadians and non-Canadians in equal numbers. The Canadians have allele 9 frequency 130/916 as in Table 3.1.2, but the non-Canadians have allele frequency 786/916. Assuming that HW holds for Canadians and non-Canadians separately and they don't intermarry, does HW hold in the combined population?

Source

Federal Bureau of Investigation, *VNTR population data: a worldwide survey*, at 461, 464-468, *reprinted in* National Research Council, *The Evaluation of Forensic DNA Evidence*, Table 4.5 at 101 (1996).

Notes

DNA profiling has been subject to searching criticism by professional groups and in the courts, but its value is now well established. The National Research Council Report is one of many discussions of this subject. See, e.g., Faigman, et al., *Modern Scientific Evidence*, ch. 47 (1997).

3.1.3   Telltale fibers

Defendant Wayne Williams was charged with the murders of two young black males in Atlanta, Georgia. There had been ten other similar murders. Critical evidence against Williams consisted of a number of fibers found on the bodies that resembled fibers taken from his environment, in particular, certain unusual trilobal Wellman 181-b carpet fibers dyed English Olive. A prosecution expert testified that this type of fiber had been discontinued and that, on conservative assumptions, there had been only enough sold in a ten-state area to carpet 820 rooms. Assuming that sales had been equal in each of the ten states, that all Georgia carpet had been sold in Atlanta, and that only one room per house was carpeted, 81 Atlanta homes had carpet containing this fiber. Because, according to the expert, there were 638,992 occupied housing units in Atlanta, the probability that a home selected at random would have such carpeting was less than 81/638,992 or 1 in 7,792. Wayne Williams's bedroom had carpet with this fiber (although defendant subsequently disputed this).

Based on this testimony, the prosecutor argued in summation that "there would be only one chance in eight thousand that there would be another house in Atlanta that would have the same kind of carpeting as the Williams home." Williams was convicted. On appeal, the Georgia Court of Appeals held that the state's expert was entitled to discuss mathematical probabilities, that counsel in closing argument was not prohibited from suggesting inferences to be drawn from the evidence, and that such inferences might include mathematical probabilities.

Questions

1. Is the prosecutor's argument correct?

2. Does the 1 in 7,792 figure imply that there is 1 chance in 7,792 that the fibers did not come from Williams's home?

3. Should the evidence have been excluded because by itself (no other evidence being considered), the probability of guilt it implies is no more than 1 in 81?

Source

Williams v. State, 251 Ga. 749, 312 S.E.2d 40 (1983).

3.1.4   Telltale hairs

State v. Carlson, 267 N.W.2d 170 (Minn. 1978).

Defendant Carlson was charged with murdering a 12-year old girl who had last been seen in his company. Investigating officers found two foreign pubic hairs stuck to the skin of the deceased in her groin area and head hairs clutched in her hand. Gaudette, an expert on hair comparisons, testified that the pubic and head hairs found on the victim microscopically matched those of the accused. Based on a study he had done a few years earlier [the Gaudette-Keeping study described below] for the pubic hair the "chances those hairs did not come from David Carlson would be on the order of 1 chance in 800 for each hair," and for the head hair the figure was 1 in 4,500. Carlson was convicted. On appeal, the Supreme Court of Minnesota found that Gaudette's testimony on mathematical probabilities was improperly received because of "its potentially exaggerated impact on the trier of fact," but affirmed the conviction because the evidence was merely "cumulative and thus nonprejudicial on the facts of the case."

State v. Massey, 594 F.2d 676 (8th Cir. 1979).

Defendant Massey was charged with bank robbery. The robber wore a blue ski mask and a similar mask was recovered from the house of Massey's associate. At his trial, an FBI expert testified that three out of five hairs found in the mask were microscopically similar to one or more of nine mutually dissimilar hairs taken from Massey's scalp.[4] Under questioning by the judge, the expert testified that he had examined over 2,000 cases and that "only on a couple of occasions" had he seen hairs from two different individuals that he "could not distinguish." He also made reference to the Gaudette-Keeping study, which found, as he described it, "that a possibility that a hair which you have done or matched in the manner which I have set forth, there's a chance of 1 in 4,500 these hairs could have come from another individual." In summation, the prosecutor argued that, assuming there were as many as 5 instances out of 2,000 in which hairs from different individuals could not be distinguished, the accuracy was better than 99.44% and thus constituted proof of guilt beyond a reasonable doubt. Massey was convicted. The court of appeals reversed the conviction, holding that the prosecutor had "confused the probability of occurrence of the identifying marks with the probability of mistaken identification of the bank robber." It also followed *Carlson* in objecting to the evidence because of its "potentially exaggerated impact upon the trier of fact."

Gaudette-Keeping study

The unwillingness of the courts to accept population frequency evidence in *Carlson* and *Massey* may have been due in part to, or at least justified by, the weaknesses of the underlying studies on which the estimates were based. The Gaudette-Keeping study to which the experts referred had been conducted several years earlier and used the following methodology. A sample of 80 to

---

[4]It is unclear how a hair could be similar to more than one of nine mutually dissimilar hairs.

100 hairs "randomly selected" from various parts of the scalps of 100 subjects was reduced to a subsample of 6 to 11 representative hairs from each subject (861 in all). Investigators examined every inter-person pair of hairs macroscopically and microscopically. Only 9 inter-person pairs were found indistinguishable. The investigators knew, however, that hairs from different people were involved. According to Gaudette, hair comparisons are somewhat subjective, and when experiments included "common featureless hairs," investigators were unable to distinguish a much higher proportion of hairs than in the original study. Nevertheless, Gaudette concluded in testimony in *Carlson* that "if nine dissimilar hairs are independently chosen to represent the hair on the scalp of Individual B, the chance that the single hair from A is distinguishable from all nine of B's may be taken as $(1 - (1/40,737))^9$, which is approximately $1 - (1/4500)$."

Questions

1. Do you see how Gaudette-Keeping derived their estimate of 1/4500 as the probability of being unable to distinguish a hair selected at random from any of 9 selected from a subject? What assumptions underlie the method of calculation?

2. Assuming the study results were accurate and representative, what two possible meanings are attributable to the expert's conclusion? Which is validly deducible without other assumptions from the study?

3. What issues would you explore on cross-examination or in rebuttal testimony with respect to the validity of the study?

Source

Gaudette's studies were reported in B. D. Gaudette and E. S. Keeping, *An Attempt at Determining Probabilities in Human Scalp Comparison*, 19 J. Forensic Sci. 599 (1974); *Probabilities and Human Pubic Hair Comparisons*, 21 id. 514 (1976); *Some Further Thoughts on Probabilities and Human Hair Comparisons*, 23 id. 758 (1978). Gaudette's work was criticized by P. D. Barnett and R. R. Ogle in *Probabilities and Human Hair Comparison*, 27 id. 272 (1982). Despite the weaknesses of the underlying studies, forensic hair analysis has rarely been rejected by the courts. For a discussion of the cases and the studies, see Clive A. Stafford Smith & Patrick Goodman, *Forensic Hair Comparison Analysis: Nineteenth Century Science or Twentieth Century Sham?*, 27 Colum. Hum. Rts. L. Rev. 227 (1996).

Notes

Carlson was not a fluke—in Minnesota. In *State v. Boyd*, 331 N.W. 2d 480 (1983), the Minnesota Supreme Court followed *Carlson*, but added some new reasons for excluding population frequency statistics. It held that a population frequency statistic of less than 1 in 1,000 should not have been admitted in evidence because of a "real danger that the jury will use the evidence as a measure of the probability of defendant's guilt or innocence, and that evidence will thereby undermine the presumption of innocence, erode the values served by the reasonable doubt standard, and dehumanize our system of justice." *Id.* at 483. *Boyd* was followed in *State v. Kim*, 398 N.W.2d 544 (1987) (population frequency less than 3.6%) and, with specific reference to DNA profiling, in *State v. Schwartz*, 447 N.W.2d 422 (1989).

The Minnesota legislature responded to the trilogy of cases ending with *Kim* by passing an act providing that: "In a civil or criminal trial or hearing, statistical population frequency evidence, based on genetic or blood test results, is admissible to demonstrate the fraction of the population that would have the same combination of genetic marks as was found in a specific human biological specimen." Minn. Stat. §634.26 (1992). In subsequent rape cases, the Minnesota Supreme Court ignored the statutes and opted for a "black box" approach: quantification of random match probabilities for DNA profiles may not be presented to the jury, although an expert may use them as the basis for testifying that, to a reasonable scientific certainty, the defendant is (or is not) the source of the bodily evidence found at the crime scene. See, e.g., *State v. Bloom*, 516 N.W.2d 159 (1994). Is this a reasonable solution to the problems of misinterpretation noted by the Minnesota Supreme Court?

The position of the Minnesota Supreme Court, as articulated in *Boyd*, is an extension to population frequency statistics of an argument by Professor Laurence Tribe against the use of Bayes's theorem in evidence. See Section 3.3.2 at p. 79. In his article, Professor Tribe objected to quantification of guilt, which Bayes's theorem could in some applications require, but did not go so far as to advocate the exclusion of population frequency statistics.[5] Most courts have not followed the Minnesota Supreme Court on this issue. The conclusion of Judge Easterbrook in his opinion in *Branion v. Gramly*, 855 F.2d 1256 (7th Cir. 1988), seems more reasonable and probably represents the dominant view:

> Statistical methods, properly employed, have substantial value. Much of the evidence we think of as most reliable is just a compendium of statistical inferences. Take fingerprints. The first serious analysis of fingerprints was conducted by Sir Francis Galton, one of the pioneering statisticians, and his demonstration that fingerprints are unique depends entirely on statistical methods. Proof based on genetic markers (critical in rape and paternity litigation) is useful though altogether statistical. So, too, is evidence that, for example, the defendant's hair matched hair

---

[5]The introduction of population frequency statistics, without Bayes's theorem, does not require jurors to come up with a numerical probability of guilt, but the Minnesota Supreme Court equated the Bayesian and non-Bayesian scenarios by focusing on the risk that jurors would misread the population statistics as just such a quantification.

found at the scene of the crime. None of these techniques leads to inaccurate verdicts or calls into question the ability of the jury to make an independent decision. Nothing about the nature of litigation in general, or the criminal process in particular, makes anathema of additional information, whether or not that knowledge has numbers attached. After all, even eyewitnesses are testifying only to probabilities (though they obscure the methods by which they generate those probabilities) – often rather lower probabilities than statistical work insists on.

Id. at 1263–1264 (citations omitted).

3.2   Selection effect

Suppose that in *Williams v. State*, section 3.1.3, a fiber found on one of the bodies was compared with fibers found in the apartment of the defendant. If the defendant did not leave the fiber, the chance of matching a pre-selected fiber is 1/100. But if there are 50 distinct fibers in defendant's apartment (each with a 1/100 chance of matching), and the fiber found on the body is compared with each, the probability of one or more matches is $1 - 0.99^{50} = 0.395$. In *Coolidge v. State*, p. 66, since the 40 pairs of particles were apparently chosen from a larger group on the basis of visual (microscopic) similarity, the probability of a random match in the data might be much higher than the 10% rate reported for studies that did not use a visual similarity screening criterion. A great deal may thus depend on the way attempts to match characteristics are made and reported, a phenomenon sometimes referred to as "selection effect."

3.2.1   L'affaire Dreyfus

In the 1899 retrial of the 1894 secret court-martial of Alfred Dreyfus, Captain in the French General Staff, the prosecution again sought to prove that Dreyfus was the author of a handwritten *bordereau* (note) that transmitted five memoranda purportedly containing secret military information to the German ambassador in Paris. This bordereau was among a package of papers that a charwoman, in the pay of French intelligence, delivered to her employers, claiming that she discovered it (the *bordereau* torn to pieces) in the ambassador's wastebasket. The famous criminologist Alphonse Bertillon testified that there were suspicious coincidences of the initial and final letters in four of the thirteen polysyllabic words in the bordereau. Evaluating the probability of such a coincidence in a single word in normal writing as 0.2, Bertillon argued that the probability of four sets of coincidences was $0.2^4 = 0.0016$ in normal writing. This suggested to him that the handwriting of the document was not normal, which connected with a prosecution theory that Dreyfus had disguised his own handwriting to conceal his authorship. A divided military court again found Dreyfus guilty of treason, but this time with extenuating circumstances.

Questions

1. Given the assumed probability of coincidence in a single word as 0.2, exactly what probability did the expert compute?

2. If a high number of coincidences in the thirteen polysyllabic words were somehow indicative of contrived handwriting, compute a relevant probability under the assumption that the handwriting was not contrived.

Source

Laurence H. Tribe, *Trial by Mathematics: Precision and Ritual in the Legal Process*, 84 Harv. L. Rev. 1329, 1333-34 (1971); Rapport de MM. Darboux, Appell et Poincaré, in *L'affaire Dreyfus: La Révision du Procés de Rennes, Enquête* 3 at 501 (1909).

Notes

The 1899 verdict was widely recognized as a manifest injustice; there was an international outcry and Dreyfus was immediately pardoned on health grounds. The Dreyfus family eventually obtained review of the court martial by the civil court of appeals. As part of its investigation, the court requested the Academie des Sciences to appoint an expert panel to examine and report on the expert evidence. The panel–which included Henri Poincaré, a famous professor of the calculus of probabilities at the Sorbonne–pronounced it worthless. They added that "its only defense against criticism was its obscurity, even as the cuttle-fish cloaks itself in a cloud of ink in order to elude its foes." Armand Charpentier, *The Dreyfus Case* 226 (J. Lewis May translated 1935). In 1906, the court exonerated Dreyfus and annulled the 1899 verdict. In the end, having endured five years on Devil's Island, Dreyfus was restored to the army, promoted to major, and decorated with the *Legion d'Honneur*.

3.2.2   Searching DNA databases

According to the 1996 Report of the Committee on DNA of the National Research Council, in criminal investigations more than 20 suspects have already been initially identified by computerized searches through DNA databases maintained by various states. As the number and size of such databases increase, it is likely that initial identifications will more frequently be made on this basis. In its report, the Committee on Forensic DNA Science of the National Research Council stated that in such cases the usual calculation of match probability had to be modified. It recommended as one of two possibilities that the calculated match probability be multiplied by the size of the data base searched.

Questions

1. Explain the theory of such a calculation by reference to Bonferroni's inequality (see Section 3.1 at p. 57).

2. What is the probability computed with such an adjustment? Is it relevant to the identification issue?

3. What is the difference between this case and the fiber matching problem referred to in Section 3.2?

4. After reading Section 3.3 on Bayes's theorem, consider the following: Suppose that after the initial identification based on a computerized search, specific evidence is discovered that would have been sufficient to have the suspect's DNA tested if the specific evidence had been discovered first. Does it make any difference to the strength of the statistical evidence that it was used before or after the specific evidence was discovered? Suppose the specific evidence was very weak or even exculpatory so that it would not have led to a testing of the suspect's DNA. How would that affect the probative force of the statistical evidence?

5. Should the adjustment recommended by the Committee be made?

Source

National Research Council, *The Evaluation of Forensic DNA Evidence* 32 (1996). For criticisms of the Committee's position see Peter Donnelly and Richard D. Freedman, *DNA Database Searches and the Legal Consumption of Scientific Evidence*, 97 Mich. L. Rev. 931 (1999).

3.3   Bayes's theorem

Bayes's theorem is a fundamental tool of inductive inference. In science, as in law, there are competing hypotheses about the true but unknown state of nature and evidence that is more or less probable depending on the hypothesis adopted. Bayes's theorem provides a way of combining our initial views of the probabilities of the possible states of nature, with the probabilities of the evidence to arrive at posterior probabilities of the states of nature, given the evidence. It is thus a way of reasoning "backward" from effects to their causes.

In mathematical notation, Bayes's theorem shows how a set of conditional probabilities of the form $P(B_i|A_j)$ may be combined with initial or prior probabilities $P(A_i)$ to arrive at final or posterior probabilities of the form $P(A_i|B_j)$, wherein the roles of conditioning event and outcome event have been interchanged.[6] In the case of discrete events, Bayes's theorem is easily derived. By definition, $P(A_i|B_j) = P(A_i \text{ and } B_j)/P(B_j)$. The joint probability $P(A_i \text{ and } B_j)$ may be written $P(B_j|A_i)P(A_i)$ and, similarly, the marginal probability $P(B_j)$ may be written as $P(B_j) = \sum_i P(B_j|A_i)P(A_i)$, the sum taken over all possible states of nature A_i (see Section 3.1). Thus we have

$$P(A_i|B_j) = \frac{P(B_j|A_i)P(A_i)}{\sum_i P(B_j|A_i)P(A_i)}.$$

In the case of only two states of nature, say A and not-A ($\bar{A}$), the result is:

$$P(A|B) = \frac{P(B|A)P(A)}{P(B|A)P(A) + P(B|\bar{A})P(\bar{A})}.$$

A more enlightening formulation is in terms of odds:

$$\underset{(1)}{\frac{P(A|B)}{P(\bar{A}|B)}} = \underset{(2)}{\frac{P(A)}{P(\bar{A})}} \times \underset{(3)}{\frac{P(B|A)}{P(B|\bar{A})}}.$$

In words, this says that (1) the posterior odds on the truth of state A as opposed to not-A given evidence B are equal to (2) the prior odds on A times (3) the likelihood ratio for B, i.e., the ratio of the probability of B given A and not-A. Thus, the probative force of evidence is an increasing function of both the prior odds and the likelihood ratio.

Bayes's theorem honors the Rev. Thomas Bayes (1702-61), whose result was published posthumously in 1762 by Richard Price in the Philosophical Transactions. In Bayes's paper, the prior probability distribution was the uniform distribution of a ball thrown at random on a billiard table. While Bayes's original example utilized a physical prior probability distribution, some more controversial applications of Bayes's theorem have involved subjective prior probability distributions. See Section 3.3.2.

Although prior odds are usually subjective, sometimes they are objective and can be estimated from data. An intriguing example of objective calculation of a prior probability was Hugo Steinhaus's computation for paternity cases. See Steinhaus, *The Establishment of Paternity*, Prace Wroclawskiego Towarzystwa Naukowego, ser. A., No. 32, at 5 (1954).

The background, or prior, probability computed by Steinhaus was the probability that the accused was the father after intercourse had been established, but before serological test results were known. The posterior probability was

---

[6]The difference between these events can be made clear from an example attributed to Keynes. If the Archbishop of Canterbury were playing in a poker game, the probability that he would deal himself a straight flush, given honest play on his part, is not the same as the probability of honest play on his part, given that he has dealt himself a straight flush. The first is 36 in 2,598,960; the second most people think would be much larger, perhaps close to 1. Would the same result apply to the cardsharping Cardinal Riario?

the probability of paternity given the test results. A significant aspect of Steinhaus's procedure was his use of population statistics to estimate the proportion of guilty fathers among those designated for the test, even though no individuals (except those subsequently exonerated by the test) could be identified as guilty or innocent. For the sake of clarifying his theory, we simplify it slightly.

Different blood types occur with different frequencies in the population. Let the type in question be called "A" and have frequency f; the frequency of those who do not have this type is $1 - f$. Consider the group of accused fathers who take the serological test because the child has blood type A, one not shared by the mother. If the mothers' accusations were always right, the serological test would show that every member of this group had type "A" blood (although the converse, of course, is not true). If the mothers' accusations were always wrong, the members of this group would constitute a random sample from the population with respect to blood type, and the expected frequency of those with blood types other than A would be $1 - f$. The disparity between the actual rate of type A blood in this accused group and the population rate measures the overall accuracy of the accusations. The higher the proportion of men with type A blood, the more correct the accusations.

Let p be the proportion of the accused group who are fathers. Then $1 - p$ is the proportion of unjustly accused men and $(1 - p)(1 - f)$ is the expected proportion of those unjustly accused whom the test will exonerate. The ratio of the expected proportion of the exonerated group to the proportion of the general population who do not have blood type A is $(1 - p)(1 - f)/(1 - f)$, or simply $1 - p$, the prior probability of a false accusation. The importance of this ratio is that both its numerator and denominator can be estimated from objective sample and population statistics.

Using the results of 1,515 Polish paternity cases in which serological tests had been administered, Steinhaus concluded that the prior probability of a true accusation was about 70 percent. (With perhaps less than complete fairness, this factor has been called "the veracity measure of women.") The 70 percent figure may be regarded as the background probability in paternity cases. It was, however, computed from a subgroup of paternity cases, including only those cases in which the child did not share the blood type of the mother, requiring a serological test to establish paternity. Nevertheless, it seems fair to test the attributes of various decision rules by this subgroup because it is probably a random sample with respect to the fact of paternity; at the very least there are not more paternities among defendants in this group than in the larger group.

3.3.1   Rogue bus

On a rainy night, a driver is forced into a collision with a parked car by a swerving bus that does not stop. There are two bus lines that travel the street: Company A has 80% of the buses; company B has 20% of the buses. Their schedules shed no light on the culprit company. An eyewitness says it was company B's bus, but eyewitness testimony under sub-optimal conditions such as those prevailing here (rainy night, speeding bus) is known to have a high error rate.

Questions

1. In a civil suit by the injured driver against company A, is the statistical evidence that company A has 80% of the buses sufficient to satisfy plaintiff's burden of proof by a preponderance of the evidence that it was company A's bus? In the *Smith* case cited below, the court held (quoting *Sargent v. Massachusetts Accident Co.*, 307 Mass. 246, 250, 29 N.E.2d 825, 827 (1940)) that the evidence must be such as to produce "actual belief" in the event by the jury, and statistical evidence could only produce probabilities, not actual belief. Do you agree that this is a valid reason for concluding that statistical evidence is per se insufficient?

2. If suit is also brought against company B, is the eyewitness's testimony sufficient to satisfy plaintiff's burden of proof that it was company B's bus?

3. If the statistical evidence is insufficient but the eyewitness testimony is sufficient, how do you reconcile those results?

4. Assume that the eyewitness testimony has a 30% error rate. Treating the statistical evidence as furnishing the prior odds, and the eyewitness testimony as supplying the likelihood ratio, use Bayes's theorem to combine the statistical and eyewitness evidence to determine the probability, given the evidence, that it was company B's bus.

Source

Cf. *Smith v. Rapid Transit, Inc.*, 317 Mass. 469, 58 N.E.2d 754 (1945). For a discussion of some other early cases on this subject, see *Quantitative Methods in Law* 60-69.

Notes

Whether "naked" statistical evidence (i.e., statistical evidence without case-specific facts) can be sufficient proof of causation in a civil or criminal case has provoked extensive academic discussion, with the professorial verdict usually being negative, at least in criminal cases. The arguments are criticized in Daniel Shaviro, *Statistical-Probability Evidence and the Appearance of Justice*, 103 Harv. L. Rev. 530 (1989). For criminal cases, the discussion has been conducted on the level of law-school hypotheticals, because in real cases there is always case-specific evidence to supplement the statistics. For civil cases, statistics have been held to be sufficient evidence of causation when that has seemed necessary to do justice. The outstanding example is the diethylstilbestrol (DES) litigation, in which the DES manufacturers were held proportionately liable, based on their market shares, to plaintiffs whose mothers had taken DES during pregnancy, even though there was no case-specific evidence of which company's DES the mother had taken. *Sindell v. Abbott Labs, Inc,* 26 Cal.3d 588, 607 P.2d 924 (1980), *cert. denied,* 449 U.S. 912 (1980); *Hymowitz v. Lilly & Co.,* 73 N.Y.2d 487 (1989).

3.3.2   Bayesian proof of paternity

A New Jersey statute criminalizes sexual penetration when the defendant has supervisory or disciplinary power by virtue of his "legal, professional or occupational status" and the victim is on "probation or parole or is detained in a hospital, prison or other institution." Defendant was a black male corrections officer at the Salem County jail where the female victim was incarcerated on a detainer from the Immigration and Naturalization Service. The victim conceived a child while in custody. If defendant was the father he was guilty of a crime, irrespective of the victim's consent.

In contested paternity proceedings, prior to the advent of DNA testing, the parties were frequently given Human Leukocyte Antigen (HLA) tests to identify certain gene-controlled antigens in the blood. After making HLA tests, an expert witness for the state testified that the child had a particular set of genes that was also possessed by the defendant, but not by the mother. She further testified that the frequency of this particular set of genes was 1% in the North American black male population. The expert assumed that the odds of defendant being the father, quite apart from the HLA tests, were 50-50 and, based on that assumption and the 1% frequency of the gene type, concluded that "the likelihood of this woman and this man producing this child with all of the genetic makeup versus this woman with a random male out of the black population . . . [results in] a probability of paternity of 96.55 percent."

Questions

1. Was the expert's testimony on the probability of paternity properly admitted?

2. Was the restriction to the rate of the haplotype in the black population warranted?

3. If the expert had proposed to give the jurors a hypothetical range of prior probabilities and the posterior probability associated with each prior, should her testimony have been admitted?

Source

State v. Spann, 130 N.J. 484, 617 A.2d 247 (Sup. Ct. N.J. 1993).

Notes

Whether Bayes's theorem should be explicitly used as suggested in Question 3 has been the subject of considerable academic and some judicial debate. On the academic side, among the first articles are Michael O Finkelstein & William Fairley, *A Bayesian Approach to Identification Evidence*, 83 Harv. L. Rev. 489 (1970)(proposing the use of Bayes's theorem); Laurence H. Tribe, *Trial by Mathematics:Precision and Ritual in the Legal Process*, 84 Harv. L. Rev. 1329 (1971) (criticizing the proposal); Finkelstein & Fairley, *A Comment on "Trial by Mathematics,"* 84 Harv. L. Rev. 1801 (responding to Tribe); and Tribe, *A Further Critique of Mathematical Proof*, 84 Harv. L. Rev. 1810 (1971) (rejoinder). A further critique appears in L. Brilmayer & L. Kornhauser, *Review: Quantitative Methods and Legal Decisions*, 46 U. Chi. L. Rev. 116 (1978). See generally two symposia: *Probability and Inference in the Law of Evidence*, 66 B. U. L. Rev. 377-952 (1986) and *Decision and Inference in Litigation*, 13 Cardozo L. Rev. 253-1079 (1991). On the judicial side, compare *Plemel v. Walter*, 303 Ore. 262, 735 P.2d 1209 (1987) and *State v. Spann, supra* (both approving an explicit use) with *Connecticut v. Skipper*, 228 Conn. 610, 637 A.2d 1104 (1994) (disapproving an explicit use).

Those approving an explicit use in a criminal case argue that jurors tend to underestimate the probative force of background statistical evidence. Such insensitivity to prior probability of outcomes appears to be a general phenomenon in subjective probability estimation. See, e.g., *Judgement Under Uncertainty: Heuristics and Biases*, at 4-5 (Daniel Kahneman, Paul Slovic & Amos Tversky, eds., 1982). Empirical studies based on simulated trials tend to support this. See, e.g., Jane Goodman, *Jurors' Comprehension and Assessment of Probabilistic Evidence*, 16 Am. J. Trial Advocacy 361 (1992). They also point to what is called the prosecutor's fallacy: the risk that the jury will misinterpret the low population frequency of the blood type as the probability of innocence. Those opposed to explicit use object that jurors would be invited to estimate a probability of guilt before hearing all the evidence, which they view as inconsistent with the presumption of innocence and the instruction commonly given to jurors to withhold judgement until all the evidence is heard. On the other hand, if the jurors wait until they hear all the evidence before estimating their priors, the statistics are likely to influence those estimates. Some scholars further object to any juror quantification of the probability of guilt as inconsistent with the "beyond a reasonable doubt" standard for criminal cases. Since conviction is proper despite some doubt, it is not clear why quantification of that doubt by a juror would be per se objectionable. There is some evidence that quantification of the burden of proof influences verdicts in an appropriate direction. Dorothy K. Kagehiro & W. Clark Stanton, *Legal vs. Quantified Definitions of Standards of Proof*, 9 L. & Hum. Behav. 159 (1985).

Perhaps the strongest case for an explicit use by the prosecution arises if the defense argues that the trace evidence does no more than place defendant in a group consisting of those in the source population with the trace in question.

Known as the defense fallacy, the argument assumes that without the trace defendant is no more likely to be guilty than anyone else in the source population. (This is an unlikely scenario since there is almost always other evidence that implicates the defendant.) The prosecution might then be justified in using Bayes's theorem to show what the probabilities of guilt would be if the jurors believed at least some of the other evidence. Conversely, the prosecutor's fallacy (that the frequency of the trace in the population is the probability of innocence) assumes that the prior probability of defendant's guilt is 50%. If the prosecutor makes such an argument, the defense should then be justified, using Bayes's theorem, to demonstrate what the probabilities would be if some or all of the other evidence were disbelieved.

Another set of issues is presented if identifying the source of the trace does not necessarily imply guilt. A thumb print on a kitchen knife, used as a murder weapon, may have been left there innocently. The complication here is that the same facts suggesting guilt that are used to form the prior probability of authorship of the print would also be used to draw an inference from authorship of the print to guilt. If this is an impermissible double use, it would be hard or impossible to partition the non-statistical evidence among uses.

Whether an explicit use of Bayes's theorem is allowed in the courtroom may stir legal academics more than jurors. In one empirical study the jurors simply disregarded the expert's Bayesian explanations of the statistics. See David L. Faigman & A. J. Baglioni, Jr., *Bayes' Theorem in the Trial Process: Instructing Jurors on the Value of Statistical Evidence*, 12 Law & Hum. Behav. 1 (1988). The more important (and often ignored) teaching of Bayes's theorem is that one need not assert that a matching trace is unique or nearly unique in a suspect population to justify its admission as powerful evidence of guilt.

3.4   Screening devices and diagnostic tests

Screening devices and diagnostic tests are procedures used to classify individuals into two or more groups, utilizing some observable characteristic or set of characteristics. Most familiar examples come from medical diagnosis of patients as "affected" or "not affected " by some disease. For our discussion we adopt the clinical paradigm, but the central ideas are by no means limited to that context.

False positives and negatives

No diagnostic test or screening device is perfect. Errors of omission and commission occur, so we need to distinguish between the *true* status (say, A = affected or U = unaffected) and the *apparent* status based on the test (say, + = test positive or − = test negative). A *false positive* diagnosis is the occurrence of a positive outcome (+) in an unaffected person (U); it is denoted by $(+, U)$. A *false negative* is the occurrence of a negative outcome $(-)$ in an affected person (A); it is denoted by $(-, A)$. The situation may be summarized in a four-fold table of events, as follows.

Test outcome

		+	−
	A	$(+, A)$	$(-, A)$
		correct	false
True			negative
status			
	U	$(+, U)$	$(-, U)$
		false	correct
		positive	

Sensitivity and specificity

Although the terms false positive and false negative are unambiguous when referring to the possible screening errors, an accuracy rate can be defined a few different ways, and these are often confused in casual or uninformed communication. The *sensitivity* of a test is the proportion of all affected individuals who (correctly) test positive, $P[+|A]$. The *specificity* of a test is the proportion of all unaffected individuals who (correctly) test negative, $P[-|U]$.

The term *false positive rate* usually (but not always) refers to the complement of specificity,

$$\text{false positive rate} = 1 - \text{specificity} = P[+|U],$$

because it measures the rate of occurrence of falsely positive outcomes (among the truly unaffected). The term *false negative rate* usually refers to the complement of sensitivity,

$$\text{false negative rate} = 1 - \text{sensitivity} = P[-|A],$$

because it measures the rate of occurrence of falsely negative outcomes (among the truly affected). These terms must always be carefully examined, as some authors use them to refer to another set of error rates with different denominators.

Positive and negative predictive values

Positive predictive value (PPV) is the proportion of all test-positive people who are truly affected,

$$\text{PPV} = P[A|+] = \frac{P[+, A]}{P[+]}.$$

Negative predictive value (NPV) is the proportion of all test-negative people who are truly unaffected,

$$\text{NPV} = P[U|-] = \frac{P[-, U]}{P[-]}.$$

A screening device is characterized by its sensitivity and specificity. These are its *operating characteristics*, which are objective quantifications of accuracy that do not depend on the prevalence rate $P[A]$. However, the recipient of a screen test (or its victim, as the case may be) is usually more concerned about the positive or negative predictive values of the test since, having the test result in hand, interest turns to the proportion of similarly classified people who are truly affected or not.

It is an important fact that predictive values do depend on overall prevalence rates. In fact, for a test with given sensitivity and specificity, the odds on being affected given a positive test are:

$$\frac{\text{PPV}}{1 - \text{PPV}} = \frac{P[A|+]}{P[U|+]} = \frac{P[+|A]P[A]}{P[+|U]P[U]}$$

$$= \frac{\text{sensitivity}}{1 - \text{specificity}} \cdot \text{prevalence odds}.$$

Compare this with Bayes's theorem on p. 75. As the prevalence of a condition becomes rare ($P[A]$ approaching zero), PPV drops too, and sometimes surprisingly so. For example, a test with sensitivity and specificity each equal to 99% is generally considered quite precise, relative to most diagnostic procedures. Yet for a condition with a not-so-rare prevalence of one per hundred, the odds on being affected given a positive test outcome are $(0.99/.01) \times (0.01/.99) = 1$, i.e., among all positive results only 50% are in fact truly affected! For a prevalence rate of one per thousand, the PPV is only about 10%. These low numbers raise serious ethical and legal questions concerning action to be taken following positive test outcomes.

One course to pursue whenever feasible to improve a screen's PPV is to repeat the test, preferably with an independent procedure. Repeated administrations of even the same screen results in much higher accuracy than does a single administration.[7] Another technique is to have some independent basis for the test, i.e., some factor associated with increased prevalence odds.

ROC curve

Although we have assumed in our discussion a natural dichotomy, affected and unaffected, one should be wary of arbitrary or forced dichotomization of an underlying continuum. Often, for convenience or simplicity, an under-

---

[7]Assuming random misclassification errors and with a "positive" outcome defined as two or more individual positives.

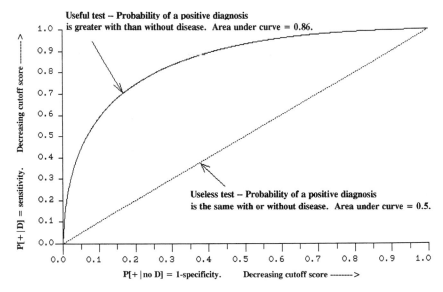

FIGURE 3.4. ROC curves for good and bad screening tests

lying spectrum of a condition (e.g., ability) is dichotomized ("fast" or "slow" learner). Aside from the dangers of stigmatization and misclassification, an ill-chosen breakpoint for the test diminishes accuracy. There is an inherent trade-off—lowering the cut-off criterion for a test-positive outcome increases sensitivity but decreases specificity, allowing more false positive errors; conversely, raising the cut-off criterion increases specificity but lowers sensitivity. In general, if a dichotomous classification is necessary, one should weigh the separate costs and benefits of each kind of error or correct decision.

To compare two competing screening procedures in the context of an underlying continuum, one should review each pair of values of sensitivity and specificity as the cut-off criterion is varied. The term "relative operating characteristic (ROC) curve" describes the locus of such pairs that trace a curve in the sensitivity-specificity plane. If the ROC curve of one procedure dominates the others for any classification criterion, then preference for that procedure is straightforward. If ROC curves cross, then only within a range of sensitivity values is one screen preferable to another. The area under an ROC curve is a commonly used one-number measure of the classificatory accuracy of the underlying continuum. (These notions developed historically from radio communications technology, where ROC originally stood for "receiver operating characteristic.") An example of ROC curves is shown in Figure 3.4.

3.4.1   Airport screening device

As of 1980, the FAA reinstituted the use of a statistically based hijacker profile program (discontinued in 1973) to help identify people who might attempt to hijack a plane using a nonmetallic weapon. Assume that approximately 1 person in 25,000 carries such a weapon and that the test has sensitivity of 90% and specificity of 99.95%. One Lopez and his companion show the wrong profile. They are searched, heroin is found, and they are arrested.

Question

On a motion to suppress the evidence in a subsequent prosecution, did the test provide either "probable cause" to justify an arrest or "reasonable suspicion" to justify a brief investigative detention?

Source

United States v. Lopez, 328 F. Supp. 1077 (E.D.N.Y. 1971); J. Monahan and L. Walker, *Social Science in Law*, 211-12 (1985).

Notes

Screening devices are used in a variety of contexts, including educational admissions, credit risk assessment, testing for drug use, and potential for criminal behavior. There are three possible methods of selection for screening: (i) attempt to identify high-risk individuals and test them; (ii) select individuals at random; and (iii) test everyone. Each of these methods has provoked litigation. In *United States v. Mendenhall*, 446 U.S. 544 (1980), three justices found a psychological profile used to identify heroin couriers sufficient to generate "reasonable suspicion" for an investigative detention, but four justices found the profile insufficient. In *Florida v. Royer*, 460 U.S. 491 (1983), a majority of the Court rejected such profiles as insufficient for probable cause, but allowed them as a basis for reasonable suspicion. In *United States v. Lopez*, 328 F. Supp. 1077 (E.D.N.Y. 1971), a hijacker screening device was upheld as sufficient for probable cause, despite "disquieting possibilities."

If psychological profiles are at the edge of acceptability, then pure chance should be an insufficient basis for a search. The Supreme Court has held that "random" (really haphazard) stops by police of automobiles, ostensibly to check licenses and registration, are unreasonable seizures within the Fourth and Fourteenth Amendments. *Delaware v. Prouse*, 440 U.S. 648 (1979). The Court found such spot checks insufficiently productive compared with action taken on the basis of observed violations, and noted that the discretion given to police officers held a potential for abuse not present in more systematic methods. *Id.* at 659-63. Query whether searches based on objective stopping rules would meet constitutional objections.

Jar Wars: Some programs of universal testing have been sustained as reasonable, while others have been rejected as too intrusive. In 1986, the Reagan administration, by executive order, declared drug use incompatible with federal employment and gave agency heads authority to require urine tests of all new applicants and of present employees suspected of drug use. The order asserted that the program was not instituted to gather evidence for criminal prosecution. Executive Order No. 12564, 3 C.F.R. 224 (1986), *reprinted in* 5 U.S.C. §7301, note at 909-911 (1998). The program as implemented by the Customs Service was upheld in *National Treasury Employees Union v. Von Raab*, 489 U.S. 656 (1989).

Bills have been introduced in many states to regulate drug testing in the workplace. In 1987, a bill introduced in the New York Senate provided that any screening test must "have a degree of accuracy of at least ninety-five percent" and "positive test results must then be confirmed by an independent test, using a fundamentally different method and having a degree of accuracy of 98%." Assuming that 0.1% of the adult working population takes drugs, and that "accuracy" refers to both sensitivity and specificity, what is the positive predictive value of a test program meeting the bill's requirements?

3.4.2   Polygraph evidence

Edward Scheffer, an airman stationed at March Air Force Base in California, volunteered to work as an informant on drug investigations for the Air Force Office of Special Investigations (OSI). He was told that from time to time he would be asked to submit to drug testing and polygraph examinations. Shortly after beginning undercover work, he was asked to take a urine test. After providing the sample, but before the results were known, he agreed to take a polygraph test administered by an OSI examiner. In the opinion of the examiner, the test "indicated no deception" when Scheffer denied using drugs after joining the Air Force. After the test, Scheffer unaccountably disappeared from the base and was arrested 13 days later in Iowa. OSI agents later learned that the urinanalysis revealed the presence of methamphetamine.

At his trial by court-martial, Scheffer sought to introduce the polygraph evidence in support of his defense of innocent ingestion, i.e., that he had not knowingly used drugs. The military judge denied the motion on the basis of Military Rule of Evidence 707, which makes polygraph evidence inadmissible in court-martial proceedings. Scheffer was convicted. On appeal, he contended that Rule 707 was unconstitutional because the blanket prohibition of polygraph evidence deprived him of a "meaningful opportunity to present a complete defense."

When the case reached the Supreme Court, it upheld Rule 707. *United States v. Scheffer,* 118 S. Ct. 1261 (1998). Justice Thomas, writing for the majority, observed that the scientific community "remains extremely polarized about the reliability of polygraph techniques" with overall accuracy rates from laboratory studies ranging from 87 percent to little more than the toss of a coin. The lack of scientific consensus was reflected in disagreements among federal and state courts over the admissibility of such evidence, with most states maintaining per se rules excluding or significantly restricting polygraph evidence. Rule 707's blanket prohibition was thus justified as in furtherance of the government's legitimate interest in excluding unreliable evidence. *Id.* at 1265–1266.

In his opinion, Justice Thomas had to deal with the fact that the government, and in particular the Defense Department, routinely uses polygraph tests in screening for personnel security matters, and in fact maintains a highly regarded Polygraph Institute for training examiners. As Justice Stevens noted in his dissent, "[b]etween 1981 and 1997, the Department of Defense conducted over 400,000 polygraph examinations to resolve issues arising in counterintelligence, security, and criminal investigations." *Id.* at 1272, n. 7. Justice Thomas responded that "[s]uch limited, out of court uses of polygraph techniques obviously differ in character from, and carry less severe consequences than, the use of polygraphs as evidence in a criminal trial." *Id.* at 1266, n. 8. The Defense Department program yields some illuminating data. For example, in fiscal year 1997, there were 7,616 individuals who were tested under the Department of Defense Counterintelligence-Scope Program. Of these, 176 individuals were evaluated as "yielding significant psychological responses, or were evaluated as inconclusive and/or provided substantive information." Out of this subgroup, at the time of the report, 6 persons had received adverse action denying or withholding access to classified information, 2 were pending adjudication, and 14 were pending investigation.

Questions

[In answering the following questions, it may be helpful to construct a two-by-two table with the rows being "guilty" and "not guilty", and the columns being the outcome of the polygraph test ("deception found" and "deception not found"). Obviously, only part of the table can be filled in from the Defense Department data.]

1. Justice Stevens's dissent cites a number of studies reporting accuracy rates for polygraph tests between 80 and 95%. Assuming that the 1997 data from the counterintelligence program are representative of the accuracy of the polygraph examinations given by the Defense Department, what is an upper bound for the positive predictive value (PPV) of the test (i.e., assuming all pending matters are resolved unfavorably to the individuals)?

2. What statistic would you want to compute to appraise the accuracy of an exonerating polygraph test, as in Scheffer's case?

3. Is it reasonable or unreasonable to assume that in the Defense Department program $PPV \geq 1 - NPV$ for its polygraph tests? If reasonable, what does this suggest about the accuracy of the test when used as exonerating evidence?

4. How is the accuracy of the test as it might have been used in Scheffer's case affected by the difference in character between criminal proceedings, such as Scheffer's, and personnel screening? Was the Court right when it declined

to infer that, since the NPV of the test was apparently deemed sufficient for counterintelligence screening purposes, it should also be sufficient for criminal trials?

Source

United States v. Scheffer, 118 S. Ct. 1261 (1998); Department of Defense, *Annual Polygraph Report to Congress* (Fiscal Year 1997).

3.5   Monte Carlo methods

Monte Carlo methods (also known as simulation methods) constitute a branch of empirical–as opposed to deductive–mathematics that deals with experiments using random numbers. That name conjures up sequences of random results, as in a gambling casino, which are at the heart of the method. With the advent of computers Monte Carlo techniques have become widely used in statistics, as they have in all fields of applied mathematics.

One of the basic problems addressed by Monte Carlo methods is to estimate a distribution or average value of some possibly complicated statistic. In the pre-computer era, problems of this sort were solved using the tools of mathematical analysis and numerical approximation, with a range of results depending on the complexity of the analysis and accuracy of the approximation. When simulation is used, the computer generates data according to the specified distribution (such as the normal distribution); calculates the statistic for each simulated data set; and estimates the average value in the obvious way from the sample average of simulated values. Thus, we replace complex analysis with reliance on the law of large numbers to ensure that the sample estimate will be close to the true value with high probability. Some mathematicians complain that mathematical insight is being replaced by mindless computing. We don't have to resolve that debate; it is unquestioned that Monte Carlo methods are highly useful in statistics.

For example, suppose we want to calculate the probability that a test statistic will reject the null hypothesis, but the mathematical distribution of the test statistic is unknown. Using Monte Carlo methods, we have the computer generate a large number of data sets, compute the test statistic each time, and count the sample proportion of times the statistic rejects the null hypothesis. Tail probabilities, or P-values, are often generated in this way. When the tail probabilities are very small, it is ineffficient to estimate them by the simple proportion of times they occur because the number of simulations required to see even one event may be enormous. In such cases, a more complicated technique known as *importance sampling* is used instead. This technique estimates very small P-values with much greater relative precision for a given number of simulations.

Simulation methods are routinely used to check the accuracy of large sample techniques, such as approximate 95% confidence intervals. Here the computer generates many data sets with known parameter values, and the proportion of these contained inside the confidence intervals constructed by the method under consideration estimates the true coverage probability.

In an interesting and relatively recent development known as the *bootstrap*, the computer generates the data for Monte Carlo methods without assuming any particular distribution. Instead, it uses the empirical distribution of an observed set of data and samples from it *with* replacement to create multiple data sets known as bootstrap samples. Statistical estimators are then calculated from each of the bootstrap samples to make robust inferences about their distribution in the population from which the original sample was drawn. For example, a famous data set consisted of the average undergraduate GPA and first year law school grades for 15 American law schools. The investigator sought to compute a confidence interval for the correlation coefficient, but confronted problems of small sample size and a nonnormal distribution for the coefficient. Using the bootstrap, he drew replicate samples of size 15 at random and with replacement. Doing this 200 times generated a Monte Carlo sample of 200 correlation coefficients. The lower 2.5% and upper 97.5% of these values provided the desired 95% confidence interval.

3.5.1   Sentencing a heroin swallower

Charles Shonubi was caught at Kennedy Airport attempting to enter the country from Nigeria with 103 condoms in his digestive tract. The condoms held 427.4 grams of white powder. Analysis of a random sample of four condoms showed them to contain heroin. Shonubi was convicted of importation and possession with intent to distribute a controlled substance. From the fact that Shonubi had made at least 8 trips to Nigeria between September 1, 1990, and December 1, 1991, without a reason or the apparent means to do so (he was earning $12,000 a year as a part-time toll collector at the George Washington Bridge), the district judge inferred that Shonubi had made seven prior trips for smuggling purposes.

Under federal sentencing guidelines, the length of sentence depends on the amount brought in as part of the same course of conduct. But there was no information about how much Shonubi had brought in on his prior seven trips. At the original sentencing, the court multiplied 427.4 grams by 8, arriving at a total of 3419.2 grams, and sentenced Shonubi to 151 months, that being the lower end of the guideline sentence for 3,000 grams. On appeal, the Second Circuit held that the district court impermissibly engaged in "surmise and conjecture" in assuming that the amounts brought in on prior trips were the same. It remanded for resentencing, requiring the prosecution to produce "specific evidence"–which it defined as "e.g., drug records, admissions, or live testimony"– of amounts brought in by Shonubi on the seven prior trips.

TABLE 3.5.1. Net weight in grams per internal smuggling trip known to DEA agents at Kennedy Airport. September 1, 1990, to December 10, 1991.

Net weight in grams	Number of occurrences
0–100	1
100–200	7
200–300	13
300–400	32
400–500	31
500–600	21
600–700	6
700–800	1
800–900	2
900–1000	2
1000–1100	0
1100–1200	0
1200–1300	1

On remand, the prosecution produced a study by an expert on drug policy, Dr. David Boyum, who obtained U.S. Customs Service data on all 117 Nigerian heroin swallowers arrested at Kennedy Airport between September 1, 1990, and December 1, 1991 (the dates of the first and last trips shown on Shonubi's passport). For each swallower, the data included the gross weight of the heroin seized (i.e., heroin plus condom or balloon). Deducting the estimated weight of the condom, Dr. Boyum produced a list of 117 net weights. The mean weight was 432.1 grams; the median was 414.5 grams; and the standard deviation was 172.6 grams. Distributing these in 100-gram ranges or bins, Dr. Boyum generated the distribution shown in Table 3.5.1 and the top panel of Figure 3.5.1.

By computer simulation, Dr. Boyum performed a Monte Carlo study by making 7 selections at random from the 117 net weights (each time including the previously selected quantity in the list before the next selection) and calculating their sum. Dr. Boyum's computer repeated this procedure 100,000 times and generated the cumulative frequency distribution of the totals shown in the bottom panel of Figure 3.5.1.

Based on the cumulative frequency distribution generated by the Monte Carlo study, the government argued that there was a 99% chance that Shonubi carried at least 2090.2 grams of heroin on the seven trips combined; a 95% chance that he carried more than 2341.4 grams; a 75% chance that he carried more than 2712.6 grams; and a 55% chance that he carried more than 3039.3 grams. Taking a conservative position, the government contended that these outcomes met its burden of proof that he carried at least 2,500 grams.

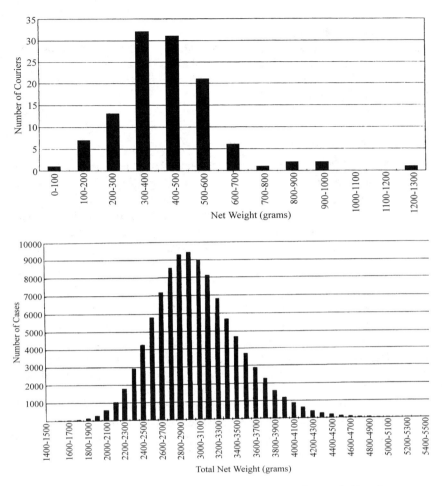

FIGURE 3.5.1. Cumulative frequency distribution of smuggled heroin

Questions

1. What objections do you have to the relevance of the government's data?

2. What additional information might be collected to assist in the estimation process?

3. The district court initially argued that, since the chemical nature of the powder as heroin in the 103 condoms was validly extrapolated from analysis of only four condoms, it should be permissible to extrapolate the amount carried on the seven prior trips from the amount carried on the eighth trip. What is the flaw in that argument?

Source

United States v. Shonubi, 895 F. Supp. 460 (E.D.N.Y. 1995) (a wide-ranging, 66-page opinion by District Judge Jack B. Weinstein that is a small treatise on statistical evidence), *vacated and remanded*, 103 F.3d 1085 (2d Cir. 1997), *on remand*, 962 F. Supp. 370 (E.D.N.Y. 1997).

3.5.2   Cheating on multiple-choice tests

In October 1993, some 12,000 police officers in New York City took a written multiple-choice examination for promotion to sergeant. Four sets of examinees, comprising 13 officers in all, were suspected of having collaborated in answering questions. The suspected officers consisted of two dyads, one triad, and one hexad. The suspicions were based on anonymous "tip" letters naming some of the officers, and the fact that two pairs of brothers managed to sit near each other despite random seating assignments. The statistician appearing for the department was given the answer sheets of the suspected officers and asked for an analysis of answer patterns to determine whether there had been cheating.

The examination consisted of morning and afternoon sessions. In each session, the officers had varying numbers of jointly wrong answers (the same questions answered incorrectly) and matching wrong answers (the same wrong answers to the same questions). Following a design first suggested by William Angoff, the statistician (Bruce Levin) counted the number of matching wrong answers in the suspected groups and compared that with the numbers of such matches in random samples of answer sheets drawn from all those answer sheets with the same numbers of wrong answers as the suspected answer sheets. For example, one member of Dyad 2 had seven wrong answers and the other had six; there were six matching wrong answers. This was compared with the rate of matching in 10,000 samples of dyads in which one answer sheet was drawn at random from all other sheets with 7 wrong answers and the other was drawn at random from sheets with 6 wrong answers. If both members of a dyad had the same number of wrong answers, selection would be made without replacement from all answer sheets having that number of wrong answers. In either case, after each group was selected, the answer sheets for that group were replaced. The same procedure was followed for the triad and the hexad except that the numbers of matching wrong answers were calculated as the total number of matching wrong answers for all possible dyads in those groups. Dr. Levin called this primary analysis a "Modified Angoff Procedure" because he had made a minor change to Angoff's procedure. (Angoff's original procedure fixed only the product of the wrong answer count per dyad; Levin fixed each person's count.)

To check his results Dr. Levin made a new "Item-Specific" analysis in which he limited the comparison groups to those who had answered incorrectly the same specific questions as had the suspected group members; when the numbers were small enough, a complete enumeration was used instead of Monte Carlo samples. The data from both procedures for the two dyads are shown in Table 3.5.2.

Consider Dyad 2 in the afternoon session. The two officers had 7 and 6 wrong answers, respectively, and 6 of them (the maximum possible) matched, i.e., were wrong in the same way (col. II). The comparison group was created from 10,000 replications in which one answer sheet was drawn from those with 7 items wrong and the other was drawn from those with 6 items wrong (in both cases including the suspected officers), with the sheets replaced after each dyad was selected. In the replications, the average number of matching wrong answers was 1.08 (col. III) and there were 0 groups with 6 matching wrong answers (col. IV). In the Item-Specific Procedure there were 283 examinees who answered the same six questions incorrectly (col. VI). The mean number of matching wrong answers for all possible dyads drawn from this group was 3.47 (col. VII), and the rate of such groups with 6 (the maximum number) matching wrong answers was 0.0396 (col. VIII).

Questions

1. (a) Assuming the matching wrong answers are due to innocent coincidence, estimate the probability of observing three matches for Dyad 1 in the afternoon session; call this the likelihood of the data under the hypothesis of innocent coincidence. (b) Assuming Dyad 1 was cheating, what likelihood would you assign to the observed data? The ratio of (b) to (a) is the likelihood ratio for these data. Use this to describe the strength of the evidence against innocent coincidence and in favor of cheating. Is this ratio sufficient proof for a finding of collaboration?

2. Why limit the comparison groups to those with the same number of wrong answers?

3. Why use the Item-Specific Procedure? What conclusion should be reached if the Item-Specific Procedure does not support the results of the Modified-Angoff procedure?

4. Why not count matching correct answers as evidence of innocence?

5. Is it fair not to include non-matching wrong answers as evidence of innocence?

6. Can the results be explained on the theory that the officers may have studied together, or studied the same material?

7. Because it cannot be determined who copied from whom, is the prosecution's evidence flawed?

TABLE 3.5.2. Wrong answers and matching wrong answers in promotion examination.

I	II	Modified-Angoff procedure		V	Item-specific procedure		
		III	IV		VI	VII	VIII
		10,000 Replications				10,000 Replications	
Group names	# WA: # MWA	Mean # MWA per group	# Groups with ≥ # MWA	# JWA: # MWA	# Examinees with same wrong items	Mean # MWA per group	Rate of groups with ≥ # MWA
Dyad 1							
a.m.	8,8:8	1.30	0	8,8:8	28	3.852	0.0005
p.m.	4,4:3	0.488	37	4,4:3	203	2.126	0.354
Dyad 2							
a.m.	1,1:1	0.0487	487	1,1:1	7,182	0.491	0.491
p.m.	7,6:6	1.08	0	6,6:6	283	3.470	0.0396

Key: #WA=# wrong answers for each officer in the group; #MWA = the sum of the # of matching wrong answer pairs; #JWA=# jointly wrong answers to the specific items.

References

Stephen P. Klein, *Statistical evidence of cheating on multiple-choice tests*, 5 Chance 23 (1992); William H. Angoff, *The development of statistical indices for detecting cheaters*, 69 J. Am. Stat. Assoc. 44 (1947). See Section 6.2.3 for further questions on these data.

3.6   Foundations of probability

Statistical inference rests on concepts of probability and randomness. It is not a trivial matter to give precise definitions for these profound ideas. In fact, the search for precise formulation has been the source of lively controversy, and of mathematical and philosophical research into the foundations of probability and statistical inference.

The current view of physics is that, at sufficiently small scales of observation, the laws of nature must be expressed probabilistically, and, indeed, any regularity or "law" of nature observed at ordinary scales is itself a consequence of statistical theorems that describe the aggregate behavior of large samples of objects. This idea is also familiar to biologists, psychologists, social scientists, and others who seek regularities in the highly complex systems of their disciplines. It is ironic, then, that when empirical scientists turn to mathematicians or philosophers for precise formulations of "probability," they meet the response that there are various types of probability (as many as five different formulations), that there is controversy over which definition, if any, should be preferred as fundamental, and even debate over whether some types of probability exist!

Objective probabilities are quantities that relate to the states of some physical system, such as energy levels of an electron or the positions and velocities of molecules in a coin at the moment it is tossed. Objective probabilities are regarded as intrinsic properties of the system under consideration, and are assumed to exist independently of an observer's personal opinion. Objective probabilities are initially unknown, but can be assigned in cases in which a physical symmetry implies equally likely outcomes, like the numbers on a roulette wheel or the spots on a fair die. These are sometimes called *logical* probabilities. More often, objective probabilities must be discovered by direct observation, when the relative frequency of outcomes consistent with a given event accumulate in increasingly larger series of identical trials. Thus, the probabilities associated with Schrödinger's wave equation in quantum mechanics are often taken as objective probabilities, because the equation is assumed to describe an intrinsic property of the physical system (although the act of observation itself may indeed alter the wave equation). These probabilities are given a "frequentist" interpretation in the sense that repeated experiments show that the relative frequency of finding a particle in a given state agrees closely with the predictions of the wave equation.

Those who argue against the objectivist, or frequentist, interpretation of probability contend primarily that its assumption of the existence of physical probabilities is too restrictive to deal with some events of interest, e.g., those that can never be repeated, or even observed. We need some concept of probability to assign degrees of certainty or belief between 1 (truth) and 0 (falsity) to a statement when its truth is unknown. Such assignments, however, are somewhat artificial in terms of objective probabilities, because there is commonly no natural frequentist interpretation for statements expressing degrees of belief, especially when unique events are involved.

A less artificial approach involves recognition that probabilities also have a subjective interpretation: *subjective* probabilities are numerical expressions of the degree of certainty about, or the degree of belief in, statements concerning the occurrence of an event. Rational observers use their probability assessments to place fair bets or to make other decisions; in fact, subjective probabilities are often elicited in terms of the amount of money one is just willing to bet on the occurrence of an event per unit reward. (Rewards should be considered in terms of utility rather than money, since money has different values to different people.) Subjective probabilities are not intrinsic properties of events, but reflect an observer's frame of mind and the amount of evidence available, which, of course, differ from observer to observer. Subjective probabilities are also called *epistemic* because they deal with a person's knowledge. Subjective probabilities are said to be *coherent* in a betting context when the assignment of probabilities to events prohibits certain loss. It has been shown that coherence suffices to ensure that all of the usual rules for manipulation of probabilities are satisfied.

There may in fact be no significant difference between the subjective definition of probability applied to unique events and the classical definition based on relative frequency. At the molecular level, two tosses of a coin are dissimilar events, yet they are treated as similar for probabilistic purposes because of our subjective belief that the chance of throwing heads is the same both times. And so, seemingly unique events may be grouped together with respect to the degree of belief they inspire, and the probability of the uncertain events in each case expressed as the relative frequency of such events over all cases in the class. This is perhaps a more precise statement of the intuitive notion that the evidence in a particular case has met a certain standard of probability or persuasion. Thus, the statement "X is more likely than not" implies that, if we affirmed a proposition (any proposition) when we had a similar degree of belief, we would be right more than half the time.

Finally, there is an *axiomatic* or formalistic approach that abandons any attempt at interpretation of the notion of probability, restricting attention to the mathematical formulations that assign numerical values to events and satisfy a few basic axioms. These axioms define probability–no more, no less. The axiomatic approach does not prescribe how to assign probabilities to events; it assumes only that some such assignment, consistent with the axioms, exists.

Each individual must decide where the assignment comes from, and what real-world interpretation it has.

What is one to make of all this? Fortunately, the diversity of interpretation is usually of no importance in practical matters, since all formulations obey the same formal rules of calculation. Differences show up most in the choice of method for statistical inference. The objective interpretation emphasizes methods with good frequentist properties, i.e., long-run averages, while the subjectivist interpretation emphasizes Bayesian methods, i.e., methods that combine prior subjective probabilities with evidence to arrive at posterior assessments. The likelihood ratio is a common thread in both interpretations.

When observed data are plentiful, frequentist and Bayesian inferences usually agree. When there is little available data, an individual's prior probability assessments are likely to make a difference, as some of the preceding problems have shown. In such cases the subjective assessment should be brought into the open and its merits debated. One reason for doing this is that studies have shown that, in estimating posterior probabilities, uninformed decisionmakers tend to misappraise the force of their prior probabilities relative to the evidence. Making the prior explicit tends to correct for such bias.

Further Reading

D. Kahneman, P. Slovic & A. Tversky (eds.), *Judgment Under Uncertainty: Heuristics and Biases* (1982).

L. Savage, *The Foundations of Probability* (1954).

3.6.1   Relevant evidence defined

Rule 401 of the Federal Rules of Evidence defines "relevant evidence" as "evidence having any tendency to make the existence of any fact that is of consequence to the determination of the action more probable or less probable than it would be without the evidence."

Questions

1. Does this definition import Bayesian or classical probability?

2. If Bayesian, reword the definition to make it classical; if classical, reword it to make it Bayesian.

3. Professor Gary Wells suggests that people's reluctance to base a verdict on "naked statistical evidence" (see Section 3.3.1) may be due to the fact that "in order for evidence to have significant impact on people's verdict preferences, one's hypothetical belief about the ultimate fact must affect one's belief about the evidence." Since statistical evidence that is not case-specific is unaffected by one's view of the case, it is seen as insufficient. Gary L.

Wells, *Naked Statistical Evidence of Liability: Is Subjective Probability Enough?*, 62 J. Personality & Soc. Psych. 739, 746-747 (1992) (suggesting the hypothesis and supporting it with some experimental results). But if statistical evidence that is not case-specific generates a 50+% prior probability in favor of a proposition, but is nevertheless insufficient, why should the addition of even weak case-specific evidence (as reflected in the likelihood ratio) make the total mix of evidence sufficient? For example, if the statistical evidence generates a prior probability of 80% in favor of a proposition, and with the case specific evidence the posterior probability is 81%, why should the latter, but not the former, be deemed sufficient in a civil case?

CHAPTER 4

Some Probability Distributions

4.1   Introduction to probability distributions

We continue the discussion of random variables that was begun in Section 1.1 at p. 1.

Discrete random variables

A *discrete* random variable is one that can take on only a finite (or at most denumerable) number of values. For example, the number of members in a sample household, X, can be thought of as a discrete random variable taking values 1, 2, 3, and 4+. If we knew the relative frequencies of different sizes of households, we could set up an assignment between these possible values of the random variable X and their relative frequencies, as shown in the following table:

x	1	2	3	4+
$P[X - x]$	0.1	0.2	0.5	0.2

The discrete random variable X is said to take the value 1 with probability 0.1, the value 2 with probability 0.2, and, in general, the value x with probability $P[X = x]$. The table is said to define the random variable's *distribution*. Nothing in mathematical theory compels any particular assignment, except that probabilities must be nonnegative numbers between 0 and 1 and the probabilities for all possible values of the random variable must sum to 1.

The probabilities associated with the various possible values of a random variable may be specified by a formula that reflects in an idealized way the process by which the values of the random variable are generated. The binomial distribution (Section 4.2) specifies the probabilities of tossing x heads with a coin out of n tosses, the number of heads being the random variable X generated by the coin tossing. The hypergeometric distribution (Section 4.3) specifies the probabilities of picking x marked chips in N selections (without replacement) from an urn that has M marked and T-M unmarked chips, the number of marked chips among those selected being the random variable X generated by the sampling without replacement. The Poisson distribution (Section 4.7) specifies the probabilities of observing x relatively rare events–such as accidents–occurring in a time period, where the random variable X is the number of such events over the period. The above distributions are for discrete random variables because they relate to counts of discrete events–numbers of heads, marked chips, or accidents. For such discrete random variables, the distributions $P[X = x]$ are also called point probability functions because they assign a probability to each discrete value of the random variable.

Probability distributions for discrete variables are usually represented by histograms, in which each value of a random variable is represented by a rectangle of unit width centered on that value and the area of the rectangle represents the probability of that value. The total area of all the rectangles is 1. The histogram for rolling a pair of dice utilizing the probability distribution in Section 1.2 at p. 6 is shown in Figure 4.1a.

Continuous random variables

A *continuous* random variable is a mathematical extrapolation of a random variable to a continuum of possible values, usually made for mathematical convenience. For example, height or weight in a human population, or time to some event, may be viewed as continuous random variables.

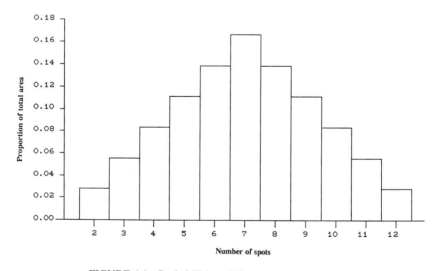

FIGURE 4.1a. Probabilities of dice outcomes 2, . . . , 12

For continuous random variables, a probability distribution is expressed as a *probability density function*, which can be viewed as a limiting form of a point probability function as the possible values of the discrete random variable become more and more continuous.[1] In a density function, the ratio of two ordinates provides the relative likelihood of observing a value near one level versus the other, and for this reason densities are also called *relative frequency functions*. The normal, or Gaussian, bell-shaped curve (see Section 4.3) is a probability density function that specifies the relative frequency with which a sample measurement will depart by x standard deviations from the true or population mean value; it also is used to approximate the probability distributions for discrete random variables referred to above. These uses give it a key role in statistical inference. The exponential density (see Section 4.8) is another important probability distribution, one that is used to specify the probabilities of waiting times to an event when the probability of the event does not change over time, and in particular is unaffected by the past pattern of events.

Cumulative distributions

Most uses of probability distributions involve the probabilities of groups of events; these are called cumulative probabilities because they involve adding together the probabilities of the elementary events that make up the group. For example, the probability of tossing x *or fewer* heads in n coin tosses, denoted by $P[X \leq x]$, is a cumulative probability, the sum of the probabilities of tossing $0, 1, 2, \ldots, x$ heads. Similarly, the cumulative probabilility of *at least x* heads, denoted $P[X \geq x]$, is the sum of the probabilities of tossing $x, x + 1, x + 2, \ldots, n$ heads.

Cumulative probabilities for continuous random variables are obtained from the relative frequency function as areas under the density curve. Relative frequency functions are non-negative and the total area beneath the curve equals 1. Thus $P[X \geq x]$ is the area beneath the relative frequency curve to the right of x. For example, if X is a random variable with a standard normal distribution, the probability that $X > 1.96$, denoted by $P[X > 1.96]$, is the proportion of total area under the curve to the right of the value $x = 1.96$ and corresponds to the shaded region in Figure 4.1b. The figure shows that the value $x = 0$ is $0.399/0.058 = 6.9$ times as likely to occur as the value $x = 1.96$. It also shows that $P[X > 1.96]$ is 0.025.

Probabilities of the form $P[X \leq x]$ are denoted by $F(x)$, and the function $F(x)$ is called the *cumulative distribution function* (cdf) of the random variable X. The cdf has an ogival shape, showing the inevitably increasing proportion

---

[1]More precisely, if we represent the discrete probability $P[X = x]$ geometrically as the area of a narrow rectangle centered at x and extending horizontally half-way to the two neighboring values of x, then the probability density function is the approximate height of the rectangle in the limit, as the neighboring values of x come closer together and $P[X = x]$ approaches 0.

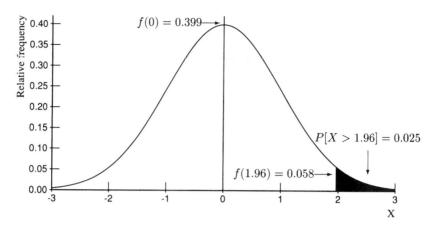

FIGURE 4.1b. Probabilities as areas under the standard normal relative frequency curve, $f(x) = (2\pi)^{-0.5} \exp(-x^2/2)$

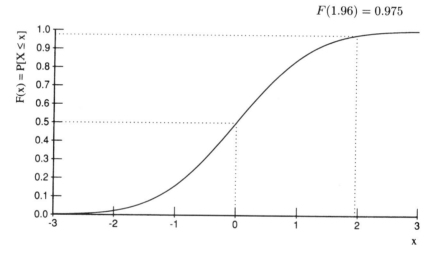

FIGURE 4.1c. Standard normal cumulative distribution function

of the population with values of X at most equal to x, as x increases without limit (see Figure 4.1c). Note that in the figure, $F(1.96)$ is 0.975, corresponding to the complementary probability illustrated in Figure 4.1b.

Further Reading

F. Mosteller, R. Rourke, and G. Thomas, *Probability with Statistical Applications*, ch. 5 (2d ed. 1970).

2 *Int'l Encyclopedia of Statistics* 1101-9 (1978) (W. Kruskal and J. Tanur, eds.).

4.2   Binomial distribution

Suppose an urn contains a certain number of chips, a proportion p of which are labeled '1,' the rest labeled '0.' Chips are withdrawn at random and replaced in the urn after each drawing, so that the contents of the urn remain constant. After n drawings, what is the probability of obtaining exactly r chips labeled '1'? An equivalent problem is to find the probability of obtaining exactly r heads among n tosses of a coin with probability of heads on a single toss equal to p. Both problems involve a sequence of n binary random variables, each identically distributed, with successive outcomes being statistically independent.[2] Being "independent" means that probabilities of subsequent outcomes do not depend on prior outcomes. "Identically distributed" here means that the probability p remains constant from one observation to the next. In such cases, the number of '1's (or heads, or generic "successes") in n trials is a random variable X having a *binomial distribution* with index n and parameter p, written $X \sim \text{Bin}(n, p)$.

Basic formula

For any integer r between 0 and n, we have the following probability formula for the event that the binomial variable takes the value r:

$$P[X = r] = \binom{n}{r} p^r (1 - p)^{n-r}.$$

To derive this formula, note that any sequence of 1's and 0's containing r 1's and $n - r$ 0's occurs with probability $p^r (1 - p)^{n-r}$ by the multiplication rule for probabilities of independent events, and there are $\binom{n}{r}$ distinct such arrangements of r 1's and $n - r$ 0's (see Section 2.1). Since each distinct arrangement occurs with the same likelihood, the total probability of obtaining r successes is as given above.

The binomial distribution may be illustrated with the following example: a balanced die is tossed 24 times. What is the probability of tossing an ace three times? The answer is given by the binomial formula for the point probability, with $n = 24$ and $p = 1/6$:

$$P[X = 3] = \binom{24}{3} \cdot \left(\frac{1}{6}\right)^3 \cdot \left(\frac{5}{6}\right)^{21} = 0.204.$$

Cumulative binomial distribution

The binomial formula for individual terms may be summed to calculate exact "tail-area" probabilities of events of the form "the number of successes is

---

[2]Trials in such a sequence are called "Bernoulli trials" in honor of Jakob Bernoulli (1654-1705), whose *Ars Conjectandi* was one of the earliest treatises on mathematical probability.

less than or equal to r," or "the number of successes is greater than r." Such probabilities, known as cumulative binomial probabilities, are calculated by summing up the individual point probabilities for all events included in the statement. To continue the prior example, the probability of tossing three *or fewer* aces, is the sum of the probabilities of tossing 0, 1, 2, and 3 aces. In symbols:

$$P[X \le 3] = \sum_{i=0}^{3} \binom{24}{i} \cdot \left(\frac{1}{6}\right)^i \cdot \left(\frac{5}{6}\right)^{24-i} = 0.416.$$

Binomial mean and variance

Calculating cumulative binomial probabilities can become laborious (without a computer), particularly when n is large and r is not close to either 0 or n. In such cases, certain approximations for these probabilities make the calculation much simpler by capitalizing on the properties of the binomial variable as a sum of independent and identically distributed random variables (see Section 4.3). These approximations require knowledge of the mean and variance of the binomial distribution, which are derived as follows.

Take a single binary variable, say Y, which might be a single toss of a coin. Heads take the value 1 and tails the value 0 with probabilities p and $1 - p$, respectively. In that case, the number of heads has an expected value equal to $EY = 1 \cdot p + 0 \cdot (1 - p) = p$, and a variance equal to $VarY = E(Y - p)^2 = (1 - p)^2 \cdot p + (0 - p)^2 \cdot (1 - p) = p(1 - p)$. Now take a binomial variable X that is the sum of n realizations of Y. Its expectation is $EX = n \cdot EY = np$. And since these realizations are statistically independent, the variances also sum to $VarX = n \cdot VarY = np(1 - p)$. Returning to our die example, the expected number of aces in 24 tosses is $24(1/6) = 4$; the variance of the number is $24(1/6)(5/6) = 10/3$; and the standard deviation is $\sqrt{3.33} = 1.82$. These results enable us to express binomial outcomes as standardized variables. For example, if we tossed 0 aces in 24 tosses, the result would be $(0 - 4)/1.82 = -2.2$ standard deviations below expectation. Such expressions are used repeatedly in making estimates of cumulative binomial probabilities.

Binomial sample proportion

Binomial variables are frequently expressed in terms of the sample proportion, $\hat{p} = X/n$, by dividing the number of successes, X, by sample size, n. The expected value of $\hat{p}$ is $E(\hat{p}) = \frac{1}{n} \cdot np = p$, i.e. $\hat{p}$ is an unbiased estimator of p. The variance of $\hat{p}$ is $Var(\hat{p}) = \frac{1}{n^2} \cdot VarX = \frac{1}{n^2}np(1 - p) = p(1 - p)/n$. (See Section 1.2.) Thus, the standard deviation of $\hat{p}$ (more commonly called the *standard error of the sample proportion*) is $\sqrt{p(1 - p)/n}$.

Binomial tables

Table B in Appendix II tabulates the cumulative binomial distribution for smaller numbers of trials ($n \leq 25$ and $n = 30, 35, \ldots, 50$) and values of p in multiples of 0.05 up to 0.50. For values of $p > 0.50$, the probability of at least X successes is computed as the probability of $n - x$ or fewer failures with $1 - p$ as the probability of failure. The probability of $n - x$ or fewer failures is the complement of the probability of $n - x + 1$ or more failures. Hence, the probability of at least X successes is equal to $1 -$ (probability of at least $n - x + 1$ failures), with $1 - p$ as the probability of failure.

Multinomial distribution

The binomial distribution generalizes to the *multinomial distribution* when there are more than two possible outcomes of a trial. The probability of obtaining cell frequencies $n_1, \ldots, n_k$ in $n = n_1 + \cdots + n_k$ independent tosses of a k-sided die with outcome probabilities $p_1, \ldots, p_k$ on each toss is given by

$$\binom{n}{n_1, \ldots, n_k} \cdot p_1^{n_1} \cdots p_k^{n_k},$$

where

$$\binom{n}{n_1, \ldots, n_k} = n! \bigg/ \prod_i^k n_i!$$

is the multinomial coefficient (see Section 2.1 at p. 43). For example, toss a pair of ordinary dice six times. What is the probability of getting three 7's, two 11's, and one 12? Rolling a pair of dice is equivalent to tossing a single eleven-sided "die" with outcomes $2, \ldots, 12$ occurring with probabilities 1/36, 2/36, 3/36, 4/36, 5/36, 6/36, 5/36, 4/36, 3/36, 2/36, 1/36, respectively. We get the set of 7's, 11's, and 12 with probability $6!/(3! \cdot 2! \cdot 1!) \cdot (6/36)^3 \cdot (2/36)^2 \cdot (1/36)^1 = 60 \cdot 0.0046296 \cdot 0.0030864 \cdot 0.027778 = 2.3815 \times 10^{-5}$.

4.2.1   Discrimination in jury selection

In *Avery* v. *Georgia*, 345 U.S. 559 (1953), a black defendant was convicted by a jury selected from a panel of sixty veniremen. Their names were drawn from a box containing tickets with the names of persons on the jury roll— yellow tickets for blacks, white tickets for whites. Five percent of the tickets were yellow, yet not a single yellow ticket was selected. Writing for the U.S. Supreme Court, Justice Frankfurter held that "[t]he mind of justice, not merely its eyes, would have to be blind to attribute such an occasion to mere fortuity." *Id.* at 564.

The Court first calculated binomial probability in a jury discrimination challenge in *Whitus* v. *Georgia*, 385 U.S. 545 (1967). In *Whitus*, blacks constituted 27% of the tax digest from which jury lists were selected. The digest indicated race. From this digest, jury commissioners selected a "revised" jury list of around 600 names on the basis of personal acquaintance. A venire of 90 was selected "at random" from this list. The venire included 7 blacks. The petit jury that convicted the defendant was selected from this venire: there were no blacks on the jury. The racial breakdown of the "revised" jury list did not appear. One jury commissioner, however, testified that his best estimate was that 25% to 30% of the list was black. Without benefit of briefs or argument, citing a law review article, the Court calculated that "[a]ssuming that 27% of the list was made up of the names of qualified Negroes, the mathematical probability of having seven Negroes on a venire of 90 is 0.000006." *Id.* at 552, n.2.

Questions

1. Using a binomial model, do you agree with Justice Frankfurter in *Avery*?

2. Is the event for which the *Whitus* court calculated a probability appropriate to test the hypothesis of random selection with respect to race?

3. Suppose the venire in *Whitus* had consisted of 25 persons, of whom 2 were black, and the revised jury list was 25% black. Using Table B in Appendix II, would you reject the hypothesis of random selection from this list?

Notes

For a discussion of this subject see Michael O. Finkelstein, *The Application of Statistical Decision Theory to the Jury Discrimination Cases*, 80 Harv. L. Rev. 338 (1966), reprinted in revised form in *Quantitative Methods in Law*, ch. 2 (1978).

4.2.2   Educational nominating panel

Under a system established in 1965, the Mayor of Philadelphia appoints nine members of the School Board, but is assisted in that task by the Educational Nominating Panel. The function of the Panel is to seek out qualified applicants and submit nominees to the Mayor. Of the Panel's 13 members, all appointed by the Mayor, four must come from the citizenry at large and nine must be the highest ranking officers of certain specified citywide organizations or institutions, such as labor union councils, commerce organizations, public school parent-teacher associations, degree-granting institutions of higher learning, and the like. Panels serve for two years. The Panels of 1965 to 1971 (all appointed by Mayor Tate) were constituted as shown in Table 4.2.2.

Throughout this period, the general population of Philadelphia was approximately one-third black and the school-age population was 60% black.

A civic organization sued, claiming that: (i) the Mayor had unconstitutionally excluded qualified blacks from consideration for membership in the 1971

TABLE 4.2.2. Educational Nominating Panels.
Philadelphia. 1965–1971

	Whites	Blacks
1965	10	3
1967	11	2
1969	12	1
1971	11	2

Panel in violation of the Fourteenth Amendment and the city charter; and (ii) the framers of the city charter intended that the nine organizational seats on the Panel, when combined with the four at-large selections, would reflect a cross section of the community. The district court rejected both claims. It held that differences between representation of blacks in the population and on the Panel were of no significance, largely because the number of positions on the Panel was too small to provide a reliable sample; the addition or subtraction of a single black meant an 8% change in racial composition.

The court of appeals reversed. It found that "the small proportion of blacks on the Panel is significant in light of the racial composition of the public schools. Because one qualification for Panel membership is interest in the public school system and because the parents of school children are likely to have this interest, a color-blind method of selection might be expected to produce that many more black Panel members." *Educational Equality League* v. *Tate*, 472 F.2d 612, 618 (3d Cir. 1973). On review, the Supreme Court found that the appropriate comparison was the population of highest-ranking officers of the designated organizations, not the general population, and that the smallness of the 13-member panel precluded reliable conclusions. For these and other reasons, it dismissed the case. *Philadelphia* v. *Educational Equality League*, 415 U.S. 605 (1974).

The record did not indicate the racial composition of the highest ranking officers of the designated organizations, nor, with one exception, did the opinions of the courts indicate whether blacks appointed to the Panel were from the designated organizations or from the general citizenry.

Questions

1. Assuming that at least one-third of the heads of the designated classes of organizations were black, does the smallness of the 13-member panel preclude reliable conclusions based on a binomial model?

2. Since selection for the Panel was not random, is the binomial model of any use in this context?

4.2.3   *Small and nonunanimous juries in criminal cases*

In Louisiana, the traditional common law unanimous 12-person jury has been replaced by a three-tier system: a unanimous 12-person jury for conviction

TABLE 4.2.3. Jury voting

Final verdict	First ballot and final verdict Number of guilty votes on first ballot				
	0	1–5	6	7–11	12
Not guilty	100%	91%	50%	5%	0%
Hung	0	7	0	9	0
Guilty	0	2	50	86	100
No. of Cases	26	41	10	105	43

From: H. Kalven and H. Zeisel, *The American Jury*, 488 Table 139 (1966).

of the most serious felonies; a 9-to-3 jury for conviction of less serious felonies; and a unanimous 5-person jury for conviction of the least serious felonies.

Assume that in Louisiana juries are selected at random from a population that is 20% minority. In addition, consider the data shown below on first ballot and final verdict collected by Kalven and Zeisel from a sample of 225 jury cases in Chicago (where unanimity was required). In the aggregate, 1,828 out of 2,700 jurors cast a first ballot for conviction.

Questions

1. As the attorney for a minority-group defendant charged with a felony in the middle category, use a binomial model to argue that the probability that a minority juror will be required to concur in the verdict is substantially reduced by the shift from unanimous 12 to 9–3 verdicts.

2. Use a binomial model to argue that a 9-to-3 conviction vote on a first ballot is easier to obtain than a unanimous 5 conviction for lesser offenses.

3. As a prosecutor, use the Kalven and Zeisel data to attack the binomial model.

Source

Johnson v. Louisiana, 406 U.S. 356 (1972); see also *Apodaca v. Oregon*, 406 U.S. 404 (1972).

Notes

In the 1970s six Supreme Court decisions dealt with the constitutionality of decisions by juries with fewer than twelve members or by nonunanimous verdicts. Five of these cases involved the "small jury" issue.

In *Williams v. Florida*, 399 U.S. 78 (1970), the Court approved (7–1) unanimous six-member juries, asserting that "[w]hat few experiments have occurred—usually in the civil area—indicate that there is no discernible difference between the results reached by ... different-sized juries." *Id.* at 101.

In *Colgrove* v. *Battin*, 413 U.S. 149 (1973), the Court held (5–4) that six-member federal civil juries meet the Seventh Amendment requirement of trial by jury. The Court bolstered its historical argument by adding that "four very recent studies have provided convincing empirical evidence of the correctness of the Williams conclusion that 'there is no discernible difference' " in verdicts between six-member and twelve-member juries (*Id.* at 159, n.15).

In *Ballew* v. *Georgia*, 435 U.S. 223 (1978), the Court drew the line and held that five-person juries in criminal cases violated the Sixth and Fourteenth Amendments. The opinion for the Court by Justice Blackmun, joined on these points by five other Justices, is striking for its extensive reliance on speculative probability models to disprove the Court's earlier opinion that there was no discernible difference between 6- and 12-member juries.

Based on various studies, Justice Blackmun found that progressively smaller juries (i) are less likely to foster effective group deliberation (*id.* at 232); (ii) are less likely to be accurate (*id.* at 234); (iii) are subject to greater inconsistencies (*id.* at 234–35); (iv) are less likely to be hung and more likely to convict (*id.* at 236); and (v) are less likely to include adequate minority representation (*id.* at 236–37).

On the issue of accuracy, Justice Blackmun cited a model by Nagel and Neef, which purported to show that the optimal jury size for minimizing error is between six and eight (*id.* at 234). The model made (i) "the temporary assumption, for the sake of calculation," that each juror votes independently (the "coin flipping" model); (ii) the assumptions "for the sake of discussion" that 40 percent of innocent defendants and 70 percent of guilty defendants are convicted; (iii) the assumption that 95% of all defendants are guilty; and (iv) the value judgment, following Blackstone, that the error of convicting an innocent person (Type I error) is ten times worse than the error of acquitting a guilty person (Type II error) (so that an error of the first type should be given ten times the weight of an error of the second type). Using the coin flipping model and the assumptions in (ii), Nagel and Neef calculated that the probability that a juror would vote to convict an innocent person was 0.926 and to convict a guilty person was 0.971; and the probability that a juror would vote to acquit a guilty person was 0.029 and to acquit an innocent person was 0.074. These are highly implausible results. Using these probabilities, the coin-flipping model, the assumption that 95% of defendants are guilty, and Blackstone's weighting scheme, Nagel and Neef calculated total weighted error for juries of different sizes. They found the point of minimum weighted errors to be 7 persons (468 errors per thousand at 7 versus, e.g., 481 errors per thousand at 12 and 470 at 5). Nagel and Neef, *Deductive Modeling to Determine an Optimum Jury Size and Fraction Required to Convict*, 1975 Wash. U.L.Q. 933, 940–48.

The Court approved the Blackstone 10 to 1 weighting of Type I and II errors as "not unreasonable" and cited the Nagel and Neef conclusion that "the optimal jury size was between six and eight. As the size diminished to five and below, the weighted sum of errors increased because of the enlarging risk of the conviction of innocent defendants." *Id. at 234.*

On the issue of consistency, the Court again cited Nagel and Neef. Based on Kalven and Zeisel's data indicating that 67.7% of jurors vote to convict on the first ballot, Nagel and Neef computed the standard error of the proportion of first ballot conviction votes as $[0.677 \cdot (1 - 0.677)/(n - 1)]^{1/2}$. Using a t-value at the two-tailed 0.5 level with 11 d.f., they computed 50% confidence intervals for the "conviction propensity," which ranged from 0.579 to 0.775 for the 12-member jury and from 0.530 to 0.830 for the 6-member jury. (For a discussion of t-tests and confidence intervals, see Section 7.1.) The Court cited these figures as evidence of significant diminished consistency, *id.* at 235, n.20. "They [Nagel and Neef] found that half of all 12-person juries would have average conviction propensities that varied by no more than 20 points. Half of all 6-person juries, on the other hand, had average conviction propensities varying by 30 points, a difference they found significant in both real and percentage terms." *Id.* at 235.

Given the arbitrariness of the assumptions, the patent inappropriateness of the binomial model, and the outlandishness of the computed probabilities, it is remarkable, and unfortunate, that the Supreme Court embraced the results of the Nagel and Neef model as expressing truths about the effect of jury size on the correctness and consistency of verdicts.

Ballew was followed by *Burch* v. *Louisiana*, 441 U.S. 130 (1979), in which the Court held that a 5-to-1 verdict for conviction in a state criminal trial was unconstitutional. The opinion by Justice Rehnquist does not cite social science studies but simply rests on arguments that (i) most states with small juries do not also allow nonunanimous verdicts, and (ii) when a state has reduced its juries to the minimum size allowed by the Constitution, the additional authorization of nonunanimous verdicts would threaten the constitutional principle that led to the establishment of the size threshold.

Modeling of jury behavior is not new to probability theory. Nicolas Bernoulli dealt with the subject at the beginning of the 18th century; Condorcet and Laplace each pursued it; and Poisson seemingly exhausted it in his *Recherches sur la probabilité des jugements. . .* published in 1837.

A modern investigation by Gelfand and Solomon, using a slight variation of Poisson's model and Kalven and Zeisel's data on initial and final votes, purports to show that about two-thirds of those accused are guilty and that the chance that a juror will not err on his first ballot (this being the measure of juror accuracy) is 90%. Interestingly, calculations from the models show virtually no differences between the probabilities of conviction for 6- and 12-member juries, the first being 0.6962, the second 0.7004. See Gelfand and Solomon, *Modeling Jury Verdicts in the American Legal System*, 69 J. Am. Stat. A. 32 (1974). For a commentary on this work, see David H. Kaye, *Mathematical Models and Legal Realities: Some Comments on the Poisson Model of Jury Behavior*, 13 Conn. L. Rev. 1 (1980); see also Michael J. Saks & Mollie

Weighner Marti, *A Meta-Analysis of the Effects of Jury Size*, 21 Law & Hum. Behav. 451 (1997).

4.2.4   Cross-section requirement for federal jury lists

The Sixth Amendment to the U.S. Constitution requires that jury panels be drawn from a source representing a "fair cross section" of the community in which the defendant is tried. The fair-cross-section requirement applies only to the larger pool serving as the source of the names and not to the petit jury itself. To make out a case of prima facie violation of the Sixth Amendment, defendant must prove that (a) the representation of a given group is not fair and reasonable in relation to the number of members of the group in the community, and (b) the underrepresentation is the result of systematic exclusion of the group in the jury selection process. A defendant need not prove discriminatory intent on the part of those constructing or administering the jury selection process.

The jury selection process in the Hartford Division of the Federal District Court for Connecticut began with a "Master Wheel" composed of 10% of the names on the voter registration lists–some 68,000 names. From those names the clerks (with judicial help in close cases) picked some 1,500 qualified people for a "Qualified Wheel." From the Qualified Wheel names were picked at random for the "Jury Clerk's Pool." When a venire was required, the jury clerk entered into a computer certain selection criteria for people in the Jury Clerk's Pool (e.g., omitting names of jurors who had served the preceding month) and from this created a "picking list" of people to be summoned for jury duty.

However, because of a programming error at Yale University (the computer read the "d" in Hartford as meaning that the person was deceased), no one from Hartford was included in the Qualified Wheel. Some other unexplained practice or mistake excluded people from New Britain. Voting-age blacks and Hispanics constituted 6.34% and 5.07%, respectively, of the voting-age population in the Division. But since about two-thirds of the voting-age blacks and Hispanics in the Division lived in Hartford and New Britain, their exclusion made the Qualified Wheel unrepresentative.

To cure that problem, a new representative Qualified Wheel was selected. But when it came time to assemble a picking list for the venire in the *Jackman* case, to save the work involved in assembling the old picking list the clerk used the remaining 78 names on that list and supplemented it with 22 names from the new Qualified Wheel.

Defendant attacked the venire selected in this way as violating the cross-section requirement of the Sixth Amendment. To analyze the underrepresentation of Hispanics and blacks, defendant created a "functional wheel" by weighting the proportions of Hispanics and blacks in the two sources (the old and the new Qualified Wheels) by the proportions in which they contributed to the picking list (i.e., 78% and 22%, respectively). With this weighting, blacks constituted 3.8% and Hispanics 1.72% of the functional wheel.

Questions

1. Appraise the representativeness of the functional wheel for blacks and His-panics by computing the probability that not more than one black would be included in a venire of 100 selected at random from the wheel. Make the above calculations using both the functional wheel and the new Qualified Wheel. Make the same calculation for Hispanics. Do the results suggest to you that the functional wheel was not a fair cross section?

2. How many Hispanics and blacks would have to be added to an average venire to make it fully representative? Do these numbers suggest that the functional wheel was not a fair cross section as to either group?

3. Which test—probabilities or absolute numbers—is the more appropriate measure of the adequacy of a cross-section?

Source

United States v. Jackman, 46 F.3d 1240 (2d Cir. 1995) (finding the functional wheel unacceptable). See Peter A. Detre, *A Proposal for Measuring Underrepresentation in the Composition of the Jury Wheel*, 103 Yale L.J. 1913 (1994).

Notes

In many district courts, the master wheel tends to underrepresent blacks and Hispanics because it is chosen primarily, if not exclusively, from voter registration lists and those groups tend to underregister in proportion to their numbers in the population. In *United States v. Biaggi*, 680 F. Supp 641, 647, 655 (SDNY 1988), *aff'd*, 909 F. 2d. 662 (2d Cir. 1990), the district court cited the following evidence of underrepresentation in the 1984 Master Wheel for the Southern District of New York:

Minority group	Eligible population	Master wheel	Percentage pt. disparity
Hispanic	15.7%	11.0%	4.7
Black	19.9%	16.3%	3.6

On these figures, the district court in *Biaggi* held that, since only 1–2 blacks and 1–4 Hispanics would have to be added to an average venire of 50–60 to make it fully representative, the jury wheel was close enough to a cross-section to pass Sixth Amendment muster. The Second Circuit affirmed, but commented that the case pushed the absolute numbers test to its limit. 909 F.2d at 678. Using probability analysis, can you distinguish *Biaggi* from *Jackman*?

4.3   Normal distribution and a central limit theorem

Normal distribution

The normal distribution is the probability law of a continuous random variable with the familiar "bell-shaped" relative frequency curve. There is a family of normal distributions whose members differ in mean and standard deviation. "The" normal distribution refers to the member of that family with zero mean and unit standard deviation. A random variable with this distribution is referred to as a standard normal random variable. Normal distributions are a family in that any normal variable may be generated from a standard normal by multiplying the standard variable by the desired standard deviation, and then adding the desired mean. This important property implies that all family members have the same general shape, and therefore only the standard normal distribution need be tabulated.

The bell-shaped relative frequency curve for the standard normal distribution is given by the formula $\phi(x) = (2\pi)^{-1/2} \cdot e^{-x^2/2}$. It is shown in Figure 4.1b. Normal tail-area probabilities are calculated by numerical integration of this function or by special approximation formulas. M. Abramowitz and I. Stegun, *Handbook of Mathematical Functions*, Chap. 26, Sec. 2, lists many of these. Normal probabilities are widely tabled, and are available on some hand calculators and in virtually all statistical software for computers. Cumulative normal probabilities are given in Table A1 of Appendix II. The chart below, excerpted from Table A2, shows the rapidity with which deviant values become improbable in normal distributions.

Distance of deviate from mean (in units of standard deviation)	Probability of deviation this extreme or more
1.64	.10
1.96	.05
2.58	.01
3.29	.001
3.90	.0001

The familiar quotation of a mean value plus or minus two standard deviations is based on the fact that a normal random variable falls within two standard deviations of its expected value with a probability close to the conventional 95% level.

The normal distribution is symmetrical around its mean, so that deviations z units above the mean are as likely as deviations z units below the mean. For example, there is a 5% chance that a normal deviate will be greater than the mean by 1.64 standard deviations or more, and a 5% chance that it will be less than the mean by 1.64 standard deviations or more, for a total of 10% for an absolute deviation of 1.64 or more.

Central limit theorems

The normal distribution derives its importance in mathematical statistics from various remarkable *central limit* theorems. In substance, these theorems describe the outcome of a process that results from the actions of large numbers of independent random factors, each of which has only a slight effect on the process as a whole. The cumulative effect of these factors is the sum $S_n = X_1 + X_2 + \cdots + X_n$, where the X_i are individual factors. The problem is to determine the probability distribution of S_n when little is known about the distribution of the component factors X_i. This might seem impossible, but the emergence of regularity in the sum from underlying disorder among the components is the essence of these far-reaching statistical laws. In particular, it can be shown that the distribution of S_n (suitably adjusted) is approximately normal. The adjustment commonly made is to subtract from S_n the sum of the expectations of the X_i, and then to divide the difference by the square root of the sum of the variances of the X_i. The result is a standardized random variable, with zero mean and unit variance, whose distribution approaches the standard normal distribution as n becomes large. The central limit theorem applies even when component variables have different distributions, provided only that their variances are not too different (in a sense that can be made quite precise). It is the central role played by the normal distribution, as the limiting distribution of standardized sums of many other distributions, that gives this particular limit theorem its name and its importance.

An equivalent formulation is that the number of standard errors between the sample mean and its expected value follows a standard normal distribution, in the limit as the sample size becomes large. The *standard error* (of the mean) is defined as the standard deviation of the sample mean regarded as a random variable. When the component variables have the same standard deviation, σ, the standard error is equal to σ divided by the square root of n; when the components differ in variability, the standard error is the root mean squared standard deviation, $\sqrt{\sum \sigma_i^2 / n}$, divided by $\sqrt{n}$.

The mathematical root of the central limit theorem, as the limit of the sum of binary variables, goes back to Abraham De Moivre (1667–1754) and Pierre Simon, Marquis de Laplace (1749–1827). An empirical root was added in the early nineteenth century, when variations in astronomical measurements led to general interest in discovering natural laws of error. Carl Friedrich Gauss (1777–1855) systematically studied the normal distribution as a description of errors of measurement, and the normal distribution is frequently called "Gaussian" in recognition of his work. Gauss used this distribution to justify his method of least-squares estimation, which he applied with spectacular success in 1801 to predict the orbits of two newly discovered asteroids, Ceres and Pallas. At the same time, the concept of a law of errors began to extend from the narrow context of measurement error to other areas of natural variation. Adolphe Quetelet (1796–1874) was prominent among those introducing the normal distribution into the investigation of biological and social phenomena. See Section 4.6.1.

Normal approximation for the cumulative binomial

Although the central limit theorem gives a precise statement only in the limit as n becomes arbitrarily large, its utility derives from the asymptotic (i.e., large sample) approximation it implies for finite sample sizes. As an example, consider n Bernoulli trials, where each trial constitutes an independent random variable that takes values 1 or 0 (respectively, "success" or "failure"), with constant probability of success denoted by p. As indicated in Section 4.1, the expected value of a single trial is p, the expected value of the sum is np, the standard deviation in a single trial is $\sqrt{p(1-p)}$, and the standard deviation of the sum is $\sqrt{np(1-p)}$. Since the sample proportion of successes equals the sample mean of the trials, the standard deviation of the sample proportion is $\sqrt{p(1-p)/n}$. The normal approximation is applied as follows.

Suppose that in n trials the number of successes is less than expected, and one wishes to assess the probability that it would be so low. If n is sufficiently large, the probability that the number of successes would be less than or equal to the observed number is given approximately by the probability that a standard normal variate would have a value that was less than or equal to z, where

$$z = \frac{\left(\begin{matrix}\text{observed num-}\\\text{ber of successes}\end{matrix}\right) - \left(\begin{matrix}\text{expected num-}\\\text{ber of successes}\end{matrix}\right)}{\left(\begin{matrix}\text{standard deviation of}\\\text{the number of successes}\end{matrix}\right)}.$$

Since the normal distribution is symmetrical, this probability is the same as the probability that a standard normal variate would have a value greater than or equal to the absolute value (i.e., magnitude) of z.

Suppose we toss a die 100 times and ask what the probability is of obtaining ten or fewer aces. Thus $n = 100$, $p = 1/6$, and $q = 5/6$. The expected number of aces is $(1/6) \cdot 100 = 16.67$, and the standard deviation of the number of aces is $\sqrt{100 \cdot (1/6) \cdot (5/6)} = 3.727$. Thus $|z| = |10 - 16.67|/3.727 = 1.79$. Interpolation in Table A1 of Appendix II shows that the probability that z would be at least 1.79 standard deviations below the mean is $1 - (0.9625 + 0.9641)/2 = 0.0367$.

How large must a sample be for the normal approximation to be reasonably accurate? Generally, the more symmetrical the distributions of the components, the more closely their sum is approximated by the normal distribution. A rule of thumb is that the mean of the sum must be at least three standard deviations from each end of its range. For example, in the binomial case, the closer p is to 0.5, the more symmetrical are the distributions of the component binary variables and the binomial distribution of their sum. If there are 10 trials with $p = 0.5$, $np = 10 \times 0.5 = 5$ and $\sqrt{np(1-p)} = \sqrt{10 \cdot 0.5 \cdot 0.5} = \sqrt{2.5} = 1.58$. Since the mean is $5/1.58 = 3.16$ standard deviations from both 0 and 10, the normal approximation is adequate. However, if $p = 0.1$, then $np = 1$ and the standard deviation is $\sqrt{10 \cdot 0.1 \cdot 0.9} = 0.949$. The mean in this case is $1/0.949 = 1.054$ standard deviations from the 0 end of the distribution; the binomial distribution is too skewed to justify use of the normal approximation. A somewhat less stringent rule is that both np and $n(1 - p)$ must be at least equal to 5 if the normal approximation is to be reasonably accurate.

In some studies (see, e.g., Section 4.7.5), one encounters data that have been truncated by the elimination of all values on one side of a certain cut-off point. When a truncation is substantial, the distribution of the remaining data becomes highly skewed, degrading the accuracy of the normal distribution.

One sometimes requires normal tail-area probability estimates for extreme deviates, e.g., those 5 or more standard deviations from the mean. A useful approximation is that the tail area to the right of x is approximately $\phi(x)/x$ for large x, where $\phi(x)$ is the standard normal density function given above. For example, if $x = 5$ the exact upper tail area is 2.86×10^{-7}; the approximation yields 2.97×10^{-7}. When extremely small tail areas are involved, the normal distribution is no longer an accurate approximation to discrete distributions such as the cumulative binomial or hypergeometric (see Section 5.4). The error becomes proportionately large, although still minute in absolute terms.

Continuity correction

Because the normal distribution describes a continuous variable while the binomial describes a discrete (integer) variable, when sample sizes are small a "correction for continuity" is generally recommended to improve the accuracy of the approximation. This correction is made by reducing the absolute difference between the observed and expected values by $1/2$ unit in computing z, so that the difference between observed and expected values is moved closer to zero.[3] See Figure 4.3. For proportions, the correction is made by reducing the absolute difference by $1/(2n)$. In the dice example, z is recomputed as follows: $z = (|10 - 16.67| - 0.5)/3.727 = 1.655$. The probability of a departure at least this far below the mean is approximately 0.05, so that the continuity correction does make a bit of a difference on these data, even with n as large as 100.

Further Reading

S. Stigler, *The History of Statistics: The Measurement of Uncertainty Before 1900* (1986).

---

[3]The correction is not applied if the observed value already differs from the expected value by less than $1/2$.

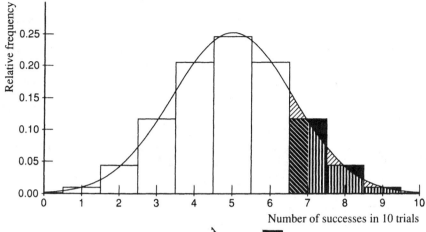

The errors in regions shaded ⧄ and ◤ tend to cancel, making the normal probability to the right of 6.5 a more accurate approximation to $P[S \geq 7]$ than the normal probability to the right of 7.

FIGURE 4.3. One-half continuity correction to approximate $P[S \geq 7]$, where $S \sim$ Bin$(10, 0.5)$

4.3.1 Alexander: *Culling the jury list*

After *Whitus* (see Section 4.2.1), the next jury challenge in which the Supreme Court noted mathematical probability was *Alexander v. Louisiana*, 405 U.S. 625 (1972). The Court summarized the facts as follows: "In Lafayette Parish, 21% of the population was Negro and 21 or over, therefore presumptively eligible for grand jury service. Use of questionnaires by the jury commissioners created a pool of possible grand jurors which was 14% Negro, a reduction by one-third of possible black grand jurors. [Of the 7,374 questionnaires returned 1,015 were from blacks.] The commissioners then twice culled this group to create a list of 400 prospective jurors, 7% of whom were Negro [27 in number]—a further reduction by one-half." *Id.* at 629.

In a footnote, the Court "took note" of "petitioner's demonstration that under one statistical technique of calculating probability, the chances that 27 Negroes would have been selected at random for the 400-member final jury list, when 1,015 out of 7,374 questionnaires returned were from Negroes, are one in 20,000." *Id.* at 630, n.9.

Questions

1. Use the normal approximation for the cumulative binomial distribution to check the probability calculation in *Alexander*.

2. Assume that in the culling process in *Alexander* the Commissioners identified a qualified pool of people from whom the 400-person list was selected at random. Use the normal approximation to estimate the largest proportion of blacks in the qualified pool that would be consistent (at the 5% level) with the 7% black representation on the list, assuming that the list was a small fraction of the qualified pool. Use this proportion to calculate a minimum ratio of the qualification rate among whites to the qualification rate among blacks.

4.3.2   Castaneda: *Measuring disparities*

In *Castaneda v. Partida*, 430 U.S. 432 (1977), the Supreme Court considered whether a criminal defendant had made a prima facie case of discrimination against Mexican-Americans in selection for grand jury service. Defendant's evidence consisted of data showing that, although Mexican-Americans constituted approximately 79.1% of Hidalgo county's population, out of 870 persons summoned for grand jury duty over an 11-year period, only 339, or 39%, were Mexican-Americans. Writing for the majority, Justice Blackmun compared this disparity with the disparities in three other cases in which discrimination had been found, and concluded, without explanation, that the disparity in *Castaneda* was sufficient for a prima facie case. In a lengthy footnote appended to that conclusion, and apparently in support of it, Blackmun accepted the idea that the racial results of nondiscriminatory jury selection should have a binomial distribution. Then, using the normal approximation for the binomial, he computed the probability that so few Mexican-Americans would have been included if selection had been made at random from the general population. In rough fashion, the opinion approves use of the conventional level of statistical significance: "As a general rule for such large samples, if the difference between the expected value and the observed number [of Mexican-Americans] is more than two or three standard deviations, then the hypothesis that the jury selection was random would be suspect to a social scientist." *Id*. at 497, n.7. (The standard deviation of a sample statistic, such as the sample mean, is frequently referred to as the *standard error* of the statistic to distinguish its sampling variability from the variability of the sampled trait in the population. It would thus be more appropriate to express the Court's social science norm in terms of standard errors rather than standard deviations. See Section 1.3, at p. 18.)

The facts in *Castaneda* put the observed disparities well beyond two or three standard errors. The Court calculated that the expected number of Mexican-Americans was 688 and the standard error for the observed minus expected difference was 12. Since the observed number was only 339, the shortfall was $688 - 339 = 349$, which represented $349/12 = 29$ standard errors. Such a disparity, the Court reported, had a probability of 1 in 10^{140}. The Court made the same computation for the 2 1/2-year period during which the state district judge who presided at Castaneda's trial was in charge of the selection process. In that period, there were only 100 Mexican-Americans out of 220 persons summoned; the disparity was 12 standard errors with an associated probability of 1 in 10^{25}. These probabilities are orders of magnitude smaller than almost any encountered in scientific pursuits.

The dissenting justices in *Castaneda* did not argue that chance could account for the underrepresentation of Mexican-Americans. Rather, they argued that the overall census figure of 79.1% Mexican-Americans was misleading because it did not take into account how many Mexican-Americans were not citizens, how many were migrant workers and not residents of the county, how many were illiterate, how many were not of sound mind and good moral character, and how many had been convicted of a felony or were under indictment or legal accusation for theft or a felony—all disqualifications for jury service.

Justice Blackmun rejected these arguments. He pointed out that none of these points was backed with evidence and that, when census figures were broken down in a more detailed way, sharp disparities remained. For example, on the literacy issue, Blackmun calculated that Spanish-surnamed people constituted 65% of those over 25 with some schooling. He found the disparity between that figure and the 39% representation to be significant, reporting that "the likelihood that such a disparity would occur by chance is less than 1 in 10^{50}." *Id.* at 489, n. 8.

Questions

1. The Court has held in some cases that, to constitute prima facie evidence of discrimination, a disparity must be "longstanding and gross." Justice Blackmun calculated P-values to buttress his conclusion that the disparities shown in *Castaneda* met that standard. What are the pros and cons of that approach?

2. The prosecution did not argue that the underrepresentation of Mexican-Americans was due to chance. Would the statistical exercise have become irrelevant if the prosecution had stipulated that the observed disparities were not caused by chance?

Notes

In the wake of *Castaneda*, the federal courts have hammered the "two or three standard deviation" social science norm into a rule of law, and frequently have relied on it, particularly in employment discrimination cases. See, e.g., Section 4.4.1.

The other issue in *Castaneda* was the so-called governing majority theory. As a discrimination case, *Castaneda* was unique in that Mexican-Americans controlled the government of the county; in particular, three of the five jury commissioners who selected jurors were Mexican-Americans. Justice Blackmun found this insufficient to rebut defendant's prima facie case because it could not be presumed, as he put it, "that human beings would not discriminate against their own kind." *Id.* at 500.

4.4   Testing statistical hypotheses

When testing an hypothesis (called a null hypothesis, denoted by H_0) against an alternative hypothesis (denoted by H_1), two types of error are possible: the error of rejecting the null hypothesis when it is true (known as Type I error), and the error of failing to reject the null hypothesis when it is false (known as Type II error).

The maximum allowable rate of Type I error is known as the *level of statistical significance* of the hypothesis test, or the α (alpha) level. The rate of Type II errors is denoted by β (beta). The complement of the Type II error rate, $1 - \beta$, is known as the power of the test, so that power is the probability of (correctly) rejecting the null hypothesis.

The level of statistical significance is a test specification that is set by the investigator in light of the Type I error rate that he or she is willing to accept. Frequently used levels are 0.05 or 0.01. By using 0.05, the investigator incorrectly rejects the null hypothesis in no more than one case in twenty; using a 0.01 level reduces the rate of such errors to one in a hundred.

Why not insist on a smaller Type I error? The reason is this tradeoff: the lower the rate of Type I error, the fewer the rejections of the null hypothesis and the greater the rate of Type II error. Thus, strengthening the level of statistical significance of a test reduces its power for fixed sample size.

On the facts in *Avery* (see Section 4.1.1), an investigator using a *one-tailed* test will decide whether or not to reject the null hypothesis by considering only the probability that so few (in this case, not any) yellow tickets would be selected in a random drawing. This probablity is variously called the "observed" or "attained" significance level, or "P-value." Since under the hypothesis of random selection the probability of selecting no yellow tickets is 0.046 (i.e., less than 0.05), an investigator using a one-tailed test and a significance level of 0.05 would reject the null hypothesis.[4]

On the other hand, an investigator using a *two-tailed* test might decide whether or not to reject the null hypothesis by computing the probability of departing from the expected number by at least the observed amount *in either direction* from the mean. Since the expected number of yellow tickets in the sample is 3, the relevant departures include zero on the low side and six or more on the high side. The P-value approximately doubles (the exact value goes from 0.046 to 0.125, somewhat more than double due to the skewness of the binomial distribution with $n = 60$ and $p = 0.05$; but see below). An investigator insisting on a 0.05 level of significance would not reject the null hypothesis on these data. The test is called two-tailed because to include in the Type I error rate departures which are either too far below *or* too far above the null expectation is to count both tail-area probabilities.

----

[4]Deciding in advance as a test specification to reject the null hypothesis when the P-value falls below 0.05 ensures that the maximum rate of Type I error is no more than 0.05.

When the normal distribution is used to approximate the cumulative binomial distribution, a one-tailed test with a 0.05 rate of Type I error rejects the null hypothesis for departures of 1.645 standard deviations or more from the expected numbers in the hypothesized direction. A two-tailed test with a Type I error rate of 0.05 rejects the null hypothesis only for departures of at least 1.96 standard deviations from the mean in either direction.

When a random variable has a normal distribution, or other symmetrical distribution, a two-tailed test doubles the P-value of the one-tailed test. If the distribution is asymmetric—as is the case in a binomial distribution with np close to 0 or n—it is no longer clear how to define the two tails. Doubling the P-value is appropriate only if the overriding concern is to equalize Type I errors, which is not a usual goal. A commonly used procedure (which we refer to as the point-probability method) takes the point probability of the observed data under the null hypothesis and adds to it the probability of any outcome in either tail with an equal or smaller probability. The resulting sum is the P-value of the two-sided test. For example, recall that in *Avery* the null hypothesis is that the probability of selecting a black juror is 0.05; since 60 selections were made, the expected number of blacks selected is $60 \times 0.05 = 3$. The probability that as few as 0 blacks would be selected is 0.046. That is the lower tail. To get the upper tail, consider in turn the probability of 4, 5, etc., blacks being selected and start adding probabilities when they are equal to or less than 0.046. When X = 6, p = 0.049, and when X = 7, p = 0.020. Thus, the probability of 7 or more blacks has to be added to 0.046. The two-tailed significance level is now 0.076, which is still not significant, although it is smaller than twice 0.046 and smaller than 0.125, the value previously computed by going equal distances on either side of the expected value.[5] In the asymmetrical situation, the point-probability method is to be preferred to the more simplistic ways of calculating P-values in a two-tailed test.

Many scientific researchers recommend two-tailed tests even if there are good reasons for assuming that the result will lie in one direction. The researcher who uses a one-tailed test is in a sense prejudging the result by ignoring the possibility that the experimental observation will not coincide with his prior views. The conservative investigator includes that possibility in reporting the rate of possible error. Thus routine calculation of significance levels, especially when there are many to report, is most often done with two-tailed tests. Large randomized clinical trials are always tested with two-tails.

In most litigated disputes, however, there is no difference between non-rejection of the null hypothesis because, e.g., blacks are represented in numbers not significantly less than their expected numbers, or because they are in fact overrepresented. In either case, the claim of underrepresentation must fail. Unless whites also sue, the only Type I error possible is that of rejecting the null

---

[5]A still smaller P-value, in which the two-tailed test remains statistically significant, is obtainable using the generalized likelihood ratio method. See Section 5.6 at p. 193.

hypothesis in cases of underrepresentation when in fact there is no discrimination: the rate of this error is controlled by a one-tailed test. As one statistican put it, a one-tailed test is appropriate when "the investigator is not interested in a difference in the reverse direction from the hypothesized." Joseph Fleiss, *Statistical Methods for Rates and Proportions* 21 (2d ed. 1981).

Note that a two-tailed test is more demanding than a one-tailed test in that it results in fewer rejections of the null hypothesis when that hypothesis is false in the direction suggested by the data. It is more demanding because it requires greater departures from the expected numbers for a given level of statistical significance; this implies that a two-tailed test has a larger Type II error rate and less power than a one-tailed test with the same significance level. Since power and statistical significance are both desirable attributes, two-tailed tests should not be used where one-tailed tests are appropriate, especially if power is an issue.

For a discussion of Type II errors and statistical power in hypothesis testing, see Section 5.4.

4.4.1   Hiring teachers

In *Hazelwood School District* v. *United States*, 433 U.S. 299 (1977), an employment discrimination case decided shortly after *Castaneda*, the data showed that of 405 school teachers hired by the defendant school district, only 15 were black. The United States, as plaintiff, claimed that this should be compared with the teacher population of St. Louis County, including the City of St. Louis, in which case the population of the pool would be 15.4% black. The defendant school district contended that the city population should be excluded, leaving a population of teachers that was 5.7% black.

In the opinion for the Court, Justice Stewart computed the disparity as less than two standard deviations using the teacher pool and excluding the City of St. Louis, but greater than five standard deviations if the city were included. *Id.* at 311, n.17. For this and other reasons, the Court remanded for a determination whether the United States' position on the labor pool was correct. In dissent, Justice Stevens noted that his law clerk had advised him (after his opinion had been drafted) that there was only about a 5% probability that a disparity that large would have occurred by random selection from even the 5.7% black teacher pool. *Id.* at 318, n. 5. Justice Stevens concluded on these and other grounds that a remand was unnecessary since discrimination was proved with either pool.

Questions

1. Compute the probability that so large a disparity would occur by random selection using each pool.

2. What is the source of the difference between the conclusions of the majority and the dissent? With whom do you agree?

3. What criticisms do you have of a two-or-three standard deviation rule?

Further Reading

Paul Meier & Sandy Zabell, *What Happened in Hazelwood: Statistics, Employment Discrimination, and the 80% Rule*, Am. Bar Foundation Res. J. 139 (1984), reprinted in *Statistics and the Law* 1 (M. DeGroot, S. Fienberg, and J. Kadane, eds., 1986).

4.5   Hypergeometric distribution

Suppose an urn contains T chips of which M are marked in a given way and $T - M$ are unmarked. A random sample of size N is drawn without replacement, i.e., at each drawing a chip is withdrawn with equal likelihood from those remaining in the urn. Let X denote the number of marked chips in the sample. Then the random variable X, has a *hypergeometric distribution* with indices T, M, and N.

Basic hypergeometric formula

The probability function of X may be derived as follows. This kind of random sampling implies that each of the possible $\binom{T}{N}$ samples of size N from the urn is equally likely (see Section 2.1). For given integer x, there are $\binom{M}{x}$ ways of choosing exactly x marked chips from among the M available, and there are $\binom{T-M}{N-x}$ ways of choosing $N - x$ unmarked chips from among the $T - M$ available: so, there are a total of $\binom{M}{x} \cdot \binom{T-M}{N-x}$ ways of choosing a sample size N containing exactly x marked chips. Thus,

$$P[X = x] = \binom{M}{x} \cdot \binom{T - M}{N - x} \Big/ \binom{T}{N}$$

is the required probability function for the hypergeometric distribution. Obviously X cannot exceed the sample size, N, or the total number of marked chips available, M. Likewise, X cannot be less than zero, or less than $N - (T - M) = M + N - T$, for otherwise the sample would include more unmarked chips than the total number of unmarked chips available, $T - M$. Consequently, the possible values of X are $\max(0, M + N - T) \leq x \leq \min(M, N)$.

It is helpful to visualize the notation in a fourfold table, illustrated below for the event $[X = x]$.

	marked	unmarked	total
sampled	x	$N - x$	N
not sampled	$M - x$	$x - (M + N - T)$	$T - N$
total	M	$T - M$	T

The hypergeometric formula can be re-expressed in a number of ways using the defintion of the binomial coefficient:

$$P[X = x] = \binom{M}{x} \cdot \binom{T - M}{N - x} \bigg/ \binom{T}{N}$$

$$= \frac{M! \cdot (T - M)! \cdot N! \cdot (T - N)!}{x! \cdot (M - x)! \cdot (N - x)! \cdot (x - M - N + T)! \cdot T!}$$

$$= \binom{N}{x} \cdot \binom{T - N}{M - x} \bigg/ \binom{T}{M}.$$

The hypergeometric formula thus yields the same result when the rows and columns of the fourfold table are transposed. This means that in applications it does not matter which of two dichotomies is regarded as "sampled" and which as "marked." For example, in a study of the influence of race on parole, we can regard black inmates as the sampled group and compute the probability of X paroles among them. Alternatively, we can regard the paroled inmates as the sampled group and compute the probability of X blacks among them. On the assumption of no association between race and parole, the hypergeometric formulas apply and lead to the same result in either case.

Mean and variance

In deriving the mean and variance of the binomial distribution, it was helpful to view the binomial random variable as a sum of independent and identically distributed binary variables (see Section 4.1). The hypergeometric random variable X is also a sum of binary outcomes ($1 = $ marked, $0 = $ unmarked), but now successive outcomes are neither statistically independent nor identically distributed as the drawing progresses. Why? This situation is typical of sampling from a finite population. The statistical dependence affects the variance of X, but not the mean: X has expectation equal to N times the expected value of a *single* drawing, which has expectation M/T, yielding $EX = MN/T$. The variance of X is equal to N times the variance of a single drawing times the finite population correction factor $(T - N)/(T - 1)$ (see Section 9.1). The variance of a single drawing is $(M/T) \cdot (1 - M/T)$ and thus,

$$\text{Var} X = \frac{M \cdot N \cdot (T - M) \cdot (T - N)}{T^2 \cdot (T - 1)}.$$

Tail area probabilities

To compute tail area probabilities, one may sum individual terms of the hypergeometric formula if they are not too numerous. Otherwise, a version of the central limit theorem for sampling without replacement from finite populations applies. As M and N both become large, with M/T approaching a limiting value p $(0 < p < 1)$ and N/T approaching a limiting value f (sampling fraction, $0 < f < 1$), the standardized variable $Z = (X - EX)/\sqrt{\mathrm{Var}\,X}$ approaches a standard normal random variable in distribution. Thus, for large M, N, and T, the distribution of X may be approximated by a normal distribution with mean Tfp and variance $T \cdot p(1-p) \cdot f(1-f)$. In the case of a fixed sample size N, with T large, the hypergeometric distribution is approximately binomial with index N and parameter p, since with N equal to a small fraction of T sampling without replacement differs negligibly from sampling with replacement.

Example. Consider the following data:

	marked	not marked	total
sampled	2	8	10
not sampled	38	52	90
total	40	60	100

The exact probability that the number X of marked chips would be $x = 2$ is given by the hypergeometric formula:

$$P[X = x] = \binom{M}{x} \cdot \binom{T-M}{n-x} \Big/ \binom{T}{N} = \binom{40}{2} \cdot \binom{60}{8} \Big/ \binom{100}{10}$$

$$= \frac{40 \cdot 39}{2} \cdot \frac{60 \cdot 59 \cdots 53}{8 \cdot 7 \cdots 2 \cdot 1} \Big/ \frac{100 \cdot 99 \cdots 91}{10 \cdot 9 \cdots 2 \cdot 1} = 0.115.$$

In a similar fashion, we find $P[X = 1] = \binom{40}{1}\binom{60}{9}/\binom{100}{10} = 0.034$ and $P[X = 0] = \binom{40}{0}\binom{60}{10}/\binom{100}{10} = 0.004$. Thus, $P[X \le 2] = 0.154$. Using the normal approximation with $1/2$ continuity correction, we find that

$$P[X \le 2] = P[X \le 2.5]$$
$$= P\left[Z \le \frac{2.5 - 40 \cdot 10/100}{[(40 \cdot 60 \cdot 10 \cdot 90)/(100^2 \cdot 99)]^{1/2}} = -1.016\right],$$

which is approximately equal to the probability that a standard normal variate is ≤ -1.016, that is, 0.156.

Non-central hypergeometric distribution

An interesting property of binomial distributions also leads to hypergeometric distributions. Suppose X has a binomial distribution with index N and parameter P, and Y has an independent binomial distribution with index $T - N$ and the same parameter P. If we are told that the sum of the two is $X + Y = M$, what is the conditional distribution of X? The answer is hypergeometric with indices T, M, and N. This property is used to derive Fisher's exact test in Section 5.1.

What happens if the two parameters are unequal, say P_1 for X and P_2 for Y? In that case, given $X + Y = M$, the conditional distribution of X is said to follow the non-central hypergeometric distribution with indices T, M, and

N, and parameter $\Omega = (P_1/Q_1)/(P_2/Q_2)$, where Ω is the odds ratio for the two Ps. Note that the distribution depends *only* on Ω and not otherwise on the individual parameters P_1 and P_2. The probability distribution for X given $X + Y = M$ has the somewhat daunting expression

$$P[X = x | X + Y = M] = \frac{\binom{N}{x}\binom{T-N}{M-x}\Omega^x}{\sum_i \binom{N}{i}\binom{T-N}{M-i}\Omega^i},$$

where the summation extends over all possible values of x (namely the larger of 0 or $M + N - T$ at the low end, up to the smaller of M and N at the high end). This distribution can be used to determine exact confidence intervals for odds ratios, (see Section 5.3 at p. 166), and point estimates for relative hazard functions in survival analysis (see Section 11.1).

Terminology

The name "hypergeometric" was given by the English mathematician John Wallis in 1655 to the terms of the sequence a, $a(a+b)$, $a(a+b)(a+2b)$, ..., with n^{th} term equal to $a(a+b)\cdots(a+(n-1)b)$. This generalizes the geometric sequence whose terms are a, a^2, a^3, ... when $b = 0$. When $b = 1$ we write the nth term in the hypergeometric sequence as $a^{[n]} = a(a+1)\cdots(a+n-1)$. The hypergeometric probability function $P[X = x]$ may then be expressed as the following ratio of hypergeometric terms:

$$\frac{(M-x+1)^{[x]} \cdot (T-M-N+x+1)^{[N-x]} \cdot 1^{[N]}}{(T-N+1)^{[N]} \cdot 1^{[x]} \cdot 1^{[N-x]}}.$$

4.5.1   Were the accountants negligent?

When conducting an audit, accountants check invoices by drawing random samples for inspection. In one case, the accountants failed to include any of 17 fraudulent invoices in their sample of 100 invoices. When the company failed, after the accountants had certified its financial statements, a creditor who had relied on the statements sued, claiming that the accountants had been negligent.

Questions

1. What is the probability that none of the 17 invoices would be included in a random sample of 100 if there were 1000 invoices in all?

2. Suppose the sample were twice as large?

3. Is the difference sufficient to justify holding the accountants liable for failing to draw a larger sample?

4. How should the accountants handle the problem if invoices are of different amounts?

Source

Ultramares Corp. v. *Touche*, 255 N.Y. 170 (1931).

Note

Statistical sampling is now widely used by accountants in audits. See, e.g., American Institute of Certified Public Accountants, Statement on Auditing Standards, No. 39, *Audit Sampling* (1981) (as amended).

See also, Donald H. Taylor & G. William Glezen, *Auditing: An Assertions Approach*, ch. 14 (7th ed. 1997).

4.5.2   Challenged election

Under New York law, as interpreted by New York's Court of Appeals, a defeated candidate in a primary election is entitled to a rerun if improperly cast votes (it being unknown for whom they were cast) "are sufficiently large in number to establish the probability that the result would be changed by a shift in, or invalidation of, the questioned votes." In an election, the challenger won by 17 votes; among the 2,827 votes cast were 101 invalid votes.

Questions

1. If X is the number of improper votes that were cast for the winner, what is the smallest value of X that would change the results if the improper votes were removed from the tally?

2. What is the expected value of X if the improperly cast votes were a random selection from the total votes cast?

3. Use the normal approximation to the cumulative hypergeometric distribution to compute the probability that X would so far exceed expectation that the results of the election would be changed.

4. Are the assumptions used in computing the probabilities reasonable?

5. From the point of view of the suitability of the model, does it matter whether the winner is the incumbent or the challenger?

Source:

Ippolito v. *Power*, 22 N.Y.2d 594, 294 N.Y.S.2d 209 (1968). See Michael O. Finkelstein and Herbert Robbins, *Mathematical Probability in Election Challenges*, 73 Colum. L. Rev. 241 (1973), reprinted in *Quantitative Methods In Law*, 120–30 (1978). A criticism of the random selection model appears in Dennis Gilliland and Paul Meier, *The Probability of Reversal in Contested*

Elections, in *Statistics and the Law*, 391 (M. DeGroot, S. Fienberg, and J. Kadane, eds., 1986, with comment by H. Robbins).

4.5.3   Election 2000: Who won Florida?

The year 2000 presidential election was one of the closest in history. In the end, the outcome turned on the result in Florida. After protest proceedings by the Democrats, on November 26, 2000, Florida election officials certified that the Bush/Cheney Republican ticket had won by 537 votes out of some 5.8 million cast. The Democrats had sought recounts by manual inspection of ballots that had registered as non-votes ("undervotes") by the punch-card voting machines in four heavily Democratic counties. Two of these counties had completed their recounts within the time allowed by Florida state officials and their results had been included in the certified totals. But two large counties–Miami-Dade and Palm Beach–did not meet the deadline and the recounted votes were not included. Palm Beach had completed the recount shortly after the deadline and Miami-Dade had completed only a partial recount and had stopped, leaving uncounted some 9,000 ballots registered as undervotes for president.

The Democrats returned to court, this time in contest proceedings. Florida law provides as a ground for contesting an election a "rejection of a number of legal votes sufficient to change or place in doubt the results of the election." The Democrats contended that the punch-card machines in particular failed to count ballots on which the intent of the voter could be discerned by manual inspection and hence were legal votes that had been rejected by the machine.

The Democrats produced evidence purporting to show that the Gore/Lieberman ticket would pick up enough votes in Miami-Dade and Palm Beach counties to win the election, given manual recounts confined to the undervotes in those counties. (The Republicans attacked this evidence as biased.) The Democrats did not offer testimony on the probability of reversal if there were a state-wide recount. Nor did the Republicans. Although the opinion is not clear, it appears that the trial court would have required such evidence. The trial judge, Sander Sauls, held that there was "no credible statistical evidence, and no other substantial evidence to establish by a preponderance of a reasonable probability that the results of the statewide election in the State of Florida would be different from the result which has been certified by the State Elections Canvassing Commission."

On December 8, the Supreme Court of Florida reversed, holding that such a showing based on a state-wide recount was not necessary. "The contestant here satisfied the threshold requirement by demonstrating that, upon consideration of the thousands of undervotes or 'no registered vote' ballots presented, the number of legal votes therein were sufficient to at least place in doubt the result of the election." The court ordered that the results of the manual recount to date be included and that a recount proceed not only with respect to the 9,000 uncounted votes in Miami-Dade, but also, as a matter of remedy, in all other counties with uncounted votes. A vote was to be included as a legal vote if "the voter's intent may be discerned from the ballot." No further specification was given.

The rate at which a voter's intent could be discerned in undervoted ballots (the recovery rate) evidently depended on the standards applied and the type of voting machine. In those counties already recounted, involving punch-card voting machines, recovery rates ranged from about 8% in Palm Beach, to 22% in Miami-Dade, to about 26% in Broward. More liberal standards, at least in punch-card counties, were thought to favor Gore. Giving effect to the votes that had already been recounted, the Bush plurality shrank to 154 votes. Immediately following the Florida Supreme Court's decision, a feverish recount began around the state. Two optical scanning counties, Escambia and Manatee, completed their recounts immediately and had in the aggregate about a 5% recovery rate.

On the next day—in a 5-4 decision—the U.S. Supreme Court stayed the recount and granted certiorari. Late in the evening on December 12, the Court reversed the Florida Supreme Court, holding (7-2) that the Florida court's opinion violated the equal protection clause of the Fourteenth Amendment by failing to identify and require standards for accepting or rejecting ballots. The Court further held (5-4) that since the Florida Supreme Court had said that the Florida Legislature intended to obtain the safe-harbor benefits of 3 U.S.C. §5 for the Florida electors, any recount would have to be completed by December 12, the safe harbor date. Since this was obviously impossible, the case was over; Bush was declared the winner of Florida and the election.

Would Bush have won if the U.S. Supreme Court had not stopped the recount? Table 4.5.3 shows the reported vote totals for each of the 67 counties in Florida, giving effect to the recount numbers as of the time of the Florida Supreme Court's decision.

Questions

1. Assume that in each county the recovery rate of legal votes from the uncounted votes is 26% for punch-card counties and 5% for optical scan counties, and the recovered votes are divided between the two candidates in the same proportions as the counted votes. If the recount had been allowed to go forward, who would have won?

2. Assuming that the machine-counted votes are a random sample from all of the ascertainable votes (i.e., the counted votes and the undervotes that would have been ascertainable upon a manual recount), test the null hypothesis that Gore won Florida among the ascertainable votes by calculating the statistical significance of the plurality computed in (1) above.

3. Had Judge Sauls granted the Democrats' request for a manual recount in the four counties, then the 195 plurality in Table 4.5.3 would be apropos, with only Miami-Dade left to finish recounting. In that case, what would have been the probability of a reversal?

TABLE 4.5.3. Counted votes and undervotes in Florida counties in the 2000 presidential election

Optical ballot counties	(1) Bush	(2) Gore	(3) Other	(4) Net for Gore	(5) Undervote[a]	(6) Expected net for Gore among undervote[b]	(7) Expected total net for Gore[c]	(8) Variance[d]
Alachua	34135	47380	4240	13245	225	1.7	13246.7	10.7
Bay	38682	18873	1318	−19809	529	−8.9	−19817.9	25.7
Bradford	5416	3075	184	−2341	40	−0.5	−2341.5	2.0
Brevard	115253	97341	5892	−17912	277	−1.1	−17913.1	13.5
Calhoun	2873	2156	146	−717	78	−0.5	−717.5	3.8
Charlotte	35428	29646	1825	−5782	168	−0.7	−5782.7	8.2
Citrus	29801	25531	1912	−4270	163	−0.6	−4270.6	7.9
Clay	41903	14668	985	−27235	100	−2.4	−27237.4	4.9
Columbia	10968	7049	497	−3919	617	−6.5	−3925.5	30.0
Flagler	12618	13897	601	1279	55	0.1	1279.1	2.7
Franklin	2454	2047	144	−407	70	−0.3	−407.3	3.4
Gadsden	4770	9736	225	4966	122	2.1	4968.1	6.0
Gulf	3553	2398	197	−1155	48	−0.5	−1155.5	2.3
Hamilton	2147	1723	96	−424	0	0.0	−424.0	0.0
Hendry	4747	3240	152	−1507	39	−0.4	−1507.4	1.9
Hernando	30658	32648	1929	1990	101	0.2	1990.7	4.9
Jackson	9139	6870	294	−2269	94	−0.7	−2269.7	4.6
Lafayette	1670	789	46	−881	0	0.0	−881.0	0.0

TABLE 4.5.3. continued

Optical ballot counties	(1) Bush	(2) Gore	(3) Other	(4) Net for Gore	(5) Undervote[a]	(6) Expected net for Gore among undervote[b]	(7) Expected total net for Gore[c]	(8) Variance[d]
Lake	50010	36571	2030	-13439	245	-1.9	-13440.9	12.0
Levy	6863	5398	468	-1465	52	-0.3	-1465.3	2.5
Liberty	1317	1017	76	-300	29	-0.2	-300.2	1.4
Manatee	58023	49226	3092	-8797	111	-0.4	-8797.4	5.4
Monroe	16063	16487	1345	424	83	0.1	424.1	4.0
Okaloosa	52186	16989	1639	-35197	85	-2.1	-35199.1	4.1
Okeechobee	5057	4589	208	-468	84	-0.2	-468.2	4.1
Orange	134531	140236	5388	5705	966	1.0	5706.0	47.4
Polk	90310	75207	3112	-15103	228	-1.0	-15104.0	11.2
Putnam	13457	12107	673	-1350	83	-0.2	-1350.2	4.0
Seminole	75790	59227	2783	-16563	219	-1.3	-16564.3	10.7
St.Johns	39564	19509	1698	-20055	426	-7.0	-20062.0	20.6
St.Lucie	34705	41560	1725	6855	537	2.4	6857.4	26.2
Suwannee	8009	4076	376	-3933	42	-0.7	-3933.7	2.0
Taylor	4058	2649	103	-1409	82	-0.8	-1409.8	4.0
Volusia	82368	97313	3992	14945	0	0.0	14945.0	0.0
Walton	12186	5643	494	-6543	133	-2.4	-6545.4	6.4
Washington	4995	2798	233	-2197	292	-4.0	-2201.0	14.1
TOTAL OPTICAL	1075707	909669	50118	-166038	6423	-38.2	-166076.2	312.5

TABLE 4.5.3. continued

Punch ballot counties	(1) Bush	(2) Gore	(3) Other	(4) Net for Gore	(5) Undervote[a]	(6) Expected net for Gore among undervote[b]	(7) Expected total net for Gore[c]	(8) Variance[d]
Baker	5611	2392	152	-3219	94	-9.6	-3228.6	23.0
Broward	177939	387760	9538	209821	0	0.0	209821.0	0.0
Collier	60467	29939	1791	-30528	2082	-179.2	-30707.2	515.4
Duval	152460	108039	4674	-44421	4967	-216.3	-44637.3	1259.2
Hardee	3765	2342	129	-1423	85	-5.0	-1428.0	21.3
Highlands	20207	14169	776	-6038	489	-21.8	-6059.8	123.4
Hillsborough	180794	169576	9978	-11218	5531	-44.8	-11262.8	1397.9
Indian River	28639	19769	1219	-8870	1058	-49.2	-8919.2	266.0
Lee	106151	73571	4676	-32580	2017	-92.7	-32672.7	506.9
Marion	55146	44674	3150	-10472	2445	-64.7	-10536.7	614.5
Miami-Dade	289708	329169	7108	39461	8845	145.0	39606.0	2271.2
Nassau	16408	6955	424	-9453	195	-20.1	-9473.1	47.7
Osceola	26237	28187	1265	1950	642	5.8	1955.8	163.1
Palm Beach	153300	270264	10503	116964	0	0.0	116964.0	0.0
Pasco	68607	69576	4585	969	1776	3.1	972.1	446.9
Pinellas	184849	200657	13017	15808	4226	43.6	15851.6	1062.4
Sarasota	83117	72869	4989	-10248	1809	-29.9	-10277.9	455.3
Sumter	12127	9637	497	-2490	593	-17.2	-2507.2	150.2
TOTAL PUNCH	1625532	1839545	78471	214013	36854	-553.1	213459.9	9324.5

TABLE 4.5.3. continued

Counties not separating over/undervote		(1) Bush	(2) Gore	(3) Other	(4) Net for Gore	(5) Undervote[a]	(6) Expected net for Gore among undervote[b]	(7) Expected total net for Gore[c]	(8) Variance[d]
Desoto	P	4256	3321	235	−935	80	−2.5	−937.5	20.1
Dixie	P	2697	1827	143	−870	14	−0.7	−870.7	3.5
Escambia	O	73171	40990	2688	−32181	665	−9.2	−32190.2	32.4
Gilchrist	P	3300	1910	185	−1390	47	−3.1	−1393.1	11.6
Glades	P	1841	1442	82	−399	62	−1.9	−400.9	15.7
Holmes	O	5012	2177	207	−2835	97	−1.9	−2836.9	4.7
Jefferson	P	2478	3041	124	563	29	0.8	563.8	7.4
Leon	O	39073	61444	2630	22371	175	1.9	22372.9	8.5
Madison	P	3038	3015	110	−23	27	0.0	−23.0	6.9
Martin	P	33972	26621	1420	−7351	133	−4.1	−7355.1	33.7
Santa Rosa	O	36339	12818	1245	−23521	153	−3.6	−23524.6	7.4
Union	O	2332	1407	87	−925	25	−0.3	−925.3	1.2
Wakulla	P	4512	3838	237	−674	49	−1.0	−675.0	12.4
TOTAL OTHER		212021	163851	9393	−48170	1556	−25.6	−48195.6	165.3
TOTAL PUNCH		1681626	1884560	81007	202934	37295	−565.8	202368.2	9435.6
TOTAL OPTICAL		1231634	1028505	56975	−203129	7538	−51.2	−203180.2	366.6
GRAND TOTAL		2913260	2913065	137982	−195	44833	−616.9	−811.9	9802.2

[a] Figures from election night reports. For counties not separating over/undervotes, figures from BDO Seidman report.
[b] Expected net for Gore among ascertained undervotes assuming a 26% recovery rate among punch ballot counties, and a 5% recovery rate in other counties.
$(6) = \{(4)/[(1) + (2) + (3)]\} \times (5) \times$ recovery rate
[c] Expected net for Gore among county totals.
$(7) = (4) + (6)$
[d] Variance of expected net for Gore for county totals.
$(8) = MR\{P[B]+P[G]-R(P[G]-P[B])^2\}$ where M = undervote = (5),
R=recovery rate, P[B]=observed proportion for Bush = (1)/[(1) + (2) + (3)],
and P[G] = observed proportion for Gore = (2)/[(1) + (2) + (3)].

Sources

Bush v. Gore, 121 S. Ct. 525 (2000).
Gore v. Harris, 2000 Fla. Lexis 2373 (2000).
County data are from a CNN website.

Notes

Monte Carlo studies show that the results given here are not sensitive to variations in recovery rates among the counties by type of machine. Contrary to general opinion at the time, higher recovery rates in punch-card counties would have served only to increase Bush's lead.

   Note, however, that the calculations here are based on county data. More detailed precinct data might show a different projected outcome if, for example, there were precincts in Bush counties that had high undercount rates and went for Gore. More significantly, no data are included on the overvotes, i.e., those ballots for which no vote was registered because the voter picked two presidential candidates. The rate of vote recovery for such ballots (based on the face of the ballot and not on demographic factors) would probably be much lower than for the undervote ballots.

4.6   Tests of normality

There are various summary measures used to appraise the shapes of distributions in general, and to check for departure from the normal distribution in particular.

Skewness

Skewness measures the asymmetry of a distribution around its central value. It is defined as the average value of the cubed deviations from the mean, often reported in standardized form by dividing by the cube of the standard deviation. For a symmetrical distribution, skewness is zero. Do you see why?

   Skewness is important to recognize because many probability calculations assume a normal distribution, which is symmetrical, but the sums of highly asymmetrical random variables approach normality slowly as their number increases, making the normal approximation inaccurate. In addition, in skewed distributions the various measures of central tendency (see Section 1.2) have different values. In unimodal distributions (those with a single hump), positive skewness means the hump is on the left and a long tail lies to the right. Negative skewness is the opposite. For example, the exponential distribution with mean $1/c$, with relative frequency function $f(x) = ce^{-cx}$ for $x > 0$, is positively skewed, with standardized third moment, i.e. skewness, equal to 2.[6]

---

[6]Exponential distributions are common waiting-time distributions; see Section 4.8.

Kurtosis

Kurtosis is a measure of the peakedness of the center of a distribution compared to the heaviness of its tails. The coefficient of kurtosis is defined as the average of deviations from the mean raised to the fourth power, standardized by dividing by the squared variance, from which result the value 3 is subtracted. For normal distributions, the coefficient of kurtosis has the value 0. For a given variance, distributions with high narrow peaks and heavy tails have positive kurtosis; distributions with low, broader peaks and thinner tails have negative kurtosis. For example, an exponential distribution has thicker tails than the normal distribution with relative frequency function proportional to $\exp(-cx)$ as opposed to $\exp(-cx^2)$ for the normal. The exponential distribution has kurtosis of 6. The uniform distribution, on the other hand, has no tails and its coefficient of kurtosis is -1.2.[7]

Kolmogorov-Smirnov tests

The coefficients of skewness and kurtosis do not make reliable tests of normality because they are too sensitive to outlying observations (due to the high power to which deviations are raised). A better test, applied to the normal and other continuous distributions, is the Kolmogorov-Smirnov test. In this test, the statistic is the maximum absolute difference between the empirical and theoretical cumulative distribution functions (cdf's). The empirical cdf in a set of data is the proportion of data points at or below any given value. (The theoretical cdf was defined in Section 4.1.) This maximum difference itself has a probability distribution that is tabled. See Appendix II, Table G1. Kolmogorov-Smirnov can also be used to test whether two empirical distributions have the same parent. See Appendix II, Table G2.

In the Silver "butterfly" straddle data (see Section 4.6.2), the entry "D Normal = 0.295232" in Figure 4.6 is the maximum absolute difference used in the Kolmogorov-Smirnov test. What does Table G1 show about the probability associated with a value at least that large under normality for sample sizes in excess of 50?

Normal probability graph paper

Normal probability graph paper is a convenient device for testing normality that also makes use of the cumulative distribution function. See Figure 4.6 for a sample. On this paper the cdf for the normal distribution is a straight line.

---

[7]A uniform distribution assigns equal probability density to all points in a given interval. The uniform distribution on the interval $[0, 1]$ has relative frequency function $f(x) = 1$ for $x \in [0, 1]$ and $f(x) = 0$ elsewhere. A uniformly distributed random variable X has mean $1/2$ and variance $1/12$. The probability that $X \le x$ is simply x for any x between 0 and 1. Its skewness is zero by symmetry.

Probability by 100 Divisions

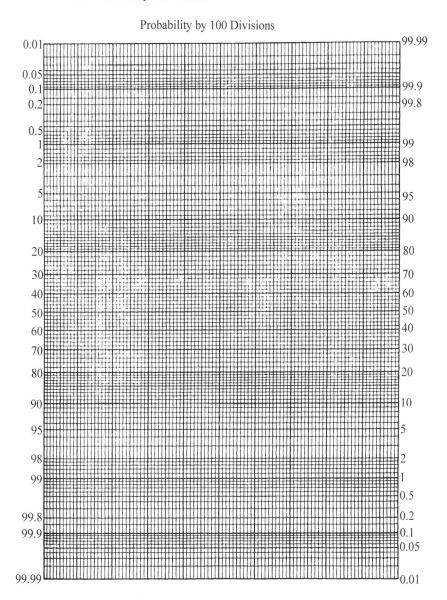

FIGURE 4.6. Normal probability paper

The empirical cdf is plotted as the ordinate against the value of the observation as the abscissa. Departures from a straight line indicate non-normality.

The student should plot the silver butterfly straddle data on the bottom of Figure 4.6.2 on the graph paper, using the scale for the data shown in the figure on the abscissa and the cdf values shown for the ordinate. Ten points should suffice to draw the curve. Compare this curve with the line for the normal distribution with the same mean (-0.11925) and standard deviation (18.1571) as in the data. This line can be drawn using only two points, e.g., the mean at the 50th percentile and 1.645 standard deviations at the 95th percentile. Are the differences between the empirical cdf and the normal cdf line consistent with positive kurtosis?

Chi-squared

For use of chi-squared to test normality, see Section 6.1.

4.6.1   Heights of French conscripts

In the first part of the nineteenth century, interest in the normal distribution as an error law led a Belgian astronomer and poet, Adolphe Quetelet, to use the normal curve to test whether data relating to human populations were sufficiently homogeneous to be aggregated as coming from a single source for purposes of social science analysis. Finding normality in many data sets led Quetelet to the view that variation in characteristics was caused by accidental factors operating on a true or ideal type—*l'homme type*—the average man. Quetelet became famous for this concept. Initially, he viewed the idea merely as a device for smoothing away random variation and uncovering mathematical laws of social behavior. But later, influenced by his conclusion that variation represented accidental imperfections, Quetelet endowed the average man with superior moral qualities emblematic of democratic virtue. Quetelet's tendency to overestimate the prevalence of the normal distribution has been dubbed "Quetelismus."

In one study, Quetelet looked at data on the chest sizes of 5,732 Scottish militiamen and the heights of 100,000 French conscripts, and compared them with the expected distribution under normality. The chest-size data appeared consistent with normality. The height data are set forth in Table 4.6.1. Conscripts below 1.57 meters (5′1.8″) were excused.

Questions

1. Test the data for normality using normal probability graph paper, Kolmogorov-Smirnov, or chi-squared.

2. What departures from normality do you find? What are the possible explanations?

TABLE 4.6.1. Heights of French conscripts—1817 data

I	II	III	IV	V
Height (in meters)	Observed # of men	Observed cumulative # of men	Expected # of men	Expected cumulative # of men
Less than 1.570	28,620	28,620	26,345	26,345
1.570–1.597	11,580	40,200	13,182	39,527
1.597–1.624	13,990	54,190	14,502	54,029
1.624–1.651	14,410	68,600	13,982	68,011
1.651–1.678	11,410	80,010	11,803	79,814
1.678–1.705	8,780	88,790	8,725	88,538
1.705–1.752	5,530	94,320	5,627	94,166
1.732–1.759	3,190	97,510	3,189	97,355
Above 1.759	2,490	100,000	2,645	100,000
Total	100,000		100,000	

3. Test Quetelet's finding that, except for the first two rows, the data are normally distributed by consolidating the first two rows and retesting for normality. What conclusion do you reach?

Source

Adolphe Quetelet, *Letters Addressed to H.R.H. The Grand Duke of Saxe Coburg and Gotha, on the Theory of Probabilities as Applied to the Moral and Political Sciences* 277-278 (Downes, tr., 1849), *discussed in* Stephen Stigler, *Measurement of Uncertainty before 1900*, at 215-216 (1986); see also, Ian Hacking, *The Taming of Chance*, ch. 13 (1990).

4.6.2   Silver "butterfly" straddles

Beginning in the late 1960s, investors began buying silver futures in certain configurations known as "butterfly" straddles. A butterfly straddle consists of a package of long and short futures contracts (10,000 oz. per contract) which is so balanced with respect to expiration dates that the value of the overall package fluctuates much less than the value of the component contracts. If properly liquidated, the straddle can create short-term capital losses in one year which are largely balanced by long-term capital gains in the next. This tax effect attracted investors, particularly those with short-term capital gains which could be offset by the short-term losses.

The IRS brought suit against investors who used straddles for this purpose. The Service claimed that, since the dominant (if not exclusive) motive was tax avoidance, the tax benefits of the transaction should be denied. As part of its case, the IRS presented a statistical model designed to show that a profit was highly improbable because the value of the butterfly moved too little to permit profitable liquidation after commissions. The data used by the

IRS commodities expert consisted of the change in closing prices of various butterfly combinations over specified periods. In a given year there might be three or four possible butterflies with different delivery months. For example, in the period 1973–1979 there were four 2–2 butterflies with delivery months as follows: January-March-May, March-May-July, May-July-September, and July-September-November. A 2-2 butterfly straddle would involve, e.g., a short position of 5 contracts for January, a long position of 10 contracts for March, and a short position of 5 contracts for May. For a given butterfly with, e.g., a four-week holding period, the weekly price change was measured as the difference between the closing price for the butterfly as a whole on Wednesday in week 1 and on Wednesday in week 4; and so on for weeks 2 and 5, etc. Each four-week price change is thus a data point. For the daily, four-week price change, the successive data points are the differences in closing prices on day 1 in weeks 1 and 4, day 2 in weeks 1 and 4, etc. The standard deviations of the daily and weekly price changes over the six-year period (1973–1979) were computed, although most contracts were held for less than one year and the maximum holding period was seventeen months. A sample of the data for butterfly spreads is shown in Table 4.6.2.

The IRS expert pointed out that, because commissions associated with buying and liquidating futures positions (approximately \$126) in most cases exceeded two standard deviations, and price changes were normally distributed, it was highly improbable that prices would fluctuate sufficiently to cover such costs and allow a profit, even if they moved in a favorable direction. In further support of that deduction, the expert produced computer analyses of samples of butterfly data. One such page appears as Figure 4.6.2.

TABLE 4.6.2. Silver butterfly analysis: August 1973–July 1979

		Width					
		1–2 months		2–2 months		2–3 months	
Holding period		Weekly	Daily	Weekly	Daily	Weekly	Daily
1 week	m	3.42	2.05	0.08	0.04	0.63	−0.30
	sd	36.36	34.91	27.40	23.30	50.78	30.09
	n	181	1150	652	3350	207	1066
4 weeks	m	13.48	8.92	1.46	0.78	0.44	−1.82
	sd	55.32	46.02	32.68	28.10	58.08	41.54
	n	178	1135	643	3505	207	1051
8 weeks	m	29.60	18.34	−2.62	−1.57	−1.64	−4.25
	sd	63.30	58.02	30.36	28.08	54.68	47.84
	n	129	899	473	2700	158	787
24 weeks	m	77.56	70.32	−5.86	−2.92	2.67	0.64
	sd	103.92	101.92	42.56	39.53	80.00	78.52
	n	70	616	271	1665	131	654

Key: m = mean (in dollars)
sd = standard of deviation (in dollars)
n = number of observations

	MOMENTS				QUANTILES (DEF=4)			EXTREMES	
N	587	SUM WGTS	587	100% MAX	170	99%	56.1987	LOWEST	HIGHEST
MEAN	-0.11925	SUM	-70	75% Q3	5	95%	15	-170	65
STD DEV	18.1571	VARIANCE	329.679	50% MED	0	90%	5	-170	65
SKEWNESS	-0.0427321	KURTOSIS	51.0753	25% Q1	-5	10%	6.00014	-65	70
USS	193200	CSS	193192	0% MIN	-170	5%	-15	-60	165
CV	-15226	STD MEAN	0.749422			1%	-55	-55	170
T:MEAN=0	-0.159123	PROB>[T]	0.073627	RANGE	340				
D:NORMAL	0.295232	PROB>D	<0.01	Q3-Q1	10				
				MODE	0				

```
BAR CHART                                        BOX PLOT               NORMAL PROBABILITY PLOT

 170.x                                      2      x    170|                                        x
    .                                                     |
    .                                                     |
    .                                                     |
    .x                                      3      x      |                                       xx
    .x                                      6      x      |                                     xxxx
    .xx                                    15      0      |                                xxxxxxxx
    .xxxxxxxxxxxxxxx                       135      _      |                    xxxxxxxxxxxxxxxxx
    .xxxxxxxxxxxxxxxxxxxxxxxxxxxxxxxxxxxxxx 402    |_|     |    xxxxxxxxxxxxxxxxxx
    .xx                                     15      0     | xxxxxxxxxxxx
    .x                                       5      x     |xxxx
    .x                                       2      x     |x
    .                                                     |
    .                                                     |
    .                                                     |
 170.x                                      2      x   -170|x
 ----|----|----|----|----|----|----|----|----|              ----|----|----|----|----|----|----|----|----|----|
                                                            -2        -1         0         1         2
        x may represent up to 9 counts
```

FREQUENCY TABLE

		PERCENTS				PERCENTS				PERCENTS				PERCENTS	
VALUE	COUNT	CELL	CUM	VALUE	COUNT	CELL	CUM	VALUE	COUNT	CELL	CUM	VALUE	COUNT	CELL	CUM
-170	2	0.3	0.3	-30	1	0.2	2.6	5	105	17.9	90.5	50	1	0.2	99.0
-65	1	0.2	0.5	-25	4	0.7	3.2	10	21	3.6	94.0	55	1	0.2	99.2
-60	1	0.2	0.7	-20	5	0.9	4.1	15	9	1.5	95.6	65	2	0.3	99.5
-55	2	0.3	1.0	15	14	2.4	6.5	20	9	1.5	97.1	70	1	0.2	99.7
-50	1	0.2	1.2	-10	20	3.4	9.9	25	3	0.5	97.6	165	1	0.2	99.8
-40	2	0.3	1.5	-5	113	19.3	29.1	30	3	0.5	98.1	170	1	0.2	100.0
-35	5	0.9	2.4	0	255	43.4	72.6	40	4	0.7	98.8				

FIGURE 4.6.2. Silver "butterfly" straddles output

Questions

1. The expert argued, on the basis of the Kolmogorov–Smirnov statistic in the computer printout [D: Normal 0.295232; Prob $> D < 0.01$] that "there was only a 0.01 chance the data were not normally distributed." What is the correct statement?

2. The expert argued that, if the data were not normally distributed due to positive kurtosis, the thinner tails made it even less likely than under the normal hypothesis that there would be opportunities for profit. What is the correct statement?

3. If the daily price changes of the 2-2 butterfly were independent (e.g., a price change from day 1 to day 2 would tell us nothing about the price change between days 2 and 3) would that support the expert's assumption that the 4-week price changes were normally distributed?

4. Are the standard deviations of the 4-week and 8-week holding period price changes consistent with independence of daily price changes?

5. Use Chebyshev's theorem (see Section 1.3 at p. 18) to argue for the IRS.

6. Does the expert's holding-period statistic reflect the potential for profitable liquidation of a butterfly straddle?

7. What alternative methods of analysis might the expert have used to demonstrate the small likelihood of a profitable liquidation of a butterfly straddle?

Source

Smith v. Commissioner, 78 T.C. 350 (1982), *aff'd*, 820 F.2d 1220 (4th Cir. 1987) (Report of Roger W. Gray).

4.7   Poisson distribution

Let X denote the number of successes in n independent trials with constant probability of success p, so that X is binomially distributed with mean np. Let n become large while p approaches zero, in such a way that np approaches the constant μ. This might occur if we consider the probability of an event—such as an accident—occurring over some time period. As we divide the time interval into progressively smaller segments, the probability p of occurrence within each segment decreases, but since the number of segments increases, the product remains constant. In such cases, the distribution of X approaches a limiting distribution known as the Poisson distribution, after the French mathematician Siméon Denis Poisson (1791–1840). Because it describes the distribution of counts of individually rare events that occur in large numbers of trials, many natural phenomena are observed to follow this distribution. For example, the number of atoms in a gram of radium that disintegrate in a unit of time is a random variable following the Poisson distribution because the disintegration of any particular atom is a rare event, and there are numerous atoms in a gram of radium. The number of traffic accidents per month at a busy intersection, the number of gypsy moth infestations per acre of forest, the number of suicides per year, all may have Poisson distributions (or mixtures of Poisson distributions to allow for heterogeneity in the average rate μ).

Basic formula

The probability function for the Poisson distribution with mean μ is

$$P[X = x] = e^{-\mu} \cdot \mu^x / x! \quad \text{for } x = 0, 1, 2, \ldots$$

The mean value is μ and the variance is also equal to μ, as a result of the limiting nature of the Poisson: the binomial variance $np(1 - p)$ approaches μ as $np \to \mu$ and $p \to 0$. The most likely outcome (modal value) in a Poisson distribution is the greatest integer less than or equal to the mean μ. The distribution is positively skewed for small values of μ, less so for large values of μ. The standardized Poisson variable $(X - \mu)/\sqrt{\mu}$ approaches the standard normal distribution as μ becomes large. This is a consequence of the following important property of the Poisson distribution: if X_1 and X_2 are independent Poisson variables with means μ_1 and μ_2, respectively, then $X_1 + X_2$ also has a Poisson distribution, with mean $\mu_1 + \mu_2$. Thus, a Poisson variable with large mean can be viewed as a sum of many independent Poisson variates with approximately unit mean, to which the central limit theorem may be applied.

As an example, suppose you drive back and forth to work each day with a per trip chance of getting a flat tire of 0.001. In one year, you risk a flat tire, say, 500 times. What are the chances of getting one or more flat tires in a year? Two or more? The chances are given by the Poisson distribution with mean $\mu = 500 \times 0.001 = 0.5 : P[X > 0] = 1 - P[X = 0] = 1 - e^{-0.5} = 0.39$ and $P[X > 1] = 1 - P[X = 0] - P[X = 1] = 1 - e^{-0.5}(1 + 0.5) = 0.09$.

Conditioning property

We mention two other useful properties of the Poisson distribution. The first is a conditioning property: let X_1 and X_2 have independent Poisson distributions with means μ_1 and μ_2. Then the conditional distribution of X_1 given fixed sum $X_1 + X_2 = N$ is binomial with index N and parameter $p = \mu_1/(\mu_1 + \mu_2)$. Here is an example of this useful relation. In an age-discrimination suit, typist A claims he was replaced by a younger typist B with inferior accuracy. A typing test is administered to both typists and the number of errors recorded as X_1 for A and X_2 for B. How do we compare typing precision? Assume X_1 and X_2 are each distributed as Poisson variables with means μ_1 and μ_2. Under the null hypothesis of equal precision, the conditional distribution of X_1 given a total of $N = X_1 + X_2$ errors observed is $\text{Bin}(N, 1/2)$ and thus H_0 can be tested in the usual manner.

Poisson process

The second property relates the number of events that occur in a given time period to the waiting time between events. An event that occurs repeatedly over time is said to follow a Poisson process if the following three conditions hold: (i) the occurrence of events in non-overlapping time periods is statistically independent; (ii) the probability of a single event occurring in a short time period of length h is approximately μh; and (iii) the probability of two or more events occurring in the same short time period is negligible. It follows from these conditions that (a) the number of events that occur in a fixed time interval of length T has a Poisson distribution with mean μT, and (b) the inter-arrival times between successive events are statistically independent random variables that follow an exponential distribution with mean $1/\mu$. The exponential waiting-time distribution is characteristic of a constant "hazard" process, where the likelihood of an occurrence does not depend on the past pattern of occurrences. The Poisson process is used in survival (follow-up) studies to analyze the effects of risk factors on mortality in a population, especially over short time periods. See Sections 11.1 and 11.2.

Compound Poisson distribution

The foregoing discussion of the Poisson distribution assumed that μ is a fixed quantity. A *compound* Poisson distribution arises when there is random variability in μ. There are many different compound Poisson distributions, depending on the population distribution of the random Poisson mean μ. For example, a random sample of n drivers will provide n realizations of a theoretical quantity, say $\mu_j (j = 1, \ldots, n)$, representing the "accident proneness" for driver j. While the parameter μ_j is not directly observable for any driver, the number of accidents, X_j, that driver j has in a unit time period is observable. If X_j has a Poisson distribution with mean μ_j, then $X_1, \ldots, X_n$ constitute a sample of size n from a compound Poisson distribution.

A remarkable theorem, due to Herbert Robbins, provides a way to estimate the mean of unobservable μ's for designated subgroups in a compound Poisson distribution. As applied to the example of drivers, Robbins's theorem states that the mean value of μ among a group of drivers with i accidents is equal to $i + 1$ times the marginal probability of $i + 1$ accidents divided by the marginal probability of i accidents; in symbols,

$$E[\mu|X = i] = (i + 1) \cdot P[X = i + 1]/P[X = i].$$

Here $P[X = i]$ is the overall probability of i accidents, which is an average (over the distribution of μ) of Poisson probabilities, $e^{-\mu}\mu^i/i!$.

The theorem is useful for what it tells us about subgroups of drivers. For example, since no driver is immune from accidents, those drivers who are accident-free in a given year would still have some accident propensity (non-zero μ's) and should attribute their good records partly to above-average driving skills and partly to luck. In the next year the group should expect to have some accidents, but fewer than the average for all drivers. We say that the better-than-average accident record of the zero-accident group has "regressed" toward the mean for all drivers. Robbins's theorem gives an estimate of this regression for each subgroup of drivers ($i = 0, 1, \ldots$) based, counterintuitively, on the performance of other drivers as well as those in the particular subgroup. The theorem is remarkable because it holds irrespective of the distribution of the random Poisson mean μ, and expresses an apparently unobservable quantity, $E[\mu \mid X = i]$, in terms of the estimable quantities $P[X = i]$ and $P[X = i + 1]$. For example, the theorem tells us that, for a cohort of drivers who had 0, 1, 2, etc., accidents in year one, if their propensities don't change, drivers with 0 accidents in year one will be expected in year two to have the same number of accidents as were had in year one by drivers who had 1 accident in that year (i.e., the number of such drivers). Do you see why?

4.7.1   Sulphur in the air

In January 1983, the Environmental Protection Agency (EPA) approved Arizona's plan for control of sulphur dioxide emissions from copper smelters. By statute, the plan had to "insure attainment and maintenance" of National Ambient Air Quality Standards (NAAQS). Under those standards, the permitted emissions may not be exceeded more than one day per year. The Arizona plan was designed to permit excess levels approximately 10 times in ten years, or an average of once per year. Discharge from smelters is highly variable, depending on meteorological conditions and other factors.

A person living near a smelter sued, contending, among other things, that there was too great a risk of bunching, i.e., two or more days exceeding the limits in a single year.

Questions

1. Use the Poisson distribution to determine the risk of bunching two or more days (not necessarily contiguous) in a single year, assuming that the occurrence of excess emissions is a simple Poisson process.

2. Is the assumption reasonable?

Source

Kamp v. Hernandez, 752 F.2d 1444 (9th Cir. 1985), *modified, Kamp v. Environmental Defense Fund,* 778 F.2d 527 (9th Cir. 1985).

4.7.2   Vaccinations

Plaintiffs sued the Israeli Ministry of Health when their child suffered irreversible functional damage after receiving a vaccination. The vaccine is known to produce functional damage as an extremely rare side effect. From earlier studies in other countries it is known that this type of vaccine has an average rate of 1 case of functional damage in 310,000 vaccinations. The plaintiffs were informed of this risk and accepted it. The child's vaccine came from a batch that was used to give 300,533 vaccinations. In this group there were four cases of functional damage.

Question

Do the data suggest that the risk of functional damage was greater for this batch of vaccine than the risk that plaintiffs had accepted?

Source

Murray Aitkin, *Evidence and the Posterior Bayes Factor*, 17 Math. Scientist 15 (1992).

Notes

In the U.S., compensation for vaccine injury is in most cases covered by the National Childhood Vaccine Injury Act, 42 U.S.C. §§300aa-1 et seq. The act authorizes compensation for specified injuries, *id.* §300aa-10(a), caused by a covered vaccine, *id.* §300aa-11(c)(1). When the time period between the vaccination and the first symptom or manifestation set in the Vaccine Injury Table of the act is met, it is presumed that the vaccine caused the injury, *id.* §300aa-11(c)(1)(c)(ii). If this time period is not met, causation must be established by a prepondenance of the evidence as a whole, *id.* §§300aa-11(c)(1)(c)(ii)(II)-13(a)(1). For a case involving the swine flu vaccine, see Section 10.3.4.

4.7.3   Is the cult dangerous?

A district attorney is considering prosecution of a cult group on the ground of "reckless endangerment" of its members. The cult's leader preaches suicide if its devotees suffer doubts about the cult, and advises long car trips with minimal sleep. Over a five-year period, 10 of some 4,000 cult members have died by suicide or in automobile accidents, while the death rate from these causes in the general population in the same age bracket is 13.2 deaths per 100,000 person-years.

Question

1. As a statistical advisor to the district attorney, use the Poisson distribution to analyze these data.

4.7.4   Incentive for good drivers

California proposes a plan to reward drivers who have no accidents for three-year periods. An incentive for such drivers is being considered on the theory that good drivers may become careless and deteriorate in performance. A study of a group of drivers for two three-year periods (in which there was no known

TABLE 4.7.4. Distribution of accidents for a cohort of drivers

	0	1	2+
Accidents in first period	0	1	2+
Drivers having this number of accidents in first period	6,305	1,231	306
Accidents by drivers having this number of accidents in first period	0	1,231	680
Accidents for these drivers in second period	1,420	421	157

change in general conditions affecting the rate of accidents) produced the data of Table 4.7.4.

Questions

1. On the assumption that each driver has a Poisson distribution for accidents, and that the accident proneness of drivers does not change between the two periods, compute the expected number of accidents in the second period for those drivers who had zero accidents in the first period. (Hint: this is an example of the conditioning property of the Poisson distribution discussed at p. 141. Write an expression for the expected contribution to the number of accidents in the second period by any driver with an aggregate n accidents in the two periods, remembering that a driver with more than zero accidents in the first period contributes nothing to the sum. Show that this is the same as the proportion of drivers with one accident in the first period and $n - 1$ in the second period. Check your result by applying Robbins's theorem, estimating marginal probabilities as needed.)

2. Assuming statistical significance, what conclusions do you draw?

Source

Ferreira, *Quantitative Models for Automobile Accidents and Insurance*, 152–55 (U.S. Dep't of Trans. 1970).

4.7.5   Epidemic of cardiac arrests

From April 1981 through June 1982, there was an unusual increase in the number of cardiopulmonary arrests and deaths in the eight-bed pediatric intensive care unit (ICU) at a large medical center hospital in San Antonio, Texas. A disproportionate number of the deaths (34 of 42) occurred during the evening shift (3:00 p.m. to 11:00 p.m.), as compared with 36 of 106 during the previous four years. A consultant believed that at least some of these events were compatible with drug intoxication. Table 4.7.5 gives the number of deaths and the number of evening shifts worked for the eight nurses who were present when at least five deaths occurred.

Other data showed that (i) only Nurse 32 was assigned to care for significantly more patients during the epidemic period at the time of their deaths (21

TABLE 4.7.5. Deaths in a pediatric ICU during shifts of selected personnel

Nurse code	Number of evening deaths ($n = 34$)	Shifts worked ($n = 454$)
13	14	171
22	8	161
31	5	42
32	27	201
43	6	96
47	18	246
56	17	229
60	22	212

of 42) than to patients in the pre-epidemic period at the times of their deaths (4 of 39); (ii) Nurse 32 had been caring for 20 of the 34 children who died during the evening shift in the epidemic period as compared with 1 of the 8 who died in the evening shift in the pre-epidemic period; (iii) for 10 of the 16 children who had coronary resuscitation, the resuscitation events were repeated on different days but on the same nursing shift, with Nurse 32 being assigned to 9 of the 10 cases; and (iv) no other nurse or non-nursing personnel had significantly elevated risk.

Questions

1. Under the null hypothesis that time of death is independent of nurse on evening shift, find the probability of a clustering of deaths as far above expectation for any nurse as that observed, assuming that (a) no additional evidence is available with respect to Nurse 32, and (b) Nurse 32 is suspected on prior grounds. (Ignore the fact that the data are truncated to five or more deaths.)

2. What is the relative risk of death when Nurse 32 is on duty vs. that when she is off duty?

3. Would the statistical evidence alone be sufficient to sustain a murder charge against Nurse 32?

4. In a prosecution based on non-statistical but circumstantial evidence with respect to the death of a particular patient, are the statistics admissible?

Source

Istre, et al., *A Mysterious Cluster of Deaths and Cardiopulmonary Arrests in a Pediatric Intensive Care Unit*, 313 New Eng. J. Med. 205 (1985).

Notes

After Nurse 32 left the hospital, she went to another smaller hospital where she killed a patient by lethal injection, was caught, convicted, and was sentenced to 99 years. She was then criminally prosecuted and convicted of injuring a child at the first hospital, one of the patients who did not die but who required recurrent cardiopulmonary resuscitations on the evening shifts. The method was overdose injection (or injections) of unprescribed heparin. The evidence was circumstantial and non-statistical, but the statistics were introduced over objection.

In another, similar case, the clustering was more extreme. Of the 52 nurses who took care of 72 patients at the time of their cardiac arrests, one nurse (Nurse 14) was the nurse at the time of 57 arrests, while the next highest total for a nurse was 5. Other statistical evidence also pointed to Nurse 14. After a bench trial, the criminal prosecution of Nurse 14 for murder and attempted murder was dismissed, the court holding that the statistics merely placed her at the scene of the cardiac arrests and there was no evidence of crime, as opposed to malpractice or mistake.

The Washington Post, June 8, 1988 at B1; Sacks, et al., *A Nurse-Associated Epidemic of Cardiac Arrests in an Intensive Care Unit*, 259 J. Am. Med. Assoc. 689 (1988). See also Buehler, et al., *Unexplained Deaths in a Children's Hospital*, 313 New Eng. J. Med. 211 (1985); Stephen Fienberg and David H. Kaye, *Legal and Statistical Aspects of some Mysterious Clusters*, 154 J. Roy. Stat. Soc. A, Part 1, p. 61 (1991).

4.8   Geometric and exponential distributions

Geometric and exponential distributions are closely related to the Poisson distribution. In a Poisson process (see Section 4.7 at p. 141), the times between successive events are statistically independent random variables that follow an exponential distribution. The exponential waiting-time distribution is characteristic of a constant "hazard" process, in which the likelihood of a new occurrence does not depend on the past pattern of occurrences. The Poisson process is used in survival (follow-up) studies to analyze the effects of risk factors on mortality in a population.

Geometric distribution

It is perhaps easiest to begin with the discrete analogue of the exponential distribution, namely, the geometric distribution. When there is a series of independent trials, with constant probability p of success and q of failure, the number of trials, n, to the first success has a geometric probability distribution defined as

$$P[X = n] = pq^{n-1},$$

where $n \geq 1$. It can easily be shown that, for this model, the mean number of trials is $1/p$, and the variance of the number of trials is q/p^2. The probability that the number of trials would exceed n is q^n. Setting this equal to $1/2$ and solving for n, the median number of trials is about $-\ln(2)/\ln(q)$. Note that p is the discrete hazard constant, i.e. the conditional probability of a success on the next trial, given no prior successes.

Exponential distribution

In the continuous exponential distribution, the analogue of p is the hazard constant β, which is the limiting value of the conditional probability of an event in a brief time period given no event up to then, divided by the length of the interval, as the period shrinks to zero. That is, the conditional probability of failure in a brief interval dt starting at any time t (given no failure up to time t) is approximately $\beta \cdot dt$. The probability density function of a waiting time x to an event is given by

$$\beta e^{-\beta x}.$$

The mean waiting time is $1/\beta$; the variance of the waiting time is $1/\beta^2$; and the standard deviation is $1/\beta$. Thus, the mean waiting time is only one standard deviation from the left end of the density (0 waiting time), while the right tail of the density extends to infinity. This skewness is reflected in the relationship of the median to the mean. The probability that the waiting time would exceed t is $e^{-\beta t}$; setting this equal to $1/2$ and solving for t, we find that the median time to failure is $\ln(2)/\beta$. Since $\ln 2$ is about 0.69 and $1/\beta$ is the mean waiting time, it follows that the median waiting time is about 69% of the mean. This result tells us that waiting times that are less than the mean are more probable than those that are greater than the mean. These relationships are depicted in

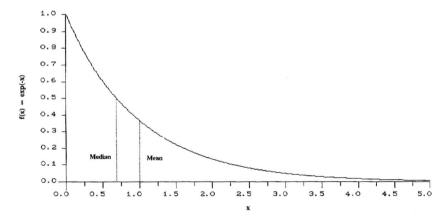

FIGURE 4.8. Probability density function for the standard exponential distribution

Figure 4.8, the curve for the standard ($\beta = 1$) exponential distribution, which has both mean and variance equal to 1.

Models that allow for changes in the hazard over time are described in Chapter 11.

4.8.1   Marine transportation of liquefied natural gas

Distrigas Corporation applied to the Federal Power Commission for permission to import Algerian liquefied natural gas (LNG), transporting it in tankers to a specially built facility on Staten Island, New York. Since a large spill of LNG in port could result in catastrophic fire, the Commissioner prepared a report on the risk of accident. An applicant barge report submitted as part of the proceedings stated that, in the case of a barge, the estimated expected time to the first accident is 7,000 years. Assume the statement is correct.

1. What probability model is appropriate for this problem?

2. Assuming that such a model was used in generating the figure quoted above, in what year is the first accident most likely?

3. What is the probability of at least one accident within the first ten years?

4. Is there any "bunching" of probability around the 7000 year mark? What is the standard deviation of the time to first accident?

Source

W. Fairley, "Evaluating the 'Small' Probability of a Catastrophic Accident from the Marine Transportation of Liquefied Natural Gas," *Statistics and Public Policy* 331 (W. Fairly and F. Mosteller, eds., 1977).

4.8.2   Network affiliation contracts

An affliation contract between a television station and a television network permits the station to broadcast network programs. Such contracts are treated as having one-year terms, but are renewed unless terminated by affirmative act of either party. Assuming that the contracts are highly valuable, may a purchaser of the station depreciate the amount allocated to the contract for tax purposes? The Internal Revenue Code provides that an intangible asset may be depreciated only if its "life" may be estimated with "reasonable accuracy." The data on terminations for 250 CBS and NBC affiliates in markets with at least three VHF stations were as shown in Table 4.8.2.

Questions

1. Use the geometric distribution and these data to estimate the life of a contract for the 1963 tax return.

TABLE 4.8.2. Contract terminations for 250 CBS and NBC affiliates

Years through which experience is determined	Number of affilation years commenced	Number of terminations
1957	944	64
1958	1110	70
1959	1274	77
⋮	⋮	⋮
1962	1767	88

2. What objections would you have to this model?

Source

W.E. Deming and G. Glasser, *Statistical Analysis of the Termination of Contracts of Television Network Affiliations* (report to the Tax Court in *Commissioner of Internal Revenue v. Indiana Broadcasting Corporation*, 41 T.C. 793, *rev'd*, 350 F.2d 580 (7th Cir. 1965)).

Notes

This tax depreciation issue has arisen in other contexts, such as the depreciation of newspaper subscriptions. See, e.g., Deming and Glasser, *A Markovian Analysis of the Life of Newspaper Subscriptions*, 14 Management Science B-283 (1968). See also Section 11.1.3.

4.8.3   Dr. Branion's case

Dr. John Branion was convicted of killing his wife by garroting her with the cord of an iron and then shooting her. Both Branion and his wife were from prominent families, and the death and trial attracted considerable public attention in Chicago, where they lived. Many years after his conviction, Branion filed for habeas corpus. He claimed that it would have been impossible for him to have driven from the hospital in the Hyde Park district of Chicago, where he was attending patients until 11:30 a.m. on the day of the murder, make two stops, go home and garrot his wife, and then call the police to report the murder by 11:57 a.m.

The statistical argument over time focused on two segments: the driving time and the garroting time. The police drove the same route six times, not exceeding the speed limit, and clocked times between 6 and 12 minutes, with the average time being 9 minutes. As for garroting time, Branion interpreted the prosecutor's pathologist as testifying that to form the observed bruises would have required pressure to his wife's neck for 15 to 30 minutes. From a graph in his brief, Branion apparently assumed that: (i) driving times and garroting times would each have normal distributions; (ii) a driving time of 6 minutes would be 3 standard deviations from the mean of 9 minutes, or 1 minute; (iii) a garroting time of 15 minutes would also be 3 standard deviations from the mean of 22.5 minutes, or 2.5 minutes. Branion argued that the P-value of each of these events (i.e., driving time less than 6 minutes and garroting time less than 15 minutes) was less than 0.01 and the P-value of the joint event was less than $0.01 \times 0.01 = 0.0001$. Branion concluded that on the basis of the driving and garroting times the chance that he was guilty was less than 1 in 9,000.

Questions

1. Are the assumed standard deviations likely to be true?

2. Are the times of the two events (driving and garroting) likely to be independent? If they are independent, is it appropriate to multiply their individual P-values together to get the P-value of their sum?

3. Referring to driving times, the court observed that "[n]othing suggests a Gaussian distribution or the absence of skewness." If the driving-time distribution was skewed, whom would that help, Branion or the state? Would the exponential distribution be appropriate here?

4. Assuming that driving and garroting times are independent, what is an upper limit of the probability that their sum would be less than 27 minutes (11:30 to 11:57) if the standard deviations were as Branion implicitly argued?

Source

Branion v. Gramly, 855 F.2d 1256 (7th Cir. 1988), *cert. denied*, 490 U.S. 1008 (1989).

Notes

According to an affidavit filed in the case, the presiding judge at the trial, Reginald Holzer, who was subsequently convicted of accepting bribes in other cases, accepted a $10,000 bribe from Branion's friends and agreed to grant a defense motion for a judgment nothwithstanding the verdict (which was not appealable) for a second $10,000. A few days later, however, the judge reneged because the district attorney had got wind of a meeting between the judge and Branion's friends and, in a private meeting with the judge, had threatened to arrest everyone concerned if the judge reversed the jury's verdict. Mindful of the money already received, Judge Holzer agreed to free Branion on bail pending appeal. Branion then fled to Africa, first to the Sudan and then to Uganda; it was said that he became Idi Amin's personal physician. When Amin fell, the new regime unceremoniously shipped him back to the United States, where he started serving a life sentence. *Id.* at 1266–1267.

The court of appeals opinion reciting these facts was caustically criticized as false by Branion's habeas corpus attorney. See Anthony D'Amato, *The Ultimate Injustice: When a Court Misstates the Facts*, 11 Cardozo L. Rev. 1313 (1990).

## CHAPTER 5

Statistical Inference
for Two Proportions

5.1 Fisher's exact test of equality for two proportions

When comparing two proportions, it is common practice simply to quote a figure representing the contrast between them, such as their difference or ratio. Several such measures of association have already been introduced in Section 1.5, and we discuss others in Section 6.3. The properties of these measures and the choice of a "best" one are topics in descriptive statistics and the theory of measurement. There are interesting questions here, but what gives the subject its depth is the fact that the data summarized in the description may often be regarded as informative about some underlying population that is the real subject of interest. In such contexts, the data are used to test some hypothesis or to estimate some characteristic of that population. In testing hypotheses, a statistician computes the statistical significance of, say, the ratio of proportions observed in a sample to test the null hypothesis H_0 that their ratio is 1 in the population. In making estimates, the statistician computes a confidence interval around the sample ratio to indicate the range of possibilities for the underlying population parameter that is consistent with the data. Methods for constructing confidence intervals are discussed in Section 5.3. We turn now to testing hypotheses.

Independence and homogeneity

The data from which two proportions are computed are frequently summarized in a fourfold, or 2×2, table. The most common null hypothesis in a fourfold table states that attribute A, reflected in one margin of the table, is *independent* of attribute B, reflected in the other margin. Attributes A and B are independent when the conditional probabilities of A given B to do not depend on B, and we have $P[A|B] = P[A] = P[A|\bar{B}]$. Thus, independence implies homogeneity of the conditional rates. For example, in Section 5.1.1, Nursing examination, the hypothesis is that whether a nurse passes the test (attribute A) is independent of ethnicity or race (attribute B). Conversely, if these two conditional probabilities are equal, then in the combined group the pass rate also equals $P[A]$. Thus, homogeneity implies independence. When A and B are independent, the joint probability is the product of the marginal probabilities, since $P[A \text{ and } B] = P[A|B] \cdot P[B] = P[A] \cdot P[B]$.

We usually formulate hypotheses in terms of independence for cross-sectional studies where both A and B are regarded as random outcomes of sampling. When the sampling design has fixed the numbers in each level of one attribute, it is customary to formulate hypotheses in terms of homogeneity, i.e., equality, of the conditional rates. The independence and homogeneity hypotheses are equivalent, and in either formulation the null hypothesis will be tested in the same way.

Exact test of equality of two proportions

One fundamental method for testing the null hypothesis is Fisher's exact test. This test treats *all* marginal totals of the 2×2 table as fixed; e.g., in Section 5.1.1 one considers possible outcomes that might have occurred keeping the numbers of black and white nurses and the numbers of those passing and failing fixed. In particular, one counts the number of ways blacks could have passed in numbers no greater than those observed. That number is divided by the total number of ways that all passing and failing outcomes could be distributed among the total number of test takers without regard to race. If this probability is small, the null hypothesis is rejected. For the hypergeometric formulas used in the test, see Section 4.5.

It may seem artificial that Fisher's exact test treats the margins of the fourfold table as fixed, when in most situations the numbers in at least two of the four margins would vary in repetitions of the sampling scheme. If men and women take a test repeatedly, the numbers passing in each group would not be fixed. In fact, where the goal is to simulate prospectively the behavior of a process–as for example in planning an experiment with adequate sample size to guarantee a certain level of statistical power– it is inappropriate to assume fixed margins. However, hypothesis tests by no means require consideration of all outcomes, for the data may already be inconsistent with the null hypothesis within a relevant subset of all the events that might have occurred. This is the key idea behind regarding all margins as fixed: if the disparity between two observed proportions is unlikely to occur under the null hypothesis among those instances when the margins assume the specific values observed, then there are already grounds for rejecting that hypothesis.[1]

To summarize the general philosophy adopted here: when testing hypotheses, we seek to embed the observed data in a probability space of other possible events. If this can be done in several ways, we generally prefer the smallest relevant space, which tends to focus speculation about "the way things might have been" closer to the facts at hand. In the case of a fourfold table, the preferred probability space consists of all those cell frequencies inside the table that are consistent with the observed margins.

There are important technical benefits from the conditional approach. First, the distribution of the cells in the table depends on fewer unknown parameters. Under the null hypothesis of homogeneity, for example, the common value of the population proportion is an unknown "nuisance" parameter that would have to be estimated or reckoned with in some manner. But once we condition on the margins, we get a single (hypergeometric) distribution with no nuisance parameters, an extremely useful simplification.

Second, among all tests at level α, the conditional test is the uniformly most powerful unbiased test.[2] That is, the conditional test is the unbiased test most likely to lead to a rejection of the null hypothesis when it is false. Occam's razor seems vindicated here, since conditioning on all margins produces these good tests.[3]

---

[1] In particular, a maximum 5% Type I error rate arranged for the test conditional on the observed margins implies a maximum 5% error rate for the unconditional test in which the margins are not fixed. The test has to produce some margins, and for any fixed set of margins the test has at most a 5% error rate by construction: since the error rate for the unconditional test is the weighted average of the error rates for each possible marginal result, the average or unconditional rate is also at most 5%. The test is conservative in that the average error rate will be less than 5%. Some authors regard this as a defect of conditional tests, but it is a necessary consequence if one decides to limit the conditional error probability to 5% under all marginal outcomes.

[2] The power of a test is the probability of rejecting H_0 when it is false (see Section 5.4). An unbiased test is one in which the probability of rejecting H_0 when it is false is not less than the probability of rejecting it when it is true. A test is uniformly most powerful unbiased if, for every value of the parameter under the alternative hypothesis, the test is more powerful than any other unbiased test with the same level of significance. Here, level of significance refers to the maximum Type I error rate over all possible parameter values satisfying the null hypothesis. Because of the discreteness of count data, the level of significance of a test may be somewhat less than the nominal α level set by the decision maker, and strictly speaking the test may be not quite unbiased or uniformly most powerful, but nearly so. The theorem asserted in the text is exactly true for a version of the conditional test that mathematically smoothes away the discreteness.

[3] Another motivation for conditioning on the margins is the "randomization model." For a given group of test-takers, the ability to pass a test may be, like one's sex, an inherent characteristic of the person taking the test. The number who will pass, like the numbers of male and female test-takers, could be regarded as predetermined given the group. The margins are fixed in this sense. Irrespective of the total number of passes, or how many men and women take the test, the only random element of interest is the distribution of that

5.1.1   Nursing examination

Twenty-six white nurses and nine black nurses took an examination. All whites and four blacks passed.

Question

Use Fisher's exact test to calculate the significance probability of the difference in pass rates.

Source

Dendy v. Washington Hospital Ctr., 431 F. Supp. 873 (D.D.C. 1977).

5.2   The chi-squared and z-score tests for the equality of two proportions

Chi-squared

A commonly used approximation to Fisher's exact test in large samples is Pearson's chi-squared test. To apply this test, one computes a statistic, somewhat confusingly called "chi-squared," that has, approximately, a chi-squared probability distribution with one degree of freedom under the null hypothesis. The chi-squared statistic is computed from the observed and expected values in the 2×2 table. The expected value for each cell, calculated under the assumption of independence, is the product of the corresponding marginal proportions, multiplied by the total sample size. The difference between the observed and expected frequencies for each cell is squared, then divided by the expected frequency; the contributions of each of the four cells are summed to obtain the chi-squared value for that table. In short, chi-squared is the sum of "observed minus expected squared over expected."

It is intuitively plausible that the chi-squared statistic is a reasonable test statistic to account for the disparity between observed and expected frequencies in a 2×2 table. The larger the disparity, the greater the value of the statistic. Squaring the difference ensures that all contributions to the total disparity are positive; a simple difference without squaring would result in positive and negative values that would sum to zero. Dividing by the expected value for each cell corrects for each disparity's dependence on its base: the difference between 52 observed and 50 expected observations is far less important (i.e., carries less evidence against H_0) than the difference between 3 observed and 1

---

number between men and women. Under H_0, the observed alignment of passes and fails is just one of many possible outcomes, all equally likely, and that forms the basis for the conditional test.

expected observation. By the same token, the difference between 30 observed and 10 expected observations contributes 10 times as much to the chi-squared statistic as does the difference between 3 observed and 1 expected, reflecting the fact that proportionate deviations from expectation become more unlikely under H_0 as sample size increases.

The chi-squared statistic is a useful test for independence of factors in the 2×2 table and for many other situations because of the remarkable fact that its distribution is approximately the same irrespective of the expected values for the cells. This means that the theoretical distribution of the chi-squared statistic under the null hypothesis can be expressed in a single table, such a Table C in Appendix II. If the chi-squared statistic exceeds the critical value given in the table, the hypothesis of independence is rejected.

We have referred above to a chi-squared distribution because there is, in fact, a family of chi-squared distributions that differ by what is called "degrees of freedom". (See Figure 5.2.) To appreciate the meaning of degrees of freedom, it is helpful to refer to the relationship between chi-squared and normal distributions. If a random variable with a standard normal distribution is squared, the squared value has, by definition, a chi-squared distribution with one degree of freedom. The sum of two such squared variables, if independent, has a chi-squared distribution with two degrees of freedom, and so on. When there are more than 30 degrees of freedom, the chi-squared distribution is approximately normal [Do you see why?], so that the normal distribution frequently is used instead.[4]

When the chi-squared statistic is computed for a 2×2 table, the absolute difference between expected and actual values is the same for each cell. Consequently, although four terms contribute to the chi-squared sum, this sum can also be expressed algebraically as the square of single approximately normal variable, so that there is really only one independent term; thus, chi-squared has only one degree of freedom.[5]

Tables for the continuous chi-squared distribution only approximate the sampling distribution of the discrete chi-squared statistic. To improve the ap-

---

[4]The expected value of a random variable with a chi-squared distribution equals the number of degrees of freedom, and its variance is twice the number of degrees of freedom.

[5]This assumes fixed margins. When the margins are viewed as random, chi-squared for the 2×2 table can be shown to have the chi-squared distribution with one degree of freedom when the population cell frequencies are unknown and estimated from the margins as described above. If the true population cell frequencies were known and used, the chi-squared statistic would have three degrees of freedom instead of one. There was an acrimonious debate between Karl Pearlson and R. A. Fisher over this point in the early 1900's. Pearson introduced chi-squared as a formal goodness-of-fit statistic. However, he thought the proper degrees of freedom for the 2×2 table should be three, even when expected cell frequencies were estimated from the margins. Fisher argued that one df is lost for each marginal parameter estimated. His calculation of one degree of freedom was ultimately proved correct, although the point was apparently never fully appreciated by Pearson.

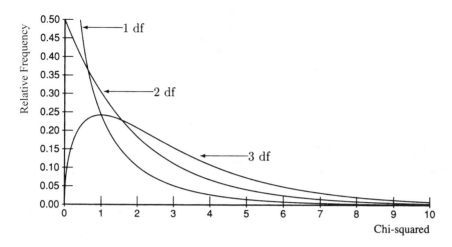

FIGURE 5.2. Members of the family of chi-squared distributions

proximation for 2×2 tables with fixed margins, the absolute difference between actual and expected values in each cell is reduced by 0.5 before squaring. This procedure is conservative in favor of the null hypothesis because it decreases the value of the chi-squared statistic. Although there has been debate over the value of this correction, it is generally used because it conforms the chi-squared approximation more closely to the results of Fisher's exact test. (The adjustment over-corrects in tables larger than 2×2 and in procedures that do not condition on the margins, so it is not applied in those instances.)

To see how chi-squared is calculated, consider *Connecticut v. Teal*, 457 U.S. 440 (1982), in which a test used to determine eligibility for promotion was challenged as discriminatory. Of 48 blacks who took the test, 26 (54%) passed, while of 259 whites who took the test, 206 (80%) passed. *Id.* at 443, n. 3. Is there a statistically significant difference in the pass rates for the two groups?

Table 5.2 shows the observed data together with the expected numbers computed from the marginal frequencies, under the equivalent hypothesis that passing is independent of race. Thus, in the upper left cell the expected number of blacks who pass is equal to the proportion of blacks in the sample (48/307) times the proportion of those who pass in the sample (232/307) times the sample size (307). This product is 36.274. A comparable calculation is made for each of the other cells.

Table C in Appendix II shows that for one degree of freedom there is a null probability of less than 0.0005 that chi-squared would exceed 12.78. Thus, we reject the hypothesis that passing is independent of race or that pass rates are homogeneous.

In the above example, if the numbers in each cell had been halved, the chi-squared statistic would have been 5.75, which would have been at about the 0.02 level of significance. The 0.02 value is the probability under the null

TABLE 5.2. Calculation of chi-squared for the *Teal* data

| | | (expected values in parentheses) | | |
		Pass	Fail	Total
	Black	26	22	48
		(36.274)	(11.726)	
	White	206	53	259
		(195.726)	(63.274)	
	Total	232	75	307

$$X^2 = \frac{(|26 - 36.274| - 0.5)^2}{36.274} + \frac{(|22 - 11.726| - 0.5)^2}{11.726}$$

$$+ \frac{(|206 - 195.726| - 0.5)^2}{195.726} + \frac{(|53 - 63.274| - 0.5)^2}{63.274}$$

$$= (9.774)^2 \cdot (36.274^{-1} + 11.726^{-1} + 195.726^{-1} + 63.274^{-1})$$

$$= 95.531 \cdot 0.1338 = 12.78$$

hypothesis that a departure this large would be observed in *either direction*. The probability that so few blacks would pass under the null hypothesis is half this, or about 0.01. This illustrates the fact that, for fourfold tables, chi-squared is automatically a two-tailed test because either positive or negative differences between proportions contribute to a significant result.[6] Since in many legal pursuits a one-tailed test is appropriate (see Section 4.4), the one-tailed significance level, which is half the two-tailed level shown in Table C, should be used instead.

Some important attributes of the chi-squared statistic follow.

1. A chi-squared test statistic has a chi-squared distribution only for *counted data*, with *independent* observations classified into a *fixed* number of *mutually exclusive and exhaustive* categories. A chi-squared test is not applicable when the units of observation are measured data, or rates or percentages. For example, to test whether the pattern of weekday vs. weekend snowfall is the same in New York and Boston, we could set up a 2 × 2 table with attribute *A* equal to city and attribute *B* equal to part of the week. However, it would *not* be valid to enter the total *number of inches* of fallen snow into the cells and then to calculate the chi-squared statistic (as one investigator essentially did in a study of snowfall—see *Snowstorms Hardest on Fridays*, N.Y. Times, Jan. 28, 1978). The reason is that depth of snow is a measured variable, not a count, with a degree of variability not provided for by the chi-squared test. Even if we agree to count inches as discrete units, they are

---

[6]Chi-squared is generally called an "omnibus" test because any pattern of departures from expected values weighs against the null hypothesis. This is of particular importance in tables larger than 2 × 2.

not independent: given that the first inch of a storm falls on a weekend, the next inch is highly likely to do so. On the other hand, it would be valid to apply the chi-squared test to the distribution of *snowstorms*, since storms (rather than inches) may be considered independent events, countable and classifiable into one of the four city/part-of-week categories.

2. The chi-squared statistic is not a good measure of the strength of association in a 2×2 table. Observe that if we double the count in each cell of a 2×2 table, the degree of association does not change, but chi-squared (without continuity correction) doubles. The value of chi-squared is determined both by the degree of association *and* by the total sample size. Dividing the (un-corrected) chi-squared statistic by the sample size, we obtain a reasonable measure of association known as phi-squared, discussed in Section 6.3.

3. Accurate approximation of the sampling distribution of the chi-squared statistic requires sample sizes that are not too small. For a fourfold table Fisher's exact test should be used instead of chi-squared if (a) the total sample size N is less than 20, or (b) N lies between 20 and 40 and the smallest expected cell frequency is less than 5. G. Snedecor and W. Cochran, *Statistical Methods* 221 (7th ed. 1980).

4. In computing expected cell frequencies one would like to use the product of the marginal proportions times the sample size, on the hypothesis that these two attributes are independent, but the population values are rarely known and usually must be estimated. Even when the null hypothesis is true, the expected values in the population will generally differ from the values de-termined by the product of the observed marginal proportions. Expected cell frequencies computed in this way are examples of *maximum likeli-hood estimates* of the population expected cell frequencies under the null hypothesis,[7] and are used because they have desirable statistical properties. For a discussion of maximum likelihood estimation, see Section 5.6.

With respect to the test for independence, these estimates conform most closely to the null hypothesis because they approximately minimize the chi-squared statistic, i.e., any other choice of expected values under the null hypothesis would result in larger values of chi-squared. This property does not hold for tables with larger numbers of cells, and different procedures are required to find the expected values that minimize the chi-squared statistic if a minimum figure is deemed appropriate. Modern computer programs to calculate chi-squared statistics in complicated situations generally estimate expected cell frequencies by the method of maximum likelihood, rather than by minimizing chi-squared.

---

[7] A maximum likelihood estimate of a population parameter is that value of the parameter that maximizes the probability of observing the sample data (regarded as fixed).

5. A convenient computing formula for chi-squared (corrected for continuity) for the fourfold table is

a	b	n_1
c	d	n_2
m_1	m_2	N

$$X^2 = \frac{N(|ad - bc| - N/2)^2}{m_1 \cdot m_2 \cdot n_1 \cdot n_2},$$

where the factor $N/2$ is the correction for continuity. This expression, multiplied by the factor $(N - 1)/N$, is equal to the squared standardized hypergeometric variable used in the normal approximation to Fisher's exact test, corrected for continuity.

6. For additional uses of chi-squared, see Section 6.1.

The two-sample z-score test

As previously stated, Fisher's exact test and the chi-squared approximation to it are based on a conditional model wherein the marginal totals are fixed, and the people in the sample who pass the test are distributed under H_0 at random with respect to the two groups taking the test. For the test of homogeneity given two groups of fixed sample sizes, there is a different model—the two-sample binomial model—that does not appear to involve conditioning on both sets of margins, at least on its face. Rather, it assumes that the members in each group are a random sample selected from large or infinite populations, wherein the probabilities of selecting a group member who would pass the test are equal under the null hypothesis. It is an interesting, but by no means intuitively obvious fact, that the square of the two-sample z-score presented below is algebraically the same as the chi-squared statistic presented above for testing the null hypothesis of independence. An advantage of the two-sample binomial model is that it can easily be adapted to test null hypotheses specifying non-zero differences in pass rates and to construct confidence intervals for the observed difference (see Section 5.6).

To implement the two-sample z-score test, note that, if the difference in pass rates in two populations is D, the corresponding difference in two large independent samples will be normally distributed, approximately, with a mean D and a standard deviation σ equal to the square root of the sum of the variances of the rates for the two samples. The variance of the sample pass rate for group $i = 1$ or 2 is $P_i Q_i / n_i$, where P_i is the pass rate for group i in the population, $Q_i = 1 - P_i$, and n_i is the sample size for group i. If the population pass rates were known, a z-score would be obtained by subtracting D from the sample difference in pass rates and dividing by $\sigma = (P_1 Q_1 / n_1 + P_2 Q_2 / n_2)^{1/2}$. This is called a two-sample z-score because it reflects the binomial variability in the samples from both groups. When the values of P_i are not known they may be estimated from each sample separately, and the estimates used to compute σ. The resulting z-score is most often used in the construction of confidence intervals for D. When testing the null hypothesis that $P_1 = P_2$, D is equal to 0 under H_0, and we require only the common value of the pass rate, say $P = P_1 = P_2$. Here it is preferable to estimate P by pooling the data from both groups, i.e., the estimate of P is obtained from the pass rate in the margin, say p, and this estimate is used to compute σ. This pooling is justified since by assumption of H_0 the probability of success is the same for each group, and the combined data provide the best estimate of that probability. Thus,

$$z = \frac{p_1 - p_2}{\left[pq(n_1^{-1} + n_2^{-1}) \right]^{1/2}},$$

where p_i is the sample pass rate for group i. Like the expression for X^2 at p. 162, this two-sample z-score test statistic is virtually identical to the standardized hypergeometric variable described in Section 4.5 (uncorrected for continuity) that could be used in Fisher's exact test; the only difference is that the standardized hypergeometric is less than the standardized difference in proportions from the two-sample binomial model by the negligible factor of $[(N - 1)/N]^{1/2}$, where $N = n_1 + n_2$, the total sample size. Thus, essentially the same numerical test procedures result whether none (chi-squared), one (two-sample z), or both (Fisher's exact) sets of margins are regarded as fixed.

The following variations are noted:

1. The correction for continuity may be applied by subtracting 0.5 from the numerator of the larger proportion and adding 0.5 to the numerator of the smaller proportion. This reduces the difference between them, and is conservative in favor of the null hypothesis. It brings the distribution of the z-score more in line with that of Fisher's test. In large samples the correction is negligible.

2. Sometimes there is only a single sample of data together with an external or a priori basis for selecting the expected pass rate. Then a one-sample z-score test is appropriate, in which the expected rate is deemed fixed and the sole source of variability is the binomial variability of the sample.

5.2.1 Suspected specialists

The administration of an options exchange has found that certain closing transactions reported on the tape of the exchange appear to be fictitious. The administration suspects that certain specialists who held inventories of the stock may have inserted the false reports to improve their positions. The specialists deny this, claiming that the fictitious transactions are mistakes. A closing transaction on an up-tic would be favorable to a specialist who had a long position in a stock; a closing transaction on a down-tic would be favorable to a specialist who had a short position in a stock. In preparation for an enforce-

TABLE 5.2.1. Options exchange transactions

Specialist position		Closing transaction		
		Tics	Fictitious	Unquestioned
Long	Plus	29	77
	Minus	6	67
Short	Plus	2	56
	Minus	45	69

ment proceeding, the staff of the exchange compared the fictitious transactions with unquestioned transactions in the same time period. The data are shown in Table 5.2.1.

Question

1. Use chi-squared to test whether the data support the specialists' claim.

5.2.2   Reallocating commodity trades

A commodities broker with discretionary authority over accounts F and G is accused by the owners of account F of siphoning off profitable trades to account G, in which the broker had an interest. The siphoning was allegedly accomplished by reallocating trades after their profitability was known. The broker denies the charge and responds that the greater proportion of profitable trades in account G is a matter of chance. There were 607 profitable and 165 unprofitable trades in account F and 98 profitable and 15 unprofitable trades in account G.

Question

Are the data consistent with the broker's defense?

5.2.3   Police examination

Twenty-six Hispanic officers and sixty-four other officers took an examination. Three Hispanic and fourteen other officers passed.

Question

1. Use the two-sample z-score method to test the null hypothesis that there is no difference between the pass rates for Hispanic and other officers.

TABLE 5.2.4. Promotions by race: Federal Reserve Bank of Richmond, 1974-1977

Grade 4	Total in grade	Total blacks in grade	Total promotions	Total black promotions
1974	85	52	47	27
1975	51	31	14	8
1976	33	21	9	3
1977	30	20	3	1
Total	199	124	73	39

Grade 5	Total in grade	Total blacks in grade	Total promotions	Total black promotions
1974	90	39	39	14
1975	107	53	28	14
1976	79	41	37	19
1977	45	24	16	5
Total	321	157	120	52

Source

Chicano Police Officer's Assn v. Stover, 526 F.2d 431 (10th Cir. 1975), *vacated*, 426 U.S. 944 (1976).

5.2.4   Promotions at a bank

The Federal Reserve Bank of Richmond is accused of discrimination against blacks because it failed to promote them equally from salary Grades 4 and 5. Table 5.2.4 shows, for each of the years 1974-77, (i) the number of employees who were employed in the pay grade at the beginning of each year, (ii) the number of black employees, (iii) the number of employees who were promoted in each year, and (iv) the number of black employees promoted.

Questions

1. Consider the aggregate data for Grade 4. Representing the employees by balls in an urn, there are 199 balls at the start, of which 124 are black and 75 are white. Seventy-three are withdrawn at random from the urn, representing promotions. The balls are not replaced. What is the probability that there would be no more than 39 black balls out of the sample of 73? Does this model correspond to a (i) one-sample binomial? (ii) a two-sample binomial? (iii) a hypergeometric model? How do things change if the balls are returned to the urn after each selection? Which model is appropriate to the problem?

2. Suppose there are two urns, one for blacks and the other for whites, each with a very large number of balls. The balls in each urn are labeled promotion and non-promotion in the proportion 73/199=0.357 for promotion. One hundred and twenty-four balls are drawn from the black urn, of which

39 are promotions, and 75 are drawn from the white urn, of which 34 are promotions. What is the probability that the difference in promotion proportions would be as large as that observed? Is this a more appropriate model than those previously described?

3. In the case, plaintiff's expert used different (and erroneous) data and applied a hypergeometric test. The court of appeals objected to the data used by the expert (the data given here were found correct by the court). The court also objected to the hypergeometric test on two grounds: (i) a statistical text stated that a binomial test was proper when sample size was at least 30, and the aggregate numbers for each grade were greater than 30; and (ii) any terminated or promoted employees during the period were presumably replaced, and thus the numbers in the sample were not "finite without replacement" as required for a hypergeometric test. Do you agree?

4. The expert testified that a one-tailed test was justified because it was reasonable to assume that if there was discrimination it was not against whites. The court of appeals rejected this approach and the one-tailed test. Do you agree?

5. For the purpose of statistical tests, is there an objection to aggregating the data over different years? Over grades and years?

Source

EEOC v. Federal Reserve Bank of Richmond, 698 F.2d 633 (4th Cir. 1983), rev'd on other grounds, sub nom. *Cooper v. Federal Reserve Bank of Richmond*, 467 U.S. 867 (1984).

5.3   Confidence intervals for proportions

Basic definition

A confidence interval for a population proportion P is a range of values around the proportion observed in a sample with the property that no value in the interval would be considered unacceptable as a possible value for P in light of the sample data. To make this more precise, consider a two-sided 95% confidence interval for a proportion P, given a sample, proportion p. The interval includes all those values P' that are consistent with the sample data in the sense that if P' were the null hypothesis it would not be rejected because of the sample, at the selected level of significance. Since we are defining a two-sided 95% confidence interval, the level of significance is 0.05, two-sided, or 0.025 in each tail. The lower bound of the 95% two-sided confidence interval is the proportion $P_l < p$ such that the sample data p would just cause us to reject the null hypothesis P_l at the 0.025 level; the upper bound of the 95%

two-sided confidence interval is the proportion $P_U > p$ such that the sample data p would just cause us to reject P_U as a null hypothesis at the 0.025 level of significance. One-sided confidence intervals are defined in the same way, except that for a one-sided 95% confidence interval (either upper or lower) the rejection value for the null hypothesis is 0.05 instead of 0.025. Figures 5.3a and 5.3b illustrate these definitions.

Example. Forty prospective jurors are selected at random from a large wheel, and eight of them are women. What is a 95% confidence interval for the proportion of women in the wheel? One way of getting an answer is to consult the cumulative binomial distribution table (Appendix II, Table B). Given the 40 prospective jurors selected and the 8 successes (women) we look for the values of P' that make the tail probabilities each equal to 0.025. If $P' = 0.10$, the Table tells us that the probability of 8 or more successes is 0.04419; since this is slightly greater than the 0.025 critical value, it follows that P_L is a little less than 0.10. If $P' = 0.35$, the Table tells us that the probability of 8 or fewer successes is $1 - 0.9697 = 0.0303$. Since the critical value of 0.025 for the lower tail is a little less than 0.0303, P_U is a little more than 0.35. Thus, based on our sample of 40 which had 20% women, we are 95% confident that the proportion of women in the wheel ranges between about 10% and 35%.

Example. In a toxicity experiment, a potential carcinogen causes no tumors ($S = 0$) when added to the diet of $n = 100$ rats for six months. What is an upper 95% confidence limit for P, the probability that a rat will develop a tumor under the same experimental conditions? The answer is given by the solution for P_U in

$$(1 - P_U)^{100} = 0.05, \text{ or } P_U = 1 - 0.05^{0.01} = 0.0295.$$

Having seen no tumors in 100 rats, we are 95% confident that the rate of such tumors does not exceed 2.95%.[8]

Figures 5.3c and 5.3d provide a graphical means for obtaining two-sided 95% and 99% (or one-sided 97.5% and 99.5%) confidence intervals for a binomial proportion. To use the charts, find the pair of curves with sample size equal to n, enter the horizontal axis with the sample proportion, and read P_U and P_L from the vertical axis. Use Figure 5.3c to confirm the 95% confidence interval for the jury selection example described above.

Why use the term "confidence" rather than "probability"? Why don't we say that "the probability is 95% that $P < 0.0295$"? The reason is that P is a

---

[8]For values of S between 0 and n, one can solve for P_U and P_L with the help of tables of the F distribution (see Table F in Appendix II). Let u_α denote the critical value of the F with $a = 2(S + 2)$ and $b = 2(n - S)$ degrees of freedom, cutting off probability α in the upper tail. Then $P_U = au_\alpha/(au_\alpha + b)$. Letting v_α denote the upper α critical value of the F distribution with $c = 2(n - S + 1)$ and $d = 2S$ degrees of freedom, the lower confidence limit is $P_L = d/(d + cv_\alpha)$. In the toxicity experiment, what would have been the upper 95% confidence limit if one mouse had developed a tumor?

The value $p = .25$ is inside the interval since tail area below .2 is greater than .05.

The value $p = .3156$ is equal to the 95% limit since tail area below .2 just equals .05.

The value $p = .40$ is not in the interval since tail area below .2 is less than .05.

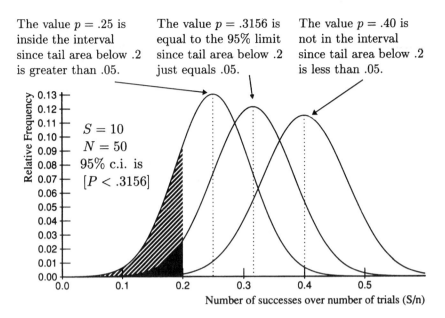

$S = 10$
$N = 50$
95% c.i. is
$[P < .3156]$

FIGURE 5.3a. Determining those values of P that lie below or above the one-sided upper 95% confidence limit

The value $p = .08$ is not in the interval since tail area above .2 is less than .05.

The value $p = .1127$ is equal to the 95% limit since tail area above .2 just equals .05

The value $p = .25$ is inside the interval since tail area above .2 is greater than .05

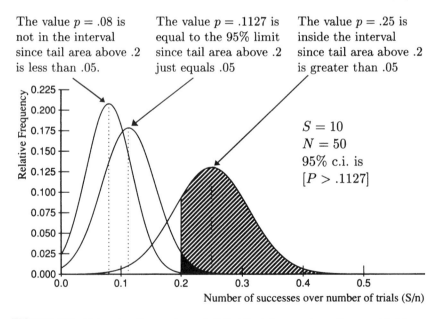

$S = 10$
$N = 50$
95% c.i. is
$[P > .1127]$

FIGURE 5.3b. Determining those values of P that lie below or above the one-sided lower 95% confidence limit

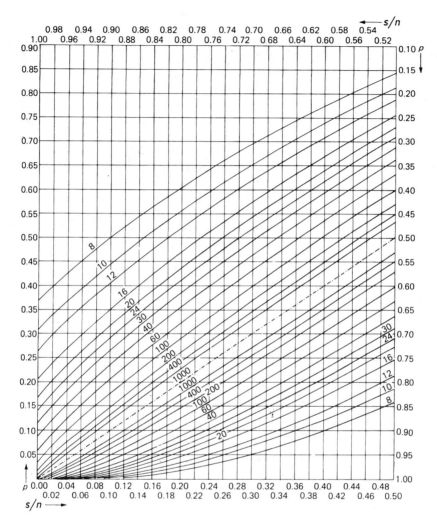

FIGURE 5.3c. Exact two-sided 95% confidence limits for a binomial proportion. (When the sample proportion s/n is 0.50 or less, read the 95% confidence limits from the left-handed vertical scale; when above 0.50, use the right-hand scale.)

population parameter and not a random variable with a probability distribution. It is the confidence limits P_L and P_U that are random variables based on the sample data. Thus, a confidence interval (P_L, P_U) is a random interval, which may or may not contain the population parameter P. The term "confidence" derives from the fundamental property that, whatever the true value of P, the 95% confidence interval will contain P within its limits 95% of the time, or with 95% probability. This statement is made only with reference to the general property of confidence intervals and not to a probabilistic evaluation of its truth in any particular instance with realized values of P_L and P_U. For

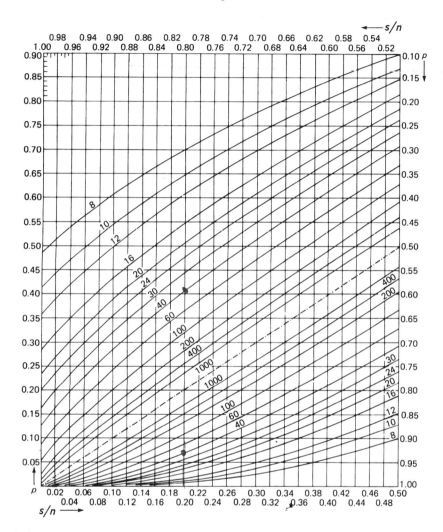

FIGURE 5.3d. Exact two-sided 99% confidence limits for a binomial proportion. (When the sample proportion s/n is 0.50 or less, read the 99% confidence limits from the left-handed vertical scale; when above 0.50, use the right-hand scale.)

the same reason it is incorrect to argue that P has a greater probability of being near the center of the confidence interval than at the edges, or that P has the same probability of being above or below the sample estimate.

A probabilistic evaluation of a confidence interval's enclosure of the population parameter would require the population parameter to have a probability distribution. This is the Bayesian approach. In the Bayesian formulation, it is the interval that is considered fixed (after the data have been realized) and the parameter P that is random and has a probability distribution. Bayesian methods are seldom used in this context, because there is no adequate basis for selecting a probability distribution for P. Still other formulations are possible, e.g., likelihood-based intervals containing all values P with sufficiently high likelihood ratios with respect to the maximum likelihood. In large samples, likelihood-based intervals yield approximately the same results as the more familiar methods we have described, although differences do arise in small samples.

Approximate intervals

When n is large and P is not too near zero or one, an approximate 95% confidence interval is given by the sample estimate p plus or minus 1.96 standard errors of p. (The standard error of p is estimated from the sample as $(pq/n)^{1/2}$.) The rationale for this approximation is simply that the normal approximation is used for the cumulative binomial. It is not an exact confidence interval because (i) the normal approximation is used, and (ii) a single value of p is used in computing the variance when obviously values of P_L and P_U should be used instead. However, in sufficiently large samples (e.g., when nP and nQ are both at least 5; see Section 5.2 at p. 157) the approximation is close enough for practical purposes.

For small n, or when P is close zero or one, the approximate interval does not closely correspond to the exact 95% confidence interval, as the following example shows.

Example. Let $n = 20$, $S = 1$, so $p = 0.05$. The approximate interval gives limits of

$$0.05 \pm 1.96[(0.05)(0.95)/20]^{1/2} = 0.05 \pm 0.0955 \text{ or } (-0.046, 0.146).$$

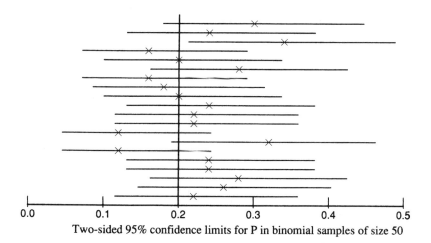

Two-sided 95% confidence limits for P in binomial samples of size 50

FIGURE 5.3e. A 95% confidence interval covers the true p in 19 out of 20 replications

The lower limit is obviously inaccurate, and the entire interval is shifted to the left of the exact interval (0.0013, 0.249).

To improve the accuracy of the approximation we may refine the approximate interval by using the 1/2 continuity correction and the null hypothesis variance $P'Q'/n$ in the test statistic. The result is a quadratic inequality whose roots are provided by the following formulas for the lower and upper critical limits (see J. Fleiss, *Statistical Methods for Rates and Proportions* 14 (2d ed.1981)):

$$P_L = \frac{(2np + c_{\alpha/2}^2 - 1) - c_{\alpha/2} \cdot [c_{\alpha/2}^2 - (2 + 1/n) + 4p(nq + 1)]^{1/2}}{2(n + c_{\alpha/2}^2)}$$

and

$$P_U = \frac{(2np + c_{\alpha/2}^2 + 1) + c_{\alpha/2} \cdot [c_{\alpha/2}^2 + (2 - 1/n) + 4p(nq - 1)]^{1/2}}{2(n + c_{\alpha/2}^2)},$$

where c is the value of a normal deviate for the selected p-value, e.g., $c = 1.96$ for $1 - \alpha = 0.95$. These limits always lie between zero and one for n and p, and come remarkably close to the exact confidence limits. In the last example, the formulas yield the interval (0.003, 0.269).

Intervals for the difference between two proportions

Approximate confidence intervals for the difference between two proportions are again the sample estimate plus or minus 1.96 standard errors for the difference. Since the variance of the difference between two proportions is the sum of the variances for each (see p. 15), the approximate confidence limits are

$$p_1 - p_2 \pm 1.96[p_1 q_1/n_1 + p_2 q_2/n_2]^{1/2}.$$

This interval will contain $P_1 - P_2$ approximately 95% of the time.

Intervals for rate ratios

For rate ratios, it is advisable first to apply a logarithmic transformation to the sample ratio in order to improve the accuracy of the normal approximation. The variance of $\ln p$ can be shown to be consistently estimated by q/np. Then since $\ln(p_1/p_2) = \ln p_1 - \ln p_2$, the standard error of $\ln(p_1/p_2)$ is consistently estimated by

$$[q_1/(n_1 p_1) + q_2/(n_2 p_2)]^{1/2}.$$

Thus, approximate 95% confidence intervals are given by

$$\ln(p_1/p_2) \pm 1.96[q_1/(n_1 p_1) + q_2/(n_2 p_2)]^{1/2}.$$

The width of the confidence interval grows as either p_1 or p_2 approaches zero, in contrast to the confidence interval for $P_1 - P_2$, which narrows as either

p_1 or p_2 approaches zero. A 95% confidence interval for P_1/P_2 is obtained by taking antilogs of the above confidence limits. One-sided limits are obtained analogously.

Example. Two hundred black employees take a test for employment and 20 pass ($p_1 = 0.10$); 100 white employees take the test and 25 pass ($p_2 = 0.25$). The point estimate for P_1/P_2 is 0.10/0.25=0.40, indicating a substantial disparate impact for the test. But is this sample difference sufficient evidence of a substantial difference in the population of test takers? Or, to put matters more specifically, does a one-sided 95% confidence interval for P_1/P_2 exclude unity at its upper limit? The value 4/5?

We proceed as follows. Taking logarithms, we have the point estimate for

$$ln(P_1/P_2) = ln(0.40) = -0.9163.$$

The estimated s.e. is

$$[0.90/(200 \cdot 0.10) + 0.75/(100 \cdot 0.25)]^{1/2} = 0.2739.$$

Thus, an upper 95% confidence limit for $ln(P_1/P_2)$ is

$$-0.9163 + 1.645 \cdot 0.2739 = -0.4657.$$

Taking the antilog, we have exp $(-0.4657) = 0.63$. The conclusion is that, at the upper limit of the one-sided 95% confidence interval, blacks pass the test at 63% of the rate at which whites pass.

Intervals for odds ratios

A similar analysis pertains to the odds ratio, $\Omega = P_1 Q_2/P_2 Q_1$. A consistent estimator of $ln \Omega$ is the sample log odds ratio $ln(p_1 q_2/p_2 q_1)$, with standard error consistently estimated by

$$s.e. = [(n_1 p_1)^{-1} + (n_1 q_1)^{-1} + (n_2 p_2)^{-1} + (n_2 q_2)^{-1}]^{1/2}.$$

Thus, an approximate 95% confidence interval for $ln\Omega$ is given by the limits

$$ln(p_1 q_2/p_2 q_1) \pm 1.96 \cdot s.e.$$

In the above example, the point estimate for the odds ratio is

$$(0.10 \times 0.75)/(0.25 \times 0.90) = 0.333,$$

i.e., the odds of a black passing are one-third the odds of a white passing. The two-sided 95% confidence interval is given by

$$ln(0.333) \pm 1.96(1/20 + 1/180 + 1/25 + 1/75)^{1/2} = -1.099 \pm 1.96 \cdot 0.330,$$

for a confidence interval of $(-1.746, -0.452)$ for $ln\Omega$. Taking antilogs, the confidence interval for Ω is given by (exp[-1.746], exp[-0.452])=(0.17, 0.64). Thus, at the upper limit of a two-sided 95% confidence interval, the odds of a black passing are about 64% of the odds of a white passing.

When sample sizes are not large, one can obtain an exact confidence interval for the odds ratio $\Omega = P_1 Q_2 / P_2 Q_1$ by following a procedure analogous to that used in computing an exact confidence interval for a binomial parameter. To illustrate the procedure, assume that n_1 blacks and n_2 whites took a test in which m_1 persons passed and m_2 persons failed. S_1 blacks passed the test, creating an odds ratio of $\hat{\Omega}$. Regarding the margins m_1, m_2, n_1, n_2 of the two-by-two table as fixed, the upper end of a two-sided 95% confidence interval, Ω_u, is the value $\Omega_u > \hat{\Omega}$ for which the lower-tail area probability, i.e., the probability of S_1 or fewer black successes, is equal to 0.025. This lower-tail area probability is the sum of the non-central hypergeometric probabilities (see Section 4.5 at p. 123) for $x = 0, 1, \ldots, S_1$. The lower end of the confidence interval, Ω_l, is the value $\Omega_l < \hat{\Omega}$ for which the upper-tail area probability, i.e., the probability of S_1 or more successes (up to the lesser of m_1 or n_1), is equal to 0.025. There is no closed form solution to the equation, so that numerical methods involving computer iteration are required to calculate the limits. In the last example, the exact two-sided 95% confidence interval for the odds ratio is (0.165, 0.670), in fairly good agreement with the large sample approximation (0.17, 0.64). In general, the exact confidence interval is somewhat wider than the approximate interval.

The reader may have noticed that the specification of 2.5% tail-area probability for the hypothesis tests at the endpoints of the confidence intervals discussed in this section has the same arbitrariness as does the doubling of a one-sided P-value to get a two-sided P-value in the case of asymmetrical distributions. Indeed, any of the more refined definitions of two-sided P-values discussed in Section 4.4 can be inverted to get a corresponding two-sided confidence interval as follows: the 95% confidence interval is the shortest interval that includes all those parameter values P' for which the two-sided P-value exceeds 0.05. Such confidence intervals are generally shorter than those that require 2.5% tail-area probabilities at the endpoints, while maintaining a guaranteed minimum coverage probability of 95%. In the binomial example of one success out of twenty, the point probability method of Section 4.4 yields an upper limit of $P_U = 0.244$, because this is the largest P' for which the two-sided P-value (in this case $P[X \leq 1|P'] + P[X \geq 9|P']$) is not less than 0.05. For the lower limit, $P_L = 0.00256$, because this is the smallest P' for which the P-value (in this case only $P[X \geq 1|P']$ because $P[X = 0|P'] > P[X = 1|P']$) is not less than 0.05. Compare the length of the interval (0.00256, 0.244) with the previously obtained interval (0.0013, 0.249). The generalized likelihood ratio method of Section 4.4 yields an even shorter interval, (0.00256, 0.227).

5.3.1   Confounders and confidence intervals

In *Brock v. Merrell Dow Pharmaceuticals*, 874 F.2d 307 (5th Cir. 1989), the court of appeals reviewed a plaintiff's verdict in one of the Bendectin cases. It held that lack of epidemiological proof was fatal to plaintiff's case. In describing epidemiology, the court wrote:

Undoubtedly, the most useful and conclusive type of evidence in a case such as this is epidemiologic studies. Epidemiology attempts to define a relationship between a disease and a factor suspected of causing it—in this case, ingestion of Bendectin during pregnancy. . . . One difficulty with epidemiologic studies is that often several factors can cause the same disease. Birth defects are known to be caused by mercury, nicotine, alcohol, radiation, and viruses, among other factors. When epidemiologists compare the birth defect rates for women who took Bendectin during pregnancy against those who did not take Bendectin during pregnancy, there is a chance that the distribution of other factors may not be even between the two groups. Usually, the larger the size of the sample, the more likely that random chance will lead to an even distribution of these factors among the two comparison groups, unless there is a dependence between some of the other factors and the factor being studied. For example, there would be a dependence between variables if women who took Bendectin during pregnancy were more or less likely to smoke than women who did not take Bendectin. Another source of error in epidemiological studies is selective recall—i.e., women who have children with birth defects may be more likely to remember taking Bendectin during pregnancy than those women with normal children. Fortunately, we do not have to resolve any of the above questions, since the studies presented to us incorporate the possibility of these factors by use of a *confidence interval*. The purpose of our mentioning these sources of error is to provide some background regarding the importance of confidence intervals.

Id. at 311-312 (emphasis in original).

Question

Do confidence intervals reflect the risk that an epidemiological study may be biased by the presence of confounders or selective recall?

5.3.2   *Paucity of Crossets*

In *State v. Sneed*, 76 N.M. 349, 414 P.2d 858 (1966), there was evidence that the accused had on occasion used the name "Robert Crosset," and that on the day of the murder someone by that name had purchased a handgun which apparently was the murder weapon. An expert witness examined telephone books in the area of the crime and found no Crosset in approximately 129 million (!) listings. He guessed the frequency of Crosset to be about one in a million and the frequency of Robert to be one in thirty. Assuming independence, he concluded that the frequency of Robert Crosset was 1 in 30 million. Reversing defendant's conviction, the Supreme Court of New Mexico objected to the use of "a positive number . . . on the basis of telephone books when the name Robert Crosset was not listed in those books."

Question

What might the expert have calculated as a reasonable, upper-bound estimate from the telephone book data for the frequency of the name Robert Crosset in the population?

5.3.3 Purloined notices

Financial Information, Inc., publishes a Daily Called Bond Service, which provides information concerning the redemption of municipal and corporate bonds. The information initially appears on index cards that are published daily and sent to subscribers. The data on the cards list the name of the issuer, redemption date, price, redemption agent, and the coupon numbers of the bonds to be redeemed. About 2,500 of these cards are published annually.

The service is provided to 500 subscribers, primarily the back offices of banks and brokerage houses. It enables subscribers to keep track of the redemption dates of portfolios of corporate and municipal bonds.

Moody's Investors Service, Inc., publishes a municipal and government manual and a municipal and government news report. These publications also provide information concerning redemption of municipal bonds, the same information provided by Financial's service with respect to those bonds. Moody's is a subscriber to Financial's service. Moody's subscribes to more sources of basic information on bond redemptions than does Financial: while both use national and local newspapers, Moody's uses more newspapers than Financial.

Sometime in 1980, Financial began to suspect that Moody's was copying information from its service. In December 1980, Financial planted an error in its index cards "redeeming" some bonds that had in fact been redeemed a year earlier. Moody's published the fake redemption. Further checking showed that in 1980 seven of ten Financial errors were reproduced by Moody's, while in 1981 all eight Financial errors were copied by Moody's. There is no serious dispute that Moody's copied the notices that repeated Financial's errors, but Moody's executives claimed that Moody's only copied about 22 notices a year.

In 1981, of 1,400 redemption notices published for municipal bonds, 358 such notices were published by Moody's before the comparable notice was published by Financial; 155 such notices were published simultaneously or within five days of Financial's first publication; 97 such notices did not appear in Financial's service; and 179 additional notices published by Moody's contained more information than appeared in Financial's service.

Questions

1. Focusing on the roughly 600 notices for which copying could not be ruled out, as an expert for Financial, compute a 99% one-sided lower confidence limit for the proportion of copied notices based on the 1981 errors. Confirm that the lower limit based on the 1980 and 1981 errors combined is 0.54.

2. As a lawyer for Moody's, how do you respond?

Source

Financial Information, Inc v. Moody's Investors, 751 F.2d 501 (2d Cir. 1984), *aff'd after remand*, 808 F.2d 204 (2d Cir. 1986). On the second appeal, the court of appeals affirmed the dismissal of Financial's claim on the ground that the notices were not copyrightable and there was preemption of state unfair competition law. In passing, it noted that the record would support a finding of no "wholesale appropriation" by Moody's, citing the facts that (i) plaintiff's expert was "statistically certain that Moody's had copied only 40-50% of the time," and (ii) a Moody's exhibit demonstrated that, of its 1,400 called bond entries in one year, 789 could not possibly have come from copying Financial's cards.

5.3.4   Commodity exchange reports

The Commodity Futures Trading Commission requires commodity exchanges to keep records of the thirty-minute interval (each such interval being referred to as a bracket) in which transactions are recorded. In December 1979, the Commission, concerned with the accuracy of the records, required each exchange to submit reports on the percentage of transactions for which bracket information was inaccurate. It advised: "The contract market may use valid random sampling techniques to estimate the degree of accuracy of the data, provided that the estimate can be expected to differ from the actual error percentage by no more than plus or minus two per cent when measured at the 95% confidence interval."

Question

If pilot studies indicate that for one exchange the error rate was about 5%, how large a sample would be required to satisfy the Commission?

Source

CFTC Letter to Commodity Exchanges dated December 7, 1979.

5.3.5   Discharge for dishonest acts

After being discharged by their employer, Pacific Northwest Bell (PNB), Charles and Linda Oliver sued PNB, contending that a company-wide policy that subjected employees to disciplinary proceedings for "dishonest acts" committed outside of employment was discriminatory because of its disproportionate impact on black employees. An expert examined personnel data of a sample of 100 employees who had been discharged for dishonest acts; of these, 18 employees, 6 of whom were black, had committed dishonest acts outside of employment. Among the six were the Olivers. The work force consisted of 975 blacks out of some 21,415 PNB employees (4.6%). The expert contrasted the 33% rate of blacks among those discharged with the 4.6% rate of blacks in the workforce and argued that PNB's policy had a disparate impact on blacks.

The appellate court found the sample too small to support a finding of disparate impact because the subtraction of even one or two blacks would shift the percentage significantly, citing *International Bd. of Teamsters v. United States*, 431 U.S. 324, 339-40 n.20 (1977) ("[c]onsiderations such as small sample size may of course detract from the value of such [statistical] evidence"); *Morita v. Southern Cal. Permanente Med. Group*, 541 F.2d 217, 220 (9th Cir. 1976), *cert. denied*, 429 U.S. 1050 (1977) (statistical evidence derived from an extremely small sample has little predictive value and must be disregarded).

Question

The court's "change one or two" analysis is a naive method of finding other rates at which blacks were discharged that are seemingly plausible due to sampling variability. Treating the sampling variability issue as one calling for a confidence interval analysis, use Figure 5.3c to determine whether you agree with the court's conclusion.

Source

Oliver v. Pacific Northwest Bell Tel. Co., 106 Wash. 2d. 675, 724 P.2d 1003 (1986).

Notes

The objection to small samples reflected in *Oliver* has surfaced in many other cases. In addition to the cases cited, see, e.g., *Wade v. New York Tel. Co.*, 500 F. Supp. 1170, 1180 (S.D.N.Y. 1980) (rejecting an inference of discrimination raised by the discharge of one or two minority employees above expected numbers); *Bridgeport Guardians Inc. v. Members of Bridgeport Civil Serv. Comm'n*, 354 F. Supp. 778, 795 (D. Conn.), *modified*, 482 F.2d 1333 (2d Cir. 1973) (rejecting an inference of discrimination "when a different result achieved by a single candidate could so drastically alter the comparative figures"). See also *Philadelphia v. Education Equality League*, 425 U.S. 604 (1974), discussed in Section 4.2.2.

5.3.6   Confidence interval for promotion test data

Compute a 95% approximate confidence interval for the ratio of pass rates for blacks and whites in the *Teal* data of Table 5.2 at p. 160.

Questions

1. Interpret your result.

2. Does the four-fifths rule of the EEOC (see Section 1.5.1) suggest a reason for preferring an interval estimate of the true ratio that you have just computed to a test of the hypothesis that the ratio is 1, as computed in Sections 5.1.1 and 5.2.3?

5.3.7   Complications in vascular surgery

Dr. Z is a vascular surgeon practicing in the Eastern United States. Administrative proceedings were brought against him to withdraw his hospital operating privileges on the ground that his rates of complication were too high. As one example, in certain peripheral vascular surgery his rate of amputation was 4 out of 61, or 6.56%, over a two-year period. This was compared with the Eastern Vascular Surgery mail-out, mail-back survey for the same time period, which reported a rate of amputation of 65/2,016, or 3.22%. The EVS is a survey conducted by mailing questionnaires to vascular surgeons; about 30-40% respond.

Questions

1. Use the Fleiss quadratic formula in Section 5.3 at p. 172 to compute an approximate 95% two-sided confidence interval for Dr. Z. What argument for Dr. Z might be based on your calculation?

2. What objections might you raise to use of the EVS as a standard?

3. What additional data might you gather in support of Dr. Z?

5.3.8   Torture, disappearance, and summary execution
in the Philippines

Victims of torture, disappearance, or summary execution at the hands of the military under the Filipino regime of Ferdinand E. Marcos brought a class

TABLE 5.3.8a. Summary of special master's damage recommendations for valid claims

	Torture	Summary Execution	Disappearance
Number of valid claims in sample	64	50	17
Sample range	$20,000–$100,000	$75,000–$220,000	$50,000–$216,667
Sample average	$51,719	$128,515	$107,853
Sample standard deviation	$26,174	$34,143	$43,103
Interquartile range	$30,000–$75,000	$95,833–$155,000	$87,222–$121,785
#Valid remaining claims in subclass	4,869	3,184	880
Subclass amounts	$251,819,811	$409,191,760	$94,910,640

action against him after he fled to Hawaii. In the compensatory damage phase of a trifurcated trial, the district court required class members to opt in by submitting claims detailing their abuse. Of the 10,059 claims filed, the court determined that 518 were facially invalid or duplicates, leaving 9,541 claims. To keep proceedings manageable, the court allowed use of a statistical sample of claims to measure damages for the entire class. A random sample of 137 claims was selected and depositions were taken of those claimants. A special master and court-appointed expert, Sol Schreiber, reviewed the claims, found that six of them were invalid, and for the rest made recommendations as to damages in three subclasses: torture, disappearance, and summary execution. He then recommended that the average award for each subclass sample be given to the remaining members of the subclass. Table 5.3.8a below gives some summary statistics for the special master's recommendations, and Table 5.3.8b gives the full data for the summary execution and disappearance cases.

The sample size of 137 was determined by a statistical expert, James Dannemiller. He testified in substance that a random sample of 137 claims would be required for a 95% confidence interval of plus or minus five percentage points (the "industry standard," as he put it) assuming that the population rate of valid claims was 90%. As a *post hoc* justification, he noted that the special master had determined that 6 of the 137 claims, or 4.38%, had been found invalid.

As to the sample claims, contrary to the Special Master's recommendations, the jury found only two claims invalid. The jury modified awards in 46 of the 135 claims and for the rest followed the recommendations. As to the claims of the remaining class members, the jury adopted the special master's recommendations. The special master recommended compensatory damages of approximately $767.5 million and the jury returned a verdict of $770 million. The court entered judgment in accordance with the jury's findings.

TABLE 5.3.8b. Special master's damage recommendations

Summary Execution		Summary Execution		Disappearance	
1	220,000	26	129,167	1	216,667
2	185,000	27	128,533	2	172,667
3	180,000	28	126,111	3	156,250
4	178,000	29	126,000	4	127,500
5	175,000	30	119,444	5	121,785
6	165,000	31	118,042	6	119,250
7	165,000	32	115,278	7	108,005
8	165,000	33	111,250	8	105,000
9	162,000	34	104,167	9	103,583
10	160,000	35	103,667	10	97,500
11	160,000	36	101,700	11	90,903
12	160,000	37	100,000	12	87,500
13	155,000	38	95,833	13	87,222
14	155,000	39	93,667	14	67,083
15	150,000	40	93,667	15	65,000
16	148,000	41	93,611	16	57,600
17	145,833	42	90,556	17	50,000
18	140,000	43	90,250		
19	140,000	44	90,000		
20	140,000	45	81,944		
21	135,750	46	81,667		
22	135,000	47	79,500		
23	135,000	48	79,042		
24	134,533	49	78,555		
25	130,000	50	75,000		

Questions

1. Does the expert's sample size guarantee that a 95% confidence interval for the percentage of invalid claims will not exceed plus or minus five percentage points?

2. Assuming that the data are normally distributed, check for overall fairness to defendant and to plaintiffs as a group by computing an approximate 95% confidence interval for the average award in each subclass. Is it valid to argue that defendant has no ground for objecting to the use of the sample average because it is as likely to be below as above the population average?

3. Check for fairness to individual plaintiffs by looking at the variation in individual awards in each subclass. Given this variation, is use of the average fair to plaintiffs within each subclass?

Source

Hilao v. Estate of Marcos, 103 F.3d 767 (9th Cir. 1996); Recommendations of Special Master and Court-Appointed Expert, Sol Schreiber, December 30, 1994.

Notes

The trial of bellwether cases in large class actions, with the results extrapolated to the class, has received approval in theory, if the cases to be tried are representative of the class. In *In re Chevron U.S.A.,* 109 F.3d 1016 (5th Cir. 1997), the court found that 15 cases selected by plaintiffs and 15 by defendants lacked the requisite level of representativeness so that results from the 30 cases, tried as bellwethers, could not be extrapolated to the class. The court laid down the requirements for extrapolation, which included confidence interval analysis, as follows:

> [B]efore a trial court may utilize results from a bellwether trial for a purpose that extends beyond the individual cases tried, it must, prior to any extrapolation, find that the cases tried are representative of the larger group from which they are selected. Typically, such a finding must be based on competent, scientific, statistical evidence that identifies the variables involved and that provides a sample of sufficient size so as to permit a finding that there is a sufficient level of confidence that the results obtained reflect results that would be obtained from trials of the whole.

Id. at 1020.

5.4   Statistical power in hypothesis testing

The power of an hypothesis test is defined as the probability of correctly rejecting the null hypothesis when it is false. The concept of power is complementary to the Type II error rate–power plus the Type II error rate equals 1. Most alternative hypotheses are composite, i.e., comprise many possible non-null parameter values. The null hypothesis that a coin is fair is the simple hypothesis that the probability of tossing heads is one-half. In the composite alternative hypothesis that the probability of tossing heads differs from one-half, the probability could be any other value. Since one can compute the power of a given test at any such parameter value, we usually refer to the test's power *function*.

The ideal hypothesis test would have a power function equal to 1 for any parameter value in the alternative hypothesis. Although this ideal cannot be attained in practice with fixed sample sizes, we may more closely approach the ideal by increasing sample size.

Power is an important concept for two related reasons. (1) When designing a sampling plan or a comparative trial, it is essential to provide for a sample sufficiently large to make it reasonably likely that any effects worth detecting will be found statistically significant. A study that has, say, only a 25% chance of finding significance in the presence of a large effect is probably not worth the effort and expense. (2) The power of a test is important to the interpretation of non-significant findings. Frequently, a finding of non-significance is interpreted as support for the null hypothesis. This may be warranted if the power to detect important effects is high, for then the failure to detect such effects cannot easily be attributed to chance. But if power is low, then the non-significant result cannot be taken as support of either hypothesis, and no inferences should be drawn from the failure to reject the null.

For example, in its final *Rules for Identification, Classification, and Regulation of Potential Occupational Carcinogens*, OSHA determined that non-positive epidemiologic studies will be considered as evidence of safety only if they are "large enough for an increase in cancer incidence of 50% above that in the unexposed controls to have been detected." 29 C.F.R. §1990. 144(a) (1997). While OSHA admitted that "[a]n excess of 50% can hardly be regarded as negligible," it determined that "to require greater sensitivity would place unreasonable demands on the epidemiologic technique." 45 Fed. Reg. 5060 (Jan. 22, 1980). This determination was based on testimony that the smallest excess risk established in an epidemiologic study was between 30% and 40%, and that epidemiologic technique cannot be expected to detect risk increases of 5-10%. Do you agree with OSHA's logic?

As another case in point, in the late 1970s a randomized clinical trial tested whether or not strictly pharmacologic therapy could lower mortality rates as much as coronary bypass surgery among patients with chronic, stable angina. The major finding was that there was no significant difference between the two treatments in terms of five-year survival. The study received much publicity, and was taken as evidence that supported a popular trend away from surgery. What was not widely reported was that (i) the observed surgical mortality rate was substantially (but not significantly) lower than the pharmacologic mortality rate, and (ii) the power of the test of the null hypothesis was low, in fact less than 30%. Consequently, there may have been great harm done to patients in this category who were advised to defer bypass surgery. See Gerald Weinstein and Bruce Levin, *The Coronary Artery Surgery Study (CASS): A Critical Appraisal*, 90 J. Thoracic & Cardiovascular Surgery 541 (1985).

The power of standard hypothesis tests has been tabled for a wide variety of parameter values and sample sizes. We illustrate the fundamental idea with a simple example for which no special tables are needed. Suppose we are testing the null hypothesis $H_0 : P_1 = P_2$ in a two-sample binomial problem with sample sizes n_1 and n_2 at level $\alpha = 0.05$. The alternative is, say, the one-sided hypothesis that $P_1 - P_2 = D > 0$. In this example there are two steps to calculating power. First, compute how large the observed difference $p_1 - p_2$ has to be to reject the null hypothesis. We call this the critical value. Then, assuming the alternative hypothesis that $P_1 - P_2 = D > 0$, compute how likely it is that the observed difference $p_1 - p_2$ would exceed the critical value. That probability is the power of the test against the specific alternative hypothesis. In detail, one proceeds as follows.

The z-score test rejects H_0 when the difference in sample proportions $p_1 - p_2$ satisfies the inequality

$$(p_1 - p_2) > 1.645[pq(n_1^{-1} + n_2^{-1})]^{1/2},$$

where p and q refer to the pooled rate of the two groups. Under the alternative hypothesis, with true proportions P_1 and P_2 and difference equal to $D > 0$, we standardize the sample difference by subtracting the true mean D and dividing by the true standard error $[(P_1 Q_1/n_1) + (P_2 Q_2/n_2)]^{1/2}$. Furthermore, in large samples the quantity pq will be approximately equal to PQ, where P is the proportion $(n_1 P_1 + n_2 P_2)/(n_1 + n_2)$. Thus, the power of the test at P_1 and P_2 is the probability that an approximately standard normal random variable exceeds the value

$$\{1.645[PQ(n_1^{-1} + n_2^{-1})]^{1/2} - D\}/[(P_1 Q_1/n_1) + (P_2 Q_2/n_2)]^{1/2}.$$

This probability can be found in tables of the normal distribution.

For example, if $P_1 = 0.6$ and $P_2 = 0.5$ and the sample sizes are 100 each, then the power of the test of $H_0 : P_1 = P_2$ equals the probability that an approximately standard normal random variable exceeds the value 0.22, which is only 0.41. The moral is: large samples are needed to detect relatively small differences with adequate power.

5.4.1   Death penalty for rape

On November 3, 1961, in Hot Springs, Arkansas, William Maxwell, a black male, was arrested for rape. He was subsequently convicted and sentenced to death.

Maxwell was 21 years old at the time of the offense. "[T]he victim was a white woman, 35 years old, who lived with her helpless ninety year-old father; ... their home was entered in the early morning by the assailant's cutting or breaking a window screen; ...and... she was assaulted and bruised, her father injured, and the lives of both threatened." *Maxwell v. Bishop*, 398 F.2d 138, 141 (8th Cir. 1968).

In a federal habeas corpus proceeding, Maxwell argued that the death sentence for rape in Arkansas was applied in a racially discriminatory manner. To support this assertion, Maxwell relied on a study by Dr. Marvin Wolfgang, a sociologist at the University of Pennsylvania. The study examined every rape conviction in a "representative sample" of 19 Arkansas counties between January 1, 1945, and August 1965.

From these Arkansas data, Dr. Wolfgang concluded that the critical variables in determining the death sentence were the race of the offender and the race of the victim. He based this conclusion on the fact that other variables that might have accounted for the disparity did not seem to be significantly associated either with defendant's race or with the sentence. Factors not significantly

TABLE 5.4.1a. Race of defendant by sentence

	Death	Life	Total
Black	10	24	34
White	4	17	21
Total	14	41	55

TABLE 5.4.1b. Race of victim by sentence

	Death	Life	Total
Black	1	14	15
White	13	26	39
Total	14	40	54

TABLE 5.4.1c. Race of defendant by prior record

	No Record	Record	Total
Black	20	14	34
White	13	8	21
Total	33	22	55

TABLE 5.4.1d. Prior record by sentence

	Death	Life	Total
Prior record	12	21	33
No record	2	20	22
Total	14	41	55

TABLE 5.4.1e. Race of defendent by type of entry

	Unauthorized	Authorized	Total
Black	13	18	31
White	3	17	20
Total	16	35	51

TABLE 5.4.1f. Type of entry by sentence

	Death	Life	Total
Unauthorized	6	10	16
Authorized	8	27	35
Total	14	37	51

associated with defendant's race or the sentence were type of entry into the victim's home, seriousness of injury to the victim, and defendant's prior record. Factors not significantly associated with defendant's race were the commission of a contemporaneous offense and previous imprisonment. A factor not significantly associated with the sentence was the victim's age. Factors that were significantly associated with defendant's race were the defendant's age and the victim's age.

Question

Excerpts from the data are shown in Tables 5.4.1a–5.4.1f. Do the data on prior record or type of entry support Dr. Wolfgang's conclusion?

Source

Maxwell v. Bishop, 398 F.2d138 (8th Cir. 1968), *vacated and remanded*, 398 U.S. 262 (1970); see also, Wolfgang and Riedel, *Race, Judicial Discretion, and the Death Penalty*, 407 Annals Am. Acad. Pol. Soc. Sci. 119 (1973). For much more elaborate studies of the relation between race and the death penalty, see Section 14.7.2.

5.4.2   Is Bendectin a teratogen?

For 23 years Merrell Dow Pharmaceuticals, Inc. manufactured Bendectin, a drug taken for morning sickness in pregnancy.

Merrell Dow estimated that some 25% of pregnant women in the United States took the drug; it was popular in other countries as well. In the 1980s, suspicions were raised that Bendectin was associated with certain types of birth defects such as limb deformities, cleft lips and palates, or congenital heart disease.

After a highly publicized trial in Florida, over 500 lawsuits were filed by plaintiffs who had been in utero when their mothers had taken Bendectin. A number of these cases were consolidated for pretrial proceedings in the U.S. District Court for the Southern District of Ohio. Merrell Dow withdrew the drug from the market and offered to settle all cases with a fund of $120 million. This offer was rejected, and a consolidated jury trial ensued solely on the issue of causation, without reference to the emotionally compelling facts relating to any particular plaintiff.

At the trial, Merrell Dow introduced the results of ten cohort studies of congenital malformations, this being the entire published literature of cohort studies at that time. (Merrell Dow also introduced some case control and other studies, but we focus on the cohort results.) The cohort studies showed that with respect to all defects there was a combined relative risk of 0.96 with a standard error for the logarithm of 0.065. The five studies with separate data for limb reduction defects showed a combined relative risk of 1.15 with a standard error of 1.22 for the logarithm. (The results for the all-defect studies are shown in Table 8.2.1a, and the method by which they were combined is described in Section 8.2 at p. 251.)

A plaintiffs' expert, Dr. Shanna Swan, contended that teratogens are usually associated only with certain types of defects, while the cohort studies in question covered all defects. She argued that the narrower effect of Bendectin might have been masked in the studies. The overall rate for all defects in the unexposed population is approximately 3%. The highest rates are 1/1000 for limb reductions, 2/1000 for major heart defects, and 3/1000 for pyloric steno-sis. If Bendectin had doubled the rate of each of these defects, but affected no others, the rate of overall defects would have increased from 3% to 3.6%, creating a relative risk of 1.2.

Questions

Were the cohort studies presented by defendants large enough to detect a rel-ative risk of 1.2? In answering this general question, resolve the following underlying questions.

1. Assuming that the rate of defects is 3% among those unexposed, what is the power in a sample of 1,000 people to reject the null hypothesis that the exposed group is at the same risk as those unexposed if the true risk for those exposed is 3.6%?

2. How many exposed people must be included in a study to give a 90% chance of rejecting the hypothesis (at the 5% one-tailed level of significance) that the rate of defects in the exposed population is no greater than the general population rate of 3%, if the actual rate is 3.6% for all defects combined in the exposed population?

3. How many exposed people must be included if the study were limited to limb reductions and Bendectin doubled the rate of such defects?

4. Assuming the all-defect data are correct and correctly combined, do they support the expert's observation that the non-significant study results have too little power to be evidence against the alternative hypothesis of a 1.2 relative risk?

5. Same question with respect to the limb reduction studies under the alternative hypothesis of a 2.0 relative risk.

Source

In re Bendectin Litig., 857 F.2d 290 (6th Cir. 1988), *cert. denied*, 488 U.S. 1006 (1989).

5.4.3   Automobile emissions and the Clean Air Act

The federal Clean Air Act, 42 U.S.C. §7521 (Supp. 1998), provides for certifi-cation of vehicles as conforming with emission standards and makes unlawful new fuels or fuel additives not used in the certification process without a waiver by the administrator of the Environmental Protection Agency (EPA). The administrator may grant a waiver if he determines that the applicant has established that the new fuel or fuel additive "will not cause or contribute to a failure of any emission control device or system (over the useful life of any vehicle in which such device or system is used) to achieve compliance by the vehicle with the emission standards with respect to which it has been certified...." *Id.*, §7545(f)(4).

The EPA allows evidence of emission effect based on sample studies and uses three different statistical tests, with respect to a fuel expected to have an instantaneous emission effect: 1) the Paired Difference Test, which determines the mean difference in emissions between the base fuel and the waiver fuel in vehicles driven first with one, and then with the other, of these fuels; 2) the Sign of the Difference Test, which assesses the number of vehicles in the test exhibiting an increase or decrease in emissions; and 3) the Deteriorated Emissions Test, which adds the incremental emissions caused by the waiver fuel to the emission levels certified for the vehicle after 50,000 miles and compares that with the standard allowed for that vehicle.

Here we consider the Deteriorated Emissions Test.[9] The EPA has provided by regulation that the Deteriorated Emissions Test should be such as to provide a 90% probability of failure of the test if 25% or more of the vehicle fleet of the type tested would fail to meet emission standards using the waiver fuel or fuel additive. In a waiver proceeding involving a proprietary fuel called Petrocoal (a gasoline with methanol additive), 16 vehicles were tested with Petrocoal and a standard fuel. Two vehicles failed the Deteriorated Emissions Test.

Questions

1. Did Petrocoal pass the Deteriorated Emissions Test?

2. What additional specifications for testing has the EPA left up to the proponents?

3. How many vehicles would need to be tested if, under the same power requirements as in the regulation, there were to be no more than a 0.10 probability of failing the test if the fleet proportion were only 5%?

Source

Motor Vehicle Mfrs. Ass'n of U.S. v E.P.A., 768 F.2d 385 (D.C. Cir. 1985); J. Gastwirth, *Statistical Reasoning in Law and Public Policy*, 616-7 (1988).

5.5   Legal and statistical significance

The fact that a difference is statistically significant does not necessarily mean that it is legally significant. When large samples are involved even small differ-

---

[9]For a discussion and application of the Paired Difference Test see Section 7.1.1. The Sign of the Difference Test is a nonparametric test generally called the sign test; it is discussed in Section 12.1.

ences can become statistically significant, but nevertheless may not turn legal litmus paper. If an employer pays men on average more than equally qualified women, the difference is not evidence of intentional discrimination unless it is large enough to imply such intent and to make it unlikely that confounding factors other than gender are responsible. Sometimes the legally required difference is given a number. In reviewing charges that a test for employment has a disparate adverse impact on a minority group, the EEOC by regulation has adopted a four-fifths rule: adverse impact will be presumed if minority group members pass at a rate that is less than 80% of the rate of the group with the highest pass rate. (See Section 1.5.1 at p. 39.) In toxic tort cases, some courts have held that, in order to prove causation from epidemiological data, the relative risk of disease from the exposure must be greater than 2.[10] (See Section 10.2 at p. 285.

The question has been raised whether statistical significance should be tested against the non-zero difference deemed legally important, rather than the zero difference required by classical statistical tests. After all, if it is only sampling error that creates a legally significant difference, it may be argued that such proof is insufficient for the same reason that leads us to test for statistical significance in the first place.

To analyze this issue, consider a group of plaintiffs that has been exposed to a toxin, contracted a disease, and sues. Define a loss as either the granting of an award to a plaintiff in the base group (who would have contracted the disease in any event), or the denial of an award to a plaintiff in the excess group (who would not have contracted the disease without the exposure). In decision theory, investigators look at the mathematically expected loss, which roughly can be translated as the average loss over a run of cases. The expected loss is a function of the true relative risk and the cutoff criterion for the point estimate.

Consider the following decision rule: if the point estimate of the relative risk is > 2 find for the plaintiff; otherwise find for the defendant. Call this a "2-rule." Under the 2-rule we ignore statistical significance. A basic argument for this rule is that it is a "minimax" solution: for any value of the "true" or long-term relative risk, the maximum expected error under the 2-rule is less than under other rules with different cutoff criteria.

To illustrate the rule, consider the case in which the true relative risk is 2. This means that half the plaintiffs should recover. Assuming symmetry in sampling (which is not unreasonable in large samples), in 1/2 of the samples the point estimate of the relative risk will be less than 2, and no award would be made. This would be an error for half the plaintiffs. Thus for $1/2 \times 1/2 = 1/4$ of the class

---

[10]This requirement is based on simple algebra. The probability that a randomly selected plaintiff would be in the base group (and hence not entitled to an award) is 1/RR, where RR is the relative risk. The probability that a plaintiff would be in the excess group (and hence entitled to an award) is $1 - 1/RR$. If RR = 2 then $1 - 1/RR = 1/2$. If RR> 2 then it is more likely than not that a randomly selected plaintiff would be in the excess group.

there will be errors in refusing to make awards. The same calculation applies to errors in making awards, so that the total expected error is 1/2 per person.

It can be shown from elementary calculus that 1/2 is the maximum expected error under the 2-rule for any value of the true relative risk. To illustrate this without calculus, consider the case in which the true relative risk is far above 2, say 8. The point estimate based on a sample would almost certainly be greater than 2, awards would be made in almost every case, and the proportion of erroneous awards would be 1/8. Thus, the rate of errors is far below 1/2. A similar calculation can be made for a true relative risk that is below 2. In fact, the curve of error frequency as against the true relative risk is bell-shaped and reaches its high point of 1/2 when the relative risk is 2 and dwindles to zero as the relative risk moves away from 2 in either direction.

Now suppose that instead of a 2-rule we require that the point estimate of the relative risk be at least 6 before an award is made. If the true relative risk is 3, there is a greater than 50% probability that the point estimate will be less than 6. In that case, no awards would be made and the rate of error would be 2/3. There is a smaller than 50% probability that the point estimate would be greater than 6, in which an award would be made, with error rate 1/3. The expected total error is the sum of these two rates weighted by their probabilities, and since it is more probable that the point estimate will be below 6 than above it, the weighted sum of the two rates will be greater than 1/2. Thus the 2-rule has a smaller maximum expected error than rules with higher cutoff points.[11]

If we adopt a 2-rule, but also require statistical significance as against 2, we would in effect be adopting rules with cutoff points above 2 in every case. How much above depends on the sample size and the degree of significance required. As we have indicated, such rules would have greater maximum expected error rates than the 2-rule.

The same line of argument indicates that using the 2-rule and requiring statistical significance even as against 1 also increases total error over the 2-rule without the significance requirement. Those cases in which a point estimate of 2 would not be significant as against 1 in effect require a cutoff criterion above 2 and this, as we have seen, increases maximum expected error. However, this would occur only in very small studies in which a point estimate above 2 would not be statistically significant as against 1, whereas testing significance against 2 would require a cutoff criterion greater than 2 in every case. The increase in maximum expected error is thus much smaller when significance is tested against 1 than it is when tested against 2. Note also that if there is no significance as against 1, there is insufficient evidence that defendant caused anyone harm; if there is no significance as against 2, the only question is whether the plaintiff is among those injured. We may therefore

---

[11]The general formula for the expected loss, which is a function of the true relative risk (R) and the point estimate criterion (r), is $1/2 + (1 - 2/R)(P(r) - 1/2)$, where $P(r)$ is the probability of no award under the point estimate criterion when $r > R$. The formula shows that the expected loss exceeds 1/2 when $R > 2$ and $r > R$ so that $P(r) > 1/2$.

justify requiring statistical significance as against 1 but not 2 on two grounds: a smaller increase in maximum expected error and a more fundamental issue.

5.5.1   Port Authority promotions

Black police officers at the Port Authority of New York and New Jersey who were candidates for promotion to sergeant brought suit, claiming that the Port Authority procedures had a disparate impact on black candidates. The promotion process had three steps: first, a written test designed to gauge a candidate's knowledge of the law, of police supervision, and of social and psychological problems at work; second, those who passed the written test took an oral test designed to measure judgment and personal qualifications; candidates who succeeded on the second step proceeded to the final step–a performance appraisal based on a supervisory performance rating and the candidate's attendance record. Candidates completing the examination process were placed on an "Eligible List" by rank according to the weighted composite score achieved on the written and oral tests and on the performance appraisal. The written examination received 55% of the weight in the composite score. After the list was issued, the Port Authority promoted candidates, starting with those with the highest scores and proceeding, as needed, down the list for a period of three years, when a new list was created.

A total of 617 candidates took part in the examination process, of whom 508 were white, 64 were black, and 45 were in other groups. The number passing the written examination was 539 of whom 455 were white and 50 were black. The mean score of whites on the examination was 79.17% and of blacks was 72.03%, yielding a difference of 5.0 standard deviations. On the oral test, blacks passed at a higher rate than whites. Whites and blacks passed the performance appraisal with approximately the same mean percentage score. Also undergoing performance appraisal were 6 white officers who were "grandfathered" from a pre-existing list.

The 316 candidates who underwent performance appraisal were listed in order of their composite scores from the three steps on the eligibility list. Promotions were made from the top of the list downward, and during its three-year life the 85th candidate was reached. The 85 candidates included 78 whites, 5 blacks, and 2 members of other groups. Eight white candidates either retired, withdrew, or were from the grandfathered group, leaving 70 whites actually promoted through the process. Thus, according to the court, 14% (70 of 501) of the white candidates who went through the process were actually promoted as against 7.9% (5 of 63) of the black candidates.[12]

One reason for these disparate results was that, on the written component, although the passing score was lower, the minimum score a candidate could

---

[12]The reason for 501 instead of $508 - 8 = 500$ white candidates and 63 instead of 64 black candidates does not appear.

achieve and still be within the top 85 candidates was 76. This level was attained by 42.2% of the black candidates but 78.1% of the white candidates, a difference that was highly statistically significant.

Questions

1. Looking at the pass-fail results of the written test alone, compute the ratio of the black to white pass rates and the *P*-value for the ratio. Do the results suggest disparate impact in light of the EEOC's four-fifths rule (see Section 1.5.1) and the Supreme Court's two- or three-standard deviations rule in *Castaneda* (see Section 4.3.2)?

   [Point of information: In the case, the calculation of significance was made for the *difference* between the two pass rates. Here we ask for a calculation of the *P*-value based on the *ratio* of rates, since it is the ratio that is tested under the four-fifths rule. Although the two null hypotheses are the same, the test statistics have slightly different properties.]

2. The district court resolved the tension between the EEOC and Supreme Court rules by arguing that the written examination did not have a disparate impact because the disparity would lose statistical significance if two additional blacks had passed the test. The court of appeals agreed. Make your own calculation. Do you agree with this argument?

3. The court of appeals looked at the overall results of the promotion process (14% of whites compared with 7.9% of blacks promoted) and concluded that there was disparate impact because the four-fifths rule was violated. But the difference was not statistically significant. The court of appeals put the lack of significance aside for two reasons: (i) "where statistics are based on a relatively small number of occurrences, the presence or absence of statistical significance is not a reliable indicator of disparate impact"; and (ii) the difference was not caused by chance, but by the written test, which, "apart from its screening out some candidates, was to cause blacks to rank lower on the eligibility test than whites." Make your own calculation of the pass ratio and its *P*-value. Do you agree with the court's reasoning?

Source

Waisome v. Port Authority of New York & New Jersey, 948 F.2d 1370 (2d Cir. 1991).

5.6   Maximum likelihood estimation

We have previously discussed two properties of the sample mean and variance: unbiasedness and consistency (see Sections 1.2 at p. 3 and 1.3 at p. 18). Although it is comforting when a statistical estimator can be proved unbiased, it is not always convenient or even possible to obtain an unbiased estimator for a given parameter. It is more important by far for an estimator to have the consistency property, which states that in large samples the estimate will be close to the true parameter with high probability, even if the expected value of the estimator is not exactly equal to the parameter it seeks to estimate. For example, while the sample variance s^2 is an unbiased estimator of a population variance σ^2, the sample standard deviation s is a slightly biased estimator of σ. This is usually of no concern, as s is a consistent estimator of σ, the bias becoming negligible as the sample size increases.

In problems more complex than estimating a mean or variance, it is not always obvious how to estimate the parameter or parameters of interest. The question then arises, how do we find an estimate for the parameters in a given problem, and can we find one that is consistent? Are there any general principles to which we can appeal? As it happens, in a surprisingly wide class of problems, we can find a consistent estimator by the *method of maximum likelihood*, which yields a special bonus: not only are maximum likelihood estimates (mle's for short) consistent, but in most cases they are also *asymptotically efficient*, which means that in large samples the mle has about the smallest variance possible among all competing estimators. The mle thus "squeezes" the maximum amount of information about the parameter from the data.[13]

Likelihood functions

The maximum likelihood method is easy to understand. First, we introduce the *likelihood function*, $L(\Theta|\underline{X})$, where Θ represents the parameter or parameters to be estimated, and $\underline{X}$ represents the observed data. The likelihood function is simply the probability of observing data $\underline{X}$, when Θ is the true parameter governing the distribution of $\underline{X}$, but here we regard the data as fixed and consider how the probability of the data varies as a function of Θ. For example, suppose we have observed the outcomes $Z_1, \ldots, Z_n$ of n tosses of a coin with probability Θ of success on a single toss. The likelihood function is then

$$L(\Theta|\underline{X}) = L(\Theta|Z_1, \ldots, Z_n) = \Theta^S \cdot (1 - \Theta)^{n-S},$$

where $S = \sum Z_i$ is the total number of successes. If only S had been recorded, but not the actual sequence of outcomes, the binomial probability formula

---

[13] In technical terms, the limiting ratio of the variance of the estimator to the theoretical minimum variance for any regular estimator approaches unity as the sample size grows large. See J. Rice, *Mathematical Statistics and Data Analyses*, ch. 8 (2d ed. 1995).

would give the likelihood function $L(\Theta|n, S) = \binom{n}{S}\Theta^S(1-\Theta)^{n-S}$, regarded as a function of Θ. In this example $L(\Theta|\underline{X})$ and $L(\Theta|n, S)$ are proportional for each value of Θ, so that the two versions are essentially equivalent for purposes of comparing the likelihood function at different values of Θ. In general, likelihood functions are only defined uniquely up to constants of proportionality, meaning constants that may depend on the (fixed) data, but not on the parameter Θ.

The mle

The maximum likelihood estimate of Θ, denoted by $\hat{\Theta}_n$, is that value of Θ which maximizes the likelihood function $L(\Theta|\underline{X})$. The mle thus has the interpretation as the parameter value that renders the observations "most likely," and as such is intuitively appealing as an estimator. In the binomial example, to find $\hat{\Theta}_n$ we ask what value of Θ maximizes the function $\Theta^S \cdot (1-\Theta)^{n-S}$, or equivalently, what value of Θ maximizes the log-likelihood function

$$\ln L(\Theta|\underline{X}) = S \cdot \ln \Theta + (n-S) \cdot \ln (1-\Theta).$$

The answer, from calculus, is that $\hat{\Theta}_n = S/n$ is the maximizer, i.e., the sample proportion S/n is the maximum likelihood estimate of the binomial parameter. In common problems like this and others involving normal, Poisson, and exponential variables, the sample mean is an mle. In more complicated problems, an iterative computing algorithm must be employed as there are no closed-form solutions like $\hat{\Theta}_n = S/n$. Because of its importance in statistical work, maximum likelihood estimation is routinely provided for in computer programs for a vast array of problems.

Here is another important property of maximum likelihood estimators: if instead of the parameter Θ interest focuses on some one-to-one transformed version of Θ, say $\Theta' = f(\Theta)$, then the value of Θ' that maximizes the likelihood function $L(\Theta'|X)$, regarded as a function of Θ', is simply $f(\hat{\Theta}_n)$, i.e., the mle of the transformed parameter is just the transformed mle of the original parameter. Thus, in the example the mle of the log odds equals the log odds of the mle of p. This invariance property means that in a given problem we are free to choose the parameter of interest however we wish; if we can find the mle for one form of the parameter, then we have found the mle for all forms.

The negative of the second derivative of the log-likelihood function is called the *observed information*, and is an important measure of the amount of statistical information contained in a sample of a data. The standard error of the mle is given by the positive square root of the reciprocal of the observed information; it is routinely calculated to determine the mle's statistical significance and confidence intervals.

Likelihood ratios

Extremely useful statistics are constructed by taking the ratio of the maximum value of the likelihood function under the alternative hypothesis to the maximum value of the function under the null hypothesis. The likelihood ratio statistic is defined as $L_1(\hat{\Theta}_1|\underline{X})/L_0(\hat{\Theta}_0|\underline{X})$, where L_1 is the likelihood under H_1, L_0 is the likelihood under H_0, and $\hat{\Theta}_i$ is the maximum likelihood estimate of any unknown parameters under either hypothesis H_i. The likelihood ratio is an important measure of the weight of statistical evidence; large values indicate strong evidence in favor of H_1 and against H_0. The likelihood ratio was used to measure the strength of evidence in the discussion of Bayes's theorem (Section 3.3), screening devices (Section 3.4), and cheating on multiple-choice tests (Section 3.5.2).

As an example, we compute the likelihood ratio for the data in the *Teal* case, at p. 160. In *Teal*, black and white officers took a test for promotion. Black officers passed the test at the rate of $26/48 = 54\%$ and failed at the rate of $22/48 = 46\%$; white officers passed at the rate of $206/259 = 80\%$ and failed at the rate of $53/259 = 20\%$. Under the null hypothesis that the pass (and fail) rates for the two groups are the same, the maximum likelihood estimates are their pooled rates, i.e., the pooled pass rate is $(26+206)/(48+259) = 232/307 = 75.57\%$, and the pooled fail rate is $75/307 = 24.43\%$. The maximum likelihood value of the data under the null hypothesis is $(\frac{232}{307})^{232} \times (\frac{75}{307})^{75} = 7.4358 \times 10^{-75}$. Under the alternative hypothesis, the pass (and fail) rates for the two groups are different. The likelihood function for the two sets of rates is the product of their individual likelihoods, using as maximum likelihood estimators the rates in the sample data. Thus, the maximum likelihood under the alternative is $[(\frac{26}{48})^{26}(\frac{22}{48})^{22}] \times [(\frac{206}{259})^{206}(\frac{53}{259})^{53}] = 4.1846 \times 10^{-72}$. The ratio of the likelihood under the alternative hypothesis to that under the null is $\frac{4.1846 \times 10^{-72}}{7.4038 \times 10^{-75}} = 564$. Thus, we are 564 times more likely to see these data under the alternative than under the null hypothesis. This is very strong evidence in favor of the alternative as against the null.

As a second example, likelihood ratios can be used to define the tails in a two-tailed test when the probability distribution of the test statistic is asymmetric. See Section 4.4 at p. 120 for a discussion of the problem in the context of *Avery v. Georgia*. The likelihood ratio (LR) method defines the two-tailed P-value as the sum of the point probabilities corresponding to those values of x with an LR statistic equal to or greater than that of the observed x. The LR statistic for testing that a binomial parameter p equals a given value p_0 is defined as $LR(x) = \hat{p}^x(1 - \hat{p})^{n-x}/p_0^x(1 - p_0)^{n-x}$, where x is the observed value of X, and $\hat{p} = x/n$ is the sample proportion and maximum likelihood estimate of p. In *Avery*, $\hat{p} = 0$ and the numerator is taken as 1, while the denominator is $0.05^0(1 - 0,05)^{60} = 0.046$, so that $LR(0) = 21.71$. Evaluating $LR(x)$ for other values of $x = 1, 2, \ldots, 60$, one finds that LR is at least 21.71 for values of x starting at $x = 9$ (with $LR(9) = 67.69$; the value for $x = 8$ just misses with $LR(8) = 21.60$). Thus the probability of 9 or more blacks (0.0028) has to be added to 0.046, yielding the two-tailed P-value 0.0488 by the LR method. The fact that the P-value by the LR method is less than the P-value by the point-probability method in this example is not a general phenomenon. The differences between them simply illustrate that different test procedure may lead to different result in borderline cases. Both methods are superior to the simpler methods. As always, the analyst should decide on the procedure before its evaluation on the data at hand.

To simplify computations and to assess the statistical significance of the likelihood ratio statistic it is common practice to compute a log-likelihood ratio statistic, G^2, given by $G^2(H_1 : H_0) = 2\log\{L_1(\hat{\Theta}_1|\underline{X})/L_0(\hat{\Theta}_0|\underline{X})\}$. It is a convenient computational fact that G^2 for a two-by-two table, as in the *Teal* example, is equal to $2 \cdot \sum(observed) \cdot \log[observed/expected \ under \ H_0]$, where the sum is over all cells in the table. The statistic is asymptotically equivalent to Pearson's chi-squared statistic, i.e., G^2 is approximately distributed as χ_1^2 under H_0, with degrees of freedom equal to the difference between the number of unknown parameters under H_1 and H_0. The likelihood ratio is obtained by dividing G^2 by 2 and exponentiating. For the *Teal* data, $G^2 = 12.67$ and $\chi^2 = 12.78$, in fairly close agreement. Using G^2, the likelihood ratio is $\exp[12.67/2] = 564$, in agreement with our previous result.

The fact that G^2 has a chi-squared distribution can be used to assess the statistical significance of the likelihood ratio. Since $\chi_1^2 = 12.67$ has a null probability of less than 0.0005, the likelihood ratio for the *Teal* data is statistically significantly greater than 1.

These results extend to larger tables, where G^2 sums over all cells in the table, and has more degrees of freedom in its chi-squared distribution. As a result, the log-likelihood ratio statistic is widely used in the analysis of multi-way contingency tables.

A virtue of the log-likelihood ratio statistic is that it can be calculated in completely general situations, e.g., with non-count distributions where the Pearson goodness-of-fit statistic would be inappropriate. In addition, its interpretation in terms of strength of evidence is regarded as superior to P-values for this purpose.

5.6.1 Purloined notices revisited

In the purloined notices problem (Section 5.3.3), Moody's executives testified that Moody's copied only about 22 notices a year out of a possible 600, or about 4%. Assuming this to be the null hypothesis and about 50% copying to be the alternative hypothesis—based on the fact that for 1980 and 1981 combined some 15 out of 18 errors were copied—compute a likelihood ratio of the alternative to the null.

5.6.2   Do microwaves cause cancer?

Concern with the effects of low-level ionizing radiation has extended to the effects of low-level microwaves. There are important biological differences between the two. As a photon or energy packet of an ionizing ray passes through a substance, the photon breaks chemical bonds and causes neutral molecules to become charged; such ionization can damage tissues. In contrast, a photon of a billion-hertz microwave has only one six-thousandth of the kinetic energy normally possessed by a molecule in the human body; consequently, such a photon would be far too weak to break even the weakest chemical bonds. Although it is still possible that weak microwave energy might directly alter tissue modecules, it is not clear by what mechanisms significant changes could occur.

A three-year animal study conducted in the 1980s involved 100 rats irradiated with microwaves for most of their lives, absorbing between 0.2 and 0.4 watts per kilogram of body weight, the latter figure being the current American National Standards Institute limit for human beings. The irradiated rats were compared with 100 control rats. The investigators examined 155 measures of health and behavior, which revealed few differences between exposed and control groups.

One striking difference was much publicized: primary malignant tumors developed in 18 of the exposed rats but in only 4 of the controls. No single type of tumor predominated. The total number of tumors in the controls was below the number expected for the strain, with the malignancy rate in the exposed group about as expected.

Questions

1. Find the maximum likelihood estimate of p_1 and p_2, the tumor rates under the two treatments. What is the mle of the relative risk, p_1/p_2?

2. Find the likelihood ratio under the hypothesis of different p_i vs. the null hypothesis $p_1 = p_2$. Is the likelihood ratio significantly large?

3. What is your assessment of the strength of the evidence that microwaves were responsible for the difference in tumor rates?

Source

Foster and Guy, *The Microwave Problem*, 225 Scientific American 32 (September 1986).

5.6.3   Peremptory challenges of prospective jurors

The distinct possibility that in certain cases lawyers will use their peremptory challenges to stack the jury by race or sex has led to new questioning of this hallowed practice.[14] One of the questions raised is the extent to which lawyers making challenges are able to identify biased jurors. To explore that question, a study used actual courtroom data (instead of simulations) and a mathematical model to estimate the extent of lawyers' consistency in strikes. The model divided prospective jurors who were struck into two conceptual groups: those who were clear choices given the facts of the case, and those who were guesses. A juror was defined as a clear choice for a strike when he or she had such characteristics (even though not justifying removal for cause) that lawyers for one side or the other (but not both) generally would agree that the juror should be struck as biased. The lawyers might be wrong, but where there is such professional agreement the strongest case is made for permitting a strike. The category of guesses included those jurors whom lawyers for both sides might strike, either because of conflicting views of the juror's bias or for some other reason, such as perceived unpredictability. The need for peremptory challenges was argued to be less cogent in the case of guesswork jurors than of clear-choice jurors.

To estimate the proportions of each type of juror, the study gathered data from the courtrooms of a few of the small minority of federal judges who require that the challenges of both sides be exercised simultaneously, instead of sequentially, as is the usual practice. The focus was on the number of overstrikes–i.e., jurors struck by both parties. The study theory was that a clear-choice juror by definition could not be overstruck, but a guesswork juror could be overstruck if the lawyers for the two sides had conflicting views of the juror's potential bias, or agreed on some other seemingly undesirable trait, such as unpredictability, or some combination of these. The numbers of clear-choice and guesswork jurors, and the numbers of strikes of such jurors were estimated by the maximum likelihood method based on the overstrike data and the key assumption that guesswork jurors had equal probabilities of being overstruck. The study gathered data from 20 selections.[15] In a common configuration, after excuses for cause, the venire consisted of 32 persons. Both sides used all their permitted strikes: the prosecution made 7 strikes and the defense 11 strikes (in both cases these strikes included alternates). The number of overstrikes ranged from none to four, averaging a little more than one per selection.

---

[14]Race and gender discrimination in the exercise of peremptory challenges has been declared unconstitutional by the Supreme Court. See, e.g., *Batson v. Kentucky*, 476 U.S. 79 (1986)(race); *J.E.B. v. Alabama*, 114 S. Ct. 1419 (1994)(gender). However, proof of discrimination is difficult because judges tend to accept weak justifications for strikes. See S.R. DiPrima, *Selecting a jury in federal criminal trials after* Batson *and* McCollum, 95 Colum. L. Rev. 888, 911 (1995).

[15]There were 16 cases and 20 selections because in four of the cases the alternates were selected separately.

Questions

1. Assume that a venire has 32 prospective jurors, the prosecution makes 7 strikes, the defense makes 11 strikes, and there is 1 overstrike. On the assumption that guesswork strikes are made at random from those jurors who are not clear choices, use the hypergeometric distribution to find the probability of one overstrike if (i) there were all clear-choice jurors save one each for the prosecution and the defense; (ii) there were no clear-choice jurors for either side; and (iii) there were three clear-choice jurors for the prosecution and five for the defense. Among these, which is your mle? [Hint: In computing the hypergeometric probabilities, assume that the prosecution first designates its 7 strikes and the defense then designates its 11 strikes, without knowledge of the prosecution's choices.]

2. Does the result suggest that lawyers may have more strikes than they can effectively use?

3. If the randomness assumption does not hold, what effect would this have on the mle of clear choices?

Source

Michael O. Finkelstein and Bruce Levin, *Clear choices and guesswork in peremptory challenges in federal criminal trials*, 160 J. Roy. Stat. Soc. A 275 (1997).

Notes

The study recognized that categorization of struck jurors as either clear choices or guesswork was too black and white. In the gray real world there would be disagreements among lawyers over clear-choice jurors and not all guesswork jurors would have the same probability of being overstruck. The authors argued, however, that any spectrum of consensus about jurors would have an expected number of overstrikes, and so could be modeled by the clear-choice/guesswork model with approximately the same expected overstrike number. Thus, the percentage of clear-choice jurors in the categorical model could be understood not literally, but as an index to express the degree of randomness in lawyers' selections. In the report of the study, the results across the 20 selections were combined, giving an mle indicating about 20% of strikes were clear-choice jurors and 80% of strikes were guesswork jurors.

  The results of the courtroom data study are broadly consistent with the results of simulation studies of lawyers' consistency and accuracy in making peremptory challenges. For a review of the simulation literature, see R. Hastie, *Is attorney-conducted voir dire an effective procedure for the selection of impartial juries?*, 40 Am. Univ. L. Rev. 703 (1991).

CHAPTER 6

Comparing Multiple Proportions

6.1 Using chi-squared to test goodness of fit

Chi-squared is a useful and convenient statistic for testing hypotheses about multinomial distributions (see Section 4.2 at p. 103). This is important because a wide range of applied problems can be formulated as hypotheses about "cell frequencies" and their underlying expectations. For example, in Section 4.6.2, Silver "butterfly" straddles, the question arises whether the price change data are distributed normally; this question can be reduced to a multinomial problem by dividing the range of price changes into subintervals and counting how many data points fall into each interval. The cell probabilities are given by the normal probabilities attaching to each of the intervals under the null hypothesis, and these form the basis of our expected cell frequencies. Chi-squared here, like chi-squared for the four-fold table, is the sum of the squares of the differences between observed and expected cell frequencies, each divided by its expected cell frequency. While slightly less powerful than the Kolmogorov-Smirnov test—in part because some information is lost by grouping the data—chi-squared is easier to apply, can be used in cases where the data form natural discrete groups, and is more widely tabled.

Chi-squared can also be used to test whether two different samples of data come from the same or different distributions when the source distribution of the data is unspecified. The cell frequencies for each sample are obtained as above, with expected cell frequencies estimated proportional to the marginal frequencies. The chi-squared statistic is applied to the resulting $2 \times k$ contingency table. Under the null hypothesis that the two samples come from the same distribution, the test statistic has a chi-squared distribution with $k - 1$ degrees of freedom.

Problems involving $2 \times k$ tables also arise when comparing k proportions. To test the homogeneity of sample proportions $p_1, \cdots, p_k$, based on samples of size $n_1, \cdots, n_k$, respectively, a convenient and equivalent computing formula for chi-squared is

$$X^2 = \sum_{i=1}^{k} n_i \frac{(p_i - \bar{p})^2}{\bar{p}\bar{q}},$$

where $\bar{p}$ is the pooled (marginal) proportion, given by

$$\bar{p} = 1 - \bar{q} = \sum_{i=1}^{k} n_i p_i \Big/ \sum_{i=1}^{k} n_i,$$

which estimates the common true proportion under the null hypothesis. In general, whenever theory or circumstance suggest expected cell frequencies, hypotheses can be tested by comparing observed with expected data using chi-squared. For example, with two categorical variables, A (with r levels) and B (with c levels), the cross-classified data form an $r \times c$ table. The hypothesis of statistical independence between A and B can be tested using chi-squared, where the expected cell frequency in the ith row and j^{th} column is estimated as the product of the ith row margin and the jth column margin, divided by the grand total. Under the hypothesis of independence, the chi-squared statistic follows a chi-squared distribution with $(r - 1)(c - 1)$ degrees of freedom. Chi-squared is also used to test hypotheses about more than two variables, i.e., is used to test goodness of fit to various hypotheses in higher-dimensional contingency tables. The expected cell frequencies are usually estimated by the method of maximum likelihood (see Section 5.6).

In many $r \times c$ tables, the numbers in one or more cells are too small for the tabled chi-squared distribution to approximate accurately the distribution of the chi-squared test statistic. For tables larger than 2×2, it is recommended that chi-squared not be used if the expected cell frequency is less than 5 in more than 20% of the cells or less than 1 in more than 10% of the cells. (A special rule for fourfold tables is given in Section 5.2 at p. 157.) In such cases one strategy is to collapse some of the rows and/or columns to form a smaller table. No change in method is required when the collapsing is done on the basis of expected values determined on an a priori basis without reference to the sample data, or from the marginal frequencies of the table.

Another reason to collapse a table is to make subgroup comparisons, e.g., to study which portions of the $r \times c$ table are contributing most to a significant overall chi-squared. When the number and type of subgroup comparisons can be pre-specified, an adjustment is required simply to control for the inflation of Type I error that occurs when making multiple comparisons. Bonferroni's method (see Section 6.2) may be applied by multiplying the attained significance level by the pre-specified number of tables tested.

On the other hand, when regrouping by inspection of the innards of the table, i.e., when post-hoc comparisons are made, an adjustment is required not only for multiple comparisons, but also for the selection effect of testing hypotheses with the same data that were used to formulate those hypotheses. Here an appropriate method is to calculate the usual chi-squared statistic for the regrouped table (even to the level of possible four-fold tables), but then to assess the significance of the result with the chi-squared distribution on $(r-1)(c-1)$ degrees of freedom.

For example, a 3×3 table has nine proper four-fold sub-tables without summing, and nine possible four-fold tables obtained by summing two rows and columns. To assess simultaneously the significance of any of these 18 four-fold tables, under Bonferroni's method, multiply the attained level of significance by 18. Alternatively, ordinary chi-squared for a collapsed table is found significant only if it exceeds the upper $\alpha = 0.05$ critical value of the chi-squared distribution on $(3-1) \times (3-1) = 4$ degrees of freedom. In so doing, one will incorrectly reject the hypothesis of independence for any of these tables with probability no greater than α.

Further Reading

J. Fleiss, *Statistical Methods for Rates and Proportions*, ch. 9 (2d ed., 1981).

K. Gabriel, *Simultaneous test procedures for multiple comparisons on categorical data*, 61 J. Am. Stat. Assoc. 1081 (1966).

R. Miller, *Simultaneous Statistical Inference*, ch. 6, § 2 (2d ed., 1981).

6.1.1   Death-qualified jurors

In a death penalty proceeding there are usually two phases: the determination of guilt or innocence and, if the defendant is convicted of a potentially capital offense, the penalty phase at which the jury decides between life imprisonment and death.

Prior to the Supreme Court's decision in *Witherspoon v. Illinois*, 391 U.S. 510 (1968), the prosecution could challenge for cause all members of the venire who expressed any degree of opposition to capital punishment. In *Witherspoon*, defendant claimed that his jury was unconstitutionally biased toward conviction and death because all individuals with scruples against the death penalty had been excluded. The Court held that the jury was biased toward death, but that the evidence did not persuade it that the jury was also biased toward conviction. It restricted the group that could be constitutionally excluded from the guilt or innocence phase to those who either (1) could not vote for capital punishment in any case ("Witherspoon excludable") or (2) could not make an impartial decision as to a capital defendant's guilt or innocence ("nullifiers"). Those permitted to serve under this test are known as death-qualified jurors.

After *Witherspoon*, challenges were made that death-qualified juries were biased toward conviction compared with juries from which only nullifiers were excluded. It was urged that Witherspoon-excludables should be included at the guilt or innocence phase, and, if defendant were found guilty, a new jury impaneled excluding them for the penalty phase.

To explore these issues, social scientists studied whether Witherspoon-excludable jurors tended to vote differently from death-qualified jurors at the guilt or innocence phase. In one such study, Cowan, Thompson, and Ellsworth obtained a sample of adults eligible for jury service from venire lists, a newspaper advertisement, and by referral. Those who said they could not be fair and impartial in deciding guilt or innocence were excluded from the sample. The remainder were divided into two groups—those who said they would be unwilling to impose the death penalty in any case (Witherspoon-excludable) and those who said they would consider imposing it in some cases (death-qualified jurors).

The subjects were then shown a videotape reenactment of an actual homicide trial. After the viewing, they were assigned to juries, some with death-qualified jurors only and others with up to 4 Witherspoon-excludables. The jurors recorded a confidential predeliberation verdict and a second verdict after one hour's deliberation. The results are shown in Table 6.1.1.

Questions

1. What does the study show?

2. Is there a statistically significant difference in conviction behavior between Witherspoon-excludable and death-qualified jurors?

Source

C. Cowan, W. Thompson, and P. Ellsworth, *The Effects of Death Qualification on Jurors' Predisposition to Convict and on the Quality of Deliberation*, 8 Law & Hum. Behav. 53 (1984); see *Hovey v. People*, 28 Cal.3d 1 (1980).

TABLE 6.1.1. Verdict choices of death-qualified and excludable jurors

Pre–deliberation ballot	Death-qualified	Excludable
First degree murder	20	1
Second degree murder	55	7
Manslaughter	126	8
Not guilty	57	14
Total	258	30

Post-deliberation ballot	Death-qualified	Excludable
First degree murder	2	1
Second degree murder	34	4
Manslaughter	134	14
Not guilty	27	10
Total	197	29

Notes

In a challenge to an Arkansas conviction, the district court found the Cowan, Thompson, and Ellsworth study and other similar studies persuasive and granted relief.[1] A sharply divided Eighth Circuit affirmed, but the Supreme Court reversed. In *Lockhart* v. *McRee*, 476 U.S. 162 (1986), Justice Rehnquist, writing for the majority, swept away the social science. In his view, the studies were defective because the individuals were not actual jurors in an actual case. Moreover, the studies did not even attempt to predict whether the presence of one or more Witherspoon-excludables on a guilt-phase jury would have altered the outcome of the guilt-phase determination.

But even if the studies had been convincing, Rehnquist regarded the social science exercise as irrelevant. The fair cross-section requirement of the Eighth Amendment for jury venires does not apply to petit juries, which cannot constitute a cross-section. In any event, the cross-section requirement applies only to "distinctive groups," and individuals whose only tie is their opposition to the death penalty do not constitute such a group. Finally, Rehnquist held that the death qualification does not deprive defendant of the impartial jury required by due process because an impartial jury consists of nothing more than "jurors who will conscientiously apply the law and find the facts," and McCree did not even claim that his jury would not do that. Justices Brennan, Marshall, and Stevens dissented.

6.1.2   Spock jurors

Dr. Benjamin Spock, author of a famous book on baby care, and others were convicted of conspiracy to cause young men to resist the draft during the Vietnam War. The defendants appealed, citing, among other grounds, the sex composition of the jury panel. The jury itself had no women, but chance and peremptory challenges could have made that happen. Although the defendants might have claimed that the jury lists (from which the jurors are chosen) should contain 55% women, as in the general population, they did not. Instead they complained that six judges in the court averaged 29% women in their jury lists, whereas the seventh judge, before whom Spock was tried, had fewer, not just on this occasion, but systematically. The last 9 jury lists for that judge, in chronological order, contained the counts shown in Table 6.1.2. Spock's jury was selected from the panel with 9 women and 91 men.

Questions

1. Use chi-squared to determine whether the proportion of women on these juries is consistent with the hypothesis of a constant probability of accepting a woman juror.

---

[1]The combined significance of the six studies was very high. See Section 8.1.

TABLE 6.1.2. Sex composition of jury panels

#	Women	Men	Total	Proportion women
1	8	42	50	0.16
2	9	41	50	0.18
3	7	43	50	0.14
4	3	50	53	0.06
5	9	41	50	0.18
6	19	110	129	0.15
7	11	59	70	0.16
8	9	91	100	0.09
9	11	34	45	0.24
Total	86	511	597	0.144

2. If so, use the pooled data for p and calculate a 95% confidence interval for that estimate. Do your results support Spock's defense on this issue?

3. Consider three post-hoc groupings of jurors: the low juries (#'s 4 and 8), the medium juries (#'s 1-3 and 5-7), and the high jury (# 9). Using this sub-grouping, test the hypothesis of constant probability of accepting a woman juror.

Source

H. Zeisel, *Dr. Spock and the Case of the Vanishing Women Jurors*, 37 U. Chi. L. Rev. 1 (1969–70); F. Mosteller and K. Rourke, *Sturdy Statistics* 206–7 (1973). The Mosteller-Rourke book misstates the outcome: the court of appeals reversed the conviction on other grounds, without reaching the jury selection issue. *U.S. v. Spock*, 416 F.2d 165 (1st Cir. 1965).

6.1.3   Grand jury selection revisited

In Orleans Parish, Louisiana, nine grand juries were selected between September 1958 and September 1962, when a grand jury was impaneled that indicted a black suspect. Each grand jury had twelve members. There were two blacks on eight of the juries and one on the ninth. Defendant contended that the number of blacks on a jury was limited by design. Blacks constituted one-third of the adult population of the Parish, but the Supreme Court of Louisiana held that the low level of black representation was due to lower literacy rates and requests for excuse based on economic hardship.

Questions

1. Test whether the lack of variation in numbers of blacks is inconsistent with a binomial selection process by using the chi-squared test of homogeneity for the observed and expected numbers of blacks on each of the juries (a 2×9 table). The test, unconventionally, requires referring the statistic to the lower tail of chi-squared. (An alternative test uses the multiple hypergeometric model, with column totals of 12 and row totals of 17 and 108, for blacks and whites, respectively.)

2. Using the binomial model, find the expected number of juries with less than 2 blacks, exactly 2 blacks, and more than 2 blacks, this collapsing being used to increase the expected numbers in each cell. Use the chi-squared goodness-of-fit statistic to compare the observed and expected number of juries in each category. What do you conclude?

Source

State v. Barksdale, 247 La. 198, 170 So.2d 374 (1964), *cert. denied*, 382 U.S. 921 (1965).

6.1.4   Howland Will contest

Hetty Howland Robinson claimed the estate of her aunt, Sylvia Ann Howland, who died in 1865. The claim was based on an alleged secret agreement between aunt and niece to leave the niece the estate. The only written evidence of the agreement was a previously undisclosed and separate "second page" inserted into a will of the aunt's. Robinson claimed that, at her aunt's direction, she wrote both copies of the second page; that her aunt signed the first copy on the morning of January 11, 1862; signed the second copy after tea on the same day; and signed the will itself that evening. A subsequent will, made in September 1863, gave Robinson only the income from the estate.

The executor rejected Robinson's claim, asserting that the signatures on the two second pages were forgeries, having been traced from the signature on the will itself. The case became a *cause célèbre*. Oliver Wendell Holmes, Sr., appeared as a witness for the executor, as did Harvard Professor Benjamin Peirce, probably the most noted mathematician of his day. Professor Peirce, assisted by his son, Charles Sanders Peirce, undertook to demonstrate by statistical means that the disputed signatures were forgeries. Their method involved counting the number of downstrokes (30 in all) that "coincided" in the authentic, and one of the disputed, signatures (the other disputed signature was not mentioned) and comparing those coincidences with coincidences in other pairs of admittedly genuine signatures of Sylvia Ann Howland. According to Charles Sanders Peirce's testimony, downstrokes were considered to coincide if "in shifting one photograph over another in order to make as many lines as possible coincide—that the line of writing should not be materially changed. By materially changed, I mean so much changed that there could be no question that there was a difference in the general direction of the two signatures." The undisputed signature on the will and one of the disputed signatures coin-

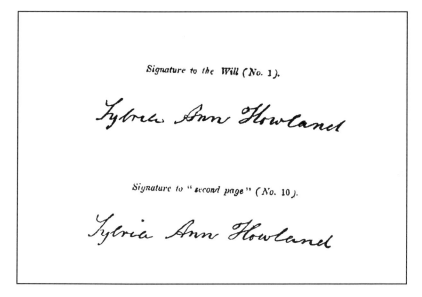

FIGURE 6.1.4. Undisputed and challenged signatures of Sylvia Ann Howland

cided with respect to all 30 downstrokes (see Figure 6.1.4). Professor Peirce compared the number of coincidences in the thirty downstrokes in every possible pairing of 42 undisputed signatures of Sylvia Ann Howland taken from various other documents. At trial he disclosed the data of Table 6.1.4 from the study. (Note that only the aggregate number of pairs and aggregate number of coinciding strokes were disclosed for pairs with between 13 and 30 coincidences.)

Assume, as appears probable, that the pair of authentic and disputed signatures was the only pair for which all 30 downstrokes coincided. Since there were 861 pairs and 30 downstrokes in each, there were 25,830 comparisons, of which 5,325, or slightly over 1 in 5, coincided. On this basis, Benjamin Peirce assumed that the probability of a coincidence was approximately 1/5 and testified, using a binomial model, that the probability of finding 30 coincidences in one pair of signatures was "once in 2,666 millions of millions of millions." (Note: his conclusion would appear to be misstated. The probability of 30 independent matches is $(5325/25, 830)^{30} = 2.666 \times 10^{-21}$, which is equal to once in 3.75×10^{20} pairs.)

Questions

1. Assuming that the 861 pairs are independent observations, use chi-squared to determine whether the data are consistent with Peirce's binomial model.

2. What reasons would you have for doubting the independence assumption?

3. What analysis of these data might be less open to dispute?

TABLE 6.1.4. Coincidence of 30 downstrokes in 861 signature pairs

Coincidences per pair	Number of pairs	Number of strokes coinciding
0	0	0
1	0	0
2	15	30
3	97	291
4	131	524
5	147	735
6	143	858
7	99	693
8	88	704
9	55	495
10	34	340
11	17	187
12	15	180
13		
⋮	20	288
30		
	861	5325

Source

The Howland Will Case, 4 Am. L. Rev. 625 (1870), *discussing Robinson* v. *Mandell*, 20 Fed. Cas. 1027 (No. 11959) (C.C.D. Mass. 1868). This case appears to be the first in which probabilities were computed in a legal proceeding. See P. Meier and S. Zabell, *Benjamin Peirce and the Howland Will*, 75 J. Am. Stat. Assoc. 497 (1980).

Notes

The Howland Will case is unusual in two respects: Professor Peirce was called to demonstrate that the differences between the questioned and exemplar signatures were too few (in fact, none) to be consistent with the natural variation of the subject's handwriting; in addition, to support his opinion, he presented a quantitative measure (albeit flawed) for the probability of the absence of such differences. In the more usual modern case, a forensic handwriting expert is called to testify that the differences between the questioned and exemplar handwritings are *too numerous* to be consistent with natural variation, and no quantitative measure of probability is given. Lack of quantification in the field of handwriting expertise is a factor that led a trial court, in a notable case, to hold that proposed testimony by a forensic document expert was not sufficiently scientific to be admitted, as such; the expert would be allowed to testify only as a person with "specialized knowledge," and the jury cautioned appropriately. *United States v. Starecpyzel*, 880 F. Supp. 1027 (S.D.N.Y. 1995) (slanting capital letter "E" in two questioned signatures found in only 5 out of 224 exemplars).

6.1.5   Imanishi-Kari's case

In 1986, Dr. Imanishi-Kari, the head of a serological research laboratory at MIT, was accused of fraud by a graduate student working in her laboratory. The student asserted that Imanishi-Kari had fabricated data behind a scientific paper that was published in the April 25, 1986, issue of the journal *Cell*. The paper was the fruit of collaborative research at Imanishi-Kari's laboratory and a Boston molecular biology laboratory headed by Dr. David Baltimore. A prominent researcher, Baltimore had won a Nobel Prize and had become president of Rockefeller University. The finding of the *Cell* paper was that the introduction of a gene from one strain of mouse to another changed the antibody repertoire in the recipient mouse. The new antibodies were found to carry certain characteristics of the donor mouse and certain characteristics of the recipient mouse.

An NIH scientific panel investigated, found some mistakes in the paper, which were corrected in subsequent publications, but cleared Imanishi-Kari of scientific misconduct. However, the case then became notorious when Representative John Dingell's subcommittee on Oversight and Investigations subpoenaed laboratory records and held hearings on the charges in 1988, 1989, and 1990. NIH's Office of Research Integrity (ORI) then reopened the investigation and in 1994 issued a 231-page report finding that Imanishi-Kari had fabricated data behind the paper. Dr. Baltimore was not accused of fraud, but was forced to resign from Rockefeller University because he had defended Imanishi-Kari.

The allegedly fabricated data were counts per minute (cpm) of gamma radiation by a gamma counter. To support the fraud charge, ORI compared the uniformity of distribution of certain "insignificant digits" in the questioned handwritten data and in the unquestioned gamma counter tapes. The theory was that digits that carried no information relating to the outcome of the ex-

TABLE 6.1.5. Frequencies of insignificant cpm digits in questioned and control data

Digit	Questioned cpm data digit frequencies	Control cpm data digit frequencies
1	0	7
2	14	6
3	2	3
4	3	10
5	4	3
6	4	2
7	6	7
8	11	7
9	3	2

periment could be expected to be uniformly distributed between 1 and 9 in an authentic data set, whereas fabricated data would tend to show nonuniformities due to the fabricator's personal and unconscious digit preferences. To obtain insignificant digits, ORI started at the right of each cpm number and moved left until reaching the first non-zero digit. That digit was then counted if it was not the left-most digit; if it was, the number was discarded. Thus, the 7 in a cpm of 7,000 would be considered significant and would not be counted, but the 8 in 28,000 would be considered insignificant and would be counted. There were always initial right-hand zeros because Imanishi-Kari rounded the cpms in the handwritten data, usually to the nearest 10. To put control group cpms on a similar basis, ORI also rounded before counting insignificant digits. This was done by rounding to the nearest 1,000 for cpms over 10,000; rounding to the nearest 100 for cpms less than 10,000 but over 1,000; rounding to the nearest 10 for cpms less than 1,000 but over 100; and discarding cpms less than 100.

Table 6.1.5 shows the digit frequencies in 47 cpms of questioned handwritten data and a similar number from one of the control data sets.

Question

Do the data support ORI's argument?

Source

Thereza Imanishi-Kari, Docket No. A-95-33, Decision No. 1582 (June 21, 1996), Dep't of Health and Human Services, Departmental Appeals Board, Research Integrity Adjudications Panel. See, in particular, Appendix B of the ORI initial decision at B-22 to B-33. The questioned cpm data are the high counts (cpm of at least 600) from page 1:121 of Imanishi-Kari's laboratory notebook; the control cpm data are the first 47 non-discarded cpms from the high counts on page 1: 102-104. On people's inability to fabricate random digits and the use of that fact in detecting fraud, see James Mosimann, Claire Wiseman, and Ruth Edelman, *Data Fabrication: Can People Generate Random Digits?*, 4 Accountability in Research 31 (1995).

Notes

The appeals panel found that ORI had not proved its charges of fraud by a preponderance of the evidence and reversed the finding of scientific misconduct. In particular, the panel found the uniformity analysis flawed because: (i) prior studies supporting uniformity analysis had used it only to demonstrate some form of human intervention and it was undisputed in this case that Imanishi-Kari had personally rounded the data; (ii) studies demonstrating uniformity of digit frequencies were only of the right-most digit (sometimes called the "terminal" or "error" digit), but ORI's study was of the right-most, non-zero digit that was not the leftmost digit and there was no authority that analysis of such digits was a commonly accepted statistical technique or that such digits would not have some information that would make them significant; (iii) the use of the control groups was suspect because the handwritten data were rounded and it did not appear that ORI's rounding protocol for the control group mimicked that used by Imanishi-Kari. Finally, the panel concluded that, even if the analysis were valid, it could not infer from a lack of uniformity alone that the data had been fabricated. Panel Op. 107-109.

6.2   Bonferroni's inequality and multiple comparisons

An important application of Bonferroni's inequality (see Section 3.1 at p. 60) lies in correcting the nominal error rate in multiple or simultaneous tests of significance. Suppose one is testing k hypotheses, not necessarily statistically independent. For example, in comparing three treatments—T_1, T_2, and T_3—we may wish to make three paired comparisons: T_1 vs. T_2, T_1 vs. T_3, and T_2 vs. T_3. The same data will be used for T_1 in each of the first two comparisons, etc. If each significance test were conducted at the nominal $\alpha = 0.05$ level, the chance of making *one or more* Type I errors is more than 5%, since there are three opportunities to err. However, the precise overall Type I error rate is often difficult to calculate because of correlations between tests. The Bonferroni inequality gives a simple solution to the problem of controlling the overall Type I error rate: for k comparisons (in the example $k = 3$), conduct each test at the nominal $\alpha = 0.05/k$ level. For if E_i denotes the event that a Type I error is made in the ith comparison, then

$$P[\text{ at least one Type I error occurs}]$$
$$= P[E_1 \text{ or } \cdots \text{ or } E_k] \leq P[E_1] + \cdots + P[E_k]$$
$$= \frac{0.05}{k} + \cdots + \frac{0.05}{k} = 0.05.$$

Thus, the overall Type I error rate is no larger than 0.05. The method is slightly conservative, in that the true error rate will be somewhat less than 0.05, but in practice the conservatism is negligible.

The second Bonferroni inequality can provide a *lower* bound to the overall Type I error if the pairwise occurrence of two Type I errors can be calculated, for then

$$P[E_1 \text{ or } \cdots \text{ or } E_k] \geq P[E_1] + \cdots + P[E_k]$$
$$- (P[E_1 \text{ and } E_2] + P[E_1 \text{ and } E_3] + \cdots + P[E_{k-1} \text{ and } E_k]).[2]$$

In practice, if k is not too large and the correlation between tests is not too strong, the terms $P[E_i \text{ and } E_j]$ will not be too far from the value they

---

[2]The possibility of errors in three or more tests also exists, but is considered negligible.

would have had if the tests had been independent, namely $(0.05/k)^2$. Since there are $\binom{k}{2} = k(k-1)/2$ of them, the correction terms to the first Bonferroni inequality are not far from $\frac{1}{2}k(k-1) \cdot (0.05/k)^2 = 0.00125 \cdot (k-1)/k$, which is fairly small. Thus $\alpha = 0.05/k$ usually provides a quick and simple control for multiple comparisons.

In particular, when comparing rates from each of k groups to a reference group, we have two possible procedures: (i) test for homogeneity of all rates (including the reference group) using chi-squared (see Section 6.1) or (ii) compare each group with the reference group and reject as significant any z-score statistic at or beyond the nominal $\alpha = 0.05/k$ level. Often, especially when k is small, the latter procedure reaches a higher level of statistical significance than the chi-squared procedure. The reason is that chi-squared is an "omnibus" test (i.e., sensitive to a wide class of alternatives), whereas pairwise comparisons are particularly powerful when the reference group is markedly different from at least one other group. On the other hand, chi-squared will be more powerful when there are diffuse but moderate differences among several groups, because each cell contributes to the chi-squared sum.

While Bonferroni's inequality is the best known method for adjusting for multiple comparisons, other procedures are more powerful, i.e., are more likely to lead to a rejection of the null hypothesis if it is false for at least one of the comparisons. The so-called Holm step-down procedure is one example. In this method, the comparisons are ranked from most to least significant. If there are k such comparisons, the first one is tested at the level α/k, as in Bonferroni's procedure; if the difference is not significant, i.e., the null hypothesis is not rejected, the procedure stops and no comparisons are significant. If the first comparison is significant, the second is tested at $\alpha/(k-1)$. If the difference is not significant, the procedure stops and only the first comparison is deemed significant. If the second comparison is significant the procedure continues by testing the next comparison at level $\alpha/(k-2)$, etc. The final criterion is simply α.

Because Holm's procedure uses a level of significance for each comparison that is equal to or larger than the level α/k used in Bonferroni's procedure, it always rejects at least as many comparisons as Bonferroni's procedure, and is therefore more powerful. However, Type I error is still bounded by α. To see this, consider the case in which the null hypothesis is true for all comparisons. If Holm's procedure is followed, what is the probability that the null hypothesis will be rejected for at least one comparison? For that to occur, the first comparison would have to be rejected, because if it is not rejected the procedure stops and no comparisons are rejected. The significance level associated with the first comparison is α/k. Thus, Type I error will occur only if the p-value for one or more of the k comparisons is $\leq \alpha/k$. The probability under the null hypothesis that a particular P-value will be $\leq \alpha/k$ is by definition α/k. By Bonferroni's inequality, the probability that one or more such events will occur is $k(\alpha/k) = \alpha$. The Holm procedure therefore bounds Type I error at $\leq \alpha$.

The attained level of significance or "Holmed" P-value for the ith comparison that preserves α as overall Type I error is determined as follows. For the first i comparisons on the list, multiply each nominal P-value by $k - i + 1$. The maximum of these values, or 1, whichever is less, is the "Holmed" P-value for the ith comparison. The Holm procedure rejects only those null hypotheses for which the Holmed P-value is $\leq \alpha$.

Further Reading

R. Miller, *Simultaneous Statistical Inference*, ch. 1, §2 (2d ed. 1981).

S. Holm, *A simple sequentially rejective multiple test procedure*, 6 Scand. J. Stat. 65 (1979).

M. Aickin and H. Gensler, *Adjusting for multiple testing when reporting research results: the Bonferroni vs. Holm methods*, 86 Am. J. Pub. Health 726 (1996).

B. Levin, *On the Holm, Simes, and Hochberg multiple test procedures, id.* at 628.

6.2.1   Wage additives and the four-fifths rule

Illinois Bell had a system of "wage additives" by which employees with certain experience or education were eligible at hire to receive salary additives that would raise their salaries one or more steps above entry level. The EEOC contended that the administration of this program with respect to the additive for nontechnical education discriminated in favor of white males and against other race/sex groups. See Section 1.5.1 for the text of the EEOC's four-fifths rule on this subject. According to an EEOC Q & A on the rule, the comparisons are to be made between males and females, and among blacks, Native Americans, Asians, Hispanics, and non-Hispanic whites; there is no obligation to compare racial groups within sexes (i.e., white males with black females, etc.) (EEOC, 8 F.E.P. 401:2306 *Q's & A's to Clarify and Provide a Common Interpretation of the Uniform Guidelines on Employer Selection Procedures*, Q & A No. 17 (1979)).

TABLE 6.2.1. Non-technical education additives

Category	Number of employees	Employees with additives
White male	143	20
Black male	38	6
Hispanic male	7	1
Asian and other male	8	0
White female	103	5
Black female	59	4
Hispanic female	38	2
Asian and other female	13	2
Total	409	40

The EEOC director involved in the case tested males and females in the aggregate and found a violation and (despite the regulation) tested each race/sex group against white males and found violations. The data are shown in Table 6.2.1.

Question

In his survey of all groups, the EEOC director found the difference in rates of wage additives between white females and white males to be statistically significant at the 5% level by the two-sample z-score test. As the attorney for Illinois Bell, what objections would you make to this calculation and what alternative calculations would you offer?

Source

K.F. Wollenberg, *The Four-Fifths Rule-of-Thumb: An EEO Problem For OTC Statisticians* (Unpubl. Memo presented at a Bell System Seminar on Statistics in EEO Litigation, Oakbrook, Illinois, October 10, 1979).

6.2.2   Discretionary parole

Native American and Mexican American inmates of the Nebraska Penal and Correctional Complex sued under 42 U.S.C. 1983 (1981), claiming that the five-member Nebraska Board of Parole denied discretionary parole to them on the grounds of race and ethnicity. The data for 1972 and 1973 are shown in Table 6.2.2.

On these data, the Eighth Circuit Court of Appeals affirmed the denial of relief. The court noted that for Native Americans the difference between the expected (35) and observed (24) number of discretionary paroles was only 11, and for Mexican Americans the difference between expected (11) and observed (5) was only 6. In each case, the court concluded that the difference represented less than two standard deviations. It computed the standard deviation by taking the square root of the product of the total number of individuals who received discretionary paroles (535), the probability of randomly selecting a Native American from the eligible inmate population (0.065), and the probability of randomly selecting someone other than a Native American (0.935). (*Id.* at 1377, n. 19). The result was 5.7. A similar computation for Mexican Americans

TABLE 6.2.2. Parole by race and ethnicity

	White	Black	Native American	Mexican American	Total
Received discretionary parole	358	148	24	5	535
Eligible for release by discretionary parole	590	235	59	18	902

yielded 3.2. Since in each case there was a departure of less than two standard deviations from expectation, the court concluded that the Supreme Court's rule in *Hazelwood* and *Castaneda* was not violated (*Id.* at 1377–78), and the statistics were "not quite sufficient" for a prima facie case. *Id.* at 1379.

Questions

1. Use chi-squared to test the null hypothesis of no association between race or ethnicity and parole against omnibus alternatives of arbitrary differences between groups.

2. What probability model underlies the court's computation?

3. If Native Americans are to be compared with all others, as the court does, what other model might be more appropriate and what are the results?

4. In using chi-squared for 2×2 tables to test significance, apply two different methods to adjust for multiple comparisons.

Source

Inmates of the Nebraska Penal and Correctional Complex v. Greenholtz, 567 F.2d 1368 (8th Cir. 1977).

6.2.3   Cheating on standardized multiple-choice tests revisited

In the police examination for promotion to sergeant described in Section 3.5.2, departmental proceedings were brought against the suspected officers. The court held that the anonymous tip letters would not be admitted in evidence. A psychometrician, testifying for the defense, argued that because the letters had been excluded, and in any event did not identify all of the suspected officers, the probabilities of matching wrong answers had to be computed on a screening basis. By that he meant that the prior identifications had to be ignored and all possible pairs of officers among the 12,570 taking the test (approximately 155 million) had to be examined to see whether the suspected officers had a statistically significant number of matching answers (in the expert's method all matching answers, both right and wrong, were counted). Statistical significance was to be measured using an alpha level of 0.001, adjusted for multiple comparisons by the Bonferroni method. With that adjustment, significance required a departure of at least seven standard errors above expectation. When this screening was done there were 48 suspicious pairs, but none of the suspected officers was included in those pairs.

The psychometrician also argued that the department's expert had overlooked the problem of multiple comparisons in his analysis. He claimed that there were 16 comparisons based on the fact that there were 4 groups of officers, two sessions (morning and afternoon), and 2 methods of analysis (Modified Angoff and Item Specific). When adjustment was made, the results lost statistical significance.

Questions

1. Is a screening appraoch to adjustment appropriate in this case?

2. Assuming a screening approach is not required, what is the number of comparisons for which adjustment should be made in determining statistical significance?

6.3   More measures of association: Phi-squared (ϕ^2) and tau B (τ_B)

The chi-squared statistic is not a good measure of association within a table because its value depends on the sample size. As stated in Section 5.2, if each cell in a 2×2 table is multiplied by 2, the association within the table does not change but chi-squared is increased by a factor of 2. A measure of association based on chi-squared that avoids this defect is phi-squared (also called the mean squared contingency), which is defined as chi-squared (without correction for continuity) divided by the sample size. Whereas the range of chi-squared is unbounded, phi-squared runs from 0, when attributes are independent, to 1, when two diagonal cells in the 2×2 table have positive values and the other two are zero (i.e., perfect correlation). Unfortunately, this characteristic does not carry over to larger tables. For the general $r \times c$ table, phi-squared ranges between zero and the smaller of $r - 1$ and $c - 1$, with the upper limit only attainable when $r = c$. See generally H. Blalock, *Social Statistics* 212–21, 228–29 (1960). A leading authority, Harald Cramér, has suggested that phi-squared be standardized by dividing by the smaller of $r - 1$ and $c - 1$. H. Cramér, *Mathematical Methods of Statistics* 282 (1946).

The positive square root of phi-squared is known as the "phi coefficient." Erratic phi coefficients for ten job groups showing the correlation between a battery of test scores and an average of two supervisory rankings were cited by the U.S. Supreme Court in *Albemarle Paper Co. v. Moody*, 422 U.S. 405 (1975), to buttress its conclusion that a test battery with a discriminatory impact was not acceptable because it was not " 'predictive of or significantly correlated with important elements of work behavior.... ' " *Id.* at 431.

In a 2×2 table phi-squared is equal to the square of Pearson's correlation coefficient (see Section 1.4) and this may help in its interpretation. In the *Teal* data (see p. 160), phi-squared is 0.0460 = 14.12 (chi-squared without the correction for continuity) divided by 307 (the sample size). The phi coefficient is thus $\sqrt{0.0460}$ = 0.214. This association appears low compared with the simple arithmetic difference in pass rates (25.4 percentage points). It also seems lower than the relative risk for pass rates of 0.68 (well below 80%) and the odds ratio of 0.30. No one of these measures is right in any absolute sense; they measure different aspects of the data.

Tau B is another measure of association between attributes on two sides of a contingency table, being one of a number of so-called "error reduction" measures. It is defined as the percentage reduction in the expected number of errors in classification into categories on one side of the contingency table that results from knowing the classification on the other side. For example, in Section 1.5.2, Bail and Bench Warrants, if the correlation between the previous issuance of a warrant and the current issuance of a warrant were perfect, knowledge of prior issuance would reduce classification errors to zero, since we would know perfectly from the issuance which defendants would fail to appear. Conversely, if prior issuance were unrelated to the need for a current warrant, knowledge of prior issuance would not help us at all and there would be no reduction in classification errors. In the more usual middle case, the computation proceeds as follows.

First, calculate the expected number of errors that would occur if we randomly allocated the 293 defendants into a group of 146 who currently failed to appear and 147 who appeared.[3] Since $147/293 = 0.502$ of the defendants appeared, we would expect this fraction of the 146 allocated to the non-appearance category to be incorrectly placed (73.25). Likewise, we would expect $146/293 = 0.498$ of the defendants allocated to the appearance group to be incorrectly placed (73.25). (These numbers are always equal.) Thus, random placement would yield a total of 146.5 errors.

Next, calculate the number of errors that would be expected if we knew whether each defendant had a bench warrant previously issued. By this logic, we expect that, of the 54 cases allocated to the non-appearance/prior warrant cell, 23/77 of them, or 16.13 cases, would be misallocated because they would appear. An equal number of misallocations (16.13 cases) would be made with respect to the appearance/prior warrant cell. The corresponding error figures for the defendants for whom no bench warrant had been previously issued are each 52.81. Therefore, by knowing whether a defendant had a bench warrant previously issued, we have reduced the expected number of errors to $16.13 + 16.13 + 52.81 + 52.81 = 137.88$.

The τ_B statistic is defined as the percentage decrease in the expected number of classification errors due to knowledge of the conditioning factor. In our example this is $(146.50 - 137.88)/146.50 = 0.059$, or 5.9%. By this measure, the reduction in errors does not appear substantial.

---

[3]These numbers are not reflective of the rate at which bench warrants were issued, but rather of the sampling design of the retrospective study. The τ_B measure (like the odds ratio) yields the same result for retrospective and prospective data in 2×2 tables, but (unlike the odds ratio) not in larger tables.

For ease of computation for 2×2 tables, τ_B may be obtained from the formula, referring to the schematic table in Section 5.2 at p. 157:

$$\tau_B = 1 - N \cdot \frac{abcd}{m_1 m_2 n_1 n_2} \left(a^{-1} + b^{-1} + c^{-1} + d^{-1}\right).$$

See generally Blalock at 232–34. This can be shown to be identical to phi-squared. For more general $r \times c$ tables with cell proportions denoted by p_{ij}, the formula becomes

$$\tau_B = \sum_i \sum_j \frac{(p_{ij} - p_{i.} \cdot p_{.j})^2}{p_{i.}} \bigg/ \left(1 - \sum_j p_{.j}^2\right),$$

where $p_{i.}$ is the marginal proportion in the ith row (the conditioning factor) and $p_{.j}$ is the marginal proportion in the jth column.

Further Reading

L. Goodman and W. Kruskal, *Measures of Association for Cross-Classifications* (1979).

6.3.1   Preventive detention

The Comprehensive Crime Control Act of 1984 authorizes denial of bail to defendants who pose a danger to "the safety of any other person and the community." 18 U.S.C. §3142(e) (1999). Among the factors that the judicial officer "shall take into account" in denying bail are "the history and characteristics of the person, including his character, physical and mental condition, family ties, employment, financial resources, length of residence in the community, community ties, past conduct, history relating to drug or alcohol abuse, criminal history, and record concerning appearances at court hearings," as well as "the nature and seriousness of the danger to any person or the community that would be posed by the person's release."

A study of 1500 defendants on pre-trial release in New York State found that four factors distinguished defendants who were rearrested before trial, or who failed to appear at trial, from those who "succeeded" on bail or personal recognizance. The factors were: 1) number of prior violent felony arrests within the previous five years; 2) number of non-felony arrests in same period; 3) months in present employment; and 4) years of education. Defendants were classified into low, medium and high risk groups based on these factors. The results were shown in Table 6.3.1.

Questions

1. Compute and interpret τ_B for these data using the "*A* or *B*" column for the outcome.

TABLE 6.3.1. Defendants' failure to appear and/or rearrest by risk level

Risk level	Percent of defendants	Failure rates (%)		
		Failed to appear (A)	Rearrested (B)	A or B
Low	61	7	12	18
Medium	20	11	16	28
High	19	16	28	37
Total	100	10	15	24

2. What is the bearing, if any, of the result on the constitutionality of the federal preventive detention statute?

Source

J. Monahan and L. Walker, *Social Science in Law* 171–72 (1985), reprinting data from Center for Governmental Research, *Final Report: An Empirical and Policy Examination of the Future of Pretrial Release Services in New York State* (1983).

Notes

The U.S. Supreme Court upheld the constitutionality of the preventive detention statute in *U.S. v. Salerno*. 481 U.S. 739 (1987). Writing for the majority, Chief Justice Rehnquist held that (i) there was no violation of substantive due process because the incarceration was not punitive, and (ii) there was no unconstitutional denial of bail because the Bail Clause does not prohibit a complete denial of bail.

The present statute is not the first preventive detention law. Under a 1970 District of Columbia law, any defendant charged with a "dangerous" or "violent" crime could be held for a pre-trial detention hearing at which a judicial officer would determine whether the defendant could be released without danger to the community; if not, he could be detained for up to 60 days. With this act as a framework, a team of Harvard law students studied pre-trial crimes committed by defendants arrested in Boston. Of 427 defendants who would have been eligible for preventive detention under the D.C. law, 62 (14.5%) were rearrested; of these, 33 (7.7%) were arrested for "violent" or "dangerous" crimes, and 22 (5.2%) were convicted of such crimes.

To determine how well these crimes could have been predicted on the basis of factors specified in the statute, investigators collected 26 items of personal data for each defendant to give substance to the general criteria of the statute. They constructed two "dangerousness scales" for each defendant, the first by combining the scores for each factor and weighting the individual items intuitively, the second by weighting each factor by weights designed to maximize predictive power of the sum. In each scale, the higher the score, the greater the probability of recidivism (in the sense of being convicted for an offense while on bail).

The results under both scales showed that whatever the cutoff point, more nonrecidivists would be detained than recidivists. In the intuitive scale, at the "best" cutoff point (30), some 70 (16.4%) of the sample defendants would be detained, of whom 18 would be recidivists and 52 would not.

The second method used linear discriminant analysis, an objective method by which a computer assigns weights to the different factors to optimize the association between scores and recidivism. The method is similar to the least squares estimation of multiple regression. See Section 13.2. The scores ranged from 52.13 to 13.5. However, even with optimal weighting, the best cutoff included more nonrecidivists than recidivists. At a 35 cutoff, a total of 63 (14.8%) defendants would have been detained; of these only 26 were recidivists. See Note, *Preventive Detention: An Empirical Analysis*, 6 Harv. Civ. Rights–Civ. Lib. L. Rev. 289 (1971).

A method of statistical prediction with important substantive consequences had been used in connection with federal parole. (Federal parole and the United States Parole Commission were abolished in 1986 by the Comprehensive Crime Control Act of 1984.) Under 18 U.S.C. 4206(d) (1982), a prisoner could not be released if there was a "reasonable probability" that he would commit any crime. To assess that probability, the U.S. Parole Commission adopted a statistical measure called a Salient Factor Score, which categorized inmates eligible for parole consideration on predictive factors. Various sets of such factors were tried from time to time. The set last used had seven factors: prior convictions; prior commitments; age at current offense; commitment within three years of current offense; probation, parole, confinement, or escape status at time of current offense; and history of heroin/opiate dependence. For each group of scores a guideline period of incarceration was specified, depending on the severity of the offense (grouped in eight categories). Departures from the guidelines were permitted, but reasons had to be given.

The accuracy of the Salient Factor Score was not impressive. A 1982 study indicated that the highest risk group (lowest Salient Factor Scores) had a failure rate of only 49%, while the lowest risk group (highest Salient Factor Scores) still had a failure rate of 12%.[4]

---

[4]Several courts suggested that strict adherence to the Parole Commission guidelines would violate a constitutional or statutory requirement of individualized determination. See, e.g., *Geraghty v. United States Parole Comm'n*, 579 F.2d 238, 259–63 (3rd Cir. 1978), *rev'd on other grounds*, 445 U.S. 388 (1980); *United States v. Cruz*, 544 F.2d 1162, 1164 (2d Cir. 1976); *United States v. Norcome*, 375 F. Supp. 270, 274, n.3 (D.C.), *aff'd*, 497 F.2d 686 (D.C. Cir. 1974). These courts rejected only exclusive reliance on the guidelines; as one court put it, the guidelines could be used "as a tool but not as a rule." *Page v. United States*, 428 F. Supp. 1007, 1009 (S.D. Fla. 1977). But other courts held that there was no inconsistency between individualized sentencing and strict adherence to parole guidelines, because each defendant was individually evaluated to obtain the guideline score. See, e.g., *Daniels v. United States Parole Comm'n*, 415 F. Supp. 990 (W.D. Okla. 1976).

Clinical and statistical predictions of future violence have been offered in sentencing proceedings involving capital punishment. In an important case, the American Psychiatric Association informed the Supreme Court, in an amicus brief, that "the unreliability of psychiatric predictions of long-term future dangerousness is by now an established fact within the profession." The APA's best estimate was that two out of three predictions of long-term future violence made by psychiatrists were wrong. The Court accepted this estimate, but held that the opinion of a psychiatrist (who had not examined the defendant) that the defendant would commit a future violent act was nevertheless admissible in a death sentencing proceeding, on the ground that cross-examination would ferret out its weaknesses. *Barefoot* v. *Estelle*, 463 U.S. 880 (1983).

In *State* v. *Davis*, 96 N.J. 611, 477 A.2d 308 (1984), defendant at a penalty trial, after a plea of guilty to murder, offered to prove by statistical data through an expert sociologist (who had never met or evaluated him) that based on "demographic features" the probability of recidivism should he be released at age 57 (after a 30-year sentence) would be extremely low. The principal demographic facts were the low rates of criminality among white males aged 57 and among first-degree murderers. Based on these factors, the expert opined that "Mr. Davis would never again commit another serious crime of any kind."

The trial court refused to allow this testimony, but on appeal the Supreme Court of New Jersey reversed, citing studies showing that age was a strong predictor of future criminal behavior, and that "clinical" predictions by psychiatrists and psychologists of future violence were accurate in no more than one case in three. The court concluded that the evidence "generally satisfied broad standards of relevancy" and was therefore admissible as a mitigating factor. The court also held that the competency of the evidence was to be determined "without strict adherence to the standards governing competency of expert testimony otherwise applicable in the guilt phase of a criminal case."

For a general discussion of the legal status of individualized ("clinical") versus statistical prediction, and some of the dilemmas inherent in this sort of prediction, see B. Underwood, *Law and the Crystal Ball: Predicting Behavior with Statistical Inference and Individualized Judgment*, 88 Yale L.J. 1408 (1979).

CHAPTER 7

Comparing Means

7.1 Student's t-test: Hypothesis testing and confidence intervals

We have seen that a hypothesis test concerning the mean of a normal population can be carried out by standardizing the sample mean, i.e., by subtracting the population mean specified by the null hypothesis and dividing by the standard error of the mean, $\sigma/\sqrt{n}$. The resulting statistic is a z-score, and may be referred to the standard normal distribution for tail area probabilities. In large samples we have also used a consistent estimate for σ when it is unknown. For example, in the two-sample binomial problem, the unknown standard error of the difference between proportions under the null hypothesis $p_1 = p_2$, $[p(1-p)(n_1^{-1}+n_2^{-1})]^{1/2}$, is estimated by $[\hat{p}(1-\hat{p})(n_1^{-1}+n_2^{-1})]^{1/2}$, where $\hat{p}$ is the pooled proportion $\hat{p} = (n_1\hat{p}_1+n_2\hat{p}_2)/(n_1+n_2)$. Use of the standard normal distribution in this case is justified as an approximation in large samples by the central limit theorem.

When the standard error of the mean is unknown in normal populations, and the sample size is small to moderate, say, less than 30, then estimation of the standard error introduces additional and nonnegligible variability in the test statistic's sampling distribution. In 1908, in a paper entitled "The Probable Error of a Mean," written under the pen name of Student while employed at the Guinness brewery in Dublin, William Sealy Gosset (1876-1937) correctly guessed the mathematical form of the sampling distribution for the statistic,

$$t = \frac{\bar{X} - \mu}{s/\sqrt{n}}.$$

Here $\bar{X}$ is the sample mean of n observations from a normal population with mean μ and variance σ^2, and $s^2 = (n-1)^{-1} \cdot \sum(X_i - \bar{X})^2$ is the usual unbiased estimate of σ^2. Student checked his work with one of the earliest examples of the Monte Carlo method; he was later proved correct by R. A. Fisher. The test statistic and its distribution now bear Student's name in his honor.

The Student *t*-distribution depends on the sample size through the degrees of freedom parameter, which equals $n-1$ for a single sample of size n. One degree of freedom is lost because, in estimating the standard deviation, one parameter, μ, has to be estimated (by $\bar{X}$). More generally, the degrees-of-freedom parameter is the total sample size decreased by the number of mean (or regression) parameters that must be estimated before one can calculate a sum of squares for the standard error. For small degrees of freedom, the distribution's relative frequency function looks similar to the bell-shaped normal curve except that it has heavier tails, reflecting the greater uncertainty introduced by the estimation of σ. As the sample size grows, the distribution approaches the normal distribution: technically, the normal distribution is a *t*-distribution with infinite degrees of freedom, although for n in excess of 30 the *t*-distribution is practically indistinguishable from the standard normal. A table of percentiles for Student's *t*-distribution appears in Appendix II, Table E. Notice how the values exceed those of the normal distribution for small degrees of freedom, and how they approach the normal percentiles as the degrees of freedom increase.

Hypothesis testing: paired samples

An important application of the single sample *t*-test is to assess the significance of the difference between the means of two *paired* variables (the "paired t-test"). For example, in a matched-pairs study (a "buddy study") female employees are paired one-to-one with male employees who most closely match them in terms of several potentially confounding factors. Then the difference, d_i, between the female and male employees' salary in the ith pair is obtained, and the sample mean $\bar{d}$ and sample variance s^2 of the differences calculated. Under the null hypothesis of no salary differences between paired employees, the test statistic $t = \bar{d}/(s/\sqrt{n})$ is referred to Student's *t*-distribution on $n-1$ degrees of freedom. The differences are assumed to be normally distributed, although mild to moderate departures from normality do not seriously affect accuracy here.

If the data are seriously non-normal, various strategies are possible to achieve accuracy. If the distribution has fat tails, a normalizing transformation can sometimes be found (e.g., a logarithmic, or a "power" transformation, such as the square root, cube root, or reciprocal). These transformations improve the approximation by symmetrizing and thinning the tails through de-emphasis of extreme values. On transformations, see generally Section 13.9. If the data are still of questionable normality, a nonparametric method should be used. See Chapter 12.

Hypothesis testing: independent samples

To compare two independent samples, the two-sample t-test is used. Assuming a sample of size n_1 from a normal population with mean μ_1 and variance σ^2, and an independent sample from another normal population with mean μ_2 and the same variance σ^2, the test statistic

$$t = \frac{(\bar{X}_1 - \bar{X}_2) - (\mu_1 - \mu_2)}{s_p(n_1^{-1} + n_2^{-1})^{1/2}}$$

has a Student t-distribution on $n_1 + n_2 - 2$ degrees of freedom. Here s_p^2 is the pooled estimate of the common variance σ^2,

$$s_p^2 = \left(\sum_{\text{sample 1}} (X_i - \bar{X}_1)^2 + \sum_{\text{sample 2}} (X_i - \bar{X}_2)^2 \right) / (n_1 + n_2 - 2).$$

Thus, to test the null hypothesis $H_0 : \mu_1 = \mu_2$, we use the test statistic

$$t = \frac{\bar{X}_1 - \bar{X}_2}{s_p(n_1^{-1} + n_2^{-1})^{1/2}}$$

and reject H_0 if $|t| \geq t_{\nu,\alpha} = $ two-tailed $100\alpha\%$ critical value for Student's t-distribution with $\nu = n_1 + n_2 - 2$ degrees of freedom.

When the assumption of equal variance in the two populations is untenable, the required distributions cannot be specified without knowledge of the variance ratio, a difficulty known as the Behrens-Fisher problem. While there is no convenient exact solution, there is a simple approximate solution known as the Satterthwaite approximation. The test statistic used is

$$t = \frac{(\bar{X}_1 - \bar{X}_2) - (\mu_1 - \mu_2)}{(s_1^2/n_1 + s_2^2/n_2)^{1/2}},$$

where s_1^2 and s_2^2 are unbiased estimates of the respective population variances σ_1^2 and σ_2^2, no longer assumed equal. While this statistic is not precisely distributed as a Student t-variable, it is closely approximated by a Student t-distribution with degrees of freedom estimated by

$$\frac{(s_1^2/n_1 + s_2^2/n_2)^2}{(s_1^2/n_1)^2/(n_1 - 1) + (s_2^2/n_2)^2/(n_2 - 1)}.$$

Critical values for fractional degrees of freedom are obtained by interpolation in Table E of Appendix II.

Confidence intervals for means

Student's t-distribution is also used to construct confidence intervals for a population mean, or difference of means. Analogous to normal–theory confidence intervals (see Section 5.3), a two-sided 95% confidence interval for the mean μ, given a sample of size n, is

$$\bar{X} - t_{\nu,.05}s/\sqrt{n} \leq \mu \leq \bar{X} + t_{\nu,.05}\, s/\sqrt{n}$$

where $t_{\nu,.05}$ is the two-tailed 5% critical value for Student's t with $\nu = n - 1$ degrees of freedom. For the difference in means, the 95% confidence interval is

$$\bar{X}_1 - \bar{X}_2 \pm t_{\nu,.05}\, s_p(n_1^{-1} + n_2^{-1})^{1/2},$$

where now $\nu = n_1 + n_2 - 2$ degrees of freedom. Note that only two changes to the normal-theory confidence intervals are needed: σ is replaced by s or s_p, and the normal critical values are replaced by those from the t-distribution.

t for 2×2

Student's t, or something analogous to it, was not needed in our discussion of tests for single-sample binomial problems because in the binomial case the variance $\sigma^2 = npq$ is completely determined once a hypothesis about the mean $\mu = np$ is specified. Thus, there is no need to account for additional variability in an independent estimator for σ^2. In the two-sample binomial case, the variance of the difference in proportions is unknown, being $pq(n_1^{-1} + n_2^{-1})$, where p is the unknown value of p_1 and p_2, assumed equal in the two populations sampled. When we use an estimator for this variance, why don't we use Student's t-distribution? An obvious answer is that Student's t assumes normal distributions, not binomial distributions, so that Student's t does not bring us toward any exact solution for small sample sizes. A more subtle reason follows from the fact that the variance of the two-sample binomial z-score (see Section 5.2 at p. 162) is not inflated when pq is estimated from the marginal data, so that there is no reason to use the t-distribution. [1] If one were to use the two-sample t-statistic in this problem, estimating the variance by s_p^2 as above, the t-distribution approximation to the exact distribution of t would tend to be not as good as the normal approximation to the exact distribution of the z-score. Of course, for truly exact distribution theory and computation in the two-sample binomial problem, we may use the hypergeometric or two-binomial distribution, thereby dispensing with approximating distributions.

---

[1] This is because for fixed marginal rate, $\hat{p}$, z is a standardized hypergeometric variable with zero mean, $E[z|\hat{p}] = 0$, and unit variance, $\text{Var}(z|\hat{p}) = 1$. The identity $\text{Var}(z) = E\,\text{Var}(z|\hat{p}) + \text{Var}\,E[z|\hat{p}]$ then shows that $\text{Var}(z) = 1$ also unconditionally.

Hypothesis testing and confidence intervals: The correlation coefficient

To test the null hypothesis that $\rho = 0$, we use the statistic

$$t = r \cdot \left(\frac{n-2}{1-r^2} \right)^{1/2},$$

where r is a sample correlation coefficient, which has a t-distribution with $n-2$ degrees of freedom (assuming that one variable is a linear function of the other plus a normal error). In the dangerous eggs data (Section 1.4.1), since $r = 0.426$, $t = 0.426 \cdot [38/(1-0.426^2)]^{1/2} = 2.903$ with 38 df, which is highly significant ($p < 0.01$, two tails).

To compute a confidence interval for ρ it is necessary to transform r into a statistic z that is approximately normally distributed for any value of ρ. On the assumption that the variables have a joint bivariate normal distribution, the required transformation, due to R.A. Fisher, is given by

$$z = \frac{1}{2} \ln \frac{1+r}{1-r},$$

with approximate expected value $Ez = 1/2 \cdot \ln [(1+\rho)/(1-\rho)]$, and variance $\mathrm{Var}\,(z) = 1/(n-3)$, where n is the sample size. This leads to the following 95% confidence interval for Ez:

$$z - \frac{1.96}{(n-3)^{1/2}} \leq Ez \leq z + \frac{1.96}{(n-3)^{1/2}}.$$

To obtain confidence limits for ρ itself, invert the transform using

$$r = \frac{e^{2z} - 1}{e^{2z} + 1}.$$

In the dangerous eggs data, $z = 1/2 \cdot \ln [(1+0.426)/(1-0.426)] = 0.455$. The confidence interval is $0.455 \pm 1.96/\sqrt{37}$, leading to an upper limit of 0.777 and a lower limit of 0.133. Inverting these values, we have $0.132 \leq \rho \leq 0.651$.

7.1.1   Automobile emissions and the Clean Air Act revisited

The federal Clean Air Act requires that before a new fuel or fuel additive is sold in the United States, the producer must demonstrate that the emission products generated will not cause a vehicle to fail to achieve compliance with certified emission standards. To estimate the difference in emission levels, the EPA requires, among other tests, a Paired-Difference Test in which a sample of cars is first driven with the standard fuel, and then with the new fuel and the emission levels compared. EPA then constructs a 90% confidence interval for the average difference; if the interval includes 0 the new fuel is eligible for a waiver (but it must pass other tests as well; see Section 5.4.3 for a further description).

TABLE 7.1.1. Emission data for NO_x

	Base Fuel (B)	Petrocoal (P)	Difference (P-B)	Sign
	1.195	1.385	+0.190	+
	1.185	1.230	+0.045	+
	.755	.755	0.000	tie
	.715	.775	+0.060	+
	1.805	2.024	+0.219	+
	1.807	1.792	−0.015	−
	2.207	2.387	+0.180	+
	.301	.532	+0.231	+
	.687	.875	+0.188	+
	.498	.541	+0.043	+
	1.843	2.186	+0.343	+
	.838	.809	−0.029	−
	.720	.900	+0.180	+
	.580	.600	+0.020	+
	.630	.720	+0.090	+
	1.440	1.040	−0.400	−
Average	1.075	1.159	0.0841	
St. Dev.	0.5796	0.6134	0.1672	

The data in Table 7.1.1 show the nitrous oxide emissions for sixteen cars driven first with a standard fuel, and then with Petrocoal, a gasoline with a methanol additive.

Questions

1. Compute a 90% confidence interval for the average difference in emissions using the *t*-distribution. Does Petrocoal pass this test?

2. Suppose the statistician ignored the fact that the data were in the form of matched pairs, but instead treated them as two independent samples, one involving the standard fuel and the other involving Petrocoal. Would Petrocoal pass the test? What is the reason for the difference in results? Do the data satisfy the assumptions required for the two-sample *t*-test to be valid in this case?

Source

Motor Vehicle Mfrs. Ass'n of U.S. v. E.P.A., 768 F. 2d 385 (D.C. Cir. 1985); J. Gastwirth, *Statistical Reasoning in Law and Public Policy* 612 (1988).

7.1.2   Voir dire of prospective trial jurors

In the California procedure for impaneling trial jurors, counsel and the judge participate in questioning. In the federal procedure (also used in many states),

TABLE 7.1.2. Test judges' voir dire impaneling times

	Minutes to impanel jury	
Judge	State	Federal
A	290	120
B	80	60
C	96	177
D	132	105
E	195	103
F	115	109
G	35	65
H	135	29
I	47	45
J	80	80
K	75	40
L	72	33
M	130	110
N	73	40
O	75	45
P	25	74
Q	270	170
R	65	89

the judge does so alone, although counsel may suggest questions. Presumably, the federal method is faster, but by how much? In a 1969 test of the two methods conducted with California state judges, the usual California procedure was followed for a two-month period and the federal procedure was followed for cases assigned to the same judges in the following two-month period. Records of impaneling times were kept. Excerpts from the data are shown in Table 7.1.2.

Questions

1. Use a t-test to assess the statistical significance of the difference in impaneling times and to construct a confidence interval for the mean difference.

2. Do the data appear normally distributed?

3. Repeat the test after making a reciprocal transformation of the data to improve normality. What conclusion do you reach?

Source

William H. Levit, *Report on Test of Federal and State Methods of Jury Selection* (Letter to Chief Justice Roger J. Traynor dated October 15, 1969).

7.1.3   Ballot position

A New York State law provided that in a particular primary in New York City incumbents be listed first on the ballot. Previously, position had been determined by lot. A non-incumbent facing an incumbent challenged the law, introducing a study of 224 two-candidate races in New York City in which position had been determined by lot. The study showed that: (i) in races in which incumbents appeared first on the ballot, they received an average 64.3% of the vote; (ii) in races in which incumbents appeared second on the ballot they received an average of 59.3% of the vote; and (iii) in races in which neither candidate was an incumbent, the average percentage of votes received by the candidate in first position was 57.6%. Of the 224 races included in the study, there were 72 two-candidate races in which no incumbent ran. In the 152 races in which incumbents ran, they appeared in second position on the ballot 31 times. The sample variance of the incumbent's percentage of the vote in those races in which they were first on the ballot was 148.19; the sample variance in those races in which they appeared in second position was 177.82.

Questions

1. Do these data show a statistically significant advantage for the first position on the ballot?

2. Is there an explanation other than ballot position?

Source

In re Holtzman v. *Power*, 62 Misc.2d 1020, 313 N.Y.S.2d 904 (Sup. Ct.), *aff'd*, 34 A.D.2d 779, 311 N.Y.S.2d 37 (1st Dep't), *aff'd*, 27 N.Y.2d 628, 313 N.Y.S.2d 760 (1970).

7.1.4   Student's t-test and the Castaneda rule

In *Moultrie* v. *Martin*, 690 F.2d 1078 (4th Cir. 1982), petitioner, a black, was indicted and convicted in 1977 and subsequently brought a habeas corpus petition attacking grand jury selection as discriminatory. The court assumed that blacks constituted 38% of the voting rolls from which the jury commissioners assembled jury lists. Each year, twelve people were selected from the list to serve as grand jurors and six of these were selected at random to serve the next year, making a total of 18 grand jurors. On the statistics shown below, the court found no discrimination under the *Castaneda* standard, particularly as adjusted using Student's *t*-distribution for small sample sizes.

The court gave the following discussion of the interrelation between the *t*-test and the *Castaneda* standard-deviation rule (citations omitted):

The Supreme Court's rule in Castaneda of course can be adjusted for small sample sizes through the use of the student's *t*-distribution. The student's *t*-distribution teaches that when the sample size is less than approximately thirty, the number of standard deviations must be increased in order to achieve the same significance level. The student's *t*-distribution, like the binomial distribution used in Castaneda, is represented by a bell shaped curve. When the sample size is small, the student's *t*-curve is flatter in the middle and plumper in the tails.

As the sample size increases, the student's *t*-curve is approximated by the binomial distribution. While a precise mathematical formula exists for computing the student's *t*-distribution, in practice it is more easily computed by the use of tables found in standard books concerning statistics.

Employing these tables, the Court's Castaneda analysis is easily adapted to small samples. The two to three standard deviations noted in Castaneda correspond to a 95% to 99.9% significance level on the two-tailed binomial distribution. Student's *t*-tables are stated in terms of significance level and degrees of freedom. The number of degrees of freedom is equal to the sample size minus one. The tables thus show that when the sample size is 15 (i.e. 14 degrees of freedom), one needs to use 2.1 standard deviations for a two-tailed 95% significance level, and 3.8 standard deviations for a two-tailed 99.9% significance level. For a sample size of five, the standard deviations become 2.8 and 7.2 respectively.

For a sample size of 18 (i.e., the number of jurors), the Castaneda range becomes 2.1 to 3.6 standard deviations.

Computing the difference between actual and expected number of jurors, in terms of standard deviation, for each of the years 1971–1977 shows the following results:

Year	# Black Jurors*	% Black Jurors	No. Std. Dev.
1971	1	6	−3.4
1972	5	28	−0.9
1973	5	28	−0.9
1974	7	39	+0.1
1975	7	39	+0.1
1976	4	22	−1.4
1977	3	17	−1.8

*This information appears elsewhere in the court's opinion.

Id. at 1084, nn. 10, 11.

Questions

1. The court recommends use of the t-distribution to approximate the binomial distribution when the sample size is less than 30. Do you agree?

2. If the sample size is 5, the court finds that a 5% level of significance requires 2.8 standard deviations. Assuming that $p = 0.5$, how many standard deviations from the mean is the most extreme result in such a sample (i.e., 0 black jurors)? Calculate the exact binomial probability of such a result.

3. Why are the thicker tails of the t-distribution appropriate in smaller samples, as compared with the normal distribution?

4. How would you analyze the data showing the numbers and percentages of black jurors?

7.2   Analysis of variance for comparing several means

We can apply the notion of analysis of variance to compare several means. Suppose there are k samples of data drawn having sample sizes n_i, sample means $\bar{Y}_i$, and sample variances s_i^2, for $i = 1, \ldots, k$. We wish to test the hypothesis that the true mean μ_i is the same in each group, $H_O : \mu_1 = \mu_2 = \cdots = \mu_k$, against the alternative hypothesis that at least two means differ. In addition to the overall test of significance, we wish to make multiple post-hoc comparisons between means, e.g., to compare mean $\bar{Y}_i$ with mean $\bar{Y}_j$ for various pairs. As in Section 6.2, the task is to limit the probability of making one or more Type I errors to a given level α. A closely related problem is to provide simultaneous confidence statements such as "μ_i lies in the interval $\bar{Y}_i \pm l_i$" or "the difference $\mu_i - \mu_j$ lies in the interval $\bar{Y}_i - \bar{Y}_j \pm l_{ij}$." The limits l_i and l_{ij} must be chosen so that for the many such statements possible, there is a probability of at least $1 - \alpha$ that each interval contains its respective parameter; i.e., that the total probability that one or more intervals will fail to contain its parameter is no more than α. We consider each of these inferential problems in the simplest case of the so-called "one-way layout," in which there is no special recognition of structure among the group categories (such as crossing or nesting of categories). More complicated designs are commonly used in experimental settings but these appear less often in observational settings most often encountered in legal disputes.

Let the jth observation in the ith sample be denoted by $Y_{ij}(j = 1, \ldots, n_i)$ and let the grand mean be denoted by $\bar{Y} = \sum_i \sum_j Y_{ij}/N$, where $N = n_1+n_2+\cdots+n_k$ is the total sample size. The analysis of variance states that the total sum of squares

$$SS_{tot} = \sum_{i=1}^{k} \sum_{j=1}^{n_i} (Y_{ij} - \bar{Y})^2$$

is equal to the sum of two components, the "between-group" sum of squares,

$$SS_b = \sum_{i=1}^{k} n_i(\bar{Y}_i - \bar{Y})^2$$

and the "within-group" sum of squares,

$$SS_w = \sum_{i=1}^{k} \sum_{j=1}^{n_i} (Y_{ij} - \bar{Y}_i)^2 = \sum_{i=1}^{k} (n_i - 1)s_i^2.$$

The between-group sum of squares, SS_b, is a weighted sum of squared deviations of each group's sample mean about the grand mean, and is a measure of how much the sample means differ from one another. The within-group sum of squares, SS_w, is a measure of inherent variability in the data. Assuming a constant variance σ^2 in each group, the expected value of SS_w is equal to $(N-k)\sigma^2$. Thus, an unbiased estimate of σ^2 is given by the within-group mean square, $MS_w = SS_w/(N - k)$.

Under the null hypothesis of equal means, the differences between the sample means $\bar{Y}_i$ also reflect only inherent variability, and the expected value of SS_b can be proven equal to $(k - 1)\sigma^2$, so that the between-group mean square $MS_b = SS_b/(k - 1)$ is a second unbiased estimate of σ^2, but only under H_0. Under the alternative hypothesis, SS_b reflects systematic group differences as well as inherent variability, and in that case MS_b tends to be larger than MS_w, with expected value that can be shown equal to

$$E(MS_b) = \sigma^2 + \sum_{i} \frac{n_i(\mu_i - \bar{\mu})^2}{k - 1},$$

where $\bar{\mu} = \sum_i n_i \mu_i/N$ is the sample-weighted average of true group means. Thus, an index of inequality between group means is given by the variance ratio, or F-statistic

$$F = \frac{MS_b}{MS_w} = \frac{SS_b/(k - 1)}{SS_w/(N - k)}.$$

Assuming the data are normally distributed, the F-statistic has an F-distribution with $k - 1$ and $N - k$ degrees of freedom under H_0. The F-distribution is defined in general as the distribution of two independent chi-squared random variables, each divided by its respective degrees of freedom.[2] Departures from H_0 in any direction (or directions) will cause MS_b to tend to exceed MS_w, and thus the null hypothesis is rejected for large values of F. Percentage points for the F-statistic are provided in Table F of Appendix II.

---

[2] An F variate with df_1 and df_2 degrees of freedom has mean $df_2/(df_2 - 1)$ and variance $2 \cdot df_2^2 \cdot (df_1 + df_2 - 2)/[df_1 \cdot (df_2 - 2)^2 \cdot (df_2 - 4)]$.

For example, suppose one is comparing scores on a proficiency test among four ethnic groups in a disparate impact case. Assume that the sufficient statistics, which summarize the relevant information, are as follows:

Ethnic group	Sample size	Group mean	Sample variance	St. dev.
Black	50	75.2	225.0	15.0
Hispanic	35	70.3	289.0	17.0
Asian	20	78.8	324.0	18.0
Other	50	84.9	256.0	16.0
	$N=155$	$Y=77.68$		

The following display, called the ANOVA table, summarizes the results of the breakdown in sums of squares and the overall test of significance:

Source of variation	Sum of squares	df	Mean square	Expected mean square	$F = MS_b/MS_w$
Between groups	4,845.27	3	1,615.09	$\sigma^2 + \sum_i n_i(\mu_i - \bar{\mu})^2$	
					6.16
Within groups	39,551.00	151	261.93	σ^2	
Total	44,396.27	154			

Since the 0.005 critical value for F on 3 and 151 degrees of freedom is less than the critical value for F on 3 and 120 degrees of freedom, which is 4.50 (see the third Table F in Appendix II), the differences among means is significant at the 0.005 level.

Which among the several groups differ significantly? The answer depends in part on how many comparisons one wishes to address. In the example above, if the Other category represented unprotected non-minorities, and only the three comparisons of the minority groups with Other were relevant, then Bonferroni's adjustment (see Section 6.2) may be used: any of the three two-sample Student t-tests that are significant at the $\alpha/3$ level will be taken as evidence of a significant difference between groups. For example, if $\alpha = 0.01$, then the t-test comparing Black to Other is 3.13 with 98 df: comparing Hispanic to Other, t is 4.04 with 83 df: and comparing Asian to Other, t is 1.39 with 68 df. From Table E, the critical value for t at the 0.01/3 level is about 3.09 on 60 df, and slightly smaller for larger dfs. Thus, two of the three comparisons are significant at the 0.01 level, adjusting for three planned comparisons.

Scheffé's method

Frequently, there are unanticipated "interesting" comparisons suggested by the data. The number of such possible post-hoc inferences grows rapidly with the number of groups. In the example, it might be of interest to see whether any significant differences exist among the minority groups; or to compare the Other group with an average of the three minority groups; or to compare the average score of Asian and Other with that of the Black group, or the Hispanic

Group, or an average of the two, etc. In fact, we might broaden our purview to include arbitrary contrasts between means, that is, a comparison of an arbitrary weighted average of one subset of means with that of another subset, even though there are infinitely many such contrasts. Clearly, Bonferroni's method cannot be used to adjust for all of these.

It may be surprising, then, that there is a method of adjustment, known as Scheffé's method, that can be used to make simultaneous comparisons and confidence statements even for a class of comparisons large enough to allow arbitrary post-hoc contrasts. A key property of the method is that with probability $1 - \alpha$, *all* of the (possibly infinitely many) confidence intervals constructed from the data will cover their respective true parameters. Another convenient feature is that, if the overall F-test of the null hypothesis of equal means is significant, then there is *some* contrast (although possibly not a simple pairwise difference) that the Scheffé method will find significant. Conversely, if the overall F-test is not significant, then *none* of the contrasts will be found significant. Thus, the Scheffé method allows us to ransack the data, examining any contrast of interest *ad libidum*.

The method is as follows. Consider an arbitrary contrast of the form $\delta = \sum_i c_i \mu_i$, where the coefficients $c_1, \ldots, c_k$ sum to zero. For example, the contrast $\delta = \mu_i - \mu_j$ has coefficients $c_i = 1$, $c_j = -1$, and the rest zero; or in the proficiency test example, the contrast $\delta = (\mu_1 + \mu_2)/2 - (\mu_3 + \mu_4)/2$ has coefficients $c_1 = c_2 = 1/2$ and $c_3 = c_4 = -1/2$. Any such contrast may be estimated by the sample version of the contrast, $\hat{\delta} = \sum_i c_i \bar{Y}_i$, which has variance $\mathrm{Var}(\hat{\delta}) = \sigma^2 \cdot \sum_i c_i^2/n_i$ that itself can be estimated by $MS_w \cdot \sum_i c_i^2/n_i$. Scheffé's method, then, constructs the intervals

$$\hat{\delta} - D \cdot \left(MS_w \cdot \sum_i c_i^2/n_i \right)^{1/2} \leq \delta \leq \hat{\delta} + D \cdot \left(MS_w \cdot \sum_i c_i^2/n_i \right)^{1/2},$$

where the multiplier D is the square root of the quantity $(k-1)$ times the upper α percentage point of the F-distribution with $k-1$ and $N-k$ degrees of freedom, $D = [(k-1) \cdot F_{k-1,N-k;\alpha}]^{1/2}$. For example, to construct a 95% Scheffé interval for the contrast $\delta = (\mu_1 + \mu_2)/2 - (\mu_3 + \mu_4)/2$ in the testing example, the sample contrast is $\hat{\delta} = (75.2 + 70.3)/2 - (78.8 + 84.9)/2 = 72.75 - 81.85 = -9.10$, with variance estimated by

$$MS_w \cdot \sum \frac{1/4}{n_i} = 261.93 \cdot 1/4 \cdot (50^{-1} + 35^{-1} + 20^{-1} + 50^{-1}) = 7.76,$$

and standard error $\sqrt{7.76} = 2.79$. The multiplier $D = (3 \cdot F_{3,151;.05})^{1/2} = (3 \cdot 2.66)^{1/2} = 2.82$. The confidence interval is then

$$-9.10 - 2.82 \cdot 2.79 \leq \delta \leq -9.10 + 2.82 \cdot 2.79,$$

or $-17.0 \leq \delta \leq -1.2$. Since zero is excluded by this interval, the contrast is declared significantly different from zero by the Scheffé criterion. The Scheffé confidence intervals for each of the six possible pairwise differences in means may be constructed similarly with the following result:

Contrast	Estimate	Standard error	Simultaneous 95% confidence interval	
Black–Hispanic	4.9	3.57	[−5.2,	15.01]
Black–Asian	−3.6	4.28	[−15.7,	8.51]
Black–Other	−9.7	3.24	[−18.8,	−0.61]
Hispanic–Asian	−8.5	4.54	[−21.3,	4.3]
Hispanic–Other	−14.6	3.57	[−24.7,	−4.51]
Asian–Other	−6.1	4.28	[−18.2,	6.0]

Thus, there are significant differences in mean scores between the Black and Other groups, and between the Hispanic and Other groups,

Note that the Scheffé method gives somewhat wider confidence intervals than the corresponding intervals would be for the difference between two means using as multiplier the critical value from Student's t-distribution at the Bonferroni-adjusted $\alpha/6$ level. This is because of the wider class of confidence statements for which the Scheffé method provides 95% confidence. In general, for a small number of planned comparisons, Bonferroni's method is preferable in the sense of yielding narrower confidence intervals. For many comparisons, or for those suggested post hoc, Scheffé's method should be used.

The Bonferroni and Scheffé methods are the most common adjustments for multiple comparisons and simultaneous confidence intervals for several means, but there are many others that are discussed in the references below.

Further Reading

R. Miller, *Simultaneous Statistical Inference* (2d ed., 1981).
H. Scheffé, *The Analysis of Variance* (1959).

7.2.1   Fiddling debt collector

The ABC Debt Collection Company [not its real name] was in the business of taking on consignment for collecting large numbers of consumer receivables. Its fees were a percentage of what it collected. Since the company had issued and sold debentures to the public, it was required under the federal securities laws to make periodic public reports of its income. At a point when collections were slow, management adopted the accounting practice of assuming that 70% of the face amount of the consigned receivables would be collected, and reported its fee income on that basis. The Securities and Exchange Commission discovered and challenged this practice. To justify it, the accountant for the company produced three allegedly random samples of receivables from three successive years. According to the accountant, in each year he selected the samples by paging through the list of receivables consigned during a year at least three years prior to the time of examination and pointing at random to those he wanted. Company clerical employees then brought the files of those selected and the accountant made the computations, producing the data given below. The accountant did not keep lists of the files he selected, but said that he trusted company employees to bring the files he designated. He also said that no files were missing and that all had sufficient data to permit him to determine what proportion of the face of the receivable was collected. For each receivable the accountant stated that he computed a proportion collected, and then averaged the proportions and computed their standard deviations.

The result was that the mean proportions collected for the three samples were 0.71, 0.70, and 0.69, respectively. The standard deviations were 0.3 in each sample. The sample size was 40 in each case.

Based on the closeness of the means to 0.70, the accountant stated that he believed he was justified in approving the company's use of the 0.70 collection figure.

Questions

Are the means of the three samples too close together to be consistent with their alleged random selection from a large pool? To answer this question, compute an F-statistic and determine a relevant probability from the F-table. To compute the F-statistic make the following calculations.

1. Compute the within-sample (within-group) sum of squares, SS_w. Compute the within group mean square, MS_w.

2. Compute the between-sample (between-group) sum of squares, SS_b. Compute the between-group mean square, MS_b.

3. Using the above-results, compute the F-statistic.

4. What is the null hypothesis? What is a plausible alternative hypothesis? Given the alternative hypothesis, which tail of the F-distribution (upper or lower) is suggested as a rejection region for the null hypothesis?

5. What is the P-value of the F-statistic? [Hint: the probability below a given value of an F-distribution with a and b degrees of freedom is the same as the upper-tail probability starting from the reciprocal of the the given value of an F-distribution with b and a degrees of freedom, $P[F_{a,b} < f] = P[F_{b,a} > 1/f]$.]

Notes

The principal of ABC was subsequently convicted of fraud in connection with the financial statements of the company, and the accountant confessed to having participated in that fraud.

CHAPTER 8

Combining Evidence Across Independent Strata

8.1 Mantel-Haenszel and Fisher methods for combining the evidence

Quite often, a party seeking to show statistical significance combines data from different sources to create larger numbers, and hence greater significance for a given disparity. Conversely, a party seeking to avoid finding significance disaggregates data insofar as possible. In a discrimination suit brought by female faculty members of a medical school, plaintiffs aggregated faculty data over several years, while the school based its statistics on separate departments and separate years (combined, however, as discussed below).

The argument for disaggregation is that pooled data may be quite misleading. A well known study showed that at the University of California at Berkeley female applicants for graduate admissions were accepted at a lower rate than male applicants. When the figures were broken down by department, however, it appeared that in most departments the women's acceptance rate was higher than the men's. The reason for the reversal was that women applied in greater numbers to departments with lower acceptance rates than to the departments to which men predominantly applied. The departments were therefore variables that confounded the association between sex and admission.[1] See Bickel, Ham-

---

[1] This is an instance of "Simpson's Paradox." In the Berkeley data, there was substantial variation in the disparity between acceptance rates from department to department, although, even if there were a fixed level of disparity, comparing aggregated rates could still be misleading. Suppose, for example, Department A had an overall acceptance rate of 50%, which was the same for men and women, while Department B had an overall acceptance rate of 10%, also the same the men and women. If 80 men and 20 women apply to Department A, while 20 men and 80 women apply to Department B, then of the 100 men, 42 will be

mel, and O'Connell, *Sex Bias in Graduate Admissions: Data from Berkeley,*
187 Science 398 (1975).

The argument for aggregation is that disaggregation deprives data of significance when there is a small but persistent tendency across strata. If blacks
are persistently underrepresented on jury panels, the difference may not be
statistically significant in any one panel, yet the aggregation would reflect
the persistence of discriminatory practices and the aggregated data would be
significant. In fact, a test based on aggregated data, when valid, is the most
powerful possible.

The Berkeley data illustrate a situation in which the comparison of a single
acceptance rate for men with one for women was thoroughly misleading. A
more subtle problem arises when a single measure of disparity is appropriate
and the question is whether the data from independent strata can be aggregated without biasing the resulting estimate of that measure. Two sufficient
conditions for an unbiased estimate of a common odds ratio are that either (i)
the outcome rate in each group (acceptance rate for men and for women) is
constant across the stratification variable (department, in the Berkeley example), or (ii) the exposure rate at each outcome level (sex, for those accepted
and rejected) is constant across the stratification variable. For an assumed constant difference or relative risk, the two sufficient conditions are that either
(i) the outcome rate overall, or (ii) the exposure rate overall are constant over
the stratification variable. These are sufficient, but not necessary, conditions
because there are special configurations in which both correlations exist, but
there nevertheless will be no aggregation bias.

A composition of the aggregation and disaggregation points of view is to
disaggregate when there is a risk of bias, but then to *combine the evidence* from
the various sources or strata. That is, having disaggregated the data to reduce
bias and increase validity, we then seek a statistic that sums up the situation in
an appropriate way. Here are some examples:

- In the Berkeley example discussed above, compute acceptance rates for men
  and women for each department separately, and then combine the evidence
  to test for a significant difference.

- In a study of advancement, compare black and white employees with respect to promotion after one year of employment. Because qualifications
  and backgrounds vary substantially, match each black employee with a
  white employee and compare their respective promotions in order to control for differences in group composition. Then combine the evidence across
  matched pairs to estimate the relative odds on promotion.

The idea behind the combination of evidence for testing hypotheses about
two groups is simple. Within each stratum we focus on the number of particular

---

accepted while of the 100 women, only 18 will be accepted. Thus, the odds ratios equal 1
for each department separately, but the odds ratio in the aggregated data is 3.7.

outcomes in a certain group, e.g., the number of offers to women or the number of black promotions. The particular choice of outcome or group is arbitrary and different choices lead to equivalent analyses. An expected value for the observed number is generated for each stratum by multiplying the probability of success under the null hypothesis for that stratum by the number of group members in the stratum. The differences between observed and expected numbers are summed for the strata, and the absolute difference is reduced by 0.5 to correct for continuity. In one form of analysis, this corrected difference is divided by the standard error of the sum. (Since the strata are independent by hypothesis, the standard error of the sum of the differences is simply the square root of the sum of the variances of those differences.) The result under the null hypothesis has an approximate standard normal distribution. In an equivalent form of analysis the corrected difference is squared and divided by the variance of the sum; the resulting statistic has a chi-squared distribution, with a single degree of freedom.

Binomial test

If it is appropriate to represent success by a binomial outcome with a given rate (e.g., selection of relatively few from a large population), the sum of the differences between observed and expected values (corrected for continuity) is divided by the square root of the sum of binomial variance terms; the resulting test statistic has an approximate normal distribution.

For example, in *Cooper v. University of Texas at Dallas*, 482 F. Supp. 187 (N.D. Tex. 1979), plaintiff charged the university with sex discrimination in hiring faculty members and submitted data comparing the hires during the 1976-77 time period with data on doctoral candidates receiving Ph.D. degrees in 1975 (from which the availability proportions were derived). The data are shown in Table 8.1.

TABLE 8.1. Hiring practices, University of Texas, 1976–1977

1 University division	2 Availability (prop. women Ph.D.s)	3 Total hires	4 Female hires	5 Expected (2 x 3)	6 Diff (4-5)	7 # of SDs	8 Exact level of signif.*
Arts-Hum.	0.383	48	14	18.38	−4.38	−1.30	0.123
Hum. Dev.	0.385	32	12	12.32	−0.32	−0.31	0.532
Manag.	0.043	26	0	1.12	−1.12	−1.08	0.319
Nat. Sci.	0.138	38	1	5.24	−4.24	−2.08	0.025
Soc. Sci.	0.209	34	6	7.11	−1.11	−0.47	0.415
Total		178	33	44.17	−11.17		

* The numbers given in the table in the court's opinion are not all correct. The correct numbers, using the binomial distribution to compute exact levels of significance, are given here.

Note that women are underrepresented in each department, but that only in Natural Science is the difference statistically significant at the 5% level. The sum of the binomial variances for each department is 30.131; the standard error is the square root of that, or 5.489. The difference between the actual and expected numbers of women hired (corrected for continuity) and divided by the standard error is $(-11.17 + 0.5)/5.489 = -1.94$. This difference has an attained level of significance of about 0.026 (one-tailed).

Mantel-Haenszel test

When complete applicant data are available (i.e., numbers hired and not hired by sex) or when promotions are considered, then a different analysis is preferable. The natural null hypothesis is that the hires or promotions were made at random with respect to sex from the combined group in each stratum. The expected value now is the number of group members times the overall success rate in each stratum.

The formula for the hypergeometric variance for each stratum is $m_1 m_2 n_1 n_2 / N^2 (N - 1)$ where m_1 is the number of successful outcomes in the stratum (e.g., aggregate promotions); m_2 is the number of non-successful outcomes in the stratum (e.g., the number of those not promoted), the n's are the respective numbers in each group (e.g., the numbers of men and women); and N is the total number in the stratum. See Section 4.5. The corrected difference between observed and expected sums divided by the standard error is a statistic that has a standard normal distribution (approximately) under the null hypothesis of no systematic influence of group on success. We refer to this as the Mantel-Haenszel z-score. In squared form, the statistic has a chi-squared distribution (approximately) with one degree of freedom, and is known as the Mantel-Haenszel chi-squared test. The Mantel-Haenszel procedure was accepted by the court in *Hogan v. Pierce*, 31 Fair Emp. Prac. Cas. (BNA) 115 (D.D.C. 1983).

As a simple illustration of the Mantel-Haenszel procedure, consider the matched-pairs advancement study. Here each pair is an independent "stratum" with $n_1 = n_2 = 1$. If both members of a pair are promoted, or if neither is promoted, then there is no departure from expectation for that pair under the null hypothesis of race-neutral promotion. Such concordant pairs are "conditionally uninformative" because their hypergeometric variance is zero, since either m_1 or m_2 is zero, and so they provide no information about the pairwise disparity in promotion rates. For discordant pairs, with one member promoted and the other not, the expected number of black promotions is $1/2$ per pair and the hypergeometric variance of the difference between observed and expected equals $1 \cdot 1 \cdot 1 \cdot 1/2^2 \cdot 1 = 1/4$ per pair. Thus if there are n *discordant* pairs, comprising b pairs with black member promoted and white member not, and $c = n - b$ pairs with white member promoted and black member not, the

Mantel-Haenszel chi-squared statistic with continuity correction is

$$X^2 = \frac{(|b - (n/2)| - 1/2)^2}{n/4} = \frac{(|b - c| - 1)^2}{b + c},$$

distributed as $\mathcal{X}^2$ with 1 df under the null hypothesis. In this form the test is called McNemar's test. Note how the Mantel-Haenszel procedure for matched pairs specializes to a binomial analysis of the number of discordant pairs in which the black member, say, is promoted, $b \sim \text{Bin}(n, 1/2)$.[2] A distinct virtue of the Mantel-Haenszel procedure is that it generalizes the buddy-study design to matched sample or stratified designs with larger strata of any size.

The Mantel-Haenszel procedure is powerful for detecting departures from the null hypothesis that are consistently in one direction; such consistent differences rapidly become highly unlikely under the null hypothesis. The Mantel-Haenszel procedure is the method of choice in such circumstances. However, the procedure is *not* powerful for detecting departures from the null hypothesis that occur in both directions, where cancellations reduce the sum of accumulated differences. Should a test that is powerful against alternatives of this form be desired, a different statistic is used: the squared difference between the observed and expected counts (without correction) in each stratum is divided by the variance term for that stratum, after which the sum is taken. This sum has a chi-squared distribution with k degrees of freedom, where k is the number of strata. In this situation, departures from expectation in either direction cause a large statistic, although if there are many strata each containing few observations, power may be modest. In the case of the Mantel-Haenszel procedure, even if the number of strata is large and sample sizes are small within strata (as in a matched-pair study), power may be adequate if the direction of effect is consistent.[3]

Fisher's method

Of several other testing methods available for combining the evidence from independent sources, the most commonly used is R.A. Fisher's method, which is easily implemented in a wide variety of circumstances. Fisher's method provides a statistic for testing whether the null hypothesis in each of k independent tests is true, against the alternative that the specific null hypothesis is false in at least one of the component problems. In principle there need be no relation

---

[2]If we assume there is a constant odds ratio Ω on promotion comparing black and white employees, then it can be shown that the random variable b has a $\text{Bin}(n, P)$ distribution with P given by $P = \Omega/(\Omega + 1)$. A maximum conditional likelihood estimate of P given n is $\hat{P} = b/n$, and a maximum likelihood estimate of the odds ratio on promotion is $\hat{\Omega} = b/c$. Binomial confidence intervals for P can be transformed into corresponding intervals for Ω via $\Omega = P/(1 - P)$.

[3]Note that if both tests are conducted with the intention to quote the more (or less) significant result, then Bonferroni's correction indicates that to limit this procedure's Type I error, each component test should be conducted at the $\alpha/2$ level.

between the component problems, hypotheses, or test statistics, but for our purposes we shall assume that the hypothesis and test statistics are of the same form in each component.

Fisher's method is based on a property of the attained level of significance. Suppose there are two independent strata, each involving a one-sided hypothesis test based on a statistic T for which large negative values lead to rejection of the null hypothesis. Since by assumption each statistic is independent, one might be tempted simply to multiply together the levels of significance, p_1 and p_2, for the statistic in each stratum to obtain a joint level of significance, $p_1 p_2$. Thus, if the test statistic has a 0.10 level of significance in each of two strata, the joint level of significance would be 0.01. However, this gives the probability that the test statistic in each of the strata would have a level of significance less than 0.10, which is artificial to the problem. The statistic of interest is actually the *probability* that the product of the two significance levels would be as small as that observed. In the example this probability will include cases in which $p_1 = 0.2$ and $p_2 = 0.05$ (for example), and not only cases in which both p_1 and p_2 are less than 0.1.

Fisher used a transformation to obtain the probability distribution under the null hypothesis of the product of the significance probabilities. He showed that if the null hypothesis were true in each stratum, then each attained level of significance would be uniformly distributed on the unit interval,[4] and as a consequence -2 times the sum of the natural logs of the attained levels of significance would have a chi-squared distribution with degrees of freedom equal to twice the number of tests. The desired probability is then given by the upper chi-squared tail area above -2 times the sum of log attained significance levels. In the example given above, the log of 0.1 = -2.3026, so that Fisher's statistic equals $-2 \cdot (-2.3026 + -2.3026) = 9.2104$. The upper 5% critical value for chi-squared with four degrees of freedom is 9.488, so that the evidence is not quite significant at that level. But note that combining the evidence has achieved a level of significance that is almost twice as high as for each stratum separately.

In the *Cooper v. University of Texas* data, the sum of the logs of the significance levels is -8.43. This times -2 equals 16.86. Since there are 5 departments, we look at chi-squared with 10 degrees of freedom and find that the upper 5% critical value is 18.31, which indicates that the data are not quite significant at the 0.05 level ($p = 0.077$). However, the data on hires are discrete, while Fisher's method assumes that the data have continuous P-values. Applying Lancaster's correction for that fact yields a value of 21.47

---

[4]In general, if X is a random variable with cumulative distribution function F, then the random variable $Y = F(X)$ is called the probability transform of X. If F is continuous, Y has a uniform distribution on the unit interval from 0 to 1, because the event $[Y < p]$ occurs if and only if X is below its Pth quantile, which occurs with probability p. In the present application, if X is a random outcome of the test statistic T with null distribution F, then the attained level of significance is $Y = P[T \leq X] = F(X)$.

for Fisher's statistic, which is significant at the 0.05 level ($p = 0.018$). This result is slightly stronger than the binomial P-value previously obtained (see p. 240) by accumulating the differences between observed and expected values and dividing by the standard deviation of the sum of the binomial variables ($p = 0.026$).[5]

The methods described above also apply to the not uncommon situation in which there are multiple independent studies of the same phenomenon and some or all of them have a low level of statistical significance. This was the situation in studies of the relationship between attitudes toward the death penalty and conviction decisions of death-qualified jurors. See Section 6.1.1. Six different studies showed that death-qualified jurors were more likely to convict, with z values ranging from 1.07 to 2.73. The z's (respectively, two-sided P-values) were: 1.90 (0.057); 2.05 (0.040); 1.40 (0.162); 1.07 (0.285); 2.58 (0.010); 2.73 (0.006). The chi-squared value ($-2 \cdot \sum \ln p$) = 37.84, which is highly significant on chi-squared with 12 df.

While simple and elegant, Fisher's method has low power when the number of studies is large and the significance levels are modest. For example, it the attained significance level were equal to $1/e = 0.37$ in each stratum, Fisher's chi-squared would be $2k$, equal to its degrees of freedom for any k, thus never much exceeding the 0.50 significance level. The Mantel-Haenszel method of combining the evidence is preferable in such cases.

Estimating a common parameter

Up to now we have combined evidence to test a null hypothesis in light of the several sources of data. In many contexts it is more important to estimate a common parameter, assumed to exist, and a confidence interval for the estimate. We discuss two methods: the first is quite general; the second applies to the Mantel-Haenszel test discussed earlier.

The more general method assumes that one has an unbiased (or at least consistent) point estimate of some common parameter of interest together with a standard error in each independent study. Under the assumption that the true parameter is the same in each study, any weighted average of the point estimates also has that same mean. To minimize the variance of the weighted average, one chooses weights for each study inversely proportional to the variance, i.e., the squared standard error, of the respective point estimates. The variance of the weighted average that results from this choice is one over the sum of the reciprocals of the variances. From this a confidence interval for the common parameter can be constructed as (weighted average) ± 1.96 s.e.

---

[5]See H.O. Lancaster, *The combination of probabilities arising from data in discrete distributions*, 36 Biometrika 370 (1949). We are indebted to Joseph L. Gastwirth for calling our attention to this adjustment. The Lancaster adjustment actually overcorrects. The exact P-value is 0.024, which is above the P-value given by Fisher's test with Lancaster's correction. In this case, the binomial P-value referred to in the text comes considerably closer to the correct figure.

For example, if study 1 has an estimated logs odds ratio of 0.5 with standard error 0.3, and study 2 has an estimated log odds ratio of 0.7 with standard error 0.4, the weighted average is $(0.5 \cdot 0.3^{-?} + 0.7 \cdot 0.4^{-2})/(0.3^{-2} + 0.4^{-2}) = 0.572$ with standard error $1/(0.3^{-2} + 0.4^{-2})^{1/2} = 0.240$. A 95% confidence interval for the weighted average is $0.572 \pm 1.96 \cdot 0.240 = 0.572 \pm 0.47$.

The Mantel-Haenszel procedure also offers an estimate of an assumed common odds ratio underlying each fourfold table. To state the estimator, together with a standard error formula for its logarithm, we use the following notation for cell frequencies in the ith table $(i = 1, \ldots, k)$:

X_i	W_i
Y_i	Z_i

T_i

Let $B_i = X_i Z_i / T_i$ denote the product of the main-diagonal cell frequencies in table i divided by the table total T_i, and let $C_i = Y_i W_i / T_i$ denote the product of the off-diagonal cell frequencies divided by the table total. Let $B = \sum_i B_i$ and $C = \sum_i C_i$ denote the sums of these products over all tables. The Mantel-Haenszel estimator of the common odds ratio is then the ratio $OR_{MH} = B/C$. It is an advantage of this method that, since we are dealing with sums of products, a study with a zero cell in its fourfold table may still be included in computing the common odds ratio; this could not be done in the first method (without arbitrarily adding a value to the zero cell) since it assumed that each study generated a finite point estimate of the log odds ratio.

Note that the estimator can be viewed as a weighted average of individual table cross-product ratios $X_i Z_i / Y_i W_i$, where the weights are given by C_i/C. From this it is easy to see that, if the number of tables k remains fixed while the table sizes T_i become large, the Mantel-Haenszel estimator is approximately a weighted average of the true underlying table odds ratios. Assuming these all have a common value, it follows that OR_{MH} is a consistent estimator of that common odds ratio.

It is also true that the Mantel-Haenszel estimator is consistent in another asymptotic (large sample) framework: when each of the table totals remains small, but the number of tables k becomes large. For example, a "buddy study" (see p. 240) may match male and female employees on important characteristics, such as experience and qualifications, to avoid confounding bias. At the end of one year, it is observed whether or not the employees are promoted. Is there an equitable promotion policy? Each pair is a two-by-two table with the total T_i for each table being 2. The Mantel-Haenszel estimator of the odds ratio on promotion for female versus male employees is given by $OR_{MH} = b/c$, where b is the number of pairs in which the woman was promoted and the man was not and c is the number of tables in which the man was promoted and the woman was not. The estimator b/c is also known as the McNemar estimate of the common odds ratio for matched pairs data. As k becomes large, OR_{MH} converges to the true odds ratio by the strong law of large numbers.

Another application of the Mantel-Haenszel estimator is given in Section 11.2. at p. 323 to estimate a hazard rate ratio in survival analysis.

Robins, Greenland, and Breslow give a formula for the large sample variance of the logarithm of OR_{MH} that is itself a consistent estimator of the variance in both asymptotic frameworks mentioned above.[6] In addition to the notation above, let $P_i = (X_i + Z_i)/T_i$ denote the fraction of total frequency T_i in the main-diagonal cells of the ith table, and let $Q_i = (Y_i + W_i)/T_i = 1 - P_i$ denote the fraction of total frequency T_i in the off-diagonal cells. Then the "RGB" estimator is given as follows:

$$V\hat{a}r\{log\,OR_{MH}\} = (1/2)\left\{\frac{\sum_{i=1}^{k}P_iB_i}{B^2} + \frac{\sum_{i=1}^{k}(P_iC_i + Q_iB_i)}{BC} + \frac{\sum_{i=1}^{k}Q_iC_i}{C^2}\right\}.$$

The square root of this variance estimate is the standard error (s.e.) of the Mantel-Haenszel log odds ratio estimator, and may be used to set confidence intervals. Thus, a large-sample approximate 95% confidence interval for the log odds ratio is given by $log\,OR_{MH} \pm 1.96\,s.e.$, and an approximate 95% confidence interval for the common odds ratio is obtained by exponentiating the endpoints of that interval, viz.,

$$exp\{log\,OR_{MH} \pm 1.96\,s.e.\} = OR_{MH} \times/\div exp\{1.96\,s.e.\}.$$

In the matched pairs example, the RGB estimate of the variance of the logarithm of OR_{MH} is $(1/2)\{(b/2)/(b/2)^2 + 0 + (c/2)/(c/2)^2\} = (1/b) + (1/c)$. We return to this subject in the discussion of meta-analysis in Section 8.2.

Further Reading

J. Fleiss, *Statistical Methods for Rates and Proportions*, ch. 10 (2d ed. 1981).

8.1.1 Hiring lawyers

The State of Mississippi Department of Justice was accused of discriminating against black lawyers and black support personnel who applied for positions with the state. The data on the numbers of applicants and hires for 1970 through 1982 for positions as attorneys are shown in Table 8.1.1. The suit was instituted in 1975 when Attorney General Winter was in office. He was followed by Attorney General Summer in 1980. (The names are not fictitious.)

---

[6]J. Robins, S. Greenland, and N.E. Breslow, *A general estimator for the variance of the Mantel-Haenszel odds ratio*, 124 Am. J. Epidemiology 719 (1986).

TABLE 8.1.1. Attorney applicants and hires at the Office of the Attorney General from 1970 to 1983

	1970	1971	1972	1973	1974	1975	1976
W	7/12	5/20	11/17	10/21	6/18	2/15	3/16
B	—	—	—	0/2	0/7	0/7	2/5

	1977	1978	1979	1980	1981	1982
W	13/19	8/18	6/15	7/27	4/21	1/4
B	1/4	0/1	1/4	3/12	1/5	3/13

A/B: The figure to the left of the slash (A) is the number of people ultimately hired from that year's applicants, even if they were not hired in the same year that they applied; the figure to the right of the slash (B) represents the number of applicants.

Question

Analyze the data for statistical significance using the Mantel-Haenszel z-score. Would Fisher's exact test and his method of combining evidence be useful in this problem? Would pooling of data be an acceptable alternative?

Source

Mississippi Council on Human Relations v. Mississippi, J76 Civ. 118R (1983) (Magistrate's decision) (unpublished).

8.1.2   Age discrimination in employment terminations

A publisher was accused of age discrimination in its terminations of copywriters and art directors. Between January 3, 1984, and January 21, 1986, there were 15 involuntary terminations of copywriters and art directors. Tables 8.1.2a and 8.1.2b show the ages of the copywriters and art directors at the date of each termination. Thus, on July 27, 1984, there were 14 copywriters ranging in age from 24 to 62 and 17 art directors ranging in age from 24 to 56; the age of the terminated copywriter was 36. Table 8.1.2c shows (1) the aggregate ages of the terminated copywriter(s) or art director(s); (2) the number terminated; (3) the average age of all copywriters and art directors at the time of a termination; (4) the average age times the number of terminations (the expected sum of ages at termination under the null hypothesis); (5) the number of persons at risk of termination at each termination; (6) the variance of the ages of those at risk of termination; and (7) the sampling variance of column (1) [equal to column $6\times$ column 2, times the finite population correction factor, which is column 5 minus column 2 divided by column 5 minus 1].

TABLE 8.1.2a. Ages of copywriters by dates of involuntary termination

	Jan 3 1984	Jan 6	Jun 29	Jul 27	Oct 5	Nov 19	Nov 20	Dec 31	Mar 1 1985	Mar 29	Jun 7	Dec 11	Jan 21 1986
1				36	36	36	36	36	36	36	37	37	37
2	52	52	52	52	52	53	53*						
3			36	36*									
4					28	28	28	28	28	28	28		
5								41	41	41	41	41	42
6			30	30	30	30	30	30	30	30	30	31	31
7	49*												
8												27	28
9	47	47	47*										
10	34	34*											
11												44	44
12			40	40	40	40	40	40	40	40	40	41	41
13	47	47	47	47	47	47	47	48	48*				
14												36	36
15										36	36	36	37
16	62	62	62	62	62	63	63	63*					
17	33	33	33	33	33	34	34	34	34	34	34	35	35
18								26	26	26*			
19	29	29	30	30	30	30	30	30	30	30	30	31	31
20	28	28	28	28	28	28	28	29	29	29	29	30	30
21	42	42	43	43	43	43	43	43	43	44			
22	24	24											
23									26	27	27	27	27
24	35	35	35	35									
25	23	23	24	24	24	24	24	24	25	25	25	25	25
26	30	30	31	31	31	31	31	31	31	31			

*involuntary termination

TABLE 8.1.2b. Ages of art directors by dates of involuntary termination

	Jan 3 1984	Jan 6	Jun 29	Jul 27	Oct 5	Nov 19	Nov 20	Dec 31	Mar 1 1985	Mar 29	Jun 7	Dec 11	Jan 21 1986
1	47	47	47	47	47	48*							
2	34	34	34	34	34	34	34	35	35	35	35	36*	
3								24	24	24	25	25	25
4	38	38	39	39	39	39	39	39	39	39	40	40	40
5	36	36											
6													52
7	55	55	55	55	55	55	55	56*					
8	55	55	56	56	56	56	56	56*					
9												24	24
10	34	34	35	35	35	35	35	35	35	36	36	36	36
11	30	30	30	31	31*								
12												28	28
13	38	38	38	38	39	39	39	39					
14													33*
15									34	35	35	35	35
16										37	37*		
17			35	35	35	35	35	35	35	35	35	36	36
18						31	31	31	32	32	32		
19	27	27	28	28	28	28	28						
20									39	39	39	39	40
21	24	24	24	24									
22	35	35	35	35	36	36	36	36	36				
23	48	48	49	49	49	49	49	49	49	49	49	50	50
24	34	34	35	35	35								
25	52	52	53	53	53	53	53	53	53	53	54		
26												25	25
27	30	30	31	31	31	31	31	31	31	32			
28	47	47	48	48	48	48	48	48	48	48	48	49	49

*involuntary termination

TABLE 8.1.2c. Combining the evidence: copywriters and art directors

	(1) $S[j]$	(2) $m[j]$	(3) $\mu[j]$	(4) 2×3	(5) $N[j]$	(6) $\sigma^2[j]$	(7) $\mathrm{Var}\,S[j]$
1.	49	1	38.68	38.68	31	106.93	106.93
2.	34	1	38.33	38.33	30	106.82	106.82
3.	47	1	39.03	39.03	31	98.16	98.16
4.	36	1	38.71	38.71	31	95.75	95.75
5.	31	1	39.14	39.14	29	97.71	97.71
6.	48	1	39.43	39.43	28	103.32	103.32
7.	53	1	39.11	39.11	27	104.32	104.32
8.	175	3	38.21	114.63	28	103.60	287.77
9.	48	1	35.44	35.44	27	58.25	58.25
10.	26	1	35.22	35.22	27	52.02	52.02
11.	37	1	35.78	35.78	23	54.08	54.08
12.	36	1	34.56	34.56	25	50.97	50.97
13.	33	1	35.27	35.27	26	60.74	60.74
Total	653	15		563.4			1276.84

Questions

1. How does the age of each terminated copywriter or art director compare with the average for all copywriters and art directors at each termination?

2. Combining the evidence at each termination, do the data show a statistically significant greater risk of termination for older employees?

We revisit these data in Section 11.2.1.

8.2   Meta-analysis

Combining the evidence is a valid and informative procedure when data are collected with similar measurement techniques in each stratum, under similar conditions of observation and study protocol. In Section 8.1, in particular, we assumed that each stratum furnished data governed in part by a common parameter value that was the target of statistical inference. For example, to test the null hypothesis of no association between group membership and employment success, the Mantel-Haenszel procedure assumes a common odds ratio of unity in each stratum. As another example, in taking a weighted average of separate point estimates, we assumed each estimate was unbiased or consistent for a single common parameter. The data were also assumed to be statistically independent across strata.

It is tempting (especially in toxic tort cases) to apply these techniques to a broader problem: to select, review, summarize, and combine the results from separate published studies on a common scientific issue. Whether Bendectin is a teratogen is an example; see Section 8.2.1. The result of such a "study of studies" is called meta-analysis. While the statistical methods used to combine the evidence across studies are those already discussed, meta-analytic results differ in their interpretability, degree of validity, and the extent to which they can be generalized.

Two different studies rarely measure precisely the same parameter. Differences in study methods and measures, subject populations, time frames, risk factors, and analytic techniques, etc., all conspire to make the "common" parameter different in each study.[7] Nor is it always clear that one wants necessarily to have the same parameter in each study: many replications of a study demonstrating the lack of a toxic effect of a drug in non-pregnant women would not be informative for pregnant women.

One must then ask what one is testing or estimating when one combines a heterogeneous group of studies. The answer is some ill-specified average of the estimated parameters from the selected studies. Whatever the average is, it depends very much on the sampling scheme by which studies are selected for combination. This is a serious problem for meta-analysis, because so often there is no "scientific" sampling of studies. Obviously, bias in the selection of studies is compounded when one limits the selection to all *published* studies, because of the well known bias that favors publication of significant over non-significant findings. This is the "file-drawer" problem for meta-analysis: how to access the unknown number of negative findings buried in researchers' file drawers to achieve a balanced cross section of findings. Finally, there is an almost inevitable lack of independence across studies. Successive studies by the same researcher on a subject are often subject to the same systematic biases and errors: a recognized authority in a field may determine an entire research program; and in extreme cases, scientific competitors may have hidden agendas that affect what gets studied and reported.

For these reasons, results of meta-analysis will be most generalizable if the protocol includes the following steps: 1) creating a preestablished research plan for including and excluding studies, which specifies criteria for the range of patients, range of diagnoses, and range of treatments; 2) making a thorough literature search, including an effort to find unpublished studies; 3) assembling a list of included and excluded studies, in the latter case with the reasons for their exclusion; 4) calculating the P-value, a point estimate of effect, and a confidence interval for each study; 5) testing whether the studies are homogeneous, i.e., whether the differences among them are consistent with random sampling error and not some systematic factor or unexplained heterogeneity; if a systematic difference is found in subgroups of studies – e.g. cohort vs. case-control studies – making separate analyses of the two groups; 6) if the studies are homogeneous, calculating a summary statistic for all of them together, with a confidence interval for the statistic; 7) calculating the statistical power curves for the result against a range of alternative hypotheses; 8) calcu-

---

[7]This point is quite apart from sampling variability. Even in large samples where sampling variability may be ignored, systematic differences between studies remain.

lating the robustness of the result, namely, how many negative studies would have to exist (presumably unpublished or perhaps in a foreign language not searched) for the observed effect to be neutralized; and 9) making a sensitivity analysis, i.e., eliminating a study or studies that appear to have more serious design flaws to measure the effect on the results.

Fixed effects

Assume that there are k independent studies each estimating a parameter, such as a relative risk (RR), and the ith study estimates RR_i, with variance of the log RR_i equal to σ_i^2, assumed to be known. The log summary relative risk (SRR) is the weighted average of the $\ln(RR_i) = b_i$ estimates, with the weights inversely proportional to σ_i^2. In symbols:

$$b = \ln(SRR) = \sum_{i=1}^{k} \sigma_i^{-2} \ln(RR_i) / \sum_{i=1}^{k} \sigma_i^{-2} = \sum_{i=1}^{k} w_i b_i / \sum_{i=1}^{k} w_i,$$

where the weight $w_i = \sigma_i^{-2}$. Taking antilogs, an approximate 95% confidence interval for SRR is obtained from

$$(SRR)e^{\pm 1.96\sqrt{V[\ln(SRR)]}},$$

with the variance (V) of the natural log of SRR given by:

$$V[\ln(SRR)] = 1 / \sum_{i=1}^{k} \sigma_i^{-2} = 1 / \sum_{i=1}^{k} w_i.$$

With respect to homogeneity, (6) above, if the studies are homogeneous in the sense of all measuring the same thing and differ merely because of sampling variation, then the squared difference between the natural log of the ith study and the natural log of the SRR divided by the variance of the ith study summed over all k studies has a chi-squared distribution with $k - 1$ degrees of freedom. In symbols:

$$\sum_{i=1}^{k} w_i(b_i - b)^2 = \sum_{i=1}^{k} [\ln(RR_i) - \ln(SRR)]^2 / \sigma_i^2 \sim \chi_{k-1}^2.$$

With respect to power, (7) above, under the assumptions already given for k studies it can be shown that an approximate expression for the power, θ_m, of the meta-analysis to reject the null hypothesis $H_0 : RR = 1$ vs. $H_1 : RR > 1$ is

$$\theta_m \cong pr\{Z < \ln(\psi)/\sqrt{V[\ln(SRR)]} - 1.96\},$$

where Z is a standard normal variable and ψ is the relative risk under the alternative hypothesis, which is assumed to be greater than 1. Given the set of $\{\sigma_i\}$ values, power curves can be constructed by plotting values of θ_m as a function of the values of ψ.[8]

Random effects

Up to now the discussion has assumed that the true value of the parameter underlying each study was the same, or that within identifiable subsets of studies, the parameter was a constant. The chi-squared for homogeneity tests this hypothesis, but it is known that this test often lacks good statistical power. Thus, even if the chi-squared test for homogeneity does not reject that hypothesis, there may be actual variation in the true odds ratios between studies, and variability in the sample odds ratios beyond what can be accounted for by sampling error within each study. In other cases, the heterogeneity may be clear, but not explainable by known, systematic factors. This is typically the case when endpoints are difficult to diagnose, or when studies are conducted under conditions with many impinging factors that are hard to control or even measure. The situation also arises when each subject in an investigation provides repeated observations over time, and, in a sense, becomes his or her own "study." The statistical methods of meta-analysis can be brought to bear to combine the evidence contributed by each subject to the overall study question, but almost always biologic variability will be a substantial source of heterogeneity between the subject (study)-specific parameters. In such cases interest shifts away from testing the hypothesis of homogeneity as a preliminary step in the analysis, and toward specifying a model that embraces the heterogeneity as a real part of the uncertainty in drawing inferences. *Random-effects* models do this, and a random-effects meta-analysis is one that explicitly permits variation in the true parameter from study to study, and seeks to make inferences about some central feature of the *distribution* of the true parameters in the population, either real or hypothetical, of all such studies.

The methods for random-effects meta-analysis are, in the simplest cases, parallel to those for fixed-effects analysis, as described above. The essential substantive difference is that the standard error of the final estimate (of the population mean parameter) in the random-effects model is larger than the standard error of the final estimate (of the assumed common parameter) in the fixed-effects model, due to the acknowledged presence of between-study heterogeneity.

Here is an explicit random-effects model. Suppose we let β_i denote the true parameter value for the ith study ($i = 1, \ldots, k$), and let b_i denote the study estimate of β_i. For example, we may let β_i be the true log relative risk for study i, and take $b_i = \ln(RR_i)$ to be the study's estimated log relative risk.[9] When individual study sample sizes are large, the central limit theorem ensures

---

[8]For values of ψ not close to SRR, V may need to be recomputed when it depends on ψ.

[9]For combining the evidence about odds ratios from several independent fourfold tables, the Mantel-Haenszel chi-squared procedure and its associated estimate of the assumed

that the sample estimate b_i will be approximately normally distributed with mean β_i and variance given by the squared standard error σ_i^2; in symbols, $b_i \sim N(\beta_i, \sigma_i^2)$. The fixed effects model assumes that all the β_i are equal, but the random effects model assumes that the β_i have a distribution, say F, with mean β, say, and variance τ^2. The object of the meta-analysis is to estimate β. (It is here that the assumption of a complete and unbiased set of studies becomes crucial for meta-analysis – to estimate β without bias, the β_i ought to be a random sample from F, or at least unbiased.) Note that to derive a summary estimator for β, we need make no distributional assumption about F other than the existence of the mean β and variance τ^2.[10]

Viewed as an estimate of β, the individual study estimates are each unbiased, because $E[b_i] = E\{E[b_i|\beta_i]\} = \beta$. Given the design of the ith study, which fixes σ_i^2, the variance of b_i as an estimate of β is $\sigma_i^2 + \tau^2$, reflecting the uncertainty of b_i as an estimate of β_i plus the variation of β_i around the population mean β. If we wish to estimate β by a weighted average of the b_i, the weights that will minimize the variance of the estimate are inversely proportional to $\sigma_i^2 + \tau^2$. Thus, we may use as the summary statistic,

$$b^* = \sum_{i=1}^{k} w_i^* b_i / \sum_{i=1}^{k} w_i^*, \quad \text{where} \quad w_i^* = 1/(\sigma_i^2 + \tau^2),$$

and the variance of the estimate will be

$$V(b^*) = \frac{1}{\sum_{i=1}^{k} w_i^*} = \frac{1}{\sum_{i=1}^{k} 1/(\sigma_i^2 + \tau^2)}.$$

Notice the similarity between these formulas and those in the fixed-effects model. Relative to the fixed-effects model, where a few studies with very large sample sizes can dominate the estimate (because their respective σ_i^2 are small and weights $w_i = \sigma_i^{-2}$ are large), in the random-effects model the presence of τ^2 in the weights w_i^* has a tendency to reduce the dominance of those large studies. There are two examples of interest. If there were no heterogeneity at all, then τ^2 would equal zero, and the analysis reduces to the fixed effects case. At the other extreme, if the between-study variance τ^2 were large compared with the squared standard errors σ_i^2 of each study, then the weights w_i^* would be essentially equal, i.c., in a random-effects meta-analysis with substantial heterogeneity, the studies contribute essentially equal weight.

---

common odds ratio are available. See Section 8.1 at p. 237. The method given here will be equivalent to the Mantel-Haenszel procedure when each of the fourfold tables has large margins.

[10]Technically, we assume only that $E[b_i|\beta_i, \sigma_i^2] = \beta_i$, $\mathrm{Var}[b_i|\beta_i, \sigma_i^2] = \sigma_i^2$, $E[\beta_i|\sigma_i^2] = \beta$, and $\mathrm{Var}[\beta_i|\sigma_i^2] = \tau^2$. The normal assumption for b_i given β_i and σ_i^2 is used when making large sample inferences about the summary estimate of β. For this purpose it is often assumed that the distribution F is also approximately normal, in which case the summary estimate of β is too. However, the assumption of normality for F is often questionable and a source of vulnerability for the analyst.

Notice, finally, that whenever τ^2 is non-zero, the weight w_i^* is less than the corresponding fixed effect weight $w_i = \sigma_i^{-2}$, and thus the variance $1/\sum_i w_i^*$ of the summary estimate b^* will be greater than the variance $1/\sum_i w_i$ of the fixed-effect summary estimate. This is the "price" to be paid for the additional uncertainty introduced by the between-study heterogeneity.

Generally τ^2 is unknown and must be estimated from the data. An unbiased estimate of τ^2 is furnished by the following.

(i) Obtain the homogeneity statistic $Q = \sum_{i=1}^{k} w_i(b_i - b)^2$ as in the fixed effects model.

(ii) Obtain the quantities $\bar{w} = (1/k) \sum_{i=1}^{k} w_i$ and $S_w^2 = \sum_{i=1}^{k}(w_i = \bar{w})^2/(k-1)$.

(iii) If $Q \leq k - 1$, then estimate $\tilde{\tau}^2 = 0$. Otherwise, estimate $\tilde{\tau}^2 =$

$$\frac{Q - (k - 1)}{(k - 1)\{\bar{w} - s_w^2/(k\bar{w})\}}.$$

Further Reading

The standard book on meta-analysis is L.V. Hedges & I. Olkin, *Statistical Methods for Meta-Analysis* (1985).

8.2.1   Bendectin revisited

In the Bendectin litigation (see Section 5.4.2) Merrell Dow, the defendant, introduced a summary of the results of published cohort studies of Bendectin and congenital malformations. The summary is shown in Tables 8.2.1a and 8.2.1b.

TABLE 8.2.1a. Cohort studies of Bendectin and congenital malformations

First author	Year	Months of pregnancy	Type of malformation
Bunde	1963	1st trimes.	All diag. at birth
Brit. GP	1963	1st trimes.	All
Milkovich	1976	1st 84 days.	"Severe"
Newman	1977	1st trimes.	"Postural excluded"
Smithells	1978	2–12 wks.[1]	"Major"
Heinonen	1979	1–4 mos.	"Uniform," "Major & minor"[2]
Michaelis	1980	1-12 wks.	"Severe"
Fleming	1981	1-13 wks.	Defined & other
Jick	1981	1st trimes.	Major
Gibson	1981	1st trimes.	"Total"

1. The most unfavorable (to Bendectin) reasonable break-point.
2. Tumors excluded.

TABLE 8.2.1b. Findings of Bendectin cohort studies

First author	Exposed # infants	Exposed # malformed	Not Exposed # infants	Not Exposed # malformed	Log RR[3]	S.E. Log RR[3]
Bunde	2,218	11	2,218	21	−0.654	0.303
Brit. GP	70	1	606	21	−0.892	0.826
Milkovich	628	14[2]	9,577	343[2]	−0.478	0.223
Newman	1,192	6	6,741	70	−0.734	0.343
Smithells	1,622	27	652[1]	8	0.307	0.332
Heinonen	1,000	45	49,282	2,094	0.058	0.124
Michaelis	951	20	11,367	175	0.307	0.195
Fleming	620	8	22,357	445	−0.431	0.296
Jick	2,255	24	4,582	56	−0.139	0.202
Gibson	1,685	78	5,771	245	0.086	0.104
Total	12,241	234	113,153	3,478		

[1] Exposed over 12 weeks. [2] Computed from the given rates. [3] Relative risk.

Questions

1. Compute an estimate for the log relative risk and a standard error for that estimate based on the combined evidence in the studies.

2. Find a 95% confidence interval for the estimate.

3. State the point estimate and confidence interval in the original scale of relative risk.

4. What magnitude of common relative risk would have an 80% probability of being declared significant at the .05 level, two-tailed?

5. The weighted variance of the $\ln(RR)$ of the individual studies (Q) is 21.46. Interpreting this as a test of the hypothesis of homogeneity of the true $\ln(RR)$'s, what conclusion do you reach?

6. Compute an estimate of τ^2 and interpret your result. (The mean and variance of the weights are 26.795 and 844.828, respectively.)

CHAPTER 9

Sampling Issues

9.1 The theory of random sampling

The objective of random sampling is to be able to draw valid statistical inferences about properties or parameters of the population from which the sample is drawn. For example, a random sample of a company's employee medical claims drawn for auditing may have the purpose of estimating the proportion of improper allowances in the entire set of claims for a given time period. Other statistical inferences that rely on sampling include hypothesis tests about purported parameter values, prediction of future events, and selection or ranking of populations along some dimension of preferability.

For valid sampling of discrete populations it is necessary to have an enumeration of the population, called the sampling frame, available either physically or in principle, so that the probability of selecting any possible sample is known in advance. A simple random sampling procedure guarantees that each possible sample of a given size is equally likely to be selected. Stratified random sampling takes a simple random sample within each of several pre-specified subsets or strata of the sampling frame.

A common misconception regarding random sampling is that the goal is to obtain a *representative* sample of the population. The term has no definite meaning. If the sole goal were representativeness in some defined sense, systematic but non-random selection of units judged to be representative might well be preferable to a simple random sample. The statistician's reason for preferring a random sample, as stated above, is to be able to make probabilistic statements, such as the following: "the estimated proportion of improper allowances in the sampled population is 0.10, and with 99% confidence the proportion is at least 0.05, in the sense that if the true (population) proportion were any less, there would have been less than a 1% chance of observing a sample with as high a proportion of improper allowances as was in fact observed." Without the probability space generated by the random sampling procedure it becomes impossible, or at best speculative, to assign quantitative statements about the reliability of inferences based on the sample.

While representativeness is not the primary goal of random sampling, it is true that random sampling does tend to produce samples generally "like" the population in the sense that the mean quantity of interest in the sample will likely be close to that parameter in the population. This is a consequence of the law of large numbers: sample means and proportions accurately approximate population means and proportions with high probability in sufficiently large samples (see Section 1.2). However, some imbalance almost always occurs. An important side benefit of random sampling is that it reduces the opportunity for bias, conscious or unconscious, to affect the resulting sample. We note that what is random, or unbiased, is the sampling *procedure*, not the sample itself. The term "random sample" is shorthand for "a sample generated by a randomization procedure."

The random sample can be selected (i) by use of a table of random digits, or appropriate "pseudo-random number generators" available on all computers and many calculators; (ii) physical randomization devices such as a roulette wheel or a lottery; (iii) systematic sampling, e.g., sampling every n^{th} item from the population ordered in suitable fashion. Of these methods, random number tables and generators provide the most reliable randomization. Physical randomization devices often produce questionable results. Systematic sampling is only random if the ordering of the population is random with respect to the attribute sampled. Hidden periodicities threaten the validity of a systematic scheme, e.g., sampling every seventh item in a daily file would give biased results if there were weekly cycles in the frequency of the sampled attribute. Note that "haphazard" and "convenience" samples are *not* random samples. Sometimes the term *probability sample* is used to distinguish random samples from these other types of samples.

Stratified and Cluster Sampling

With *stratified* random sampling we reinstate controlled features into the sampling design. The relative proportion of the total sample size devoted to a given stratum may be chosen to give that stratum its "fair share" according to its size in the population. In this way, finely stratified random sampling can be both adequately "representative" and still support statistical inferences *within* strata. Inferences can then be combined *across* strata to arrive at general statements about the population as a whole. For example, mortality rates are often estimated by stratified random sampling within specific age- and sex-based strata. The specific mortality rates may then be applied to the population age/sex structure to arrive at an estimated average mortality rate, or they may be applied to a reference age/sex structure to arrive at a "standardized" rate.

Stratified random sampling has another important property that often makes its use imperative. The sampling units within a stratum may be more homogeneous in terms of the properties being measured than they are across strata. In such cases more precise estimates are obtained from a weighted average of stratum-specific estimates than from a simple random sample. An extreme case makes the point obvious. Suppose there are two strata: stratum A has a property occurring 100% of the time; stratum B has the property occurring 0% of the time. Even a small random sample from each stratum can be used to infer correctly that the property occurs $p\%$ of the time in the population, where p is the percentage share of stratum A. A simple random sample, on the other hand, will provide an unbiased estimate of p, but with extra variability due to the random mixture of stratum A and B units in the sample.

The optimal sampling scheme–i.e., the one that produces estimates with the smallest sampling error–allocates sample sizes in proportion to the relative stratum size times the stratum standard deviation. Most often, however, the stratum standard deviations will be unknown before sampling, in which case one usually allocates the sample sizes in proportion to the stratum sizes. This is less efficient than the optimal allocation unless the stratum standard deviations are equal, in which case the methods are equally efficient. To get a more precise idea of how much more efficient stratified sampling can be, the variance of the stratified estimator of a sample mean ($\hat{\mu}$) when the sizes of the strata are known and sampling is proportional to stratum size can be written as

$$\mathrm{Var}(\hat{\mu}) = \sum_i p_i \sigma_i^2 / n,$$

where p_i is the proportional size of the ith stratum, σ_i^2 is the variance of the population in the ith stratum, and n is the sample size. By contrast, if strata sizes are unknown so that a simple random sample must be used, the variance of the mean of the sample ($\bar{X}$) can be written as

$$Var(\bar{X}) = \left\{ \sum_i p_i \sigma_i^2 + \sum_i p_i (\mu_i - \mu)^2 \right\} / n = \sigma^2 / n,$$

where σ^2 is the marginal variance of a single randomly selected observation X without regard to stratum, μ_i is the mean of the population in the ith stratum, and the other terms are as defined above. Notice that the first term in the braces (divided by n) is the variance of the mean of a stratified sample. If the strata are ignored, the use of a simple random sample will increase the variance of the sample mean by the second term within the braces (also divided by n). The equations make clear that if X varies widely among strata, so that the second term is large relative to the first, the variance of the stratified estimator $\hat{\mu}$ can be substantially smaller than the variance of the simple mean $\bar{X}$.

Cluster sampling uses a random sample of special units. These units may be large, e.g. cities, in which case there is subsequent random selection from them, or quite small, e.g. city blocks, in which case all members are interviewed.

Cluster sampling often provides an economical strategy for surveys in human populations. When deciding upon the type of cluster to be sampled, e.g., census tracts, blocks, or households, consideration should always be given to the cost of sampling the units in relation to the variance of the estimate. Thus, one may select the type of unit as the one which gives smallest variance for a given cost, or smallest cost for a given variance. The variance of an estimate in cluster sampling is typically greater than in simple random sampling with the same total sample size, although the trade-off in terms of cost economy is often well worth the inflation in variance.

The variance of a sample mean obtained from a single-stage cluster sample is given by the variance that would obtain if the sample had been drawn by simple random sampling, multiplied by a quantity called the *variance inflation factor* (VIF). The VIF depends on two quantities: the size of the clusters sampled and the homogeneity of the units within a cluster relative to that between clusters. The latter quantity is most conveniently measured by the *intracluster correlation coefficient* (ICC), which is defined to be the correlation that would appear between pairs of observations sampled at random from the same randomly selected cluster. If two subjects within a given cluster tend to be more similar than two subjects from two different clusters, the ICC is positive. This reduces the information per subject sampled and increases the variance of estimated means. (It is possible for the ICC to be negative, in which case cluster sampling can be more efficient than random sampling in terms of both cost and precision, although this situation occurs only occasionally.) The variance inflation factor is given by the expression

$$VIF = 1 + (\bar{m} - 1)\rho.$$

where ρ is the ICC and $\bar{m} = \sum m_i^2 / \sum m_i = \sum m_i(m_i / \sum m_i)$ is a weighted average of the individual cluster sizes m_i, weighted in proportion to the cluster sizes themselves.

When the size of the cluster may be chosen by the survey designer, and the ICC is positive and does not vary substantially across the different choices of cluster size, the VIF formula shows that for a given total sample size it is better to sample more clusters of smaller size than it is to sample fewer clusters of larger size.

Because cluster sampling produces observations that are not statistically independent, special methods are required to estimate the population variance and ICC before one can provide a correct standard error for estimated means or proportions. Software packages such as SUDAAN (Survey Data Analysis) are generally required in complex surveys to allow for the design effects introduced by clustering and multistage sampling.

Stratified and cluster sampling are almost always needed when sampling a large natural population. A good example, described in *Zippo Manufacturing Company v. Rogers Imports, Inc.*, 216 F. Supp. 670 (S.D.N.Y. 1963), was a survey to determine whether Rogers lighters were confused with Zippo lighters. The universe was the adult (over 18) smoking population (then 115 million people) of the continental United States. Judge Fienberg described the stratified cluster sampling method as follows:

> Three separate surveys were conducted across a national probability sample of smokers, with a sample of approximately 500 for each survey. The samples were chosen on the basis of data obtained from the Bureau of Census by a procedure which started with the selection of fifty-three localities (metropolitan areas and non-metropolitan counties), and proceeded to a selection of 100 clusters within each of these localities—each cluster consisting of about 150–250 dwelling units—and then to approximately 500 respondents within the clusters. The manner of arriving at these clusters and respondents within each cluster was described in detail. The entire procedure was designed to obtain a representative sample of all smoking adults in the country. The procedures used to avoid [or more properly, reduce] sampling error and errors arising from other sources, the methods of processing, the instructions for the interviewers, and the approximate tolerance limits for a sample base of 500 were also described.

Id. at 681.

Ratio estimates

In estimating the size of some subgroup of a population from a sample, it is sometimes more efficient to estimate the ratio of the subgroup to the population in the sample; the ratio thus estimated is multiplied by a figure for the total population obtained from other sources to arrive at an estimate of the subgroup. This is called ratio estimation. Its advantage lies in the fact that, if the sizes of the subgroup and the population are highly correlated in samples, their ratio will vary less from sample to sample than either quantity separately; use of the ratio will thus reduce the sampling error of the estimate.

Sampling with and without replacement

Random sampling from a finite population of size N *with* replacement means that each possible sample of n units is equally likely, including those in which units may be re-sampled two or more times. Sampling *without* replacement means that only samples with n *distinct* units may be drawn. Usually survey sampling is done without replacement for practical reasons, although when the population sampling frame is large there is little difference between the two methods. For finite populations of small size, however, sampling without replacement produces estimates that are somewhat more precise than analogous estimates obtained with replacement. This is because, as the sample size increases to an appreciable fraction of the population, it becomes increasingly unlikely that the sample mean will vary by a given amount from the population mean. The effect on the variance of the sample mean is quantifiable as a multiplicative correction factor, known as the *finite population correction factor*. The factor is $(N - n)/(N - 1)$, where N is the size of the population, and n is the size of the sample. Thus, the sample mean has a variance equal to $(\sigma^2/n)(N - n)/(N - 1)$, which is approximately equal to $(\sigma^2/n)(1 - f)$ when N is large, where $f = n/N$ is the sampling fraction.

Nonsampling variability

In practice, sampling variation is rarely the only source of variability. The other principal sources are (i) defects in the sampling frame involving incomplete or inaccurate enumeration of the target populations; (ii) defects in methods of selection that result in unequal probabilities of selection; and (iii) defects in collection of data from the sample (including such matters as nonresponse, evasive answer, and recall bias).

A common misconception is that a sample's precision depends on its size relative to the size of the population; in this view, samples that are a tiny fraction of the population do not provide reliable information about that population. The more correct view is that if a population can be sampled correctly (a difficult task for large and diverse populations), the precision of the sample depends principally on the variability of the population, to a lesser extent on the sample size, and perhaps most importantly on the avoidance of defects that create nonsampling variation. A drop of blood is a good sample, even though it is a tiny fraction of the body's blood supply, because the heart is a good randomization device and blood is a relatively homogeneous fluid. Skepticism about the validity of small samples from large populations is justified when it is difficult or impossible to compile a complete list. For this reason, public opinion surveys are likely to have greater variability than that indicated by formal calculation of sampling variability. For example, "random digit dialing" is a widely used method for conducting telephone surveys. While the sampling procedure is random with respect to telephone numbers, systematic differences do exist between those with and without telephones. Many systematic errors or sampling biases have a potential for far more serious distortion than does sampling variability. Some statisticians believe that an estimate of nonsampling variability should be integrated with sampling variability to derive confidence intervals for sampling estimates that reflect both sources of error, but this is not yet the general practice.

Direct assessment of sampling and non-sampling error

Complex, multi-stage sampling plans often make it impossible to provide formulas for sampling variability. One technique to measure such variability involves splitting the sample in various ways and computing the sample results for each of the subsamples. The variation in results provides a direct assessment of variability, which is usually expressed as the root mean squared difference between each subsample and the average of the subsamples. These techniques have the virtue of measuring all variability, both sampling and nonsampling.

Sample v. census

In many cases, a well organized and well executed sample may be more accurate than a census. The reason is that nonsampling errors can be minimized when working with a smaller number of items or respondents, and such errors are often more important than sampling errors. The Census Bureau itself plans to use post-census samples to correct the census for the undercount of some minorities, a proposal that has drawn much litigation. See Section 9.2.1.

Acceptance of sampling

After some initial rejections, sampling is now widely accepted in judicial and administrative proceedings, and required in numerous contexts by government regulations. The use of sampling is expressly approved by the *Manual for Complex Litigation, Third*, 21.493 (1995). For some sampling guidelines developed for law cases see, e.g., G. Glasser, *Recommended Standards on Disclosure of Procedures Used for Statistical Studies to Collect Data Submitted in Evidence in Legal Cases*, 39 The Record of the Association of the Bar of the City of New York 49 at 64 (Jan./Feb. 1984).

Further Reading

W. Cochran, *Sampling Techniques* (3rd ed., 1977).
W. Deming, *Some Theory of Sampling* (1950).

9.1.1   Selective Service draft lotteries

President Nixon's executive order for the 1970 draft lottery provided that the sequence for inducting men into military service would be based on birthdays selected at random. The order provided:

> That a random selection sequence will be established by a drawing to be conducted in Washington, D.C., on December 1, 1969, and will be applied nationwide. The random selection method will use 366 days to represent the birthdays (month and day only) of all registrants who, prior to January 1, 1970, shall have attained their nineteenth year of age but not their twenty-sixth. The drawing, commencing with the first day selected and continuing until all 366 days are drawn, shall be accomplished impartially.

> On the day designated above, a supplemental drawing or drawings will be conducted to determine alphabetically the random selection sequence by name among registrants who have the same birthday.

> The random selection sequence obtained as described above shall determine the order of selection of registrants who prior to January 1, 1970,

TABLE 9.1.1a. 1970 Draft—Random selection sequence, by month and day

	Jan.	Feb.	Mar.	Apr.	May	June	July	Aug.	Sep.	Oct.	Nov.	Dec.
1	305	086	108	032	330	249	093	111	225	359	019	129
2	159	144	029	271	298	228	350	045	161	125	034	328
3	251	297	267	083	040	301	115	261	049	244	348	157
4	215	210	275	081	276	020	279	145	232	202	266	165
5	101	214	293	269	364	028	188	054	082	024	310	056
6	224	347	139	253	155	110	327	114	006	087	076	010
7	306	091	122	147	035	085	050	168	008	234	051	012
8	199	181	213	312	321	366	013	048	184	283	097	105
9	194	338	317	219	197	335	277	106	263	342	080	043
10	325	216	323	218	065	206	284	021	071	220	282	041
11	329	150	136	014	037	134	248	324	158	237	046	039
12	221	068	300	346	133	272	015	142	242	072	066	314
13	318	152	259	124	295	069	042	307	175	138	126	163
14	238	004	354	231	178	356	331	198	001	294	127	026
15	017	089	169	273	130	180	322	102	113	171	131	320
16	121	212	166	148	055	274	120	044	207	254	107	096
17	235	189	033	260	112	073	098	154	255	288	143	304
18	140	292	332	090	278	341	190	141	246	005	146	128
19	058	025	200	336	075	104	227	311	177	241	203	240
20	280	302	239	345	183	360	187	344	063	192	185	135
21	186	363	334	062	250	060	027	291	204	243	156	070
22	337	290	265	316	326	247	153	339	160	117	009	053
23	118	057	256	252	319	109	172	116	119	201	182	162
24	059	236	258	002	031	358	023	036	195	196	230	095
25	052	179	343	351	361	137	067	286	149	176	132	084
26	092	365	170	340	357	022	303	245	018	007	309	173
27	355	205	268	074	296	064	289	352	233	264	047	078
28	077	299	223	262	308	222	088	167	257	094	281	123
29	349	285	362	191	226	353	270	061	151	229	099	016
30	164		217	208	103	209	287	333	315	038	174	003
31	211		030		313		193	011		079		100

shall have attained their nineteenth year of age but not their twenty-sixth and who are not volunteers and not delinquents. New random selection sequences shall be established, in a similar manner, for registrants who attain their nineteenth year of age on or after January 1, 1970.

The random sequence number determined for any registrant shall apply to him so long as he remains subject to induction for military training and service by random selection.

Random Selection for Military Service, Proc. No. 3945, 34 Fed. Reg. 19,017 (1969), *reprinted in* 50 U.S.C. 455 app. (1970).

The sequence was important because Selective Service officials announced that those in the top third would probably be called, those in the middle

TABLE 9.1.1b. 1971 Draft–Random selection sequence, by month and day

	Jan.	Feb.	Mar.	Apr.	May	June	July	Aug.	Sep.	Oct	Nov.	Dcc.
1	133	335	014	224	179	065	104	326	283	306	243	347
2	195	354	077	216	096	304	322	102	161	191	205	321
3	336	186	207	297	171	135	030	279	183	134	294	110
4	099	094	117	037	240	042	059	300	231	266	039	305
5	033	097	299	124	301	233	287	064	295	166	286	027
6	285	016	296	312	268	153	164	251	021	078	245	198
7	159	025	141	142	029	169	365	263	265	131	072	162
8	116	127	079	267	105	007	106	049	108	045	119	323
9	053	187	278	223	357	352	001	125	313	302	176	114
10	101	046	150	165	146	076	158	359	130	160	063	204
11	144	227	317	178	293	355	174	230	288	084	123	073
12	152	262	024	089	210	051	257	320	314	070	255	019
13	330	013	241	143	353	342	349	058	238	092	272	151
14	071	260	012	202	040	363	156	103	247	115	011	348
15	075	201	157	182	344	276	273	270	291	310	362	087
16	136	334	258	031	175	229	284	329	139	034	197	041
17	054	345	220	264	212	289	341	343	200	290	006	315
18	185	337	319	138	180	214	090	109	333	340	280	208
19	188	331	189	062	155	163	316	083	228	074	252	249
20	211	020	170	118	242	043	120	069	261	196	098	218
21	129	213	246	008	225	113	356	050	068	005	035	181
22	132	271	269	256	199	307	282	250	088	036	253	194
23	048	351	281	292	222	044	172	010	206	339	193	219
24	177	226	203	244	022	236	360	274	237	149	081	002
25	057	325	298	328	026	327	003	364	107	017	023	361
26	140	086	121	137	148	308	047	091	093	184	052	080
27	173	066	254	235	122	055	085	232	338	318	168	239
28	346	234	095	082	009	215	190	248	309	028	324	128
29	277		147	111	061	154	004	032	303	259	100	145
30	112		056	358	209	217	015	167	018	332	067	192
31	060		038		350		221	275		311		126

third would possibly be called, and those in the bottom third were unlikely to be called.

Pursuant to this order, 366 birth dates were marked on slips and put into a bowl (beginning with 31 for January, etc.). After stirring, the slips were drawn one by one from the bowl. All men with the given birth dates were called to military service in the order of selection. Thus, since September 14 was the first date selected, all men born on September 14 were selected first. Only the first 100 were to be drafted from any birth date. The data appear in Table 9.1.1a.

In the 1971 draft lottery, two bowls were used; in one were slips for 365 days (omitting February 29) and in the other were slips numbered from 1 to 365. A date drawn from one bowl was given a selection number by drawing from the second bowl. The data from this drawing are given in Table 9.1.1b.

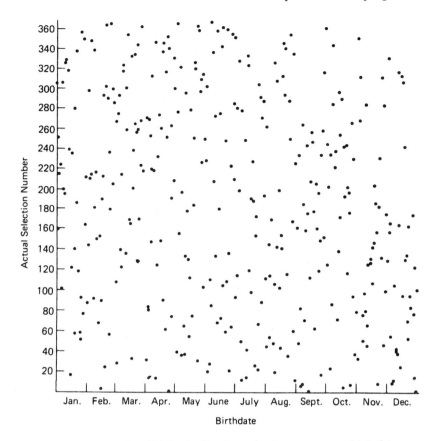

FIGURE 9.1.1. The 1970 Draft—Random selection sequence vs. birth date

Questions

1. Looking at Figure 9.1.1, a scatterplot for the 1970 lottery, do the data appear uniformly distributed?

2. Classifying sequence numbers into two or more appropriate subsets, find the tally for each birth month and use chi-squared to test for randomness in the 1970 lottery. Use the same method to test for randomness in the 1971 draft lottery.

3. What conclusions do you draw about the randomization mechanisms in the two lotteries?

Source

S. E. Fienberg, *Randomization and Social Affairs: the 1970 Draft Lottery* 171 Science 255 (1971); see also, 172 Science 630 (1971); M. Rosenblatt and J. Filliben, *Randomization and the Draft Lottery*, 171 Science 306 (1971).

9.1.2   Uninsured motor vehicles

When Michigan required "no fault" insurance coverage of motor vehicles, the law was challenged on constitutional grounds. One challenge was that the law compelled purchase of insurance by all motorists, including "many poor people who had the right to drive but had not the means to buy insurance." The question of how many uninsured motorists would be coerced by the law to buy insurance became a key issue in the trial. To resolve this question, the state proposed drawing a random sample of registered vehicles to investigate their insurance status.

At the time of trial, the state's computer file listed 4,569,317 authorized license plate numbers; of these, 63,652 plates had been stolen, were missing, or had never been manufactured, leaving 4,505,665 passenger vehicles registered in Michigan. A two-stage sample was drawn: first, a systematic sample of every 1,126th number from the total file, starting with the randomly selected number 577, to produce a sample of just under 4,000 vehicles; second, a subsample of every 16th number, starting with the random number 9, a subsample of 249 vehicles. This sample was the largest that could be investigated in the time allowed. Investigation determined that 233 vehicles were still being operated in the state, and that of this number 226 were insured. The statistician estimated from these data that 226/233, or 97% of the vehicles in the state were insured.

Questions

1. Compute a 95% confidence interval for this estimate.

2. Assuming the sampling methodology was correct, is it fair to argue that the sample (233 vehicles) is too small to "represent" the universe of motor vehicles?

3. What would be the basis for arguing that the statistician's estimate understates the true percentage of uninsured vehicles?

Source

Katz, *Presentation of a confidence interval estimate as evidence in a legal proceeding*, Dept. of Stat., Mich. St. U. (1974).

9.1.3   Mail order survey

A mail-order house in Colorado sells stationery and novelty items in California and other states. Because the company had a "substantial presence" in the state, California claimed a sales tax on the company's California sales for the period 1972–1976. The company responded that part of its sales were at wholesale, for which no tax was due.

TABLE 9.1.3a. Percentage distribution of orders by size (by number not value)

	Respondents	Nonrespondents
Under $10	44	49
$10–$29	48	42
$30-$49	6	6
$50 and up	2	3

TABLE 9.1.3b. Percentage distribution of orders by shipping address

	Respondents	Nonrespondents
San Francisco	23	24
Los Angeles	27	25
Other	50	51

To substantiate its claim, the company made a sample survey of its mail orders. The survey procedure, agreed to in advance by the California Board of Equalization, consisted of a questionnaire mailed to the California entity that put in every 35th California order placed in 1975. The questionnaire asked whether the purchase was for resale or personal use. Of 5,121 entities in the sample, 2,966 responded to the questionnaire, a typical response rate for such mail surveys. Among survey respondents, the aggregate value of orders for resale was $8,029 out of a total (for respondents) of $38,160. The value of respondents' resale orders was thus $8,029/$38,160, or 21% of their total orders. The total value of orders in the sample (respondents and nonrespondents combined) was $66,016. The value of respondents' orders for resale was thus $8,029/$66,016, approximately 12% of all orders in the sample. The company's total mail-order sales in California for the period were $7.01 million.

Question

The company proposed allocating 21% of its business to the resale category, for which no tax is due. The board proposed allocating 12% to that category. Which has the better case?

Source

D. Freedman, A *Case Study of Nonresponse: Plaintiff v. California State Board of Equalization* (unpublished ms.). For a similar survey that was not allowed by the court, see R. Clay Sprowls, *The Admissibility of Sample Data into a Court of Law: a Case History*, 4 UCLA L. Rev. 222 (1956-57).

Notes

Missing information and nonresponse are endemic problems in statistical samples and surveys. When the sampling technique is proper and there is indication from a covariate that the sample is not skewed, studies with substantial missing data and nonresponse have been accepted by the courts. Compare, *Vuyanich v. Republic National Bank*, 505 F. Supp. 224, 255–58 (N.D. Tex. 1980) (party challenging data should demonstrate that errors and omissions are not distributed randomly and bias the results; despite challenges, data base was accepted); and *Rosado v. Wyman*, 322 F. Supp. 1173 (E.D.N.Y. 1970) (due to the passage of time only 62.6% of the welfare records in a random sample of 5,344 could be found; the court accepted the sample after noting that the average payment and family size approximated those known characteristics of the whole population) with *E.E.O.C v. Eagle Iron Works*, 424 F. Supp. 240, 246–47 (S.D. Ia. 1946) (data for 60% of current and former employees rejected where all the missing racial data were from former employees) and *Bristol Meyers v. F.T.C.*, 185 F. 2d 258 (4th Cir. 1950) (survey with 20% response rate rejected; no followup study of the nonrespondents). See also Section 13.5 at p. 383.

The courts have generally been reluctant to accept estimates of taxes due based on samples taken by taxing authorities when the taxpayer has kept required records, but have permitted such estimates when required records are missing. See Bright, Kadane, and Nagin, *Statistical Sampling in Tax Audits*, 1988 Am. Bar Foundation Res. J. 305; *Narmac, Inc. v. Tracy*, 614 N.E.2d 1042 (Ohio 1993); *Wallach v. Tax Appeals Tribunal*, 614 N.Y.S.2d 647 (3rd Dep't 1994). Pennsylvania's Department of Revenue has been authorized by statute to determine tax liabilities "based upon a reasonable statistical sample or test audit" when the taxpayer's records are incomplete or when review of each item "would place an undue burden upon the Department." Act of June 30, 1995 (P.L. 139, No. 21), Section 2915-A.

9.1.4   Domino effect

Amstar Corp. sued Domino's Pizza, Inc., claiming that defendant's use of the name "Domino" on its pizza infringed on the Domino Sugar trademark. Both sides presented survey evidence on whether defendant's use of the name "Domino" tended to create confusion among consumers.

Plaintiff surveyed 525 persons in 10 cities in the eastern United States (two of which had Domino's Pizza outlets). The persons interviewed were women reached at home during the day who identified themselves as the household member responsible for grocery buying. Shown a Domino's Pizza box, 44.2% of those interviewed indicated their belief that the company that made the pizza made other products; 72% of that group (31.6% of all respondents) believed that the pizza company also made sugar.

Question

What criticisms do you have of the sampling frame for this study?

Source

Amstar Corp. v. Domino's Pizza, Inc., 205 U.S.P.Q. 128 (N.D. Ga. 1979), *rev'd*, 615 F. 2d 252 (5th Cir. 1980).

Notes

The use of sample surveys in trademark infringement cases to establish or refute claims of consumer confusion is now well established. *Zippo Manufacturing Co. v. Rogers Imports, Inc.*, 216 F. Supp. 670 (S.D.N.Y. 1963) was a seminal case in which a finding of confusion was based on such a survey. See p. 257 for a description of the stratified cluster samples used in that case. A much-followed statement of the requirements for admissibility of surveys was given by Judge Glasser in *Toys "R" Us, Inc. v. Canarsie Kiddie Shop, Inc.*, 559 F. Supp. 1189, 1205 (E.D.N.Y. 1983):

> The trustworthiness of survey evidence depends upon foundation evidence that (1) the 'universe' was properly defined, (2) a representative sample of that universe was selected, (3) the questions to be asked of interviewees were framed in a clear, precise and non-leading manner, (4) sound interview procedures were followed by competent interviewers who had no knowledge of the litigation or of the purpose for which the survey was conducted, (5) the data gathered was accurately reported, (6) the data was analyzed in accordance with accepted statistical principles, and (7) objectivity of the entire process was assured.

Do you have any objection to Judge Glasser's formulation? The survey in *Toys "R" Us* was conducted by "random interception" of customers at shopping areas, such as shopping malls. Would that method of selection satisfy the described standard?

Surveys purporting to show confusion are frequently rejected, perhaps more frequently than those purporting to show little or no confusion. See, e.g., *Alltel Corp. v. Actel Integrated Communications, Inc.*, 42 F. Supp. 2d 1265, 1269, 1273 (S.D. Ala. 1999) (survey purporting to show confusion rejected); *Cumberland Packing Corp. v. Monsanto Company*, 32 F. Supp. 2d 561 (E.D.N.Y. 1999) (see Section 9.1.5; same); *Levi Strauss & Co. v. Blue Bell, Inc.*, 216 U.S.P.Q. 606, *modified*, 78 F.2d 1352 (9th Cir. 1984) (defendant's survey showing that an overwhelming majority of shirt purchasers were not confused found more persuasive than plaintiff's survey, which showed 20% consumer confusion); but see, e.g., *Sterling Drug, Inc. v. Bayer AG*, 14 F. 3d 733, 741 (2d Cir. 1994) (survey showing confusion accepted).

Further reading

Shari Seidman Diamond, *Reference Guide on Survey Research* in Federal Judicial Center, *Reference Manual on Scientific Evidence* 229 (2d ed. 2000).

9.1.5   NatraTaste versus NutraSweet

The manufacturer of NatraTaste, an aspartame sweetener, brought suit against the maker of NutraSweet and EQUAL, other aspartame sweeteners, for trade dress infringement in violation of the Lanham Act. A product's trade dress is its total image, including features such as size, shape, color or color combinations, texture, or graphics. Plaintiff contended that the NutraSweet box in which the packets were sold too closely resembled the trade dress of NatraTaste's box. To be protected, a trade dress must be so distinctive as to point to a single source of origin. A generic trade dress, i.e., one pointing to a category, but not to a particular product, is ineligible for protection under the act. The court found that both the NatraTaste and NutraSweet trade dresses were distinctive and then turned to the question of confusion.

As the court described it, plaintiff's NatraTaste box is rectangular with over-all blue coloring. The name "NatraTaste" appears in large cursive font on the front and back panels of the box. The letters are green with white lining and stand out against the blue background. There is a photograph of a coffee cup with a saucer in the center right. Resting on the saucer is a photograph of an individually wrapped paper packet marked "NatraTaste." Another coffee cup to the lower left and saucers to the upper left cast shadows in the background. In the center right is a bright pink burst containing a comparative advertisement stating "Same Sweetener AS EQUAL. . . At A Sweeter Price." The top and side panels also say "NatraTaste" in the same style and format as in the front and back panels, but in slightly smaller font. Plaintiff introduced NatraTaste to the table-top market in 1993, shortly after defendant's patent on aspartame expired.

Defendant's NutraSweet box is also rectangular, but not as wide along the front and the side panels are thinner than the NatraTaste box. Although the box is light blue, there is less variation in blue tones than on the NatraTaste box. At the top of the front panel of the box is the red and white NutraSweet swirl logo, which stands out against the blue background. Below the logo is the trademark "NutraSweet" in big, black, block print. The bottom half of the front panel is largely taken up by a picture of a coffee cup on a saucer. An individual packet of NutraSweet is on the left side of the saucer, tilted against the cup so that the NutraSweet swirl logo and the NutraSweet trademark printed on the packet are visible. The back panel is identical to the front panel. The two side panels also display prominently the logo and the NutraSweet trademark. Defendant introduced and advertised Nutrasweet as a branded ingredient of foods and soft drinks in the 1980s, entered the table-top market with the aspartame sweetener EQUAL in 1982, and in 1997 introduced NutraSweet into that market.

Plaintiff claimed that points of similarity likely to cause confusion were (1) blue lighting and "the pronounced shadowing and gradation of the blue shades;" (2) a cup of black coffee (although NutraSweet's coffee is dark brown); (3) a packet resting on a saucer; and (4) a cup and a saucer resting on an undefined surface. Plaintiff did not claim that similarity of names was a source of confusion. The court commented that "The relevant inquiry is not how many details the two boxes share. It is whether the similarities create the same general, overall impression such as to make more likely the requisite confusion among the appropriate purchasers."

As part of its case that the boxes were confusingly similar, plaintiff introduced studies of consumer confusion. Respondents were selected from users or buyers of sugar substitutes in the previous six months. In the Two-Room, Array-Methodology study, respondents were shown a NatraTaste box in one room and were then escorted into a second room in which five other boxes of sweeteners, including NutraSweet, were displayed in a row on a table (the order was rotated). The four "control" boxes were Sweet Servings, Equal, Sweet One, and Sweet Thing. Only one of the control boxes had a blue background, and its blue was indigo. They were then asked: "Do you think any of these sugar substitutes is made by the same company as the first box you saw?" Those who answered affirmatively were then asked to identify the boxes and to give their reasons. No instruction was given against guessing.

The results were these: of the 120 people surveyed, 43% picked the NutraSweet box as made by the same company that made the NatraTaste box, while the percentages for the four control boxes were 13%, 18%, 7%, and 8%, respectively. The average for the control boxes was thus 12%, which the expert deducted from the 43% to arrive at 31% as the level of consumer confusion attributable to trade dress with regard to the NutraSweet box.

Among its other findings, the court held that the color blue signified aspartame in the sweetener industry (pink was for saccharin); being generic, it could not be a protected element of NatraTaste's trade dress.

Questions

1. What objections do you have with regard to the entry criteria and the method of conducting the study?

2. What criticisms do you have of the expert's analysis of the responses?

Source

Cumberland Packing Corp. v. Monsanto Company, 32 F. Supp. 2d 561 (E.D.N.Y. 1999).

9.1.6   Cocaine by the bag

In *People v. Hill*, 524 N.E. 2d 604 (Ill. App. 1998), defendants were convicted of possessing more than 30 grams of substance containing cocaine. They had been caught with some 55 plastic bags containing white powder. Although the opinion is not perfectly clear, it appears that the government chemist selected two or three bags at random, definitively tested them, and found that they each contained cocaine. The total weight of substance in those bags was 21.93 grams. The other bags were subjected to a color test, the results of which led the chemist to conclude, and she so testified, that they probably also contained cocaine. However, since the color test was only preliminary, the chemist conceded that it would not prove to a reasonable scientific certainty that the substance being examined was cocaine.

The appellate court allowed the conviction to stand only for the offense of possessing less than 30 grams of cocaine, since only 21.93 grams had been established by definitive tests. It held that "where separate bags or containers of suspected drugs are seized, a sample from each bag or container must be conclusively tested to prove that it contains a controlled substance." *Id.* at 611. In reaching this conclusion the court distinguished *People v Kaludis*, 497 N.E.2d 360 (Ill. App. 1996), a case decided two years earlier, in which the issue was whether random sampling was sufficient to sustain a conviction for possession of 100 tablets suspected of being methaqualone. The tablets were of the same size, shape, and hardness, and all were falsely marked "Lemmon 714." The chemist conducted tests on several randomly selected tablets, which conclusively established the presence of methaqualone. On appeal, the court concluded that where the substance is homogeneous, random sampling provides a sufficient basis for proving beyond a reasonable doubt that all of the tablets contained a controlled substance. *Id.* at 365–366.

Questions

1. Assuming that the chemist selected three bags at the random for definitive tests, use a confidence interval analysis to explore whether the sample data should be deemed sufficient for a jury to find beyond a reasonable doubt that at least some of the unsampled bags contained cocaine.

2. Use the hypergeometric distribution to demonstrate that the lower limit of a 95% confidence interval is 21 bags.

3. Do you agree that *Hill* was properly distinguished from *Kaludis*?

Notes

Estimating the amount of drugs for purposes of defining the offense or for sentencing has become an important issue in enforcement of the drug laws. Random testing has generally been approved by the courts. See Richard S.

Frank, et al., *Representative Sampling of Drug Seizures in Multiple Containers*, 36 J. Forensic Sci. 350 (1991) (listing cases and discussing construction of a lower-tail confidence limit for the number of drug containers in a seizure based on analysis of a sample of such containers). For another example of the problems, see Section 3.5.1.

9.1.7   ASCAP sampling plan

The American Society of Composers, Authors & Publishers ("ASCAP") is a performing rights society that issues licenses for the performance of music composed and published by its members. ASCAP has over 17,000 writer-members, 2,000 publisher-members, and 39,000 licensees. Its fees are collected primarily from local television and radio stations and from television networks pursuant to "blanket" licenses, which give the licensees the right to perform any piece in ASCAP's repertoire. Revenues are distributed among ASCAP's members, approximately one-half to the publishers and one-half to the writers.

The amount distributed to each member is based upon how often and how prominently his or her music is played. Music on network television is censused, but plays on local radio and local television—which account for the bulk of ASCAP's revenue—are sampled. The general features of the sampling plan are controlled by the provisions of an amended antitrust consent decree obtained by the United States in proceedings against ASCAP. The provisions of the decree, as amended in 1950, are that ASCAP is required "to distribute to its members the monies received by licensing rights of public performance on a basis which gives primary consideration to the performance of the compositions of the members as indicated by objective surveys of performances ...periodically made by or for ASCAP." *U.S. v. ASCAP*, 1950–1951 Trade Cas. (CCH), ¶62,595 at p. 63,755 (consent injunction filed March 14, 1950). In 1960, this general provision was implemented by a more detailed decree. *U.S. v. ASCAP*, 1960 Trade Cas. (CCH), ¶69,612 at p. 76,468 (consent injunction filed Jan. 7, 1960).

ASCAP implemented the amended decree by designing a random, stratified, and disproportionate sample for non-network television and radio.

The sample was stratified first by media (local TV, local radio, wired music, and other services); second, by major geographic regions (New England, Middle Atlantic, etc.); third, by type of community (standard metropolitan, other metropolitan, non-metropolitan, rural); fourth, by size of station measured by the ASCAP license fee (through categories); and a separate stratum was set up for solely FM stations. Within each cell a sample was drawn at random. Sample results were then multiplied by statistical multipliers to blow the sample up to the universe and by economic multipliers to reflect the value of the performances to ASCAP.

In ASCAP's report to its members, ASCAP's expert made these statements about the sampling plan:

Randomness: What it is and how achieved

(a) Scientific process of selection so that sample differs from universe it represents only by chance.

(b) Which days of the year, which times of day, and which stations are included in the sample are determined by statistical laws.

(c) Each playing has a knowable chance of inclusion in the sample.

Sampling Precision

(a) Sampling precision is the degree to which the sample represents the universe.

(b) The less the sampling precision, the greater the overstatement or understatement of a member's playings.

(c) Sampling precision is greater for a composition with many playings and is least when playings are few.

(d) To increase sampling precision for works with few playings requires expensive, overall increases in sample size and leaves less money for distribution.

Questions

1. To what extent are the above statements correct?

2. What types of data should the court require in order to support estimates of the accuracy of the samples?

9.1.8   Current Population Survey

The Census Bureau conducts a monthly sample survey of the population called the Current Population Survey (CPS). Among other things, the Bureau of Labor Statistics uses these data to generate unemployment figures. The CPS is re-designed every ten years to take advantage of population counts in the decennial census, but the basic features of the sampling plan remain fairly constant.

The Bureau begins the CPS by dividing the United States into 3,141 counties and independent cities, which are grouped together to form 2,007 Primary Sampling Units (PSUs). Each PSU consists of either a city, a county, or a group of contiguous counties. PSUs are then further grouped into 754 strata, each within a state. Some of the largest PSUs, like New York or Los Angeles metropolitan areas within their respective states, are thought to be unique, and constitute strata all by themselves. There are 428 PSUs that are in strata all by themselves. The remaining 326 strata are formed by combining PSUs that are similar in certain demographic and economic characteristics, like unemployment, proportion of housing units with three or more people, numbers of people employed in various industries, and average monthly wages for various industries.

The sample is chosen in stages. In the first stage, one PSU is selected from each stratum. PSUs that are in strata all by themselves are automatically included. PSUs that are combined with others in a stratum are selected using a probability method which ensures that, within each stratum, the chance of a PSU getting into the sample is proportional to its population.

Within each PSU, working with census block data, housing units are sorted by geographic and socioeconomic factors and are assembled into groups of four neighboring households each. At the second and final stage, a systematic sample of every nth group in the PSU is then selected. Every person aged sixteen or over living in a selected group is included in the sample. The 1999 CPS included about 50,000 households and about 1 person in 2,000 (from the civilian noninstitutional population aged 16 and over).

Because the sample design is state-based, the sampling ratios differ by state and depend on state population size as well as both national and state reliability requirements. The state sampling ratios range roughly from 1 in 100 households to 1 in 3,000 households. The sampling ratio used within a sample PSU depends on the probability of selection of the PSU and sampling ratio for the state. In a sample PSU with a probability of selection of 1 in 10 and a state sampling ratio of 1 in 3,000, a within-PSU sampling ratio of 1 in 300 groups achieves the desired ratio of 1 in 3,000 for the stratum.

One important use of the CPS is to measure unemployment. Suppose that in one month the Bureau's sample consisted of 100,000 people, 4,300 of whom were unemployed. The Bureau could estimate a national unemployment figure by multiplying 4,300 by the factor 2,000: $4{,}300 \times 2{,}000 = 6{,}800{,}000$. However, the Bureau does not apply the same weight to everyone in its sample. Instead, it divides the sample into groups (by age, sex, race, Hispanic origin, and state of residence) and weights each group separately.

One group, for example, consists of white men aged 16 to 19. (We ignore state of residence and use national figures in this example.) The Bureau uses census data to estimate the total number of these men in the civilian noninstitutional population of the country, correcting for known trends in population growth. Suppose this figure was about 6.5 million. If in one month 4,400 people turned up in the sample, of whom 418 were unemployed, the Bureau would weight each such unemployed person in the sample by the factor 6.5 million/4,400 = 1,477 instead of 2,000. The Bureau therefore estimates the total number of unemployed white men aged 16 to 19 in the civilian noninstitutional population as $1{,}477 \times 418 = 617{,}386$.

Each group gets its own weight, depending on the composition of the sample. The total number of unemployed people is estimated by adding up the separate estimates for each group.

The Bureau estimates that the current sample design maintains a 1.9% coefficient of variation (CV) on national monthly estimates of unemployment, using an assumed 6% level of unemployment as a control to establish consistent estimates of sampling error.

Questions

In making the CPS the Census Bureau uses stratified sampling, cluster sampling, and ratio estimation.

1. How is each method used?

2. What is the purpose of each method?

3. What is each method's effect on sampling error compared with a simple random sample of the same size from the entire population?

4. What factors should influence the sampling ratios used in the systematic samples within PSUs?

5. Assuming the given value for the CV, what would be a 90% confidence interval for a national unemployment estimate?

Source

Bureau of Labor Statistics, U.S. Dep't of Labor, *How the Government Measures Unemployment*, available online at <www.stats.bls.gov/cps—htgm>; and *Technical notes to household survey data published in Employment and Earnings*, *("A" tables, monthly; "D" tables, quarterly)*, available online at <www.stats.bls.gov/cpstn3.htm>; D. Freedman, R. Pisani, and R. Purvis, *Statistics*, ch. 22 (3rd ed. 1998).

9.2   Capture/recapture

Capture/recapture is a method for estimating the unknown size of a population. The method was originally developed for estimating the number of fish in a lake, or the size of a wild animal population. We use the fish example. The method is predicated on the properties of random sampling. It involves picking a random sample of m fish from the lake, tagging them, and then returning them to the lake. One then picks a second sample of n fish and counts the number of tagged fish. From this procedure, we can immediately write down a fourfold table as follows:

	Tagged	Untagged	Totals
In second sample	X		n
Not in second sample			
Totals	m		T

We observe $X = x$, the number of tagged fish in the second sample. If being caught in the second sample is independent of being caught and tagged in the first sample, then the expected value of X is

$$EX = \left(\frac{m}{T}\right)\left(\frac{n}{T}\right)T = \frac{mn}{T}.$$

Notice that $EX/n = m/T$, so that the expected proportion of tagged fish in the second sample equals the proportion that the first sample is to the total population. Replacing EX by its unbiased estimator x and solving for T yields $\hat{T} = mn/x$.[1] We emphasize that this estimate assumes that (i) all fish have an equal probability of being caught in the first sample (i.e., it is not the case that some fish are more likely to be caught than others), and (ii) the fact of being caught in the first sample does not affect the probability of being caught in the second sample (e.g., fish don't like being caught and become more wary). If either of these assumptions is violated, there is what is known as correlation bias: the population will be underestimated if the correlation is positive (which would occur if some fish are more prone to being caught than others), or overestimated if the correlation is negative (which would occur if the event of being tagged made fish more wary).

9.2.1 Adjusting the census

Since 1980 the U.S. Census has been the subject of intense litigation. The root of the problem is the recognition that there is a net undercount of the population and that some groups–notably blacks and Hispanics–are undercounted to a greater degree than others. In 1990, the census counted 248.7 million people. A post-census survey estimated that there were 19 million omissions and 13 million erroneous enumerations; the Census Bureau estimated that these numbers would grow in the 2000 census. The solution studied for many years in various permutations and now favored by the Bureau is to use a post-census sample survey as a basis for adjusting the census. The proposed adjustments probably would alter the population shares of states sufficiently to shift the apportionment of a few congressional seats. It would also affect the distribution of federal funds, and probably also intrastate redistricting. The matter is political because the undercounted groups tend to vote Democratic.

When New York State and New York City and others first demanded adjustment, the Census Bureau rejected it as infeasible. In the ensuing litigation, the Supreme Court held that adjustment for the undercount of minorities was not constitutionally required.[2] But after this victory, the Bureau changed its position and announced that improvements in technology and the growing difficulty of conducting the traditional census made adjustment both feasible and

---

[1] This estimator is slightly biased because $E\hat{T} > T$. To substantially eliminate this bias one uses the slightly modified estimator $T = (m + 1)(n + 1)/(x + 1) - 1$.

[2] *Wisconsin v. City of New York*, 517 U.S. 1 (1996).

desirable. Subsequently, the Bureau made public its plan to adjust for the 2000 census.

In the Bureau's plan, there were two types of sampling. First, there was a Nonresponse Followup Program in which follow-up visits would be made to a random sample of housing units that did not respond to the mailout-mailback census questionnaire. The sample was designed so that responses (either from the questionnaire or a visit) would be obtained from a total of 90% of the housing units in a census tract. Information from the visited nonresponding units would be used to impute the characteristics of the unvisited 10% of units.

Second, the Bureau would conduct a Post Enumeration Survey ("PES") to adjust census figures. The PES would be a cluster sample of 60,000 blocks containing some 750,000 housing units and 1.7 million people. After the census has been completed, the people in these blocks would be interviewed and the list created from the interviews would be matched against the census for those blocks. A person found in the PES, but not in the census, would presumptively be an erroneous omission from the census, although follow-up might be necessary to confirm that the person was properly included in the PES (e.g., did not move to the block after census day). A person found in the census, but not in the PES, would presumptively be an erroneous enumeration, although once again follow-up might be necessary (e.g., the person moved out of the block after census day). The estimation of the undercount is based on capture/recapture technology: the census is the capture and the PES is the recapture.

In drawing the PES and in estimating adjustment factors from it, the Bureau would divide the population into post-strata defined by demographic and geographic factors; one post-stratum might be Hispanic male renters age 30-49 in California. For each state there are six race-ethnicity groups, seven age-sex groups, and two property tenure groups—renters and owners. No differentiation is made between urban, suburban, and rural areas; the Bureau assumes that the undercount rate is the same everywhere within each post-stratum. The "raw" dual system estimator for each post-stratum is computed from the (slightly simplified) formula

$$DSE = \frac{Census}{M/N_p} \cdot \left[1 - \frac{EE}{N_e} \right].$$

In this formula, DSE is the dual-system estimate of the population in a post-stratum; *Census* is the census count in that stratum; M is the estimated total number of matches obtained by weighting up sample matches; N_p is the estimated population obtained by weighting up post-census sample counts; EE is the estimated number of erroneous enumerations obtained by weighting up the number of erroneous enumerations in the sample blocks; and N_e is the estimated population obtained by weighting up the census counts in the sample blocks. The ratio of the DSE to the census count is the "raw" adjustment figure for the post-stratum. It is called a raw figure because the Bureau statistically "smoothes" the estimates to reduce sampling error. The adjusted counts from the post-strata are then combined to produce state and local population figures.

Matching two large files of data without unique identifiers, like social security numbers or fingerprints, is a complex and error-prone process. Even after follow-up there are relatively large numbers of unmatched persons for whom it cannot be determined whether the error is in the census or the PES. Statistical models whose error rates are themselves unknown are planned to resolve those cases. Even the determination of match status can be problematical. For example, if a person picked up in a block in the PES cannot be matched to that block in the census, nearby blocks are searched for a match, but a search of the entire file would be impossible. There is a trade-off here: the smaller the area searched, the larger the number of false omissions; but enlarging the area searched creates operational problems. People who move between census day and the PES are another complication in determining match status. To match people who move out of the block after census day, information has to be collected from "proxy" interviews–with neighbors, current occupants of the household, etc. The chance for error here is large. Given the manifold problems, it is not surprising that studies of the matching process have shown that the number of errors is relatively large in comparison to the effect being measured. Moreover, as would be expected, false non-matches systematically exceed false matches, creating an upward bias in the estimates of gross omissions.

Would the proposed DSE account for the disproportionate undercount of minority groups? A test of capture/recapture estimation is to compare the figures resulting from adjustment aggregated to a national level with national figures obtained by demographic analysis. Demographic analysis gives figures for the total U.S. population by age, race, and sex that are derived from administrative records independent of the census, including birth and death certificates, and Medicare records. Some figures are shown in Table 9.2.1. Figure 9.2.1 shows the share changes, by state, from the proposed adjustment to the census as of July 15, 1991.

TABLE 9.2.1. Estimated net census undercounts, by race and sex, from the Post-Enumeration Survey and from demographic analysis. Figures as of 15 July 1991. "Other" is non-black, including whites and Asians.

	Post enumeration survey	Demographic analysis	Difference
Black males	804,000	1,338,000	−534,000
Black females	716,000	498,000	+218,000
Other males	2,205,000	2,142,000	+63,000
Other females	1,554,000	706,000	+838,000

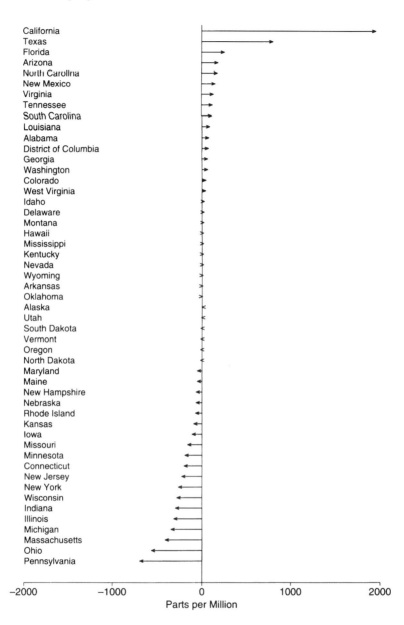

FIGURE 9.2.1. Share changes from the proposed adjustment to the census, as of July 15, 1991

Questions

1. Justify the DSE formula given above in terms of capture/recapture methodology.

2. What do Table 9.2.1 and Figure 9.2.1 suggest as to the validity of the assumptions underlying capture/recapture?

3. If adjusted figures cannot be shown to create more accurate proportionate shares down to the state or intrastate district levels, should adjustment nevertheless be made to increase numeric accuracy and eliminate systematic error against minorities?

4. If national demographic figures by race are more reliable, would it be preferable simply to "scale up" each counted minority person so that the national totals were correct?

Source

L.D. Brown, et al., *Statistical Controversies in Census 2000*, 39 Jurimetrics 347 (1999).

Notes

The Bureau's plans provoked a new flurry of congressional activity and lawsuits, this time to prevent adjustment. When the cases reached the Supreme Court, it held that the Census Act prohibited sampling for congressional apportionment purposes, but allowed it for other purposes.[3] This round of litigation has left two important open questions. One is whether adjustment would be constitutional for apportionment under the "actual Enumeration" clause, if the census statutes were amended to permit it. The other is whether adjusted or unadjusted figures are constitutionally required for drawing intrastate district lines.

The issues have become moot, for the moment, by the recommendation of the Bureau, made after the 2000 census, not to adjust the count. The advisory committee considering the question found that the national adjusted figures were significantly different from demographic estimates for some population groups, suggesting that there was an unidentified error in the adjustment method. President Bush's Secretary of Commerce, to whom the decision whether to adjust had been delegated, adopted the recommendation.

---

[3] *Department of Commerce v. United States House of Representatives*, 525 U.S. 316 (1999).

CHAPTER 10

Epidemiology

10.1 Introduction

Statistical data are frequently used, in the law and elsewhere, to establish or measure causal relations based on associations. The gold standard for doing this is the randomized, double-blind experiment. Here are the key features of such an experiment.

First, the subjects are divided into treatment and control groups by some random mechanism, such as a random number table. This feature has important advantages. Random selection: (i) does not depend on the personal preferences of the investigator, whose choices might bias the selections; (ii) tends to eliminate any systematic differences between the treatment and control groups that might account for, or "confound," an effect associated with treatment status; and (iii) is the model used in calculating the statistical significance of the results, and thus provides the firmest basis for such calculations.

Second, the study is further strengthened by requiring that it be double-blind, i.e., neither the investigators nor the subjects know to which group the subjects have been assigned. This precaution is aimed at preventing conscious or unconscious bias from affecting the investigators' behavior, e.g., in the evaluation of subject responses or outcomes; it also distributes any placebo effect equally between treatment and control groups.

Third, the groups are followed for predetermined periods by a prespecified protocol and preselected outcomes are studied for differences. Predetermination is important to prevent an opportunistic cutoff from biasing the results; prespecification assures that both groups are treated in as similar a manner as possible, except for treatment, so that differences in outcome can be attributed to treatment effect; and preselection of outcomes prevents a searching for "interesting" effects, which, if found, cannot be reliably judged because the search itself will compromise calculations of statistical significance.

Finally, important studies are then replicated with varying protocols and subjects to reduce the chance of mistaken conclusions.

Gold standard studies are seldom encountered in the law. In most cases, we cannot ethically or practically by randomized experiment replicate the setting or the conduct that produced a lawsuit. When residents of southern Utah sued the federal government contending that fallout from atmospheric A-bomb tests in Nevada caused the leukemia that killed their children, the question of causation could not be addressed by randomly selecting children for comparable exposures. And so when the law asks causation questions like this, it has to be content with less—sometimes far less—than the gold standard of the double-blind, randomized, controlled trial.

Below the gold standard we have what are called observational studies. In such studies, as the name suggests, we do not create the data but assemble and examine what already exists. Perhaps the prime example is the epidemiologic study, which from careful study of statistical associations attempts to unearth the causes of disease. In epidemiologic studies, assignment to study or comparison groups is not made by chance, but by the vicissitudes of life. Children exposed to fallout may be compared with those living in the general area who were born and grew up before the tests had been started or after they had been stopped.

When there is both an exposed group, and a control group that has not been exposed, the groups usually are compared by computing a *relative risk* (*RR*) or *odds ratio* (*OR*). The relative risk is the rate of disease in the exposed group divided by rate in the unexposed control group. The odds ratio is the same, except using odds instead of rates. See Section 1.5 for a discussion of these measures. A variant of these measures is the *standardized mortality* (*or morbidity*) *ratio* (*SMR*), defined as the observed number of deaths (or disease cases) in the study group divided by the expected number given the observed follow-up time of the study group and the incidence rates from a reference population. The denominator of the *SMR* is the number of deaths (or cases) one would expect in the study population if it had the same rates as the reference population. Usually the numerator of an *SMR* is estimated from sample data drawn from the study population, while the denominator uses rates in health census data taken from the reference population. Where age-specific rates are available, these can be used to provide more accurate expectations for the deaths or cases in the study group.

Epidemiologic studies can be highly persuasive, but are weakened by the possibility of confounding factors or other sources of bias. Because of the prominence of such potential weaknesses, inferences of causation tend to be more intensely scrutinized in epidemiology than they may in other statistical contexts, where the same problems may lurk unexamined in the background. Studying causation issues in epidemiology is thus useful, not only of itself, but also because it sheds light on the general problems of causal inference in statistics. In this chapter we introduce the subject, using epidemiologic examples as illustrations.

Further Reading

Kenneth J. Rothman & Sander Greenland, *Modern Epidemiology* (2d ed. 1998).

10.2   Attributable risk

The concept of *attributable risk*[1] is used to measure the potential reduction in occurrence of a given disease that would result by complete elimination of a particular risk factor. There are two related concepts: attributable risk in the population exposed to the risk factor and attributable risk in the entire population (both exposed and unexposed). We describe each in turn.

Attributable risk in the exposed population

In most cases in which attributable risk calculations are made, the disease under study has a greater incidence or prevalence in the exposed population.[2] Consider first the group of people who were exposed and subsequently contracted the disease. They can be divided conceptually into a "base" subgroup, consisting of those who would have contracted the disease in any event, i.e., without the exposure, and an "excess" subgroup, whose disease was caused by the exposure. The attributable risk in the exposed population (AR_e) is the proportion that the excess subgroup bears to the whole exposed diseased group. This proportion is also the probability that a person selected at random from the

---

[1] The term is something of a misnomer because what is computed is not a risk but the fraction of a risk attributable to a cause. Sometimes the term *etiologic fraction* is used, which is more accurate but less common. For convenience of reference we use the more common term.

[2] The *incidence* rate of a disease is a velocity, like the speed of a car, except that it is measured, not in miles per hour, but in the proportion of the population that newly contracts the disease in a relatively short time period. For example, the incidence of heart disease among male smokers 65 to 70 years old is about 0.1% per month, or 1.2% per year, assuming the incidence of heart disease is fairly constant over the year. If incidence varies over time, calculations become more accurate as the unit time interval shrinks to zero and the rate approaches the instantaneous incidence, a limit concept equivalent to the hazard concept discussed in Section 11.1, and analogous to the instantaneous velocity of a car.

The concept of incidence should be distinguished from that of *prevalence*, which is the proportion of the population that has the disease at a given time and thus includes the effect of current and past incidence and disease duration. Attributable risk calculations can be made in terms of incidence or prevalence, although incidence rates are more appropriate to disease prevention.

whole group would be in the excess subgroup, i.e., would not have contracted the disease without the exposure. The statement that 80% of all smokers who get lung cancer could have avoided the disease if they had not smoked is a description of the fraction of lung cancers among smokers that is attributable to smoking. The equivalent statement is that smoking caused the lung cancer in 80% of cancer victims who smoked. Simple algebra shows us that the attributable risk in the exposed population is related to the rate ratio (or relative risk) by the formula

$$AR_e = 1 - 1/RR, \tag{1}$$

where RR is the ratio of the rate of disease given the exposure relative to the rate given no exposure. AR_e is sometimes referred to as the probability of causation.

To illustrate the formula, the relative risk of colon cancer given exposure to asbestos has been estimated at 1.55. Hence $AR_e = 1 - 1/1.55 = 0.35$, approximately. Thus, 35% of colon cancers in asbestos workers might have been avoided if they had not been exposed. Equivalently, there is a 35% chance that a randomly selected asbestos worker's colon cancer was caused by asbestos.

It is of some consequence to the legal issue of causation that if the relative risk is greater than two, it is more likely than not that a random exposed worker's disease was caused by the exposure. Some courts have in fact held that if there are no facts to distinguish the specific plaintiff's case from other exposed cases, then causation for the particular plaintiff is not proved by a preponderance of evidence unless the relative risk from exposure is greater than two.[3] When there *are* specific facts, inferences based on relative risk calculations must be modified accordingly. For example, the relative risk for the exposed group may be greater than two, but this applies to the average exposure, and the plaintiff's exposure may have been below the group's average. On the other hand, if the relative risk is less than two, but other evidence suggests causation, the combination may be sufficient for an expert to conclude that the exposure more likely than not caused plaintiff's disease. The specific evidence may even be a differential diagnosis, i.e., the attribution of cause by eliminating other risk factors.[4] However, valid differential diagnosis requires both proof of general causation and affirmative evidence of an absence of other risk factors for a disease with a reasonably well established etiology.[5]

---

[3]See, e.g., *Daubert v. Merrell Dow Pharmaceuticals*, 43 F.3d 1311 (9th Cir. 1995) (Bendectin and birth defects); *Hall v. Baxter Healthcare Corp.*, 1996 U.S. Dist. Lexis 18960 at p. 15 (D. Ore. 1996) (silicone breast implants and connective tissue disease).

[4]See *Landrigan v. Celotex Corp*, 127 N.J. 404, 605 A.2d 1079 (1992) (1.55 relative risk of colon cancer from exposure to asbestos plus an absence of other causal factors is sufficient for plaintiff's proof of causation).

[5]*In re Joint Eastern and Southern District Asbestos Litig. (John Maiorana)*, 758 F. Supp. 199 (S.D.N.Y. 1991) (expert opinion in asbestos/colon cancer case found insufficient because the assumption that the deceased had no other risk factors was unsupported).

Attributable risk in the population

The second concept is that of attributable risk in the population AR_p. This is the proportion of disease cases in the general population that would be eliminated if the exposure were eliminated. The statement that "20% of all lung cancers could be avoided by smoking cessation" is a statement of attributable risk in the population. AR_p is a function of both the relative risk and exposure prevalence, which makes it a reasonable indicator of the importance of the risk factor as contributing to disease. Even if AR_e is high, because the disease is strongly associated with the exposure, AR_p may still be low if the exposed proportion of the population is low. If AR_p is high, because the exposure is widespread, the absolute number of people affected by the exposure may still be small if the disease in question is rare. For example, in the United States, the attributable risks relating smoking to lung cancer are far larger than the attributable risks relating smoking to heart disease. However, because the incidence of heart disease is much greater than that of lung cancer, a smoking-cessation policy would save many more people from heart disease than from lung cancer.

The illustration below depicts a population in which 20% of unexposed subjects and 50% of exposed subjects have the disease, and 10% of the population has been exposed. Assuming 20% of all those exposed would have contracted the disease in any event, the excess risk is $0.50 - 0.20 = 0.30$, which is 60% of the proportion of exposed disease cases. Here $RR = 0.5/0.2 = 2.5$ and $AR_e = 1 - 1/2.5 = 0.6$. Because only 10% of the population has been exposed, the excess cases attributable to the exposure account for only $0.30 \times 0.10 = 0.03$ of the total population. The proportion of all cases in the population is: $0.20 \times 0.90 + 0.50 \times 0.10 = 0.23$. Hence AR_p (the proportion of excess cases to total population cases)$= 0.03/0.23 = 13\%$.

There are three equivalent ways to calculate AR_p. The numerator can be shown to equal the difference between the *overall* disease rate and the disease rate among the unexposed, with the denominator being the overall disease rate. The overall disease rate reflects both the heightened risk due to the exposure and the prevalence of the exposure. Thus,

$$AR_p = \{P(d) - P_0(d)\}/P(d) = 1 - P_0(d)/P(d), \qquad (2)$$

where $P_0(d)$ is the risk (prevalence or incidence) of disease in the unexposed population, and $P(d)$ is the overall risk of disease in both the exposed and unexposed populations combined.

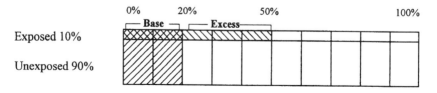

FIGURE 10.2. Attributable risk in the population

Alternatively, and equivalently, AR_p may be calculated as

$$AR_p = X/(1 + X), \tag{3}$$

where $X = P(e)(RR - 1)$ and $P(e)$ is the proportion exposed.

A third method uses the formula

$$AR_p = (1 - 1/RR)P(e|d), \tag{4}$$

where $1 - 1/RR$ has the same value given for AR_e and $P(e|d)$ is the proportion exposed among those with the disease. In a comparative cohort study, data from which to estimate $P(e|d)$ might not be available, in which case formula (4) could not be used and formulas (2) or (3), with some estimate of $P(e)$, would have to suffice. Case-control studies provide direct estimates of $P(e|d)$, but not of RR, so that to use formula (4), RR must be approximated by the odds ratio (which is acceptable if the disease is rare).

If data are available there is an advantage to formula (4). All of the formulas given so far are idealizations in the sense that counterfactually eliminating the exposure would not necessarily reduce the incidence of disease in the exposed subjects to the level of those unexposed. There may be confounding factors that correlate with exposure, contribute to the disease, and would remain even if the exposure were eliminated. Smokers may have characteristics other than smoking, e.g., sedentary lifestyles, that contribute to their higher rates of heart disease. To take these factors into account one has to compute an adjusted relative risk (RR_a) in which the numerator is, as before, the rate of disease among those exposed, but the denominator is the rate of disease among those unexposed if the exposure alone were eliminated, all other characteristics of those persons remaining unchanged. To compute AR_e and AR_p it is necessary to substitute RR_a in formulas (1) and (4), respectively. (This substitution is not valid for the other formulas for AR_p.) Part of an epidemiologist's job is to identify confounding factors, if any, and to estimate an adjusted RR that takes account of them.

An adjustment in RR may also be justified if, as is usually the case, interventions to eliminate exposure are only partially successful or if the exposed population would tend to substitute another risk factor if compelled to give up the exposure. If smoking were eliminated, smokers might turn for consolation to greater consumption of fatty foods. In that case calculations of potential health benefit should take into account the risks of the substituted behavior by computing an adjusted RR, as described above. However, for purposes of determining causation in a lawsuit, RR should not be adjusted for substitute behavior because the risk of substituted behavior does not affect the probability that the original behavior caused the disease.

10.2.1   Atomic weapons tests

Between January 1951 and the end of October 1958, at least 97 atomic devices were detonated in above-ground tests in the Nevada desert at a site located 160 km west of the Utah–Nevada border. Fallout from at least 26 tests (amounting to over half the total kiloton yield) was carried by wind eastward to Utah (see Figure 10.2.1). Testing continued underground after 1961, with some tests venting fallout into the atmosphere.

There were 17 counties in southern and eastern Utah that were identified from U.S. Government fallout maps as "high fallout" counties. These counties were rural and had about 10% of the state's population.

Two epidemiologic studies of leukemia deaths in children were conducted using high-exposure and low-exposure cohorts. Deaths were assigned to the high-exposure, age-time cohort if they occurred before age 1 for the years 1951 through 1958, at age 1 for the years 1951 through 1959, and so on up to deaths at age 14 for the years 1951 through 1972. All remaining deaths of children up to age 14, that is, all deaths through the end of 1950 and all deaths subsequent to the upper-year limits of the previously defined high-exposure cohorts for each age at death, were assigned to the low-exposure cohort. For purposes of the person-years calculation, for each year the number of persons who contribute a person-year in each cohort is the number of persons who, if they had died in that year, would have had their deaths assigned to that cohort for that year.

The two different teams of investigators used different data. Lyon, et al., used data from 1944 through 1975. A summary is shown in Table 10.2.1a.

Land, et al., critical of the Lyon study, did not use the earlier (1944–49) data because they believed these to be unreliable, but did use data from the later years. They also added three "control" areas (Eastern Oregon, Iowa, and the United States as a whole). The study used the same high-exposure, low-exposure defined cohorts as in the Lyon study. An excerpt from the Land data on leukemia deaths is in Table 10.2.1b.

Questions

Peggy Orton, born on January 12, 1946, was 5 years old when testing began in 1951. She lived in St. George, Utah. On November 11, 1959, at the age of 13, she was diagnosed as having acute lymphoblastic leukemia. She died six months later, on May 29, 1960. Orton's estate brought suit for wrongful death against the United States under the Federal Tort Claims Act, 28 U.S.C. §§2671–80 (1994).

In your computations made below, assume that the age distribution of all cohorts is the same; in the actual studies the mortality rates were age-adjusted by the "direct" method to the U.S. 1960 white population.[6]

---

[6]In comparing two populations, adjustments may sometimes be needed for differences in age distribution. In the direct method of adjustment, a weighted average is computed from age-specific rates, where the weights are the proportions of people in each age bracket in the reference population.

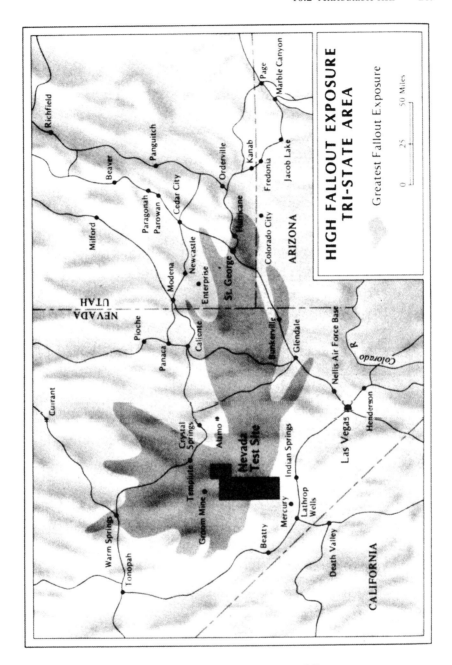

FIGURE 10.2.1. Atomic weapons test fallout

1. What is the relative risk of leukemia death in the high-exposure and combined low-exposure cohorts in Table 10.2.1a?

2. Assuming that the entire excess risk is caused by the exposure, what is the attributable risk of leukemia due to the exposure in (i) the exposed population, and (ii) the general population?

3. As attorney for Orton's estate, what argument would you make from the above calculations and other facts that, by a preponderance of the evidence, Orton's exposure caused her leukemia?

TABLE 10.2.1a. Distribution of childhood leukemia deaths. Person-years and age-specific mortality rates by three age classes and exposure

| | Cohort: Males—Southern Utah | | | | | | | | |
| | Low-exposure I (1944–50) | | | High-exposure (1951–58) | | | Low-exposure II (1959–75) | | |
Age	Cases	Person-years	Rate*	Cases	Person-years	Rate*	Cases	Person-years	Rate*
0–4	4	62,978	6.35	6	92,914	6.46	1	115,000	0.87
5–9	1	55,958	1.79	4	128,438	3.10	2	74,588	2.68
10–14	0	49,845	0.00	6	151,435	3.96	0	40,172	0.00
0–14	5	168,781	2.96	16	372,787	4.29	3	229,760	1.31

| | Cohort: Females—Southern Utah | | | | | | | | |
| | Low-exposure I (1944–50) | | | High-exposure (1951-58) | | | Low-exposure II (1959-75) | | |
Age	Cases	Person-years	Rate*	Cases	Person-years	Rate*	Cases	Person-years	Rate*
0–4	0	59,214	0.00	2	86,518	2.31	4	109,675	3.65
5–9	1	53,696	1.86	6	121,302	4.95	3	71,828	4.18
10–14	1	48,486	2.06	8	143,924	5.56	0	39,645	0.00
0–14	2	161,396	1.24	16	351,744	4.55	7	221,148	3.17

* Age-specific leukemia mortality rate expressed in cases per 100,000 computed from the respective observed cases and person-years of cohort experience.

TABLE 10.2.1b. Distribution of leukemia deaths and person-years for southern and northern Utah, eastern Oregon, and Iowa

| | S. Utah | | N. Utah | | E. Oregon | | Iowa | |
Exposure	High	Low	High	Low	High	Low	High	Low
Deaths	30	16	150	127	28	16	415	348
Person-years (in thousands)	691.3	590.1	3,768.8	4,237.0	910.7	770.4	11,958.1	10,288.8

4. As attorney for the government, what argument would you make that the association between the high-exposure cohort and childhood leukemia was not causal, and even if generally causal, was not shown to be the cause of Orton's leukemia?

5. Instead of an all-or-nothing result, should recovery be allowed more generally in cases of excess risk, but the amount thereof proportionally reduced by the probability that decedent's leukemia would have occurred in any event, without the exposure?

Source

Allen v. United States, 588 F. Supp. 247 (D. Utah 1984), *rev'd*, 816 F.2d 1417 (10th Cir. 1987), *cert. denied*, 108 S. Ct. 694 (1988) (district court judgment for the plaintiffs reversed; the testing activity was a discretionary function for which recovery against the federal government was not sanctioned by the Federal Tort Claims Act).

Notes

For the conflicting studies on the causation issue in the *Allen* case, compare Joseph L. Lyon, et al., *Childhood leukemias associated with fallout from nuclear testing*, 300 New Eng. J. Med. 397 (1979) (concluding that the data show an excess of leukemia) with Charles E. Land, et al., *Childhood leukemia and fallout from the Nevada nuclear tests*, 223 Science 139 (1984) (concluding that the data were not sufficient to support a finding of such an excess). However, a later study by Land, et al., confirmed the Lyon, et al., finding. See S.G. Machado, C.D. Land, & F. W. McKay, *Cancer mortality and radiation fallout in Southwest Utah*, 125 Am. J. Epidemiology 44 (1987).

In 1997, a National Cancer Institute study released after long delay found that radioactive iodine dispersed from the atmospheric tests had delivered varying doses of radioactivity, principally through milk, to large segments of the U.S. population east of the test site. See the NCI website at <www.nci.nih.gov.> It has been estimated that thousands of cases of thyroid cancer were caused as a result.

The story of the "downwinders," the medical controversy surrounding the issue of causation, and the legal battle for compensation are given in H. Ball, *Justice Downwind* (1986).

10.3   Epidemiologic principles of causation

Epidemiology is the study of the distribution and causes of disease in mankind. It uses statistical tools, but generally is regarded as a separate specialty. Its purview includes, but is much broader than, infectious disease epidemiology and the study of epidemics. Three types of studies are commonly used in epidemiology. The most reliable are *cohort studies* in which two or more highly similar unaffected groups are selected for study and control; the study subjects are exposed while the controls are not. The cohorts are followed for a predesignated period of time after which the health or disease outcomes in the two groups are compared, usually by computing the relative risk of disease associated with the exposure. See Section 10.2.1 (Atomic weapons tests) for examples of birth cohort studies. Cohort studies are hard to manage when the follow-up is long or the disease is rare. In such situations, an alternative is a *case-control study* in which the cases are people with the disease and the controls are without the disease but are highly similar with respect to factors related to both disease and the exposure of interest. The antecedent exposure histories in the case and control groups are then compared. The results are reflected in the computation of an odds ratio, which for rare diseases is approximately equal to a relative risk. Careful selection of controls and measurement of confounding factors are of critical importance in case-control studies. See Section 10.3.1 (Dalkon Shield) for a case-control study. A third type of study is called *ecologic*. These are generally descriptive, looking at relations between average levels of risk factors and disease in several population or sampling units. Section 1.4.1 (Dangerous eggs) is an example of an ecologic study.

Epidemiological studies may involve observations over time, or in cross-section at a particular time, or both.

General and Specific Causation

Regardless of the type of study, epidemiology follows a two-step process: first, determining whether there is an association between the exposure and a disease; and then, if there is an association, determining whether it is causal. Judicial opinions sometimes refer to an epidemiologic finding of causation as "general causation," meaning that the exposure has caused at least some disease in the exposed group. In a law case in which causation is an issue, the question is whether the exposure caused the particular plaintiff's disease. This is known as specific causation. Specific causation is not the special province of epidemiology, although epidemiologic data may be important to the answer.

Does negative epidemiology "trump" other evidence? The issue arises when studies show no association between exposure and disease, but plaintiffs propose to have experts testify that there is a causal relation. Sometimes the proposed testimony is based on laboratory *(in vitro)* studies, or animal *(in vivo)* studies, or clinical observations, or on an isolated epidemiologic study that shows a low level or weak association. When the epidemiology is strong and negative, such testimony has rightly been viewed skeptically, and in a number of cases judges have not permitted plaintiffs' claims to go to a jury.[7]

Biases and Confounding

Epidemiology is not a razor-sharp tool for detecting causal relationships. Case-control studies are particularly prone to bias because of the difficulty in identifying control subjects who match the cases in everything except the disease (but not the antecedent risk factor that may have caused the disease). Cohort studies tend to be better because they offer the possibility of composing groups differing only in exposure (i.e., before disease is detected), but even in these studies a finding of association may be produced by study biases, errors, confounding, or chance, and not by causal connections. This means that in any epidemiologic study positive findings cannot be accepted without a searching look at factors that may weaken or vitiate the conclusions. Some of these are described below.[8]

Confounding—An apparent causal relation between a supposed hazard and a disease may be due to the effect of a third variable that is correlated with both. An example would be a study which showed that people living in areas where jet airplane noise was greater than 90 decibels had a significantly shorter life expectancy. Did the noise shorten their lives? Perhaps, but the study is questionable because people who live in devalued housing close to airports tend to be poorer than the general population. Recent evidence suggests that exposure to jet exhaust pollution may be another confounder.

Selection bias—The cases or the controls may be selected in a way that makes them significantly different from the populations they are supposed to represent. Suppose that at an Alaskan hospital the rate of complications for women in labor is higher in winter than in summer. Could changes in diet and reduced light be responsible? Again, perhaps. But it may be that during the winter when travel is difficult women tend to come to the hospital only for more difficult pregnancies; otherwise they deliver at home. In short, the women appearing at the hospital during the winter were subject to self-selection that is biased with respect to the population of deliveries. The "healthy worker" effect is an example of selection bias.

Response bias—This bias arises when subjects respond inaccurately to an investigator's questions. A study of the effects of preconception x-rays of fathers on the incidence of childhood leukemia in their subsequently born

---

[7]See, e.g., cases cited in notes 3, 5 at p. 285; see also *In re Agent Orange Liability Lit.*, 597 F. Supp. 740, 787-795 (E.D.N.Y. 1984) and 611 F. Supp. 1223, 1231–1234 (1985), *aff'd in rel. part*, 818 F.2d 145, 171–173 (2d Cir. 1987) (Agent Orange and a variety of illnesses).

[8]Some of the hypothetical examples given below are from Max Mitchell III, W. Thomas Boyce, & Allen J. Wilcox, *Biomedical Bestiary: Epidemiological Guide to Flaws and Fallacies in the Medical Literature* (1984).

offspring suggests an association. But the value of the study was discounted because the information was based upon the recollections of the mothers, who may have been more likely to recall such x-rays if the child had contracted leukemia than if the child's development had been normal. Evasive answer bias is another response bias that is particularly serious when sensitive areas are probed.

Under- and over-matched controls—Under-matching is the failure to se-lect controls that are sufficiently like the cases. When that occurs, differences between the cases and controls other than the exposures may be responsible for the differences in outcomes. To avoid this, controls should match the cases in everything but the exposure. A more precise statement of what is required of control selection in a case-control study is that the probability of inclusion into the study must not depend on the exposure under study. A study of the relation between chronic bronchitis and air pollution would be under-matched if the cases were chosen from a city and the controls from rural areas. There are too many differences between city and country life that might cause bronchitis to have much confidence that an observed association was causal. A study of the relation between pregnancy outcomes and drug use would be under-matched if the cases were chosen from women patients at city hospitals with adverse pregnancy outcomes while the controls were chosen from women visiting a hospital clinic for other reasons. Women who use drugs would probably be less likely to visit a clinic and therefore the exposure to drugs would diminish the probability of their selection as controls. Since the controls would have less drug use than the cases, the study would show an effect, even if there were none.

Over-matching occurs when the controls are selected so that they excessively resemble the cases, as when controls are selected under criteria that give them the same exposure as the cases. When that occurs, the association between the disease and the exposure may be concealed. A study of the association between bronchitis and air pollution would be over-matched if the matched controls were from the same census tract as the cases because the air quality within the small area of a census tract is likely to be much the same. Since the exposure of bronchitic and non-bronchitic subjects would be essentially the same, one would conclude, falsely, that air pollution had no effect.

Cohort effect—The cohort effect is the tendency for persons born in certain years to reflect a relatively higher (or lower) risk of a disease throughout their lives. For example, a cross-sectional study of tuberculosis antibodies may show a higher prevalence in older people. This would not necessarily mean that older people suffer a higher incidence of tuberculosis, but only that the incidence of tuberculosis was greater in earlier years than it is today. In that case, a sample of tuberculosis patients would show a disproportionate representation of older people.

Observer variability—In clinical judgments, it is well recognized that ob-servers may differ markedly in their subjective evaluations. This may produce spurious effects. A study that has a single observer is particularly vulnerable to this possible bias. Unreliability (i.e., lack of reproducibility) almost always causes loss of efficiency and precision.

Hawthorne effect—The name comes from the fact that officials at a Westinghouse plant in Hawthorne, Illinois, found that merely consulting workers about conditions improved productivity. Analogously, simply participating in a study may change results, particularly when the study is of attitude or behavior. Related to this is social desirability bias, when subjects give responses they believe the investigator wants to hear.

Ascertainment—Changes in reported incidence of a disease associated with some environmental or risk factor may in fact be due to changes in the accuracy of diagnosis or classification of the disease, or the case definition, as when the CDC changed the criteria defining the Acquired Immune Deficiency Syndrome (AIDS).

Regression to the mean—Regression to the mean is the tendency of individuals who score either very high or low on one measurement to move closer to the mean on a second one. An apparent improvement in scores following some treatment may be due wholly or in part to regression. See Section 13.1 for a further discussion.

Ecological fallacy—In ecological studies the investigator attempts to detect associations between risk factors and disease from the behavior or experience of groups. The ecologic fallacy is the assumption that such associations apply at an individual level. The relation between egg consumption and ischemic heart disease in various countries (Section 1.4.1) is an example of an ecological association that may be fallacious as applied to individuals because, as noted, factors other than egg consumption are producing the association.

Association v. Causation

A well designed study will not have features that suggest bias or confounding. Yet there may be hidden defects. To minimize the possibility of false conclusions, epidemiologists have developed certain common-sense criteria of validity for a causal inference. These are referred to as the "Bradford-Hill" criteria.[9] They are described below.

Strength of association—The larger the relative risk associated with exposure, the less likely it is to be due to undetected bias or confounding. But even large relative risks may be weak evidence of causation if the number of cases is so small that chance alone could explain the association. To measure sampling uncertainty, confidence intervals for the point estimates of the relative risk should be computed.

Consistency—The more consistent the findings from all relevant epidemiologic studies, the less likely it is that the association in a particular study is due

---

[9]After A. Bradford-Hill, *The environment and disease: association or causation?*, 58 Proc. Roy. Soc. Med. 295–300 (1965).

to some undetected bias. All relevant studies should therefore be examined. What is sufficiently relevant is a matter of judgment.

Dose-response relationship—If risk increases with increasing exposure, it is more likely that the corresponding association would be causal. This is a strong indication of causality when present, but its absence does not rule out causality.

Biological plausibility—There must be a plausible biological mechanism that accounts for the association. This is sometimes said to be a weaker criterion because a biological scenario can be hypothesized for almost any association, and plausibility is often subjective.

Temporality—Cause must precede effect. This *sine qua non* for causation is usually satisfied. However, some patterns of temporality may affect the plausibility of the causal inference, as when the time between the exposure and the onset of disease is inconsistent with causality.

Analogy—Similar causes should produce similar effects. Knowledge that a chemical is known to cause a disease is some evidence that a related chemical may also cause the same disease. This is related to the plausibility argument, and similar caution is warranted.

Specificity—If risk is concentrated in a specific subgroup of those exposed, or in a specific subgroup of types of diseases under scrutiny, the association is more likely to be causal. This is a strong indication when present.

Experimental evidence—Experimental evidence in man (i.e., random allocation to exposure or removal of exposure) would constitute the strongest evidence of causation. However, such data usually are not available.

10.3.1   Dalkon Shield

The Dalkon Shield was an oval-shaped plastic intrauterine device (IUD), slightly larger than a dime, with fin-type projections on each side. It was designed to conform to the shape of the uterine cavity; the projections were added to improve resistance to expulsion and to cause the device to seek the fundus with normal uterine movements. A tailstring was tied to the bottom of the Shield by a knotted loop as an indicator of proper placement of the device and as a means for removal of the Shield. When this IUD was placed properly in the uterus, the tailstring projected through the cervix into the vagina. Because the removal of the Dalkon Shield required greater force than removal of other IUDs, the string had to be stronger and this was accomplished by the use of multiple strands of suture material encased by a nylon sheath open at both ends.

A.H. Robins, Inc. ("Robins"), bought the Shield for $750,000 plus a 10% royalty on June 12, 1970. In January 1971, Robins began a dual campaign of marketing and testing the Shield. After the Shield had been on the market for over a year (with more than one million sold) Robins began to receive complaints of high pregnancy rates, acute pelvic inflammatory disease ("PID"), uterine perforations, ectopic pregnancies, and others. By June 1974, some 2.8

million Shields had been distributed in the U.S.; this was about 6.5% of the women in the 18-44 age group. On June 28, 1974, Robins voluntarily suspended domestic distribution of the Shield at the request of the U.S. Food and Drug Administration and on August 8, 1975, announced that it would not remarket the Shield. Throughout this period there was extensive negative publicity concerning the Shield. There were studies indicating that the multifilamented tail of the Shield might "wick" bacteria from the vagina to the otherwise sterile uterus. There was evidence that the risk of infection increased with the length of time the Shield was worn. On September 25, 1980, Robins wrote a "Dear Doctor" letter recommending removal of the Shield from asymptomatic patients because of possible association between long-term use and PID.

The Women's Health Study was a multi-center, case-control investigation that included information on the relationship between the use of contraceptive methods and hospitalization for certain gynecologic disorders. Data were collected at 16 hospitals in 9 U.S. cities between October 1976 and August 1978. Women aged 18-44 were eligible. Cases consisted of women who had a discharge diagnosis of PID; controls were women who had a discharge diagnosis of a non-gynecological condition. Excluded were cases and controls who had had a history of PID or conditions that might lower risk of pregnancy, e.g., lack of sexual activity, sterility, or illness. Current IUD type was identified for cases and controls. In total, 1,996 potential cases and 7,162 potential controls were identified. After exclusions, 622 cases and 2,369 controls remained for analysis. The data are shown in Table 10.3.1.

Other studies have shown that IUDs in general raise the risk of PID, while barrier contraceptives have a protective effect.

Questions

Plaintiff is a woman who wore an IUD, and switched to the Dalkon Shield in 1972. It was removed in 1974, when she developed PID. She now sues Robins on various theories.

1. What analysis of the above data should be made in support of the contention that plaintiff's PID was caused by the Dalkon Shield?

2. Should an epidemiologist be permitted to testify as to causation (i) in general or (ii) in this particular case?

TABLE 10.3.1. Method of contraception for PID cases and controls

Current contraception method	Women with PID	Controls
Oral contraceptive (OC)	127	830
Barrier (diaphragm/condom/sponge)	60	439
Dalkon shield	35	15
Other IUDs	150	322
No method	250	763
Total	622	2,369

3. Is there reason to believe that the proportion of Dalkon Shield users among the controls was different than their proportion in the population at the time plaintiff wore the Shield? If so, what bias would that create in the study?

4. Is there reason to believe that the proportion of Dalkon Shield users among the cases was different than their proportion in the case population at the time plaintiff wore the Shield? If so, what bias would that create in the study?

Source

Lee, et al., *Type of intrauterine device and the risk of pelvic inflammatory disease*, 62 J. Obs. & Gyn. 1 (1983) (finding increased risk for the Dalkon Shield); compare Snowden, et al., *Pelvic infection: a comparison of the Dalkon Shield and three other intrauterine devices*, 258 British Med. J. 1570 (1984) (finding no increased risk for the Dalkon Shield).

Notes

An extensive recital of the facts surrounding the Dalkon Shield is given in *Hawkinson v. A. H. Robins Co., Inc.*, 595 F. Supp. 1290 (D. Colo. 1984). The *Hawkinson* case (12 women) was part of a flood of lawsuits by some 14,000 women who had worn the Dalkon Shield. Settlements cost the company hundreds of millions of dollars. Of the cases it litigated, Robins won about half, usually those tried by judges and usually on the ground that causation had not been proved. In many cases, epidemiologic data were not introduced directly, although the data possibly influenced the opinions of the clinicians who testified. In most cases of substantial recovery, punitive damages far exceeded compensatory damages. Robins claimed that it was being repeatedly punished for the same act and sought to consolidate the cases in a single tribunal. When this failed, in August 1983, the company filed for bankruptcy reorganization under Chapter 11.

Robins made $10 million in profits from the device. By 1985, when it filed for protection under Chapter 11, Robins and Aetna Casualty & Insurance Company, its insurer, had disposed of 9,500 injury suits at a cost of $530 million; more than 2,500 suits were pending and new ones were coming in at a rate of 400 a month. N.Y. Times, Dec. 13, 1987, at F17, col. 1.

In September 1985, Johnson & Johnson stopped selling its 20-year old Lippes Loop after 200 lawsuits. In January 1986, G.D. Searle & Company withdrew its Copper-7 IUD after 775 lawsuits. This was the last major manufacturer of intrauterine devices in the U.S. and its withdrawal marked the end of their production here. N. Y. Times, Feb. 1, 1986 at A1, col. 5.

10.3.2 *Radioactive "cocktails" for pregnant women*

Our story begins more than 50 years ago at the dawn of the atomic era in the fall of 1945. World War II had just ended after the atomic bombing of Hiroshima and Nagasaki. And not coincidentally, radioactive isotopes of a number of elements—such as sodium, phosphorous, and iron—produced in cyclotrons at MIT or the University of California at Berkeley, had become available for medical uses. Their potential for treatment of disease caused great excitement in the medical community. There was also research. It had become possible for the first time to trace the metabolism of elements in the body by tagging compounds with radioactivity and recording their passage by measuring the emissions. Iron was a particularly intriguing subject. Its metabolism is complex and was not perfectly understood. George Whipple won a Nobel Prize for experiments on dogs using iron 59, a radioactive isotope of iron. Paul Hahn was a young scientist working with Whipple. In 1944, Hahn was brought to Vanderbilt Medical School in Nashville, Tennessee, as a radiobiologist to continue studies using radioactive tracers.

In the meantime, and quite independently, the state of Tennessee and Vanderbilt University had set up a project to study nutrition in the South. This was known as the Tennessee Vanderbilt Nutrition Project (TVNP). In the summer of 1945, Hahn proposed that the TVNP study the metabolism of iron in pregnant women by giving them a radioactive tracer. Among other things, it was thought that knowledge of the passage of iron across the placenta would be useful in delivering dietary iron supplements to prevent maternal anemia in pregnancy. The study was approved by Tennessee's Commissioner of Health and the administration of Vanderbilt Medical School.

That these approvals were forthcoming was not surprising given the medical practices of the day. Nontherapeutic experiments on humans involving tracer doses of radioactivity were not unusual in the 1940s and 1950s, and a number of those studies included pregnant women as subjects. It was recognized that high levels of radioactivity created health risks, but exposure to low level radiation was thought to be harmless. It was common at that time to x-ray pregnant women to determine the position of the fetus (especially when twins were suspected) and the first study of the health effects of fetal x-rays, which showed an elevated risk of cancer in the children exposed in utero, was not to be published until 1957.

On the other hand, the TVNP study involved over 800 white women visiting a prenatal clinic, which made it the largest such study by far. Moreover, the women were not told that the vitamin "cocktail" they were given to drink was radioactive and had no therapeutic purpose. The extent to which the absence of informed consent departed from standard research procedure at that time was much debated by the experts in the lawsuit that followed. Informed consent as a developed concept required in human experimentation did not emerge until a later era, but the general advisability of obtaining patient consent was recognized. What was practiced was another story. There was testimony that the need to obtain consent was more opportunistic than principled: procedures believed to be harmless were performed on patients in hospitals when those procedures could appear to be part of the treatment or hospital protocol. Moreover, non paying patients were sometimes so used on the theory that they were paying back the medical establishment for their free treatment.

Whatever the standards were, it is clear that the experimenters did not believe that their project was nefarious and had to be concealed. Vanderbilt issued a press release about the study and stories appeared in Nashville newspapers. Hahn and others wrote scientific papers recording their results and delivered them at scientific meetings. Nothing happened; there was no outcry or objection of which we have any record.

But then the study became unique in another respect. About 20 years later, some researchers at Vanderbilt got the idea to do a follow-up study. We refer to this as the Hagstrom study. Their purpose was to see whether the low level radioactivity had caused any adverse effect on either the women or their children. By this time, studies beginning in the late 1950s had documented some increase in childhood cancer in children x-rayed in utero. The relative risk was about 1.4 to 2.0, depending on age, with the greatest excess at ages five to seven. After age eight the excess risk appeared to be exhausted. On the other hand, studies of children in utero in Hiroshima and Nagasaki—who had received much larger doses—showed no increase in childhood leukemia. The state of scientific knowledge suggested a possible connection, but the degree of fetal sensitivity was still unknown. The Vanderbilt follow-up study was seen as an unusual opportunity to add knowledge of this subject by using a large and well defined group.

The results of the study were as follows. Hagstrom evaluated an index group of 751 pregnant white women who were fed the isotope and a control group of 771 pregnant white women visiting the clinic who were not known to have been fed the radioactive iron. Some 679 of the index group and 705 of the control group responded to questionnaires. Follow-up of the children born to these women was completed for 634 index children and for 655 control children covering a period that averaged 18.44 years. For the mothers, no difference in health effects appeared. For the children, however, there were four cancer deaths in the index group and none in the control group. The cancers were not the same. One was a leukemia, another a sinovial sarcoma, a third a lymphosarcoma, and the fourth a liver cancer. The child with leukemia died at age five; the children with sarcomas died at age eleven. Hagstrom put the liver cancer to one side because it appeared to be familial: two older siblings of the child had died of the same disease in their twenties.

The dose of radiation received in utero was the subject of debate among the experts. A separate study made at the time of the Hagstrom study estimated the exposures as follows: for the sinovial sarcoma, the exposures were 0.095 rads[10] for the fetal dose and 1.28 rads for the fetal-liver dose; for the lymphosarcoma, the figures were 0.036 rads for the fetal dose and 0.160 for the fetal-liver dose; for the leukemia case, the exposure was 1.78 rads for the fetal dose and 14.2 rads for the fetal-liver dose. The first two were far below, and the third was above, the median for the index group. These estimates were made many years after the fact and were subject to considerable uncertainty. As a point of comparison: in the 1950s, when fetal x-rays were not uncommon, the average in utero exposure from a fetal x-ray was in the neighborhood of one rad.

The discovery of three cases of malignancy in the exposed group (excluding the liver cancer) and none in the control group led Hagstrom, et al., to conclude, in words carefully chosen, that the result "suggests a cause and effect relationship." Hagstrom, et al., also calculated that, on the basis of cancer rates in the statewide, white population of Tennessee up to age 14, the expected number of childhood cancers in the index group was 0.65 (averaging figures for 1950 and 1960).

The Hagstrom study was published in 1969. Again, nothing happened. More than 20 years passed. Then, in the 1990s, public attention was drawn to human experimentation because of revelations regarding the infamous Tuskegee experiment, in which black men with syphilis were recruited as patients and then deliberately denied treatment because researchers were interested in tracking the long-term effects of the disease. A presidential commission was appointed to investigate and report on such experiments. The commission's report, issued in December 1993, listed a number of such studies; the Vanderbilt study achieved a notoriety of sorts by being first on the list. Emma Craft, whose child had died an excruciating death from the sinovial sarcoma, testified before Congress. Very shortly thereafter, a class action was brought on behalf of the women and children in the study against the State of Tennessee, Vanderbilt, and others. Causation, both general and specific, was a major issue in the case.

Questions

1. What is the relative risk of childhood cancer in the index group vs. (i) the control group? (ii) the general population up to age 14?

2. Using a two-tailed test (see p. 120) and the Poisson distribution, are the relative risks statistically significant?

3. Assuming that the log of the estimated Poisson parameter, $\hat{\mu}$, is approximately normally distributed with variance $1/\hat{\mu}$, what is a 95% confidence interval for the relative risk in 1(ii) above? What is its relevance?

---

[10] A unit of absorbed radiation in which 100 ergs of energy are absorbed per gram of tissue. By way of comparison, background radiation delivers about 0.3 to 0.4 rads per year to humans, most of it from radon.

4. What criticisms would you make of the 0.65 figure as the expected number of childhood cancers in the exposed group?

5. What arguments can be made for and against including the liver cancer?

6. Do the results satisfy the Bradford-Hill criteria for inferring causation?

Sources

P. F. Hahn, et al., *Iron metabolism in human pregnancy as studied with the radioactive isotope Fe 59*, 61 Am. J. Obs. & Gyn. 477 (1951); R. M. Hagstrom, et al., *Long term effects of radioactive iron administered during human pregnancy*, 90 Am. J. Epidemiology 1 (1969); N.C. Dyer & A.B. Brill, *Fetal radiation doses from maternally administered ^{59}Fe and ^{131}I*, in Sikov & Mahlum, eds., *Radiation Biology of the Fetal and Juvenile Mammal* 73 (U.S. Atomic Energy Commission 1969). The most important study of the effects of fetal x-rays was Brian MacMahon, *Prenatal x-ray exposure and childhood cancer*, 28 J. Nat'l Cancer Inst. 1173 (1962). The litigation was styled *Craft et al. v. Vanderbilt et al.*, No. 3-94-0090 (M.D. Tenn 1994).

Notes

Highly qualified epidemiologists for the parties disagreed sharply over whether the epidemiology was sufficient to support the conclusion that it was more likely than not that the in utero exposure had caused the childhood cancers. Dr. Richard Monson, of the Harvard School of Public Health, treated the Hahn study as a unique case because it involved fetal exposure from an internal emitter. On the basis of the fact that there were 0.65 expected cases and 4 observed cases (he included the liver cancer), and without consideration of dose, he concluded that it was more likely than not that the malignancies were caused by the exposure. Dr. Geoffrey Howe, of the Columbia School of Public Health, believed that whether the emitter was internal or external to the fetus affected only dose. He compared the Hahn study with a number of other fetal radiation studies, primarily those involving the atomic bomb and fetal x-rays, and derived a relationship between dose and childhood cancer risk. From this he concluded that the doses involved in the Hahn study were too small to have created an appreciably increased childhood cancer risk. The case was settled before trial when Vanderbilt agreed to pay plaintiffs $10 million. The court never ruled on the adequacy of the scientific evidence of causation.

10.3.3   Preconception paternal irradiation and leukemia

In 1983, a television program reported a cluster of leukemia among young people in Seascale, a village about 3 km south of the Sellafield nuclear reprocessing plant in West Cumbria in northwest England. There were 5 such cases when, based on national rates, only 0.5 would have been expected. The program and a subsequent report caused an uproar and many investigations were launched. A subsequent study by Professor Martin Gardner proposed an explanation (the "Gardner Hypothesis"): preconception paternal irradiation (ppi) of male workers at the Sellafield plant caused mutations in their spermatagonia which, in turn, caused a predisposition to leukemia and/or non-Hodgkins lymphoma in their offspring. On this theory, two lawsuits were brought against British Nuclear Fuels, the owner of the plant.

In one lawsuit, the only one considered here, plaintiff was the mother of Dorothy Reay, who was born in 1961 in Whitehaven, West Cumbria, contracted childhood leukemia, and died in Whitehaven at the age of 10 months. Whitehaven is a village some 15 km from Seascale. Dorothy Reay's father, George Reay, had been employed at Sellafield prior to her conception. It was agreed at the trial that his total ppi was 530 milliSieverts (mSv) and his dose in the six months immediately preceding conception was 12 mSv. A Sievert is a measure of the effective dose of radiation in man. By way of comparison, natural background radiation delivers up to 3-4 mSv per year.

(During the trial, another study was published that revealed a significant excess of leukemia in the ward of Egremont North about 7 km north of Sellafield. There were four cases of childhood leukemia diagnosed in the period 1968–1985, but none of the fathers of these cases had any recorded ppi despite the fact that the collective dose of ppi associated with children born in Egremont North was higher than that associated with children born in Seascale. Whitehaven is closer to Egremont North than to Seascale; see Figure 10.3.3.)

In the epidemiologic studies, the subjects were 46 cases of leukemia occurring in young people (under the age of 25) born in the West Cumbria health district (including the 5 in Seascale) in 1950-1985. Dorothy Reay was among these cases. After the data collection had begun, a decision was made to limit cases to those not merely born but also diagnosed in West Cumbria. However, a young man was included whose home was Seascale but who was diagnosed in Bristol, where he was attending the university.

There were two groups of controls that were matched to each case: area and local. For the area controls, searches were made backwards and forwards in time from the case's entry in the West Cumbria birth register until the nearest four appropriate controls of the same sex in each direction were found. For the local controls, the residence of their mothers was matched for residence (civil parish) of the case; otherwise the procedure was the same as for the area controls. (Thus, a local control could also be an area control.) As an example of exclusions: eight potential controls for the young man diagnosed in Bristol were excluded, apparently because the physician registry in those cases was outside of West Cumbria; in seven of those exclusions the fathers had worked at Sellafield.

The data for cases and local and area controls are shown in Table 10.3.3.

About 1 in 2,000 English children contract childhood leukemia and it is estimated that about 5% of the cases are due to a genetic link. From 1950

TABLE 10.3.3. Leukemia in young persons (age under 25 years) born and diagnosed in West Cumbria, 1950-1985, by father's occupational exposure to ionizing radiation.

Father's preconception dose (in milliSieverts)	Leukemic children W. Cumbria	Local controls	Area controls
Total preconception			
0	38	236	253
1 – 99	4	37	30
≥100	4	3	5
Previous 6 months			
0	38	246	262
1 – 9	4	27	21
≥10	4	3	5

to 1989, of the 9,260 births to Sellafield employees, there were 779 births to Seascale residents and 8,482 in the rest of West Cumbria, with total collective ppi doses of 38 and 501.4 man-Sv, respectively. The number of childhood leukemias in West Cumbria (excluding Seascale) did not exceed expectation based on national rates.

Questions

1. Consider, a priori, and then compute, the odds ratio you deem most relevant to the issue of causation.

2. Based on the data provided, what arguments would you make for and against a finding that ppi caused childhood leukemia in Seascale (general causation) or in Dorothy Reay's case (specific causation)?

Source

Martin J. Gardner, et al., *Results of case-control study of leukemia and lymphoma among young people near Sellafield nuclear plant in West Cumbria*, 300 Brit. Med. J. 423 (1990); R. Doll, H.J. Evans & S.C. Darby, *Paternal exposure not to blame*, 367 Nature 678 (1994).

Notes

Application of the Bradford-Hill criteria (see Section 10.3) is illustrated by the court's resolution of the claims in the Sellafield case.

Strength of Association. The court agreed that there was a cluster of leukemia in Seascale creating a high relative risk, but for several reasons declined to find that the point estimate of the relative risk was evidence of causation.

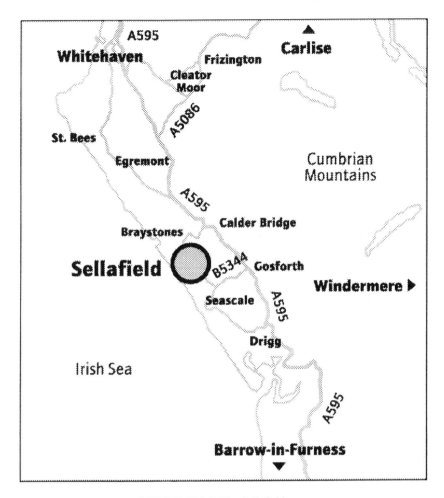

FIGURE 10.3.3. The Sellafield area

First, the number of leukemic children was only 5. Such small numbers raise the possibility of chance association and increase the risk of error if a case is wrongly included.

Second, the excess of leukemia that was investigated was to some extent identified in advance by a television program and prior studies. While neither side claimed the excess was due entirely to chance, the degree of association might have been increased by boundary tightening around Seascale.

Third, a decision was made after the Gardner team had started to collect data to limit cases to those not merely born, but born and diagnosed in West Cumbria. This switch, if made with knowledge of the data, would have compromised the statistical significance (P-values) of the results.

Fourth, contrary to the rules of the study, a young man was included whose home was Seascale, but who was diagnosed in Bristol, where he was attending university. This departure from protocol was important because seven of the eight potential controls who were excluded (apparently because their physician registry was outside of West Cumbria) had fathers who had worked at Sellafield and so would have been candidates for exposure.

Fifth, the Gardner study examined multiple hypotheses and this would have reduced the statistical significance of the results for any one comparison.

Consistency. The court examined other studies of ppi cited by the plaintiffs' experts, but found that they did not support the ppi hypothesis. In one study, the ppi hypothesis was examined for men who had received diagnostic radiation before conception of at least some of their children. The court rejected this study because, in a subsequent publication, the authors admitted that bias may have crept into their work. A study of the children of uranium miners showed an excess of leukemia, but the excess was not statistically significant and the doses were not comparable to the Sellafield doses. An excess of childhood leukemia around Dounreay, another nuclear facility, could not be explained by ppi because, of the 14 cases involved there, only 3 of the fathers had worked at Dounreay and none had received a large preconception dose. A study of the effect of ppi from diagnostic x-rays in Shanghai was found inherently unreliable because most of the information on the x-rays was obtained from the mothers.

On the other hand, studies of A-bomb survivors followed 30,000 offspring of over 70,000 irradiated parents over 45 years and found no ppi effect. These studies were particularly important because they were prospective cohort studies, the most reliable, and were conducted by highly qualified scientists under international supervision.

Dose Response. The court noted that the only evidence of positive association between dose and risk was between children and fathers who had received more than 1000 mSv ppi and that in any event the numbers involved were so small that the data, while not inconsistent with a dose-response relationship in that region, fell short of demonstrating the existence of such a relationship.

Biological Plausibility. The court addressed two questions: whether leukemia had a heritable component and, if so, whether ppi could explain the Seascale excess.

The court found that there was a heritable component in leukemia, but that it was very small (about 5% of leukemia). Moreover, to explain the Seascale excess one would have to assume transmissible gene mutation rates that were far in excess of those shown in insect (Drosophila), mouse, and human in vitro studies.

Recognizing this difficulty, plaintiff argued a synergy theory: the radiation caused a mutation that predisposed the offspring to leukemia, but the leukemia only developed if the mutation was activated or acted in synergy with an X-factor, such as a virus or environmental background radiation.

The court rejected this argument, pointing out that the theory presupposed a high rate of mutations that was inconsistent with the cited studies. Moreover, it was not clear why the X-factor could not operate on its own. In that connection, the synergy theory did not explain another cluster of leukemia in children from Egremont North (a nearby community), none of whose fathers had more than negligible ppi. Nor did the theory explain the absence of any excess of other heritable diseases in the children. To reconcile the excess of leukemia with normal rates of other heritable diseases would require either an exquisitely sensitive leukemia gene or an X-factor that was specific to leukemia. While plaintiff's experts offered some theories to that effect, the court rejected them as speculative.

An alternative hypothesis was proposed by Professor Kinlen. He concluded on the basis of studies of oil refineries in remote places that the Sellafield cluster was caused by chance and by the influx of a large number of outside workers into a previously isolated rural community; some unknown infectious agent was responsible. The court quoted Sir Richard Doll, a world famous epidemiologist, who wrote that the association with ppi was caused by chance and the post hoc selection of an atypical subgroup of the young people in West Cumbria. The court stated that it found the Kinlen hypothesis no less plausible than the Gardner hypothesis.

Analogy. Five studies of the offspring of fathers in motor-related occupations or in paint and pigment-related occupations showed relative risks of 1-5 or more. The court said that, in view of the criticisms of these studies, it gave the evidence from analogy very little weight.

Specificity. The parties were agreed that radiation had a "scatter gun" effect on the germline. Plaintiffs' experts argued that the variety of leukemia was consistent with this. Defendants argued that, if radiation were responsible, one would expect to see an excess of a wide range of diseases with a heritable component in the offspring. The court held that the plaintiffs were right to concede that this criterion afforded them little assistance.

Temporal Association. There was no dispute that ppi preceded leukemia.

Experimentation. There was no suggestion that there were any human experimental data that affected the issues in the case.

The court concluded that plaintiffs had failed to prove by a preponderance of the evidence that ppi was a material contributory cause of the excess childhood leukemia in Seascale. The court also stated that, if it had determined that ppi was the cause of the excess childhood leukemia in Seascale, it would still have found that Dorothy Reay's leukemia was not part of the Seascale excess. Reay had lived in Whitehaven, which was much closer to Egremont North than to Seascale, and so she would have been regarded as part of the Egremont North excess, as to which there was no proof of ppi.

A subsequent study of cancer in some 46,000 children of men and women employed in nuclear establishments in the U.K. found that the leukemia rate in children whose fathers had a preconception dose of > 100 mSv was 5.8 times that in children conceived before their fathers' employment in the nuclear industry (95% c.i. 1.3 to 24.8). However, this result was based on only three exposed cases, and two of those cases had previously been identified in the West Cumbrian case-control study involved in the *Reay* case. No significant trends were found between increasing dose and leukemia. Eve Roman, et al., *Cancer in children of nuclear industry employees: report on children aged under 25 years from nuclear industry family study*, 318 Brit. Med. J. 1443 (1999).

10.3.4   Swine flu vaccine and Guillain-Barré Syndrome

In 1976 there was a cluster of swine flu cases among soldiers in Fort Dix, New Jersey, and one soldier died. The virus resembled the virus in the pandemic that killed 20 million people world-wide in 1918 and 1919. Fearing an outbreak, the federal government undertook a massive immunization effort through the National Swine Flu Immunization Program Act of 1976, 42 U.S.C. 247(b)(1978). The program began on October 1, 1976, but a moratorium was declared on December 18, 1976, eleven weeks after it started, in part because the epidemic did not materialize and in part because there were sporadic reports of Guillain-Barré Syndrome (GBS) following immunization with the vaccine. GBS is a rare neurological disorder that is characterized by a destruction and loss of the myelin sheath, the insulation surrounding nerve fibers. Demyelinated nerve fiber will not conduct impulses from the central nervous system in a normal manner. A typical case of GBS begins with tingling, numbness, or weakness in the limbs, which is followed by paralysis of the extremities, with maximum weakness reached in about two months. The acute phase is followed by a much longer period in which most patients recover the use of their limbs. GBS is sometimes triggered by vaccination or infection, but it is not a well defined organic disorder and consequently is difficult to diagnose.

When insurance companies declined to issue coverage for adverse effects resulting from vaccination, the drug companies declined to produce the vaccine without coverage. To resolve this impasse, the federal government accepted liability for adverse effects under the Federal Tort Claims Act, 28 U.S.C. §1346(b) (1999). All cases of adverse reaction following vaccination were to be reported to the federal government to begin the compensation process.

Louis Manko received a vaccination on October 20, 1976. He testified that, about ten days later, he began to feel muscular weakness, which became progressive. On January 15, 1977, some thirteen weeks after vaccination, he was diagnosed as having GBS. He brought suit against the federal government. The principal issue was causation and the court seemed to assume that causation was established if the rate ratio (*RR*) for GBS comparing vaccinated and unvaccinated subjects was greater than 2.

The experts for the parties were agreed that the *RR* for GBS most relevant to Manko's case was the ratio of observed to expected numbers of new GBS cases 13 weeks after vaccination. Manko's experts computed those numbers by treating as a separate cohort those vaccinated in each time period (using periods of about a week) from the beginning to the end of the program, and then computing for each cohort the observed and expected numbers of new GBS cases 13 weeks after vaccination. The observed number of new GBS cases was simply the number of new GBS cases among the cohort appearing in the 13th week after vaccination. The expected number of new GBS cases (under the null hypothesis that vaccination had no effect on the incidence of GBS) was obtained by multiplying the size of the cohort by the background rate of GBS in the unvaccinated population in that 13th week. The numbers for the first, second, third, etc., cohorts were then summed to arrive at an overall number of observed and expected cases. Since the numbers were still very small, Manko's experts averaged the *RRs* thus computed between 11 and 16 weeks after vaccination. By their computation, the *RR* was greater than 2; one calculation gave it as $29/7.40 = 3.92$.

Following the moratorium, there was a sharp drop in the incidence of GBS among *un*vaccinated persons, which experts for both sides attributed to underreporting. But there was also a drop in rate of GBS among vaccinated persons. Manko's experts took the position that this was also due to underreporting comparable to that for unvaccinated persons.

The government's experts took the position that the decline in reported GBS following the moratorium was due to the passage of time after the vaccinations had stopped. They argued that, if prior estimates of the background rate of 0.24 per million person-weeks of exposure continued to apply, the relative risk would be $29/17.5 = 1.66$.

In Figure 10.3.4, Panel A shows the incidence rate of GBS among the vaccinated, by week since vaccination. The government's argument is reflected in the dashed line, the rate of GBS among the unvaccinated, which is truncated, after the moratorium, to the background rate of 0.24 per million person-weeks of exposure. Panel B shows, in greater detail, the incidence rate among the unvaccinated, by week since the start of the vaccination campaign.

Manko had an infection with respiratory and gastrointestinal symptoms a week or two before his hospitalization for acute GBS. Prior infection is a risk factor for GBS: about 62% of the unvaccinated GBS cases had some illness in the month before onset; for the vaccinated cases, only 33% had been ill. As a first cut, to account for this confounder, the *RR* computed above was multiplied by 33/62. Manko's experts objected that the *RR* so adjusted would be skewed by selection effect, because vaccination was contraindicated immediately after illness. To focus on cases in which illness occurred after vaccination, but within a month of the onset of GBS, they argued that the data should be restricted to late-onset cases like Manko's. In that category, the adjustment would be $53/62 = 0.85$, because 53% of the vaccinated, late-onset GBS cases had a prior illness, while, as stated, 62% of the unvaccinated GBS cases had illness in the month before GBS. Note that, in the above calculation of adjustment, it is assumed that the *RR* of illness for vaccinated vs. unvaccinated persons is 1. This is so because the formal relation between the *RR* of GBS and the *RR* of GBS for people with prior illness is given in general by:

$$[RR \text{ of } GBS \text{ given illness}] = \frac{[RR \text{ of } GBS] \times [RR \text{ of illness given } GBS]}{[RR \text{ of illness}]},$$

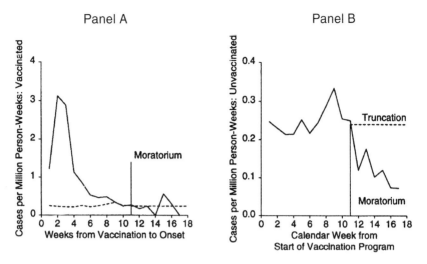

FIGURE 10.3.4. Panel A shows the incidence of GBS among the vaccinated, by week since vaccination. Panel B shows the incidence of GBS among the unvaccinated from the start of the vaccination program.

where the *RR* for GBS or for illness are for vaccinated vs. unvaccinated subjects.

Manko's experts argued that the *RR* for illness would be less than 1 because people were advised against vaccination immediately following illness. Another reason is that vaccinated persons with GBS may have been less willing to have reported any illness. On the other hand, the *RR* of illness could be greater than 1 if people prone to illness tended to seek vaccination and ignored the advice in contraindication.

Questions

1. If the government's experts were correct in their interpretation of the decline of GBS among the vaccinated, explain why the *RR* calculation of Manko's experts would be biased. Which interpretation, Manko's experts or the defendant's experts, is more reasonable?

2. Use Bayes's theorem to derive the relation shown for the *RR* of GBS given illness. Is there a problem with the definition of illness?

3. Was the adjustment for prior illness properly made?

Source

Manko v. United States, 636 F. Supp. 1419 (W.D. Mo. 1986), *aff'd in part*, 830 F.2d 831 (8th Cir. 1987); D.A. Freedman and P.B. Stark, *The swine flu vaccine and Guillain-Barré syndrome: a case study in relative risk and specific causation*, 23 Eval. Rev. 619 (1999).

Notes

On the causation issue, the district court found that the vaccine caused plaintiff's GBS on two separate theories: (i) plaintiff's symptoms of muscle weakness occurring shortly after vaccination amounted to "smouldering GBS," which became acute in January 1977; and (ii) even if plaintiff's GBS did not occur until January 1977, some 13 weeks after vaccination, causation was proved because the relative risk was greater than 2. The court of appeals affirmed on the first ground without reaching the second.

10.3.5   Silicone breast implants

Between 650,000 and one million women received silicone breast implants during the 1970s and 1980s. The legal disputes started shortly afterwards as women claimed that leaks from the implants had damaged their health. Claims were made that, besides localized complications, the implants caused connective tissue diseases (CTD), such as rheumatoid arthritis, lupus, and scleroderma. In December 1991, a California jury awarded a woman $7.34 million for CTD. There was much publicity of the verdict and the causal link. Shortly thereafter, probably influenced by the verdict and the public outcry, the Commissioner of the Federal Food and Drug Administration sharply limited the future sales of the implants pending submission of additional safety data. It was not until June 16, 1994, two years after the FDA had taken breast implants off the market, that the first reliable observational epidemiologic study was published. In the meantime, the FDA action was a major factor in causing a tidal wave of litigation against the manufacturers.

The federal cases were consolidated for pre-trial proceedings before District Judge Sam C. Pointer, Jr., of the Northern District of Alabama. Availing himself of his power to appoint experts, Judge Pointer selected an expert panel to examine the evidence that silicone breast implants caused CTD. The panel took two years and in December 1998 issued its report finding no connection between implants and CTD. Chapter III of the report, entitled "Epidemiological Analysis of Silicone Breast Implants and Connective Tissue Disease" (Principal Author, Barbara S. Hulka), did not present new research, but rather consisted of a meta-analysis of the available epidemiologic studies. The principal features of the meta-analysis were as follows.

The report analyzed the connection between implants and five conditions with established diagnoses. These were rheumatoid arthritis, lupus, scleroderma, Sjögren's syndrome, and dermatomyositis/polymyositis. There was also a diagnostic category called "definite CTDs combined," which included the five definite diagnoses plus what the individual study authors regarded as "definite CTD," even though the diagnoses did not fit into the established categories. This last category was designed to allow for some uncertainty and change in diagnoses. In addition, the report used a category called "other autoimmune/rheumatic conditions" to include ill-defined diagnoses, such as undifferentiated connective tissue disease.

The report used as the exposure of interest *any* implant, not merely silicone implants, because information on type of implant frequently was missing or could not be verified. Separate analyses were also made of known silicone implants.

The panel searched data bases of published materials, and a data base of dissertation abstracts for unpublished sources. Studies published only in abstract or letter form in unrefereed journals were also included. The search was limited to English-language sources and to human subjects. Submissions were also received from a plaintiffs' law firm. This search yielded 756 citations, most of which were not epidemiologic studies. All potentially relevant papers were reviewed independently by the investigators. The criteria for inclusion were: (1) an internal comparison group; (2) available numbers from which the investigators could identify the numbers of implanted women with and without the disease and the numbers of nonimplanted women with and without the disease; (3) the exposure variable being the presence or absence of breast implants; and (4) the disease variable being some type of CTD. For each study, the investigators calculated an *OR* or an *RR* by the exact method based on the non-central hypergeometric distribution. The *OR*s and *RR*s thus obtained were unadjusted for confounders.

In producing an overall summary estimate, it was assumed that the studies were homogeneous, i.e., they constituted repeated measurements of the same *OR* or *RR*. To test this assumption, the investigators used a chi-squared test and assumed that heterogeneity might exist if the *P*-value were less than 0.10. When that occurred they attempted to stratify the studies to achieve homogeneity within the strata. The chief available stratification variables (i.e., variables for which there was information for all studies) were study design (cohort or other), medical record validation of disease (yes or no), and the date of data collection on disease diagnosis ($<$ 1992 or $\geq$ 1992). If homogeneity could not be achieved through stratification, the investigators resorted to visual inspection of individual studies in search of outliers. Studies were removed individually or in pairs to achieve homogeneity among the remaining studies. The final set of studies represented the largest number of studies and subjects that were homogeneous.

The investigators also prepared meta-analyses of effect estimates that were adjusted by the authors of the individual studies to reflect possible confounders. The primary confounders were age, with controls frequently being older than cases, and secular time, because implant frequency and implant type varied by calendar time period. Additionally, length of follow-up sometimes varied between implanted and nonimplanted women within individual studies. Because the exact methods previously used for unadjusted results could not be used for

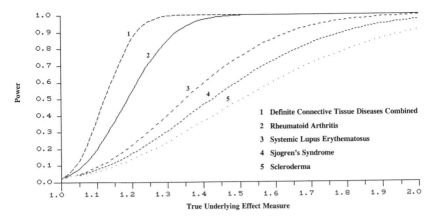

FIGURE 10.3.5. Power versus underlying effect measure (including Hennekens)

adjusted results, large-sample, meta-analytic methods were used to combine adjusted *OR*s and *RR*s obtained from the individual studies.

Table 10.3.5a. shows the individual study results for the category "Definite CTDs combined," for the 16 studies with unadjusted odds ratios. Eleven of these studies had adjusted odds ratios, which are also given. Table 10.3.5b shows some derived statistics for the Table 10.3.5a studies with adjusted odds ratios.

Figure 10.3.5 shows power calculations for Definite CTDs combined and for the conditions separately.

Questions

1. Using Tables 10.3.5a and 10.3.5b and the weighted-average method of Section 8.1 at p. 243, compute a common adjusted *OR* and a 95% confidence interval for the estimate. What conclusion do you reach?

2. Compute a *P*-value for the homogeneity statistic. The chi-squared statistic used in the Mantel-Haenszel test for homogeneity is $\sum_i w_i(l_i - \bar{l})^2$, where l_i and $\bar{l}$ are, respectively, the log odds ratio for the ith study and weighted average log odds ratio for the studies, and $w_i = 1/[se(l_i)]^2$ is the weight used for the ith study in averaging the studies. For ease of calculation, use the alternative formula for the variance (see section 1.3 at p. 18), giving

$$\sum_i w_i(l_i - \bar{l})^2 = \sum_i w_i l_i^2 - \left(\sum_i w_i\right) \bar{l}^2.$$ What conclusion do you reach?

3. What systematic difference is there between the unadjusted and adjusted estimates in the studies? What could cause that? Which set of results appears to be more valid?

4. What are the arguments for and against excluding the Hennekens study?

TABLE 10.3.5a. Sum of "Definite CTDs" (Rheumatoid arthritis, Systemic Lupus, Scleroderma, Sjögren's Syndrome, Dermatomyositis/Polymyositis, and those so designated by author) by study

Study by first author	Implant[1]		No implant		Crude OR[3,4] (95%CI)	Ln(OR)	SE[ln(OR)]	Cond MLE[4] (Exact 95%CI)	Adj OR[5] (95%CI)	Dsn[6]	Dx[7]	Yr[8]
	D[2]	D̄	D	D̄								
Burns	2	14	272	1170	0.61 (0.14, 2.72)	−0.4870	0.7589	0.61 (0.07, 2.70)	0.95 (0.21, 4.36)	0	1	1
Dugowson, 1992	1	12	299	1444	0.40 (0.05, 3.11)	−0.9102	1.0428	0.40 (0.01, 2.74)	0.41 (0.05, 3.13)	0	1	1
Edworthy	19	1093	16	711	0.77 (0.39, 1.51)	−0.2582	0.3427	0.77 (0.37, 1.62)	1.00 (0.45, 2.22)	1	1	0
Englert	4	4	527	249	0.47 (0.12, 1.90)	−0.7497	0.7113	0.47 (0.09, 2.56)	–	0	1	1
Friis	10	2560	25	10998	1.72 (0.82, 3.58)	0.5414	0.3748	1.72 (0.74, 3.71)	–	1	1	1
Gabriel	5	744	10	1488	1.00 (0.34, 2.94)	0.0000	0.5496	1.00 (0.27, 3.22)	1.10 (0.37, 3.23)	1	1	1
Goldman	6	144	709	3370	0.20 (0.09, 0.45)	−1.6192	0.4187	0.20 (0.07, 0.44)	–	0	1	0
Hennekens	231	10599	11574	373139	0.70 (0.62, 0.80)	−0.3529	0.0672	0.7026 (0.6133, 0.8016)	1.24 (1.08, 1.41)	0	0	0
Hochberg, 1996b	11	31	826	2476	1.06 (0.53, 2.13)	0.0617	0.3532	1.06 (0.48, 2.19)	1.07 (0.53, 2.13)	0	1	1
Lacey	2	13	187	1030	0.85 (0.19, 3.79)	−0.1656	0.7637	0.85 (0.09, 3.79)	1.48 (0.34, 6.39)	0	1	1
Nyren	16	7426	11	3342	0.65 (0.30, 1.41)	−0.4237	0.3922	0.65 (0.28, 1.56)	0.80 (0.50, 1.40)	1	1	1
Park	1	316	1	215	0.68 (0.04,10.94)	−0.3851	1.4170	0.68 (0.01, 53.65)	–	1	1	0
Sanchez-Guerrero	3	1180	513	85805	0.43 (0.14, 1.32)	−0.8551	0.5798	0.43 (0.09, 1.25)	0.60 (0.02, 2.00)	1	1	1
Strom	1	0	132	100	–	–	–	(0.02, ∞)	–	0	1	1
Teel	6	24	421	1553	0.92 (0.37, 2.27)	−0.0810	0.4597	0.92 (0.31, 2.34)	0.90 (0.40, 2.30)	0	1	1
Wolfe	3	4	634	1130	1.34 (0.30, 5.99)	0.2902	0.7654	1.34 (0.20, 7.93)	1.35 (0.30, 6.06)	0	1	1

[1] All types of implants were included. [2] Only cases in which disease followed implant were included, when that information was available. [3] Mantel-Haenszel estimate with Robins–Breslow–Greenland (RBG) confidence limits. [4] Obtained with exact non-central hypergeometric algorithms. [5] Reported by author. [6] 1 if cohort, 0 if not. [7] 1 if by medical record validation, 0 if not. [8] 1 if data on disease diagnosis < 1992 in at least 90% of the cases, 0 if not.

TABLE 10.3.5b. Silicone breast implants: Weighted least squares meta-analysis using studies with adjusted odds ratios.

	Study	(1) adjusted ln(OR)	(2) Se(1) (a)	(3) Weights (b)	(4) (3) × (1)	(5) (3) × (1)²
1.	Burns	−0.051	0.774	1.670	−0.085	0.004
2.	Dugowson, 1992	−0.892	1.055	0.898	−0.801	0.715
3.	Edworthy	0.000	0.407	6.033	0.000	0.000
6.	Gabriel	0.095	0.553	3.273	0.311	0.030
8.	Hennekens	0.215	0.068	216.152	46.473	9.992
9.	Hochberg, 1996b	0.068	0.355	7.942	0.540	0.037
10.	Lacey	0.392	0.748	1.786	0.700	0.274
11.	Nyren	−0.223	0.263	14.495	−3.232	0.721
13.	Sanchez-Guerrero	−0.511	1.175	0.725	−0.370	0.189
15.	Teel	−0.105	0.446	5.022	−0.527	0.055
16.	Wolfe	0.300	0.767	1.701	0.510	0.153
	Total			259.697	43.518	12.170

(a) Se{adjusted ln(OR)} approximated by half-width of (log endpoints of 95% confidence interval for adjusted OR) divided by 1.96.

(b) Weights are squared reciprocals of column (2).

5. Interpret the power curves of Figure 10.3.5.

6. In *Daubert v. Merrell Dow Pharmaceuticals*, 509 U.S. 579 (1993), the Supreme Court held that trial judges must be more vigilant gate keepers, assuring that "scientific testimony or evidence admitted is not only relevant but reliable." Should proffered expert testimony that silicone implants can cause CTD based on case reports, clinical studies, in vitro, or in vivo studies be deemed sufficiently reliable to be admissible under *Daubert* despite the negative epidemiology?

Source

Betty A. Diamond, Barbara S. Hulka, Nancy I. Kerkvliet, and Peter Tug-well, *Silicone Breast Implants in Relation to Connective Tissue Diseases and Immunologic Dysfunction: A Report by a National Science Panel to The Honorable Sam C. Pointer Jr., Coordinating Judge for the Federal Breast Implant Multi-District Litigation* (November 17,1998), available at <www.fic.gov/breimlit/mdl926.htm>. See also, Esther C. Janowsky, Lawrence L. Kupper & Barbara S. Hulka, *Meta-Analyses of the relation between silicone breast implants and the risk of connective-tissue disease*, 342 New Eng. J. Med. 781 (2000).

Notes

Despite the publication of this and other studies by scientific bodies reaching the same conclusion, litigation and the process of settlement continued. In particular, Dow Corning, which was forced into bankruptcy, went ahead with hearings before a bankruptcy court for approval of a planned $3.1 billion settlement with a class of some 170,000 women who had received breast implants. Three other manufacturers also reached class action settlements totaling some $3 billion. These were, at the time, the largest settlements in the history of American tort law.

The performance of the legal system in these cases was sharply criticized by Marcia Angell, the executive editor of the New England Journal of Medicine, in a notable book, *Science on Trial: The Clash of Medical Evidence and Law in the Breast Implant Cases* (1996). Some legal scholars defended the results, arguing that, even if the implants did not cause CTDs, the manufacturers were at fault (and presumably were properly punished) for selling a product that turned out to be safe, but before there had been sufficient studies with which to prove that safety. Rebecca S. Dresser, Wendy E. Wagner & Paul C. Giannelli, *Breast Implants Revisited: Beyond Science on Trial*, 1997 Wisc. L. Rev. 705. As they put it: "If well-designed studies had been conducted sooner, there would have been no need for drastic FDA intervention and no basis for the implant litigation." *Id.* at 708. Do you agree?

On question (6), compare *Vassallo v. Baxter Healthcare Corp.*, 696 N.E.2d 909 (Mass. 1998)(testimony allowed despite absence of supporting epidemiology) with *In re Breast Implant Litigation*, 11 F.Supp.2d 1217 (D. Colo. 1998) (testimony excluded).

CHAPTER 11

Survival Analysis

11.1 Death-density, survival, and hazard functions

There are many situations in law that require estimation of the time to failure, e.g., how long someone will live or something will last. The event of failure may be anything—death, filing a claim after exposure to a carcinogen, peeling of defective paint on house sidings, or termination of an at-will relationship. If there are comparable populations in which all or almost all members have already failed, the probabilities of survivorship and death at various ages can be estimated directly from data and presented in the form of a life table. The classic life table tracks an initial cohort of people year by year from birth, showing in each year how many people die and how many survive, until the entire cohort has died. If there are no comparable populations with complete survival data, the problem of estimation is more difficult. The average life of those who have already failed would seriously underestimate the average life of those still surviving. An estimate of the likelihood of early failure is still possible; but at later times, to take account of the fact that many in the population have not yet failed, one must assume that the force of mortality has a certain mathematical form or model as a function of time and estimate the model's parameters from the data. The future rate of mortality is then extrapolated from the model.

In survival analysis, four functions related to the failure time, viewed as a random variable T, are ordinarily defined:

1. The *death-density* function $f(t)$, or unconditional failure rate, is the probability that an individual will die within a given time interval divided by the length of that interval. Where this probability changes in the interval, a limit is taken as the interval shrinks to zero, at which point we have the instanta-

neous death rate, which is the probability density function for the failure time. Because death must occur sometime, the total area under the death-density function from zero to infinity equals one.

2. The *survival function* $S(t)$ is the probability that an individual fails at a given time t or later, $S(t) = P[T \geq t]$. If $F(t)$ denotes the cumulative distribution function of T, $F(t) = P[T \leq t]$, then $S(t) = 1 - F(t) + P[T = t]$. When failure times are measured precisely in time (the "continuous-time" case), so that $P[T = t] = 0$, $S(t)$ is simply one minus the cumulative distribution function of the failure time. The survival function never increases, and generally decreases, with time. It is often estimated from follow-up data by the "product-limit" estimator, also known as the Kaplan-Meier estimator, $\hat{S}(t) = \prod_t (1 - d_j/n_j)$, where d_j is the number of observed deaths at the moment t_j of the jth death time, n_j is the number at risk of death just prior to time t_j, and where the product $\prod_t$ is taken over all death times prior to time t, i.e., for all j with $t_j < t$. The product-limit estimator thus equals the product of sample survival proportions up to time t. In the continuous-time case, there are no ties among the observed times of failure, and d_j equals 1. Occasionally failure time data are measured only in discrete units or at the end of fixed periods, so that failure can only occur, or be observed to occur, at the end of each time period. In this "discrete-time" case, d_j may be greater than one. $\hat{S}(t)$ is particularly useful because it is defined even when losses to follow-up occur, so that not all failure times can be observed ("censored" failure times). Such losses are subtracted from the number at risk, n_j, after the time of censoring. In this way we utilize the survival information provided by those lost to follow-up, up to the time of censoring.

3. The *hazard function* $\theta(t)$, also known as the "force of mortality," is a rate measured in units of deaths per person per time period, as, e.g., 5 deaths per 100,000 persons per year, or 5 deaths per 100,000 person-years. The hazard function generally varies with time. For example, the hazard function for death from all causes in human beings is higher around birth and during old age than during late childhood through middle age.

In the continuous-time case, $\theta(t)$ is an instantaneous rate, defined as the limiting value of the ratio of the conditional probability that an individual dies during a short interval starting at time t, given his survival before time t, to the length of the interval as that length approaches zero. From the definition it follows that the hazard function is equal to the death-density function at time t divided by the survival function at time t, $\theta(t) = f(t)/S(t)$.

We emphasize that in the continuous-time case, the hazard function is a rate, i.e., a proportion per time unit, not a probability. In particular, the hazard rate can be greater than 1 whenever the rate of death, if continued long enough, would exhaust the population before the unit time period elapsed. For example, if 2,000 deaths occur in a population of 12,000 during one month, the person-years at risk are $12,000 \cdot (1/12) = 1,000$ person-years, so that the hazard rate is 2,000 deaths/1,000 person-years $= 2$ deaths per person-year. For a small time interval of length h, the quantity $\theta(t) \cdot h$ does approximate the conditional probability of death in the time interval from t to $t + h$. Thus, in the above example with $h = 1/12$, the quantity $\theta(t) \cdot h = 2 \cdot 1/12 = 1/6$, which is the conditional probability of death in the first month.

In the discrete-time case, with time periods indexed by $j = 1, 2, \ldots$, the hazard function is defined slightly differently as the conditional probability of death in period j given survival past period $j - 1$. The units are still deaths per person per period, but since the period is fixed and indivisible, the hazard rate never exceeds 1 per person per period. In the above example, with monthly periods fixed, the discrete hazard function would equal 1/6 for the given period.

4. The *cumulative hazard* function, $H(t)$, at time t is the integral of the hazard function θ from 0 to t. In continuous time, the cumulative hazard function is equal to minus the natural logarithm of the survival function; taking antilogs, the survival function is equal to e raised to the power of minus the cumulative hazard function. In symbols, $H(t) = \int_0^t \theta(u)\,du = -\log S(t)$, and $S(t) = \exp[-H(t)]$. Thus, one way to estimate $H(t)$ is to take the negative log of the product-limit estimator, $\hat{H}(t) = -\log \hat{S}(t)$. Another way to estimate $H(t)$ is by summing the discrete hazard estimates d_j/n_j up to time t. The two estimates agree closely except at very large times where data become sparse. In discrete time, $H(t)$ is defined as the cumulative sum of the hazard function, and is estimated by summing the discrete hazard estimates. The cumulative hazard function is useful for plotting procedures designed to check on specific parametric forms for the survival distribution, e.g., the Weibull failure time discussed below.

Specific forms of the foregoing functions that are widely used are the Weibull, exponential, and geometric functions.

The *Weibull* hazard function with parameters c and θ is a widely used continuous-time hazard rate function given by $\theta(t) = \theta_0 \cdot c \cdot t^{c-1}$ where θ_0 is $1/c$ times the hazard rate at $t = 1$. If $0 < c < 1$, then the hazard declines with time; if $c = 1$ the hazard remains constant (this is the exponential distribution); if $c > 1$, the hazard increases with time. Using the definition of the cumulative hazard rate for the Weibull distribution, $H(t) = \int_0^t \theta_0 \cdot c \cdot x^{c-1}\,dx = \theta_0 \cdot t^c$. Taking logarithms of both sides yields $\log H(t) = \log \theta_0 + c \cdot \log t$. Thus, if the values of the cumulative hazard function fall on a straight line against time when graphed on log-log paper, this relation can be used to estimate the parameters of the Weibull hazard rate function and, from these, the survival function. For the Weibull distribution, $S(t) = \exp(-\theta_0 \cdot t^c)$. The median survival time for a Weibull distribution is $\{(\log 2)/\theta_0\}^{1/c}$, and the mean survival time is $\Gamma(1+1/c)\theta_0^{-1/c}$, where $\Gamma(x)$ is the gamma function satisfying $\Gamma(x+1) = x\Gamma(x)$: if $1/c$ is an integer, $\Gamma(1 + 1/c) = (1/c)!$

In the case of the *exponential* distribution the cumulative hazard function is $\theta_0 \cdot t$ and the survival function is $\exp(-\theta_0 t)$. The mean time to failure is equal to the reciprocal of the hazard constant θ_0, and the standard deviation of the time to failure is equal to the mean. The term "half-life" refers to the median life in an exponential survival distribution. The half-life is given by $(\log 2)/\theta_0$. Do you see why? See Section 4.8.

The *geometric* distribution is a discrete time analogue of the exponential distribution and shares many of its properties. Like the exponential distribution, it is characterized by a constant hazard, p, and is without memory, i.e., the expected residual lifetime is constant no matter when it is evaluated. What is the median lifetime (half-life) in a geometric survival distribution? See Section 4.8.

Further Reading

J. Kalbfleisch and R. Prentice, *The Statistical Analysis of Failure Time Data* (1980).

J. Lawless, *Statistical Models and Methods for Lifetime Data* (1982).

11.1.1   Valuing charitable remainders

Humes v. United States, 276 U.S. 487 (1928). Della R. Gates died leaving one-half her residuary estate of $11,783,072.30 in trust for her 15-year old niece, who lived in Texas. The niece was to receive income and portions of the corpus during her lifetime. If she died without issue before reaching the age of 40, any principal not given to her was to be distributed to charity. There was a parallel provision for the testatrix's brother involving the other half of the estate. Actuaries determined that present value of the contingent interest was $482,034, or 4.0909% of the residue. This amount was deducted from the gross estate for estate tax purposes.

The probability that the charity would take was calculated using a standard experience table of mortality with two other "relatively little known" tables not previously used in legal proceedings. These tables were described by Justice Brandeis as follows:

> Both of these tables are based on data contained in volumes of Lodge's Peerage. The first table, which may be found in the Transactions of the Faculty of Actuaries in Scotland, Vol. 1, pp. 278–79, and is called Lees' Female Peerage Tables, was constructed by M. Mackensie Lees. It deals with 4,440 lives, of whom 2,010 died during the period of observation. The second of the tables, which may be found in an article entitled "On the Probability that a Marriage entered into by a Man of any Age, will be Fruitful," in the Journal of the Institute of Actuaries of Great Britain, Vol. 27, pp. 212–13, was constructed by Dr. Thomas Bond Sprague. It deals with the experience of 1,522 male members of the Scotch [sic] peerage and purports to show the probability that a marriage will be childless both as respects men married as peer or heir apparent and men who did not marry as peer or heir apparent. In order to apply the latter table to females certain assumptions and adjustments are necessarily made.

Id. at 493.

The Revenue Act of 1918 provided that "the amount of all bequests to or for" a charity was to be deducted from the gross estate. Was the value of the charitable bequest sufficiently ascertainable to be deductible?

Commissioner v. Marisi, 156 F.2d 920 (2d Cir. 1946). When Marisi divorced his wife in 1931, she became entitled to alimony until her death or remarriage. In 1940 Marisi died and his estate claimed a deduction for estate tax purposes of the value of that obligation. Marisi's wife was then 49 and unmarried. To determine the expected time to remarriage, the Tax Court, over the Commissioner's objection, accepted actuarial tables prepared by the Casualty Actuarial Society of America, based upon the remarriage experience of American widows entitled to workmen's compensation in states where compensation benefits were lost upon remarriage. A deduction was allowed accordingly.

The Commissioner objected that no deduction should be allowed because the tables were not relevant from an actuarial point of view. Was the Commissioner right?

Commissioner v. Sternberger, 348 U.S. 187 (1954). Sternberger established trusts to provide income to his wife and daughter, with remainder to charity if the daughter had no descendants living at her death. The estate claimed a charitable deduction based on the actuarial value of the conditional gift to charity. The probability of the daughter's marriage was computed from the American Remarriage Table (referred to in *Marisi* above). A specially devised table was used to compute the probability of the daughter's having issue. It was assumed that any child would survive its mother. On that basis, the present value of the charitable remainder was computed as 24.06% of the principal value of the trust funds (some $2,000,000).

The Commissioner objected that the tables were not relevant from an actuarial point of view. Was the Commissioner right?

11.1.2   Defective house sidings

Whittaker claimed that Sherwin-Williams breached its express and implied warranties by supplying Crown with defective Superclad paint, which Crown in turn sold to consumers under a 5-year warranty. When warranty claims were made against Whittaker, it sued Sherwin-Williams. Assume that at the time of trial only about 2 years of the 5-year warranty period had run. The court required proof of damages arising from all past and future failures, not only those failures occurring up to the time of trial. Set forth in Table 11.1.2 are data showing the cumulative hazard by months to failure for the color laurel green. A plot of cumulative hazard vs. months to failure on log-log paper (the logs used here are common logs) shows that the points fall close to a straight line. The slope of that line (as determined by least squares regression) is 2.484 and the intercept is -3.742.

TABLE 11.1.2. Failure data for laurel green paint

Age (months)	Cumulative hazard	$\log_{10}$ Age	$\log_{10}$ Cum haz
9	0.02459	0.95424	−1.60924
9	0.04919	0.95424	−1.30812
12	0.07380	1.07918	−1.13194
12	0.09841	1.07918	−1.00696
13	0.12303	1.11394	−0.90999
14	0.14765	1.14613	−0.83077
14	0.17228	1.14613	−0.76377
15	0.19692	1.17609	−0.70571
17	0.22156	1.23045	−0.65451
19	0.24621	1.27875	−0.60869
20	0.27086	1.30103	−0.56726
22	0.29552	1.34242	−0.52941

Questions

1. Is the life of laurel green house sidings exponentially distributed?

2. What is the probability that a laurel green siding will last for at least five years?

11.1.3 "Lifing" deposit accounts

When Trustmark National Bank acquired Canton Exchange Bank on December 14, 1983, it allocated a part of the purchase price to Canton's regular savings deposit accounts, and then sought to amortize the cost for tax purposes. The willingness of depositors to continue their accounts is regarded as an intangible asset, the cost of which can only be amortized if it has a limited life that can be estimated with reasonable accuracy. The IRS denied any deduction by amortization, taking the position that the life of the asset could not be so estimated. The taxpayer brought suit in the Tax Court and at the trial estimated a Weibull model from data on the lives of accounts. These are shown in Table 11.1.3.

Questions

1. Referring to Table 11.1.3, is the hazard of termination increasing, decreasing, or staying the same with age? Is the pattern reasonable?

2. Plot the logarithms of the estimated cumulative hazard, $H(t)$, on the vertical axis of a graph against the logarithms of the average age at the end of 1983. From this graph estimate the slope coefficient c and the intercept term $\log \theta_0$. This provides a graphical estimation procedure for the parameters c and θ_0 of a Weibull distribution.

TABLE 11.1.3. Observed closures in 1983: regular savings accounts

Year open	Avg age acc'ts at end of 1983	Log of avg age	No. acc'ts at start of 1983	No. of closures	Prop. of closures	Est. cum. hazard	Log est. cum. hazard
1983	0.5	−0.693	409[1]	133	0.325	0.3252	−1.123
1982	1.5	0.405	347	110	0.317	0.6422	−0.443
1981	2.5	0.916	257	55	0.214	0.8562	−0.155
1980	3.5	1.253	312	52	0.167	1.0229	0.023
1979	4.5	1.504	310	29	0.094	1.1164	0.110
1978	5.5	1.705	249	35	0.141	1.2570	0.229
1977	6.5	1.872	197	28	0.142	1.3991	0.336
1976	7.5	2.015	152	20	0.132	1.5307	0.442
1975	8.5	2.140	129	17	0.132	1.6625	0.508
1974	9.5	2.251	108	13	0.120	1.7828	0.578
1973	10.5	2.351	89	7	0.079	1.8615	0.621
1972	11.5	2.442	82	8	0.098	1.9590	0.672
1971	12.5	2.526	65	4	0.062	2.0206	0.703
1970	13.5	2.603	50	3	0.060	2.0806	0.733
1969	14.5	2.674	21	2	0.095	2.1758	0.777

[1] Accounts opened during 1983.

3. Use the above parameters to estimate the median and mean of a Weibull distribution (rounding $1/c$ to the nearest integer).

4. Should a depreciation deduction be allowed?

Source

Trustmark Corp v. Commissioner of Internal Revenue, T. C. Memo 1994-184 (1994).

11.2   The proportional hazards model

The Weibull and similar models purport to estimate the probability of "death" as a function of time. Such models can be used to compare hazards in two or more populations or groups. However, a more non-parametric approach, known as a proportional hazards model, can be used to estimate the increase in risk in a population exposed to some condition or treatment. See Cox, *Regression Models and Life Tables*, 34 J. Roy. Stat. Soc. (Series B) 187 (1972). In such models the hazard function for the exposed population is postulated to be a multiple (or proportionality factor) times the baseline hazard associated with the unexposed population. The proportionality factor may be a function of the degree of exposure, usually assumed to be of the form $\exp(\beta X)$, where X is the degree of exposure and β is a coefficient to be estimated from the data. The hazard function of the exposed population is $\theta(t \mid X) = \theta_0(t)\exp(\beta X)$, where $\theta(t \mid X)$ is the hazard for a population with exposure X and $\theta_0(t)$ is the baseline hazard function for the unexposed population. Under this model, each unit increase in exposure corresponds to an e^β-fold increase in risk over baseline; if β is small, this increase in exposure corresponds approximately to a 100β percent increase in risk over the baseline.

A distinct advantage of this model is that the coefficient β can be estimated from data without any assumptions regarding the specific form of the baseline hazard $\theta_0(t)$. The key element in the analysis is the conditional probability that a person with a given degree of exposure X_i (possibly 0) would die at a certain moment given that someone in the risk group has died at that moment; this probability is ascertained at each observed death time. (The risk group at time t includes all persons alive and at risk just prior to time t, both exposed and unexposed.) This conditional probability is proportional to the hazard function $\theta(t|X)$ among those at risk at the time of each death. Because the baseline risk $\theta_0(t)$ appears in both numerator and denominator, it cancels out and the conditional probability simply equals $\exp(\beta X_i)$ divided by the sum of such terms over all individuals in the risk set. The product of the conditional probabilities is known as the partial likelihood function.[1] The factor β is then estimated by maximum likelihood, namely, it is the value that maximizes the partial likelihood function.

When $\beta = 0$ there is no excess risk associated with the exposure. In that case exposure is irrelevant and the conditional probability that a death would occur to a person with any exposure, given that a death has occurred among those at risk, is simply one divided by the total number of persons in the risk group. For dangerous exposures, the conditional probability that a death would involve a person with high exposure is large, and β is estimated to be correspondingly greater.

When testing the hypothesis that $\beta = 0$, a central limit theorem usually applies to the sum of the exposure variates for the observed deaths, provided there are enough of them. This is true even if the exposure variate is not normally distributed. The difference between the sum of the exposures for the people dying and their expected exposures divided by the square root of the sum of the variances for such exposures tends to a normal distribution with zero mean and unit standard deviation. The statistic is called a score test,[2] and it is powerful against proportional hazards alternatives.

To estimate β, a computer iteration is generally required. A closed-form approximation to the maximum likelihood estimator of β (really the first step

---

[1]Multiplying these conditional probabilities together, even though they are not strictly independent, can be justified by partial likelihood theory, the properties of which are similar to standard likelihood theory. See Section 5.6.

[2]Score tests are generally based on the statistic that results from differentiating the logarithm of the likelihood function. In this instance, the score test is the derivative of the log partial likelihood, evaluated at $\beta = 0$.

in the iteration process) is given by the sum of the differences between the observed and expected exposures divided by the sum of the variances of the exposures under the null hypothesis. To be accurate, the full iteration should be carried out for three or four steps.

If the exposure variate is a binary indicator, simplifications occur, which we examine in some detail. For any time interval, a fourfold table can be constructed with rows giving the number of deaths during, and survivals beyond, the interval (among those still alive and at risk at the beginning of the interval), and columns corresponding to exposure status. Suppose the full study period is divided into a sequence of equal subintervals and corresponding fourfold tables are constructed. Some tables may indicate no deaths in a subinterval: these are non-informative about β and may be ignored. The remaining informative tables will show the occurrence of one or more deaths. Say the ith such table has n_i persons at risk, with m_i exposed and $n_i - m_i$ not exposed, and with d_i who died and $n_i - d_i$ who survived. (If the subinterval length is short, d_i will equal 1 for most of the informative tables.) The situation may be summarized thus:

	exposed		
	yes	no	
died	d_{i1}	d_{i0}	d_i
survived	l_{i1}	l_{i0}	$n_i - d_i$
at risk	m_i	$n_i - m_i$	n_i

The proportional hazards model implies that whatever odds on death there may be during any short interval for those unexposed, the odds on death for those exposed are greater by a constant factor e^β, i.e., there is a common odds ratio $\Omega = e^\beta$ underlying each table.

To test the hypothesis $H_0 : \beta = 0$, we may combine the evidence from the fourfold tables constructed above using the Mantel-Haenszel statistic (this score test is also called the log-rank test),

$$z = \frac{\left(\sum_i d_{i1}\right) - \left(\sum_i d_i m_i / n_i\right)}{\left\{\sum_i d_i(n_i - d_i)m_i(n_i - m_i)/\left[n_i^2(n_i - 1)\right]\right\}^{\frac{1}{2}}}.$$

The justification for using this statistic, however, is different from that used in Section 8.1. There, the fourfold tables were statistically independent, whereas here they are not: outcomes d_{i1} and l_{i1} in table i affect the margins in subsequent tables, and hence the conditional distribution of the quantities inside those tables. In fact, in the present case there is a conceptual problem in conditioning on both sets of margins in all tables simultaneously that does not arise when the tables are truly independent. This is because once we know the numbers of exposed and unexposed at each stage (and the numbers lost to follow-up in each group), then we know how many deaths and survivals occurred inside each table, leaving no room for variation.

The two situations differ also respecting the quantities that are fixed or random in the test statistic. In the independent case, when all margins are fixed, the sum of the reference cells over all tables is the pivotal random variable, while the sum of the cell expectations and variances are fixed. In the survival application, it is still appropriate to regard the marginal number of deaths d_i in each table as fixed, since these depend on our choice of time intervals; the number of losses to follow-up between points of observation may also be regarded as fixed. One may further condition on the initial number of people exposed and unexposed, m_1 and $n_1 - m_1$, and even on the final total of exposed and unexposed deaths, $\sum_i d_{i1}$ and $\sum_i d_{i0}$. However, margins m_i and $n_i - m_i$, the number of those exposed and unexposed and at risk at any stage after the first, are random. Thus, the sum of reference cells, $\sum_i d_{i1}$, may be regarded as fixed, while the other terms in the expression for z are random.

The situation may be likened under the null hypothesis to a series of selections at random without replacement from an urn containing at the outset m_1 "exposed" chips and $n_1 - m_1$ "unexposed" chips. A sampling plan is adopted pursuant to which d_i chips, corresponding to deaths, are selected at the ith stage. Drawing single chips, $d_i = 1$, is an example of such a plan. What is random is the composition of the urn after each selection, which corresponds to the number of exposed and unexposed chips in the urn at each stage.

So the circumstances for applying z to a two-group comparison of survival are different, but in one respect not: it can still be shown that under H_0 the numerator of z has zero expectation, and that z has an approximately standard normal distribution in large samples. A fact used in the proof of this assertion is that the terms summed in the numerator, $(d_{i1} - d_i m_i / n_i)$, while not independent, are nevertheless uncorrelated, so it is correct to sum their variances to find the variance of the sum. The random variable in the denominator of z turns out to be a consistent estimator of the actual null variance of the numerator, so z is approximately of unit variance. That the asymptotic distribution of z is normal follows from special limit laws for certain stochastic processes, which generalize the notion of sums of independent random variables to the dependent case. These are studied in the theory of "martingales."

One important difference that does emerge as a consequence of the role reversal between fixed and random quantities is that the one-half continuity correction is inappropriate to use in the score test statistic for survival comparison. Unlike the case of independent tables, where the correction is valid and improves accuracy, the random variable in the numerator of z is no longer integer-valued, but takes on fractional, unequally spaced values. As a result, the distribution of the score statistic is much better approximated by the normal distribution without continuity correction.

Returning to the estimation problem, one may estimate $\Omega = e^\beta$ simply with the Mantel-Haenszel estimate of a common odds ratio. The estimate is given by

$$\Omega_{MH} = \sum_i \frac{d_{i1} l_{i0}}{n_i} \bigg/ \sum_i \frac{d_{i0} l_{i1}}{n_i},$$

where sums are taken over all informative tables. In the present application, assuming $d_i = 1$ for each table, the estimator reduces to

$$\Omega_{MH} = \sum_1 (1 - f_i) \bigg/ \sum_0 f_i,$$

where $f_i = m_i/n_i$ is the fraction exposed among those at risk at the time of the ith death, and where the summation in the numerator is over those tables corresponding to an exposed death, and the summation in the denominator is over those tables corresponding to an unexposed death. The logarithm of Ω_{MH} is then a consistent estimator of β. The standard error for $\log \Omega_{MH}$ may be obtained from the RGB formula at p. 245 of Section 8.1.

Further Reading

N. Breslow & N. Day, *The Design and Analysis of Cohort Studies, 2 Statistical Methods in Cancer Research*, ch. 4 (1987).

J. Kalbfleisch & R. Prentice, *The Statistical Analysis of Failure Time Data*, ch. 4 (1980).

11.2.1 Age discrimination in employment terminations revisited

Question

Using the proportional hazards model, compute a "one-step approximation" for the increase in risk of termination for each year of age from the data of Section 8.1.2.

11.2.2 Contaminated wells in Woburn

Woburn, Massachusetts, a community of some 37,000 residents located 12 miles from Boston, has been an industrial site for more than 130 years. The town was a major chemical and leather processing center, a producer of arsenic compounds for insect control, and a producer of textiles, paper, TNT, and animal glues.

The town's drinking water was supplied by eight municipal wells, two of which (designated "G & H") were contiguous and operated as a single source to supply water to sections of eastern Woburn. The chance discovery of toxic wastes near wells G & H led to their testing in May 1979. The tests showed concentrations of toxic chemicals in the wells and they were shut down. The types and amounts of contamination in wells G & H before 1979 were not known. All other town wells met state and federal drinking water standards.

Independently, abandoned waste sites were found near wells G & H, which led to testing of groundwater. Pollution was discovered at 61 sites in eastern Woburn.

These events led to a series of studies in 1980 and 1981. A review of mortality statistics for 1969 to 1979 showed a significantly elevated rate of childhood leukemia (ages 19 and under) in Woburn between 1969 and 1979, with 12 cases diagnosed when only 5.3 were expected ($O/E = 2.3$, $p = 0.008$). The study noted that the leukemia excess was attributable to six cases occurring in one of the town's six census tracts (No. 3334, see Figure 2.1.4). A new investigation, by the Environmental Protection Agency (EPA) and the Harvard School of Public Health, identified 20 cases of childhood leukemia diagnosed in Woburn between 1964, the year wells G & H began pumping, and 1983. Based on national rates, only 9.1 cases were expected over this period ($O/E = 2.2$, $p = 0.001$).[3]

The town's water pipes were interconnected, so that each residence received a blend of water from several of the town's eight municipal wells. The specific blend varied with the location of the residence and over time. The Massachusetts Department of Environmental Quality and Engineering estimated, on a monthly basis, which zones of Woburn received none, some, or all of their water from wells G & H. These results (obtained in August 1983) were used to estimate the percentage of each household's annual water supply from wells G & H. The study assigned to each leukemic child an exposure history, consisting of his or her set of annual exposure scores, beginning from birth. The score for each year was the percentage of water received by the zone of residence from wells G & H. For example, a leukemic child born in 1967 and residing in the intersection of zones 1 and B for the first four years of life would generate cumulative exposures of 0.51, 1.23, 1.98, and 2.25 over this period. However, this kind of detailed information was not available for the population at large. In determining its exposure distribution, the investigators surveyed a sample of the risk-set population. As the investigators described their procedure: "For each case we identified all surveyed children who were born in the same year and were residents at the same time as the case and then computed the average and variance of their exposure values for the period of residency of the case."

The EPA/Harvard study used two measures of exposure: (i) cumulative exposure from birth until age t, and (ii) a binary indicator of whether there had been any G & H exposure by age t. In models utilizing the first of these measures, risk increases steadily with cumulative exposure. With the latter, an individual's hazard function jumps upon exposure.

The observed and expected exposure scores for each leukemic death are shown in Table 11.2.2.

---

[3]The study also identified elevated levels of birth defects and other health problems, but these are not covered here.

TABLE 11.2.2. Observed and expected exposures to wells G and H for twenty childhood leukemia cases

Case	Year of diagnosis	Year of birth	Period of residency	Observed cumulative exposure	Size of risk set sample	Expected cumulative exposure	Var	Proportion of risk set exposed
1	1966	1959	1959-1966	1.26	218	0.31	0.26	0.33
2	1969	1957	1968-1969	0.00	290	0.34	0.36	0.26
3	1969	1964	1969	0.75	265	0.17	0.10	0.25
4	1972	1965	1965-1972	4.30	182	0.90	2.23	0.36
5	1972	1968	1968-1972	2.76	183	0.58	0.88	0.32
6	1973	1970	1970-1973	0.94	170	0.20	0.20	0.19
7	1974	1965	1968-1974	0.00	213	0.56	1.04	0.29
8	1975	1964	1965-1975	0.00	239	0.99	2.78	0.38
9	1975	1975	1975	0.00	115	0.09	0.03	0.25
10	1976	1963	1963-1976	0.37	119	1.18	3.87	0.40
11	1976	1972	1972-1976	0.00	132	0.24	0.32	0.18
12	1978	1963	1963-1978	7.88	219	1.41	6.23	0.40
13	1979	1969	1969-1979	2.41	164	0.73	2.56	0.31
14	1980	1966	1966-1980	0.00	199	1.38	6.00	0.39
15	1981	1968	1968-1981	0.00	187	1.14	4.20	0.35
16	1982	1979	1979-1982	0.39	154	0.08	0.02	0.23
17	1983	1974	1974-77/80-83	0.00	84	0.25	0.45	0.23
18	1982	1981	1981-1983	0.00	0	0.00	0.00	0.00
19	1983	1980	1980-1982	0.00	0	0.00	0.00	0.00
20	1983	1980	1981-1983	0.00	0	0.00	0.00	0.00
Totals				21.06		10.55	31.53	5.12
			Score test statistic			1.87		2.08
			Significance level			$P = 0.03$		$P = 0.02$

Questions

The parents of 20 Woburn children who were diagnosed as having leukemia between 1964 and 1983 sued W.R. Grace & Co. for contaminating wells G & H, claiming that the well water caused the disease. The Harvard study was accepted in evidence.

1. What is the composition of the risk set at the point of diagnosis of leukemia for Case No. 1?

2. What are the definitions, under the null hypothesis, of the expected exposure and variance of exposure for a leukemic child?

3. What number of excess leukemia cases does the study attribute to wells G & H?

4. Using the one-step approximation given in Section 11.2 at 324, 325, what is the relative risk of leukemia per child-year of exposure to water from wells G & H?

5. Compute the Mantel-Haenszel estimate from Section 11.2 for the odds ratio on leukemia, across all deaths for exposed vs. unexposed children.

6. Is there anything in the data to cast doubt on the connection between leukemia and wells G & H?

7. Are the data sufficient to establish causation, prima facie?

Source

S.W. Lagakos, B.J. Wesson & M. Zelen, *An analysis of contaminated well water and health effects in Woburn, Massachusetts*, 81 J. Am. Stat. Assoc. 583 (1986); *Anderson v. Cryovac, Inc.*, Civ. A. No. 82–1672-S (D. Mass. 1982).

Notes

A subsequent study found the greatest elevation in childhood leukemia ($OR = 8.33$; c.i. 0.73, 94.67) in children whose mothers were exposed to water from wells G and H during pregnancy. See Massachusetts Department of Public Health, Bureau of Environmental Health Assessment, *Woburn Childhood Follow-up Study* (July, 1997), available on the Massachusetts Department of Public Health website at <www.state.ma.us/dph/beha>.

The Woburn lawsuit was the subject of a popular book by Jonathan Harr, *A Civil Action* (1995), and a movie of the same title, starring John Travolta.

11.3   Quantitative risk assessment

Under the Occupational Safety and Health Act of 1970, 29 U.S.C. §§651-78 (1994), the Secretary of Labor is delegated broad authority to promulgate occupational safety and health standards, defined as standards that are "reasonably necessary or appropriate to provide safe or healthful employment and places of employment." *Id*. at §652(8). Acting pursuant to this authority, the Occupational Safety and Health Administration (OSHA), the responsible "Agency" within the Department of Labor, has issued standards regulating workplace exposure limits for chemicals believed to be toxic. OSHA's limits frequently have been subjected to legal challenge.

The leading case relating to risk assessment is *Industrial Union Dept., AFL-CIO v. American Petroleum Inst.*, 448 U.S. 607 (1980). In that case, the Secretary of Labor issued a standard that reduced the exposure limit on airborne benzene from 10 parts per million (ppm) averaged over an eight-hour period to 1 ppm. There was evidence that benzene in concentrations as low as 10 ppm caused leukemia and some expert opinion that there was a risk even below that level, but there were no statistical data in that range or quantification of that risk. The Agency argued that, since substantial evidence supported the view that there was no safe level for a carcinogen, the burden was on industry to establish that a safe level exists. Rejecting an industry witness, the Agency found that since there was no safe level of exposure and it was impossible to quantify the benefits of the proposed reduction in exposure, the industry "must select the level of exposure which is most protective of exposed employees." 448 U.S. at 654 (quoting OSHA).

The plurality opinion of the Court did not accept OSHA's position. In remanding for further proceedings, the Court held that "[a]lthough the Agency has no duty to calculate the exact probability of harm, it does have an obligation to find that a significant risk is present before it can characterize a place of employment as 'unsafe.' " A reasonable person might consider that a risk of one in a thousand was significant. This does not mean that "anything approaching scientific certainty is required, but only the best available evidence....Thus, so long as they are supported by a body of reputable scientific thought, the Agency is free to use conservative assumptions in interpreting the data with respect to carcinogens, risking error on the side of overprotection rather than underprotection." OSHA has interpreted this mandate as requiring it to make a quantitative assessment of risks when issuing exposure limits.

The multistage extrapolation model

OSHA makes quantitative assessments by extrapolating from animal studies in which the animals are exposed to much higher doses of the chemicals being tested than would be encountered by humans in the workplace. The animals are then sacrificed and examined for disease. This approach requires OSHA to make two extrapolations: from animals to humans and from high-dose to low-dose effect; we deal here with the second extrapolation. To make the extrapolation from high to low doses OSHA uses a "multistage" model first proposed in the 1950s by Armitage and Doll,[4] and now in general use. The multistage model assumes that: (1) a normal cell is transformed into a cancer cell in a fixed and irreversible sequence of steps; (2) the waiting times in the various stages are statistically independent and exponentially distributed; and (3) the mean time spent in each stage is inversely related to the dose, with higher doses shortening the time by a factor specific to that stage. If there are m steps, the total risk of cancer from all sources within a specified time period, $P_{total}(d)$, in the presence of dose d is usually approximated by the equation

$$P_{total}(d) = 1 - \exp[-(q_0 + q_1 d + q_2 d^2 + \cdots + q_m d^m)],$$

where the q_i are parameters estimated from the data, but constrained to be not less than zero. This equation is a convenient mathematical approximation to the exact form of $P_{total}(d)$ under assumptions (1)–(3) above.

The first term of the exponent, q_0, represents the natural or base rate of cancer (without the carcinogen); the second term, q_1, represents the linear part of the dose-response curve (which would be a straight line if there were no higher power terms); and the remaining q_i's allow for a nonlinear response. The q_i's are generally estimated by the maximum likelihood (mle) method (see Section 5.6).

Two aspects or assumptions of this model should be noted. First, the model makes no allowance for a threshold effect: *any* dose, no matter how small, increases risk. That there is no absolutely safe dose for a carcinogen is an assumption generally made by regulators in the United States. Second, the model is consistent with the assumption generally made that the dose-response curve is linear at low doses. This will be true under the model if the mle estimate of q_1 is greater than zero. In such cases, since d is much greater than d^2 and higher powers of d at low doses, the linear term, $q_1 d$, will predominate and the model will be essentially linear. However, where the rate of cancer rises rapidly at higher doses (more rapidly than a linear model would predict), as it tends to do in these studies, the sharp rise in the curve leads to a small linear term and larger estimates of the nonlinear terms. When the maximum likelihood estimate of q_1 in the multistage model is very small or zero, so that the linear term does not predominate, regulators have made ad hoc adjustments, to implement the assumption of low-dose linearity. One approach simply assumes that at low doses the multistage model is approximated by a simple linear model in which the terms with d^2 and higher powers of d are omitted. Another approach uses the upper end of a 95% confidence interval for q_1 instead of the mle. Since the standard error of q_1 is usually quite large, the linear term becomes substantial and predominates.

---

[4]See P. Armitage & R. Doll, *Stochastic models for carcinogenesis*, Proceedings of the Fourth Berkeley Symposium of Mathematical Statistics and Probability 19 (1961).

Both the no-threshold and linear dose-response assumptions are "conservative" in the sense that they predict greater risk at low doses than most other models. It is probably fair to say that no empirical evidence either supports or refutes these assumptions for most exposures, nor does any appear possible at present. It is also quite clear that neither the multistage, nor any of the other models discussed below, fully reflect the complexity of the carcinogenic process. In particular, the linearity assumption assumes that the number of cells at risk of undergoing the first transformation is constant and thus the normal processes of cell division, differentiation, and death are not taken into account by the model.

Other extrapolation models

Other models used to make low-dose extrapolations are the "one-hit," the "multi-hit," the Weibull, the logit, and the probit. The formulas for the one-hit, multi-hit, and Weibull models are as follows. Let $P_{\text{excess}}(d)$ represent the conditional chance of getting cancer because of exposure at dose level d, given that the animal escapes cancer from other causes. This may also be interpreted as the risk within the specified time period when there is zero independent background risk. A little algebra shows that $P_{\text{excess}}(d)$ for the multi-stage model is the same as above without the term q_0.

Under the assumption that hits follow a Poisson distribution with intensity parameter k such that, given dose d, the hits have a mean kd, then

$$P_{\text{excess}}(d) = 1 - \sum_{i=0}^{m-1} e^{-kd}(kd)^i / i!,$$

the summation term being the probability of fewer than m hits. For the one-hit model, which assumes that a cell turns malignant after a single hit, $m = 1$ and this becomes $P_{\text{excess}}(d) = 1 - \exp(-kd)$. At low doses, this expression is approximately $P_{\text{excess}}(d) = kd$, so that the one-hit model may be characterized as linear at low doses. For the multi-hit model, which assumes that more than one hit is required to turn a cell malignant, $m > 1$ and at low doses $P_{\text{excess}}(d)$ is approximately a polynomial with leading term of order higher than linear. For the Weibull model with parameters m and k, $P_{\text{excess}}(d) = 1 - \exp(-kd^m)$. For a discussion of the logit and probit models, see Section 14.7. Both the logit and the probit models assume that individuals have a threshold for disease: a dose above the threshold is certain to result in disease; a dose below that is certain to have no effect. In the logistic model, the thresholds are assumed to have a logistic distribution (which is like a normal distribution except that the tails are fatter) among individuals; in the probit model they are assumed to be normally distributed. The multi-hit model also admits of an individual threshold model interpretation, where the threshold distribution is gamma with shape parameter m and scale parameter $1/k$.

While regulators tend to rely primarily on multistage models, they compare the risk estimates thus obtained with those from other models and derive comfort from estimates that are consistent within a range. In one case it was said that a factor of three was well within a range of consistency. This is not always obtainable. In general, the one-hit model gives the highest risk estimates and the multi-hit or probit the lowest, with Weibull and multistage in the middle. The differences between estimates can be large. For example, for aflatoxin, the virtually safe dose (defined as a dose that yields a risk of one in a million) is 30 times greater for the multistage model than for the one-hit model, 1,000 times greater for the Weibull model, and 40,000 times greater for the multi-hit model.

Further Reading

Bernard D. Goldstein and Mary Sue Henifin, *Reference Guide on Toxicology* in Federal Judicial Center, *Reference Manual on Scientific Evidence* 181 (2nd ed. 2000).

National Research Council, *Science and Judgment in Risk Assessment* (1994).

D. Krewski and J. Van Ryzin, *Dose response models for quantal toxicity data* in *Current Topics in Probability and Statistics* (North Holland, Amsterdam, 1981).

11.3.1   Ureaformaldehyde foam insulation

In 1982, after extensive hearings, the U.S. Consumer Products Safety Commission found that ureaformaldehyde foam insulation (UFFI) presented an unreasonable risk of cancer and irritant effects and banned its future use in residences and schools. On appeal, various petitioners challenged in particular the commission's finding of unreasonable risk of cancer. As required by the Administrative Procedure Act, the court of appeals reviewed the agency's action under the "substantial evidence" standard, which requires that an agency's actions be supported by substantial evidence on the record taken as a whole.

In estimating the cancer risk to humans, the commission had several types of data: (i) two sets of animal studies, which showed that formaldehyde gas at high concentrations (e.g., an average of 14.3 ppm with concentrations of 17 to 20 ppm, not uncommon) caused cancer in rats; (ii) data from 1,164 homes with UFFI; and (iii) test data from measurements of formaldehyde gas in simulated UFFI wall segments. Both home and test data were similar and showed concentrations of gas on the order of 0.12 ppm. Petitioners challenged the concentrations to which the rats had been exposed (the maximum was far greater than the average), the sample from which home levels of exposure had been calculated (they were not a random sample, but were mostly "complaint" homes), and the validity of the simulated tests (materials simulated unheated, unairconditioned homes). In addition, petitioners challenged the method of

TABLE 11.3.1. Chemical Industry Institute of Toxicology rat formaldehyde study

Average exposure (ppm)	0	2.00	5.60	14.30
Equivalent continuous exposure (ppm)	0	0.34	0.94	2.40
Number of rats at risk	216	218	214	199
Number of carcinomas	0	0	2	103

extrapolation from the high-dose response in rats to the low-dose response in humans. Only this last point need concern us here.

In the high-dose study, rats were exposed to formaldehyde gas in varying concentrations for six hours a day, five days a week. After 24 months the animals were sacrificed and examined for nasal carcinomas. Table 11.3.1 shows the results, with the administered dose converted into an equivalent continuous-exposure dose.[5]

For purposes of extrapolation, an industrial worker receives an equivalent lifetime continuous exposure of 0.13 ppm and a homeowner with formaldehyde insulation in his or her house receives 0.004 ppm. The concentration of formaldehyde in the Los Angeles ambient air was 0.015 ppm in 1979.

To estimate the risk of cancer at low doses, the agency used a multistage model; for a description see Section 11.3. Because of the sharp rise in cancer at higher doses, the estimate of q_1 was 0. The commission, however, implemented its assumption of low-dose linearity by assessing the risk at low doses from the upper end of a 95% confidence interval for q_1 instead of the maximum likelihood estimate. The value of the linear term was substantial because the confidence interval was wide.

In response to an objection that the CPSC should not have used the upper 95 percent confidence interval as opposed to the maximum likelihood estimate, the commission noted:

"The 'most likely' estimates from the multistage and other models do not consider that formaldehyde can interact with other environmental carcinogens and ongoing carcinogenic processes within the human body. The ability of formaldehyde and many other carcinogens to interact with the genetic material makes it likely that they also interact with ongoing carcinogenic processes within the human body. Thus, the true risk is probably closer to a predicted risk based on the upper 95 percent confidence limit [of the model used by the CPSC] than to one predicted from any other model. If formaldehyde could not interact with the genetic material, the predicted risk would be low. However, the Commission believes this is unlikely."

CPSC, *Ban of Urea-Formaldehyde Foam Insulation*, 47 Fed. Reg. 14366 (April 2, 1982).

---

[5]An equivalent continuous dose that would yield the same total exposure (the actual numbers differ slightly from those that would be produced by a simple pro-rating).

Questions

The court of appeals rejected the rat study evidence because (i) "[I]n a study as small as this one the margin of error is inherently large. For example, had 20 fewer rats, or 20 more, developed carcinomas, the predicted risk would be altered drastically"; (ii) there was only a single experiment; and (iii) the administered dose was only an average; "the rats in fact were exposed regularly to much higher doses."

1. Do you agree that the rat study evidence should have been rejected for these reasons?

2. Does the CPSC justification for using the upper 95 percent confidence interval instead of the maximum likelihood estimate satisfy the grounding in science required of an agency, as described by the Supreme Court in *American Petroleum Institute*?

Source

Gulf South Insulation v. U.S. Consumer Product Safety Commission, 701 F.2d 1137 (5th Cir. 1983); S. Fienberg, ed., *The Evolving Role of Statistical Assessments as Evidence in the Courts* 46 (1989).

11.3.2   Ethylene oxide

In 1983, the federal Occupational Safety and Health Administration (OSHA) proposed to reduce the permissible exposure limit for ethylene oxide (EtO) from 50 parts per million (ppm) of air as an eight-hour time-weighted average to 1 ppm. EtO is a colorless gas with an ether-like odor which is primarily used as an intermediate in the manufacture of ethylene glycol, a major component of anti-freezes and an intermediate in the production of certain polyester fibers, bottles, and films. More than five billion pounds of EtO are produced domestically each year.

In making a quantitative risk assessment, OSHA relied primarily on a rat study. In the study, rats were exposed to airborne concentrations of 100, 33, or 10 ppm of ethylene oxide vapor for 6 hours a day, 5 days per week, for approximately two years. Two control groups were exposed to air only, under similar conditions. Initially, 120 rats per sex per group were exposed, and at each 6-month interval some animals were sacrificed to determine treatment-related effects. The results are shown in Table 11.3.2a and Table 11.3.2b below.

Questions

1. OSHA proposed reducing permissible exposures from 50 ppm to 1 ppm. Using the two-stage model, estimated from male rat data, confirm that

TABLE 11.3.2a. Pathologies in rats after exposure to ethylene oxide

Tumor Type	100 ppm(a)	33 ppm	10 ppm	Controls
Peritoneal	22/96(b)	7/82	3/88	4/187
mesothelioma	(22.9%)	(8.5%)	(3.4%)	(2.1%)
(Males)	$[< 0.0001](c)$	[0.0213]	[0.3983]	
Mononuclear	26/98	25/77	21/77	38/193
cell	(26.5%)	(32.5%)	(27.3%)	(19.7%)
leukemia	[0.1192]	[0.0202]	[0.1163]	
(Males)				
Mononuclear	28/73	24/72	14/71	22/186
cell	(38.4%)	(33.3%)	(19.7%)	(11.8%)
leukemia	$[< 0.0001^*]$	$[< 0.0001^*]$	[0.0791]	
(Females)				

(a)—Dose used in risk assessment was expressed in mg/kg of body weight/day. Dose was calculated as $d(\text{mg/kg/day}) = d(\text{ppm}) \times S$, where S is a scaling factor. For male rats $S = 0.1930324$; for female rats $S = 0.2393601$; for humans $S = 0.129898$. Almost certainly the measurement of scaling factors was not accurate to the number of decimals shown.

(b)—Number of tumor-bearing animals/effective number of animals at risk. The effective number of animals at risk was the number of rats alive when the first tumor was observed (since the animals were sacrificed at several intervals before the first tumor was observed).

(c)—P-value from Fisher's exact test (upper-tail probability) when compared to controls. A Bonferroni correction was used in evaluating significance at the 0.05 level. * indicates P less than $0.05/r$, where r is the number of test doses.

the excess cancer rates per 10,000 workers for 50 ppm and 1 ppm are approximately 634 and 12, as shown in Table 11.3.2b.

2. OSHA used the one-stage model to extrapolate the data for female rats on the ground that the data fit the one-stage model. Plot the data for female rats. Are they consistent with the use of the one-stage model?

3. Using the Weibull model for excess risk estimated from female rat data, compute the excess cancer rates for 10,000 workers for 50 ppm and 1 ppm. The mle's for k and m in the model risk are approximately 0.0829 and 0.4830, respectively. Are these extrapolations consistent with the extrapolations for males?

4. What goodness of fit is chi-squared measuring here? To what extent do the values of the chi-squared statistic constitute evidence in favor of the estimated risks at low doses given by the models?

TABLE 11.3.2b. Excess lifetime risk of cancer per 10,000 workers (a).

Exposure level(ppm)	Two-stage(b)		One-stage(c)		One-stage(d)	
	MLE(e)	UCL(f)	MLE	UCL	MLE	UCL
50	634	1,008	746	1,018	1,093	1,524
10	118	211	154	213	229	326
5	58	106	77	107	115	164
1	12	21	15	21	23	33
0.5	6	11	8	11	12	16
$\chi^2(g)$	0.0579(1)		0.2360(2)		4.3255(2)	
$P(h)$	0.81		0.89		0.12	

(a)—Excess risk per 10,000 workers, $P[E] = (P(d) - P(0))/(1 - P(0))$. This is known as Sheps' relative difference index. Lifetime exposure is assumed to be 8 hours per day, 5 days per week, 46 weeks per year for 45 years in a 54-year lifespan since initial exposure.
(b)—Two-stage model with 3 parameters: $q[0] = 0.020965$; $q[1] = 0.008905$; and $q[2] = 0.000184$. Extrapolated from male rats; data for females not available.
(c)—One-stage model with parameters: $q[0] = 0.019682$; $q[1] = 0.011943$. Extrapolated from male rats.
(d)—One-stage model with parameters: $q[0] = 0.14700$ and $q[1] = 0.017841$. Extrapolated from female rats.
(e)—Maximum likelihood estimate of excess risk.
(f)—95% upper confidence limit on excess risk.
(g)—One-tailed chi-squared goodness-of-fit test. Degrees of freedom are in parentheses.
(h)—P-value associated with (g).

Sources

Occupational Safety and Health Administration, *Docket H-200, Occupational Exposure to Ethylene Oxide*, 48 Fed. Reg. 17284 (April 21, 1983); *id., Preliminary Quantitative Risk Assessment for Ethylene Oxide* (Exhibit 6-18); *Public Citizen Health Research Group v. Tyson*, 796 F.2d 1479 (D.C. Cir. 1986).

CHAPTER 12

Nonparametric Methods

12.1 The sign test

A nonparametric test is one for which the stated Type I error rate holds no matter what distribution may underlie the observations. Nonparametric tests are thus broadly applicable and valid for testing, relying only on the assumption of random sampling, and generally requiring only simple calculation. The sign test is the simplest example of this class of procedures.

Since the underlying distributions are no longer assumed symmetrical, hypotheses are often formulated in terms of population medians instead of means. Suppose we wish to test the hypothesis H_0 that the median μ of a continuous but otherwise unknown distribution equals zero, against the one-sided alternative that the median is, say, a negative number, $\mu < 0$.[1] For example, we may wish to test the hypothesis that the median lifetime of some product equals or exceeds an advertiser's claim of μ_0. Then given sample lifetime data $X_1, \ldots, X_n$ we test the hypothesis that the distribution of $X - \mu_0$ has zero median against the alternative that the median is negative. Another example arises when comparing two distributions in a matched pairs study: the null hypothesis of equal distribution implies that the median of the paired difference distribution is zero. We can test H_0 without knowledge of the shape, or parametric form, of the distribution by noting that, under the hypothesis of zero median, each observation will be positive or negative with probability one-half. Thus, in a sample of size n, the number S of observations with a plus "sign" will follow a binomial distribution with index n and $p = 1/2$. Under the alternative hypoth-

---

[1] For discrete distributions, there could be several candidates for population median. In fact, a median in such cases is any number μ satisfying $P[X < \mu] \leq 1/2 \leq P[X \leq \mu]$.

esis, S follows a binomial distribution with $p < 1/2$, so that the strategy of counting positive outcomes (or outcomes in excess of μ_0) reduces the matter to a simple binomial problem. Table B in Appendix II may be used to assess significance of the binomial tail probability, or in large samples the normal approximation may be applied (see Section 4.2). Thus, in the lifetime example, a skewed sample of size $n = 10$ such as $\{1, 1, 1, 1, 1, 1, 1, 1, 1, 11\}$ provides evidence against a claimed median lifetime of $\mu_0 = 2$ that is significant at the $P = P[S \le 1 | n = 10, p = 1/2] = \left[\binom{10}{0} + \binom{10}{1}\right] \cdot (1/2)^{10} = 0.011$ level.

A two-sample cousin of the sign test arises when comparing two distributions, say F_1 and F_2, with independent samples of size n_1 and n_2. To test the hypothesis $H_0 : F_1 = F_2$, we find the sample median in the combined sample of size $n_1 + n_2$, and then count the number of observations in each group that exceeds the median or that falls below it. The results may be cast in a fourfold table with fixed row margins (n_1, n_2) and column margins (m_1, m_2), where m_1 is the number of observations in the combined sample that exceeds the sample median, and m_2 is the number that falls at or below it (if $n_1 + n_2$ is even and there are no ties in the data, then $m_1 = m_2 = (n_1 + n_2)/2$, but m_1 or m_2 may be less than this if $n_1 + n_2$ is odd or there are observations tied at the median). Under H_0, the cells of the fourfold table follow Fisher's exact distribution, i.e., the number of group 1 observations exceeding the combined median has a hypergeometric distribution, and so H_0 may be tested with the methods of Sections 4.5 and 5.1. Clearly, these methods may be generalized to quantiles other than the median, or to more than the two categories above or below the reference quantile.

The price one pays for the general applicability of a nonparametric test is a loss of some efficiency: comparisons of the power of the nonparametric test with the power of a given parametric test of the same hypotheses generally show that, when the parametric test is valid, it is more powerful than the nonparametric test. The trade-off, then, is power for robustness against departures from distributional assumptions. We illustrate this in the case of the sign test, which throws away all information in the observed values except whether they are positive or negative. If the data were known to come from a normal distribution with unit variance, one could test $H_0 : \mu = 0$ with the z-score $z = \sqrt{n} \cdot \bar{X}$, rejecting H_0 at level $\alpha = 0.05$ when $z < -1.645$. When H_0 is false, e.g., when the true parameter is $\mu = -1$, the event $[z < -1.645]$, which is equivalent to $[\bar{X} < -1.645/\sqrt{n}]$, occurs with probability $P[\sqrt{n}(\bar{X} - \mu) < \sqrt{n}\{-1.645/\sqrt{n} - (-1)\}] = \Phi(\sqrt{n} - 1.645)$, where Φ denotes the standard normal cumulative distribution function. Then the power to reject H_0 when the true mean is $\mu = -1$ is given by $\Phi(\sqrt{n} - 1.645)$. For example, when $n = 10$, power $= \Phi(1.517) = 0.935$. For the sign test, we reject H_0 if the number of positive observations is 2 or less, with type I error rate $P[S \le 2 | n = 10, p = 1/2] = 0.0547$. If $\mu = -1$, then the probability that any observation is positive becomes $p = 0.16$, so that the power of the sign test is $P[S \le 2 | n = 10, p = 0.16] = 0.79$. To achieve the same power as the z-score, we would require a sample size of about 16, for a relative efficiency (the ratio of sample sizes) of about 62.5%. Of course, if the data were markedly non-normal, the z-score could not be relied upon to give an accurate result.

For example, suppose that the true distribution of X is that of the variable $X = (\log 2) - Y$, where Y has a standard exponential distribution. The median of X equals 0 and the variance of X is 1. This distribution satisfies the null hypothesis, but because of its skewness, the probability that $\sqrt{n} \cdot \bar{X} < -1.645$ is substantially in excess of 0.05. For $n = 10$, for example, the Type I error already exceeds 0.20, and grows larger as n increases.

Fortunately, there are nonparametric tests that make better use of the data than the sign test, and which have remarkably high efficiency relative to the best available parametric tests—90% or better—so that the loss of efficiency is of lesser concern, and the trade-off worthwhile with data of uncertain distribution. We study some of these tests below.

Further Reading

E. Lehmann, *Nonparametrics: Statistical Methods Based on Ranks*, ch. 3 (1975).

12.1.1   Supervisory examinations

The Board of Education of the City of New York requires teachers seeking supervisory positions in the school system to take qualifying examinations for the various positions. These examinations were challenged as discriminatory against black and Hispanic teachers. Approximately 10% of the teaching force eligible to take the examinations was black or Hispanic. The data showed that out of 6,201 candidates taking most of the supervisory examinations in the previous seven years, including all such examinations in the preceding three years, 5,910 were identified by race. Of the 5,910 so identified, 818 were black or Hispanic and 5,092 were Caucasian.

Of 50 different examinations given, 32 were examinations where blacks or Hispanics took the examination and at least one person in any group passed. Among these 32 examinations, the Caucasian group had a larger percentage passing in 25 examinations and the black and Hispanic group combined had a larger percentage passing in 7 examinations. For most examinations, the number of blacks and Hispanics taking the test was very small (e.g., 41 out of 50 examinations were taken by an aggregate of 83 blacks and Hispanics).

Questions

1. Do the data from the 32 examinations show a statistically significant disparate impact on black and Hispanic teachers?

2. What objections would you have to a test of statistical significance in this context?

Source

Chance v. Board of Examiners, 330 F.Supp. 203 (S.D.N.Y. 1971), *aff'd*, 458 F.2d 1167 (2d Cir. 1972).

12.2   Wilcoxon signed-rank test

The most important applications of the Wilcoxon signed-rank test arise when pairs of subjects from a treatment group and a control group are compared to determine whether there is a statistically significant difference between treatments. Although there are two groups, the test is a one-sample test because the quantity of interest is the difference between paired observations from the two groups. Under the null hypothesis of no treatment effect the median of the differences would be zero. As the treatment becomes more effective, the magnitude of the observations in the treated group is systematically raised (or lowered) relative to the control group, and the median difference is correspondingly raised or lowered.

To apply the test: (1) rank the absolute values of the data (in this case, the differences) from 1 to sample size n; (2) give the ranks the signs of the items to which they relate; (3) sum the negative or positive ranks. This is the Wilcoxon signed-rank statistic (W). Under the null hypothesis each rank has a 50% chance of being positive (or negative), so that each particular assignment of $+$'s and $-$'s to the ranks has probability $(1/2)^n$. Computation of the W distribution for any particular value involves counting the number of assignments with a W value equal to or more extreme than that observed. The distribution of the W statistic has been tabulated for small sample sizes. As an example, supposing there are $n = 20$ items and $W = 52$, $P[W \leq 52]$ is about 0.025 by Table H1 of Appendix II. This figure would double for a two-tailed test (i.e., doubling gives the probability that W would be either as far below expectation as that observed or as far above expectation).

Under the null hypothesis, the signed rank statistic has mean $EW = n(n + 1)/4$ and variance $\text{Var } W = n(n + 1)(2n + 1)/24$. In large samples, the z-score $z = (W - EW)/\sqrt{\text{Var } W}$ may be used to assess significance. In the example above, we would calculate $EW = 20(21)/4 = 105$, $\text{Var } W = 20(21)(41)/24 = 717.5$, and then using the $1/2$ continuity correction, $P[W \leq 52] = P[W < 52.5] = P[(W - EW)/\sqrt{\text{Var } W} < (52.5 - 105)/\sqrt{717.5} = -1.96] \approx 0.025$.

We have assumed above that there are no zeros among the absolute differences. If zeros do occur, they are dropped from the calculations *after* assigning ranks to all the differences. If there are d_0 zero differences, the expected value of W is modified to $EW = [n(n + 1) - d_0(d_0 + 1)]/4$. Occasionally there are other tied values among the absolute differences, and when they occur, the average rank is assigned to each member in a group of tied differences. In the presence of tied values, the variance of W is $\text{Var } W = [n(n + 1)(2n + 1) - d_0(d_0 + 1)(2d_0 + 1)]/24 - [\Sigma d_i(d_i^2 - 1)]/48$, where the sum is taken over the tied groups, with d_i items tied in the ith group. (The $1/2$ continuity correction is not used when ties are present since the statistic is no longer integer-valued.) The modifications for ties are usually negligible, except when sample size is small or there are many tied values (which generally occurs only when the measures being ranked are discrete).

Further Reading

E. Lehmann, *Nonparametrics: Statistical Methods Based on Ranks*, §§3.2 and 4.2 (1975).

12.2.1   Voir dire of prospective trial jurors revisited

Questions

1. Use the Wilcoxon signed-rank statistic to test the statistical significance of the median difference in impaneling times under state and federal methods for voir dire, using the data in Section 7.1.2.

2. What advantages are there in this method of testing statistical significance over the t-tests used in Section 7.1.2?

12.3   Wilcoxon rank-sum test

Assume that there are m "treatment" and n "control" items ranked from 1 to $m + n = N$ (rank 1 representing the best result). There are no ties, and the two groups are statistically independent.

A plausible measure of treatment effect is the sum of the ranks, S, of the m treatment items. The probability under the null hypothesis that S would be as low as the observed sum can then be computed by counting the number of possible rank distributions whose sum is less than or equal to the observed sum (practical only if m and n are quite small) divided by the total number of possible different rankings. Table H2 in Appendix II gives critical values for the rank-sum statistic for small values of m and n. For larger values, use the normal approximation, as follows.

Under the null hypothesis of no treatment effect, the expected value of a rank is $(N + 1)/2$; for the m treatment subjects the expected value of their sum is $ES = m(N + 1)/2$. The variance of a rank selected at random from N ranks is $\sigma^2 = (N^2 - 1)/12$. The variance of the sum of m randomly selected ranks out of $m + n = N$ ranks, without replacement, is thus $(m/12) \cdot (N^2 - 1) \cdot (N - m)/(N - 1)$, or $\mathrm{Var}\,S = mn(N + 1)/12$. Despite the lack of independence of rank assignments, when m and n are sufficiently large the sum of ranks is approximately normally distributed.

When ties are present, the average rank among tied items is assigned to each item in the tied group. These adjusted ranks still have mean $ES = (N + 1)/2$, but their variance becomes

$$\sigma^2 = \{N^2 - 1 - (1/N) \cdot \sum d_i(d_i^2 - 1)\}/12,$$

where the sum is taken over the tied groups, with d_i items tied in the ith group. The adjusted variance of the sum of m ranks selected without replacement is then

$$\mathrm{Var}\, S = (1/12) \cdot mn(N + 1) - \{mn/[12N(N - 1)]\} \cdot \sum_i d_i(d_i^2 - 1).$$

The rank-sum statistic is statistically equivalent to the Mann–Whitney U statistic, defined as the number of item pairs that can be formed with one item from each group, such that the rank of the item from the treatment group is less than the rank of the item from the control group. The relation between the two statistics is

$$U = mn + \frac{1}{2}m(m + 1) - S.$$

The quantity $U/(mn)$ estimates the probability that a treatment outcome will be better than a control outcome, when applied to two items chosen at random.

Further Reading

E. Lehmann, *Nonparametrics: Statistical Methods Based on Ranks*, §§1.2, 2.2 (1975).

12.3.1   Sex discrimination in time to promotion

K & B, a Louisiana corporation domiciled in New Orleans, operates a chain of drug stores in Louisiana, Mississippi, and Alabama. The EEOC and a female pharmacist sued K & B, alleging discrimination in the promotion of women from pharmacist to chief pharmacist. The data below show, for a certain cohort of employees, the time in months from hire to promotion to chief pharmacist for male and female employees.

Plaintiff's statistician, using the Wilcoxon rank-sum test, concluded that the median time to promotion was greater for women than for men. Defendant's statistician computed the probability under the null hypothesis that both female time-to-promotions would exceed the male median time-to-promotion and concluded that the difference was not statistically significant.

TABLE 12.3.1. Time, in months, from hire to promotion

Females:	229, 453
Males:	5, 7, 12, 14, 14, 14, 18, 21, 22, 23, 24, 25, 34, 34, 37, 47, 49, 64, 67, 69, 125, 192, 483.

Defendant's expert explained the difference in results by saying that he was "looking at the typical time to promotion," whereas plaintiff's expert, by using the Wilcoxon rank-sum test, tested the similarity of the shapes of the two distributions—male and female—and "statistical significance will be obtained if the two distributions do not match, regardless of the manner one distribution did not match the other."

Questions

1. Make the calculations for plaintiff's and defendant's experts.

2. Do you agree with the statement of defendant's expert?

3. Which expert used the more appropriate test?

Source

Capaci v. Katz & Besthoff, Inc., 525 F. Supp. 317 (E.D. La. 1981), *affirmed in part, reversed and remanded in part*, 711 F. 2d 647 (5th Cir. 1983); Gastwirth & Wang, *Nonparametric tests in small unbalanced samples: application in employment discrimination cases*, 15 Canadian J. Stat. 339 (1987)

12.3.2   Selection for employment from a list

Persons are selected for employment, in order, from an eligibility list. Rank on the list is determined by the person's score on a competitive examination. The Puerto Rican Legal Defense & Education Fund contends that the test has an adverse impact on blacks and Hispanics. Of 2,158 test takers, 118 were black and 44 Hispanic; 738 passed the test, 27 of whom were black, and 11 Hispanic. See Table 12.3.2 for the scores and rankings of blacks and Hispanics who passed. An applicant who scored below a certain cutoff point is not included on the list.

Questions

1. Use the Wilcoxon rank-sum test to determine whether minorities have significantly lower average scores than non-minorities. The sum of the list nos. in Table 12.3.2 is 19,785.

2. Would a two-sample t-test on these truncated data also be valid to answer that question?

3. Compute the probability that a minority person randomly selected from the list would have a lower rank (higher score) than a randomly selected non-minority person.

TABLE 12.3.2. Selection for employment: Minority list

Blacks		Hispanics	
List No.	Adj. final score	List No.	Adj. final score
058	91.753	162	90.051
237	88.223	273	87.372
315	86.371	369	85.479
325	86.138	432	84.293
349	85.799	473	83.327
350	85.769	493	80.323
371	85.459	529	82.083
375	82.878	605	79.780
484	83.081	631	79.277
486	82.965	692	77.263
527	82.137	710	76.114
551	81.550		
575	80.818		
584	80.452		
587	80.419		
615	79.598		
642	78.854		
645	78.762		
654	78.419		
688	77.270		
693	77.184		
694	77.137		
713	75.935		
716	75.779		
725	74.887		
726	74.683		
731	74.168		

12.3.3   Sentencing by federal judges

In a Federal Judicial Center study of sentencing practices by federal judges, 50 federal judges were given case descriptions and asked to impose sentence. The sentences were ranked by severity for each case (1 being the most severe and 50 being the most lenient). Data showing the average ranks given by the 50 judges in 13 model cases chosen to represent the range of types to come before them are given in Table 12.3.3.

Questions

Do the data support the common beliefs that some judges are more severe than others? Consider whether the average sentence rankings of the judges are more dispersed than they would have been if ranks had been assigned on a random basis, i.e., each possible permutation of rankings is equally likely. Test that proposition using the average ranking for all 13 cases, as follows.

TABLE 12.3.3. Ranks of sentences of individual judges in thirteen model cases

Judge	Average rank	Judge	Average rank
1	5.4	26	25.9
2	10.6	27	26.0
3	12.1	28	26.0
4	15.3	29	26.1
5	19.2	30	26.7
6	19.2	31	26.7
7	19.6	32	26.8
8	19.6	33	27.0
9	20.8	34	27.6
10	22.7	35	27.8
11	22.8	36	27.8
12	23.0	37	27.9
13	23.4	38	28.3
14	24.3	39	29.3
15	24.5	40	30.0
16	24.5	41	30.1
17	24.6	42	31.5
18	24.6	43	31.8
19	24.6	44	32.1
20	24.7	45	32.7
21	25.0	46	33.0
22	25.2	47	33.4
23	25.5	48	34.7
24	25.7	49	36.1
25	25.8	50	36.9

1. Ignoring the variance adjustment for ties, determine the mean and variance of an average of 13 independent rankings for a single judge, under the null hypothesis that all judges sentence with equal severity.

2. It can be shown that $N - 1$ multiplied by the ratio of the mean squared deviation to its expected value under the null hypothesis has a chi-squared distribution with $N - 1$ df as the number of cases becomes large. (One df is lost because the ranks have a fixed sum.) Test the null hypothesis using this statistic. What do you conclude? How would you describe the degree of departure from the null hypothesis?

The sum of the squared deviations of the 50 average ranks is 1,822.47.

Source

Partridge & Eldridge, *The Second Circuit Sentencing Study: A Report to the Judges of the Second Circuit* (Federal Judicial Center, 1974) (Appendix C).

12.4   Spearman's rank correlation coefficient

To measure the correlation between two variables, X and Y, there is available a nonparametric version of Pearson's product-moment correlation coefficient that is based on ranks rather than on actual numerical values. Whenever there is risk of distortion due to outliers, or when Y depends on X non-linearly, the non-parametric version should be used. Known as Spearman's rank correlation coefficient, it is always applicable to measured data, and is uniquely suited to assess information coming only in the form of relative rankings or ordered qualitative ratings. An example is a concordance study of two or more raters, where each rater arranges a list of items in order of preference. Like other nonparametric procedures discussed above, rank correlation trades a slight loss of efficiency relative to the ordinary sample correlation coefficient (for making inferences about true correlation in bivariate normal situations), in exchange for applicability to a much broader array of problems for which rank correlation alone is valid, since it does not require specific distributional assumptions. The procedure is as follows.

Given N data pairs (X_i, Y_i), $i = 1, \ldots, N$, replace each X value by its corresponding rank, from 1 to N, from among the N X-values; similarly, replace each Y value by its rank from among the N Y-values; then, calculate the ordinary product-moment correlation between the paired ranks.

Since the rank values $(1, \ldots, N)$ are known in advance, the formula for the correlation coefficient simplifies appreciably: letting d_i denote the difference between the X-rank and the Y-rank in the ith pair, Spearman's rank correlation coefficient is

$$r_S = 1 - 6 \sum d_i^2 \Big/ (N^3 - N).$$

A significance test of a correlation based on r_S may be obtained by referring $z = r_S \cdot \sqrt{N - 1}$ to tables of the standard normal distribution because, in large samples, under the null hypothesis of no association between X and Y, r_S has an approximate normal distribution with mean 0 and standard deviation $1/\sqrt{N - 1}$. Tables of the exact distribution of r_S for small N are widely available.

Further Reading

E. Lehmann, *Nonparametrics: Statistical Methods Based on Ranks*, ch. 7 (1975).

12.4.1   Draft lottery revisited

Questions

1. Compute Spearman's rank correlation coefficient for the 1970 draft lottery (see Section 9.1.1). The sum of squared rank differences (between birthday rank and sequence rank) is 10,015,394 with $N = 366$.

2. Do the same for the 1971 draft lottery. The sum of squared rank differences is 7,988,976 with $N = 365$.

CHAPTER 13

Regression Models

13.1   Introduction to multiple regression models

Multiple regression is a statistical technique for estimating relationships between variables that has become a key tool in mathematical economics (econometrics) and in the social sciences. By these routes it has invaded the law. The principal applications have been in antidiscrimination class action litigation, but the technique has also been applied in a variety of other legal contexts—antitrust price-fixing, securities market manipulation, litigation over capital punishment, attacks on bail, and others. It is now so easy to fit models to data by computer that multiple regression and related techniques are likely to become even more widely used—and probably also abused—in cases involving statistical proof.

Here are some examples of regression models in law:

- In a class-action antidiscrimination suit, plaintiffs assert that after accounting for differences in productivity factors between men and women, there is a residual difference in salaries that should be attributed to discrimination. To quantify that relationship, plaintiffs estimate a regression relationship in which salary is the dependent variable and sex and productivity factors (such as education and experience) are explanatory factors. The argument is that, if no discrimination exists, then sex would not be a significant explanatory factor in the regression equation. To the extent that it is a factor, its importance is a measure of the extent of discrimination. The weight assigned to the sex explanatory variable in the regression equation represents an average dollar shortfall between equally productive men and women. Often the focus is on testing for the presence of discrimination by assessing the statistical significance of the sex coefficient.

- In a price-fixing case, plaintiff is entitled to recover three times the difference between the price it paid and the price it would have paid in the absence of the conspiracy. To estimate the "but-for" price, a multiple regression model is used in which the dependent variable is price and the explanatory variables are those factors of demand, supply, and inflation that are believed to influence it. The regression equation is estimated from data in the non-conspiratorial period, and the but-for price in the conspiracy period is projected by inserting the values of the explanatory variables for that period in the equation to obtain the expected price in the absence of conspiracy. The regression estimate of the competitive price (the predicted value of the dependent variable) is compared with the conspiracy price to determine damages.

- A regulatory agency may need to determine the effect of a rate it sets on the regulated industry. Regulators formulate regression equations in which industry revenue is the dependent variable and the regulated rate and other variables are explanatory factors; they use this equation to estimate the effect on revenue of various possible rates given certain levels of the other explanatory factors.

There are many types of regression models. In the linear multiple regression model, the type most commonly encountered, the dependent variable is portrayed as a weighted sum of the explanatory factors plus a random error term. The weights are called regression coefficients. A large positive coefficient implies that small increases in the explanatory factor are associated with large increases in the dependent variable, all other factors being held constant; a large negative coefficient implies an inverse relation. Conversely, a coefficient near zero implies that changes in the explanatory variable are not associated with changes in the dependent variable, all other factors being held constant. In some contexts the quantity of interest is the size of the coefficient of a particular explanatory factor, while in others it is the expected value of the dependent variable as determined by the regression equation. That is, the regression equation may be used for explanation (analysis) or prediction (synthesis).

Regression analysis does not require that the relation between response and explanatory variables be exact or perfectly deterministic. All that is necessary is that the mean of the error term be zero for any fixed values of the explanatory factors, so that it is the *average* value of the dependent variable which is systematically related to the values of the explanatory factors by the regression model.

We have described the concept of regression in terms of "model-plus-error," where the model has a stimulus-response or input-output nature implying a causal relation. In the model-plus-error view, explanatory factors need not be regarded as random variables with distributions, but may be deterministic values, possibly fixed by design. The rate-setting example is of this type. Another important perspective considers the joint distribution of two or more random variables and inquires about the mean value of one of them for given values of the others. No causality need be assumed for there to be a systematic relation between the average value of the dependent variable and the values of the other variables. That is, an increase of one unit in factor X may not lead instrumentally to any corresponding change in Y. The regression model merely states that, on average, different subsets of the data have different means. Analysis of scores on tests given before and after training is in this noncausal category. The score X on the first test is not an input that causes the score Y on the second test (leaving aside learning from test-taking). Here one must consider an important phenomenon known as "regression to the mean," which we describe below. Whereas in the model-plus-error view the regression model is postulated from the start, in the joint-distribution view it is the regression phenomenon that gives us the (sometimes surprising) regression model.

Regression to the mean

Regression to the mean was discovered by Sir Francis Galton in the late 1880s, while he was studying data on the heights of parents and their adult offspring, sons in particular. Galton noticed that if he divided parents into groups by height and then plotted the average height of the sons in each group against the midparent height,[1] the averages tended to lie on a straight line. That linear regression relation constituted a regression model for Galton, even though a son's height in any particular case could not be predicted with great accuracy from his parents' heights. The line of averages represented the component of sons' heights "explained" by the model, while the difference between each individual son's height and the group-specific average was the error, or unexplained, component of that son's height.

Galton called the relation one of *regression* because, as he further noticed, for any given group of parents of fixed height there is an apparent shift toward the mean in the average height of their sons. That is, for the group of parents whose height is, say, one standard deviation above the mean for all parents, the average height of their sons will be above average, but by less than one standard deviation above the mean for all sons, thus "regressing" toward the grand mean for all sons. See Figure 13.1a. The degree of regression turns on the tightness of correlation. At one extreme, if midparent heights perfectly predicted sons' heights, then for parents whose height was d standard deviations above the mean for midparent height, their sons' mean height would also be d standard deviations above the mean for sons; in such cases there would be

---

[1] Galton used the "midparent" height, defined as one-half the sum of father's height and 1.08 times mother's height, to summarize the joint contribution of parents. (The factor 1.08 scales up height of mothers to be equal in mean and variability to that of fathers.) Galton satisfied himself that midparent height was an adequate summarization by finding no systematic variation in son's height based on the *difference* between father's and mother's height, after fixing midparent height.

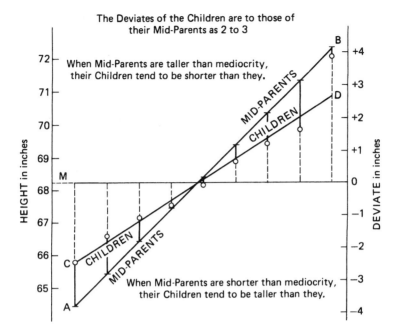

FIGURE 13.1a. Galton's illustration of regression to the mean

no regression to the mean. At the other extreme, if there were no correlation between heights of parents and sons, the mean height of sons of any given midparent height would be the same as the grand average for all sons, i.e., it would regress all the way to the mean. In intermediate cases, if there is a correlation r between the height of parents and the height of their sons, then parents whose height is d standard deviations above the mean midparent height will have sons whose mean height will be rd standard deviations above the mean for sons; or, to put it differently, they would have regressed by a factor of $1 - r$ toward the mean for sons (in standard deviation terms).

Specifically, Galton found a correlation of $r - 0.5$ between midparent height and height of sons, so that parents with a height d standard deviations above or below the mean midparent height had sons whose height was, on average, $d/2$ standard deviations above or below the average of sons' heights. To put this in units of inches, note that the standard deviation of midparent height is smaller than that of son's height by a factor of about $\sqrt{2}$, since the former is an average of two quantities, each with the same variance as the latter.[2] Therefore, parents with height h inches above or below the mean had sons whose height

---

[2]This ignores the correlation between mother's and father's heights, which was about 1/4. Taking this into account yields the slightly smaller factor $\sqrt{1.6} = 1.26$.

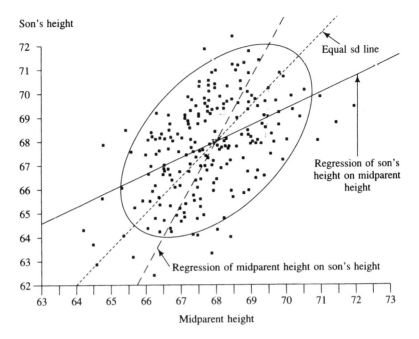

FIGURE 13.1b. Heights of sons and midparents, showing two regression lines

was, on average, $\frac{1}{2}h \cdot \sqrt{2} = h/\sqrt{2}$ inches above or below the average son's height. Hence, the regression equation of son's height on midparent height is

$$Y = EY + (X - EX)/\sqrt{2} + e,$$

where Y = son's height, EY = average height among sons, X = midparent height, EX = average midparent height, and where e represents the error term.[3] The equation may be expressed equivalently as $Y = a + bX + e$ where $a = EY - bEX$ and $b = 1/\sqrt{2}$.

In Galton's example, both the outcome factor (son's height) and the explanatory factor (midparent height) are random variables in the population. The regression equation is an asymmetrical relation, however, describing the conditional mean of the outcome variable given fixed levels of the explanatory factor(s). This asymmetrical approach is a key feature of regression analysis. It implies, for example, that there are two regression relationships, the regression of son's height on midparent height, and the regression of midparent height on son's height. See Figure 13.1b.

The former regression is the average height of all sons whose parents are a given height; the latter regression is the average midparent height of all par-

---

[3]Using the smaller factor 1.26 yields the regression coefficient $\frac{1}{2}(1.26) = 0.63$, close to the value 2/3 preferred by Galton. See Figure 13.1a. Our value of $1/\sqrt{2} = 0.71$ is slightly higher than Galton's.

ents whose sons are a given height. It is often surprising to learn that these two regression relations are not the same. For example, as we have seen, parents whose heights are h inches above average have sons whose mean height is $h/\sqrt{2}$ inches above average. But sons whose heights are $h/\sqrt{2}$ inches above average do not have parents whose mean height is h inches above average. (Their mean height is $h/4$ inches above average.) The regression of midparent height on son's height obviously does not represent a causal connection from son to parent, but merely the correlation induced by the contribution of genetic and environmental factors from parent to offspring. The regression of midparent height on son's height could nevertheless be used, for example, to predict an unseen couple's midparent height on the basis of their son's height.

Further Reading

D. Freedman, R. Pisani & R. Purves, *Statistics*, 169–74 (3d ed. 1998).
S. Stigler, *The History of Statistics*, ch. 8 (1986).

13.1.1   Head Start programs

Head Start programs provide compensatory education for disadvantaged children. Do they work? To answer that question, investigators compared a group of disadvantaged children eligible for compensatory education (the Experimental Group) with a group of children selected at random from the community who were not eligible for such education (the Control Group). Both groups were pre-tested, the Experimental Group was given a program of compensatory education, and then both groups were post-tested. Children from both groups with the same pre-test scores were then compared with respect to their post-test scores. In addition, children were matched by post-test scores, and their gain from pre-test scores compared. The results are shown in Tables 13.1.1a and 13.1.1b (simulated data). In these data, the correlation between pre-test and post-test scores for both groups is about 0.5; the mean pre-test score is approximately 50 for the Experimental Group and 70 for the Control Group; the post-test means are 60 and 80, respectively; the standard deviations of the scores are each about 16.6.

Congressional hearings were held on funding of the program. Opponents claimed the data show that compensatory programs have a negative effect, because Experimental Group children frequently have lower post-test scores than Control Group children with the same pre-test scores. Thus, for pre-test scores with the midpoint 55 (the largest group), the post-test mean for the Experimental Group is 62.93, while for the Control Group it is 70.29 (Table 13.1.1a).

On the other hand, the director of the program claimed it is beneficial. He pointed out that for children scoring the same on the post-test (Table 13.1.1b) the Experimental Group shows a greater gain from the pre-test than the Control

TABLE 13.1.1a. Means on post-test according to pre-test score

Pre-test score (Midpoint)	n	Experimental group Post-test mean	n	Control group Post-test mean
110	0	—	1	110.98
105	0	—	5	95.58
100	0	—	13	103.75
95	2	77.74	21	92.71
90	2	69.54	27	86.99
85	6	87.11	41	84.13
80	10	76.25	48	84.85
75	23	72.52	64	83.20
70	26	67.73	55	78.92
65	33	67.16	60	76.31
60	57	64.78	45	78.50
55	64	62.93	47	70.29
50	56	61.38	30	74.12
45	50	56.51	25	66.11
40	54	51.87	7	61.14
35	42	53.91	4	70.58
30	35	54.17	4	43.44
25	20	53.69	2	59.24
20	6	43.57	0	—
15	5	41.32	1	37.95
10	6	34.36	0	—
5	3	44.34	0	—

Group. Thus, for one of the two largest post-test groups (midpoint 60), the Experimental Group increased from 49.52, while the Control Group showed no gain over its pre-test mean.

Questions

1. For experimental and control groups with given pre-test or post-test scores, illustrate the fact that the data exhibit regression to the mean and compute the magnitude of this effect. What would be the explanation for the regression phenomenon?

2. Explain the fallacy in the arguments of the opponents and the director in terms of regression to the mean.

3. Compare the average difference in pre-test and post-test scores for all the experimental and control groups combined. What does this difference suggest about the efficacy of the program?

This case raises subtle points, some of which are taken up again in Section 14.5.

TABLE 13.1.1b. Means on pre-test according to post-test score

Post-test score (Midpoint)	n	Experimental group pre-test mean	n	Control group pre-test mean
125	0	—	1	78.77
120	0	—	3	94.59
115	0	—	4	94.06
110	1	82.66	14	81.53
105	0	—	16	87.46
100	2	70.36	25	77.62
95	3	67.19	51	77.19
90	14	64.20	41	74.07
85	21	67.90	52	72.56
80	29	51.64	75	69.16
75	48	53.40	52	68.62
70	50	55.05	57	61.52
65	57	56.23	37	63.38
60	57	49.52	27	61.43
55	52	48.99	21	62.09
50	54	42.81	15	58.41
45	42	44.94	4	38.98
40	26	37.19	5	42.29
35	22	33.77	0	—
30	15	37.41	0	—
25	2	30.29	0	—
20	4	16.75	0	—
15	1	30.34	0	—

Source

Campbell and Erlebacher, *How regression artifacts in quasi-experimental evaluations look harmful*, in *Compensatory Education: A National Debate*, 3 *The Disadvantaged Child* (J. Helmuth 1970).

Notes

In a study sponsored by the American Pediatrics Institute, an investigator gave propranolol, one of a class of heart drugs called beta blockers, to 25 high school students. The students were selected because they suffered from unusually severe anxiety about the Scholastic Aptitude Tests and evaluations suggested that "they had not done as well as they should have on the S.A.T." When the students took the S.A.T. again an hour after taking propranolol, their scores improved by a mean of 50 points on the verbal section and 70 points on the mathematics section. Each part of the multiple-choice exam is scored on a scale of 200–800. Students who retake the test without special preparation typically increase their verbal scores by 18 points and their math scores by 20 points. As a *New York Times* lead on the story reported, "A drug used to control high blood pressure has dramatically improved Scholastic Aptitude Test scores for students suffering from unusually severe anxiety, according to a preliminary study." N. Y. Times, October 22, 1987 at A27. What reservations would you have about the validity of this assessment?

13.2   Estimating and interpreting coefficients of the regression equation

In the usual case, a regression model is postulated, but the mathematical form of the relation, the explanatory factors, and their coefficients are unknown. Both the form of regression relation (linear or nonlinear) and the choice of explanatory factors involve substantive expertise, which of course depends on the case, while the assignment of coefficients to those factors is one of statistical technique, for which there are general solutions. We describe first the statistical technique for estimating coefficients in a linear regression model. Choice of explanatory factors is taken up in Section 13.5 and alternate forms of regression models are discussed in Sections 13.9, 14.3, and 14.7.

Each possible set of regression coefficients for a given equation produces a set of "fitted" values for the dependent variable by ignoring the error component. The differences between the actual values of the dependent variable in the data and the fitted values are known residuals. See Figure 13.2. When the actual value of the variable lies above the regression estimate the residual is positive; when below, the residual is negative. One cannot select the equation that minimizes the algebraic sum of the residuals because that sum can equal zero for many different sets of coefficients as positive and negative residuals cancel. Selecting the equation that minimizes the sum of the absolute values of the residuals gives a unique solution in most cases, but absolute values are awkward to manipulate and lack certain other advantages discussed below. The procedure most widely followed is to pick the set of coefficients that minimizes the sum of the squared residuals. This is known as the "ordinary least squares" (OLS) method of estimation of the regression.

Consider a simple linear regression specified by the model

$$Y = a + bX + e,$$

where the error term e has zero mean for any fixed value of X. Given a sample of n pairs (X_i, Y_i), $i = 1, \ldots, n$, the least squares estimate of the constant term a is given by $\hat{a} = \bar{Y} - \hat{b}\bar{X}$, where $\bar{X}$ and $\bar{Y}$ are the sample means of X and Y, respectively. Here $\hat{b}$ is the least squares estimate of b, given by $\hat{b} = rs_Y/s_X$, where s_X and s_Y are the sample standard deviations of X and Y, respectively, and r is the sample correlation coefficient between X and Y (see Section 1.4).

It is instructive to re-express the fitted model in terms of standardized variables, $x = (X - \bar{X})/s_x$ and $y = (Y - \bar{Y})/s_Y$, with zero sample mean and unit sample variance, as follows:

$$y = rx + e',$$

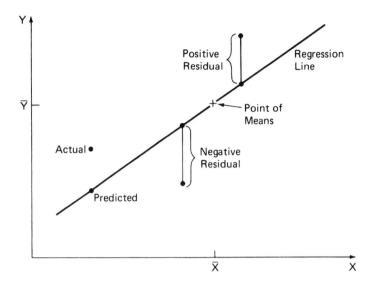

FIGURE 13.2. Residuals from a fitted regression model

where e' is the regression residual, $Y - (\hat{a} + \hat{b}X)$, divided by s_Y.

Thus, in terms of standardized variables, the estimated regression coefficient equals the sample correlation coefficient. When the model is expressed in this form, the regression coefficient is called a *standardized regression coefficient*, also known as a *path coefficient*. In general, a standardized regression coefficient is equal to the ordinary least squares estimate of a regression coefficient corresponding to an explanatory factor X, divided by the ratio s_Y/s_X.

When there are two explanatory factors, the multiple linear regression model becomes

$$Y = a + b_1 X_1 + b_2 X_2 + e,$$

where the error term has zero mean for all fixed values of X_1 and X_2. In terms of standardized variables, the fitted model is given by

$$y = p_1 x_1 + p_2 x_2 + e'.$$

The standardized regression coefficients p_1, p_2 satisfy the pair of simultaneous linear equations (the "normal" equations)

$$r_{1Y} = p_1 + r_{12} \cdot p_2$$
$$r_{2Y} = r_{12} \cdot p_1 + p_2$$

where r_{1Y} and r_{2Y} denote the correlations between Y and X_1 or X_2, respectively, and r_{12} is the correlation between X_1 and X_2. The solution to these equations is given by

$$p_1 = (r_{1Y} - r_{12} \cdot r_{2Y})/(1 - r_{12}^2)$$
$$p_2 = (r_{2Y} - r_{12} \cdot r_{1Y})/(1 - r_{12}^2).$$

Then, the least-squares estimates of the coefficients of the original are given by

$$\hat{a} = \bar{Y} - \hat{b}_1 X_1 - \hat{b}_2 \bar{X}_2$$
$$\hat{b}_1 = p_1 \cdot s_Y / s_{X_1}$$
$$\hat{b}_2 = p_2 \cdot s_Y / s_{X_2}.$$

Notice that if X_1 and X_2 are uncorrelated, $r_{12} = 0$ and the regression coefficients are equal to the values that would obtain if Y were regressed on either variable alone. This means that, in the single variable case, adding to the model a second variable uncorrelated with the first will not change the estimated coefficient of the first variable. This general statement also holds when there are more than two explanatory factors, in the sense that a new variable, when added to the model, will not change the coefficient of an existing variable if the two are uncorrelated, given fixed values of all remaining variables.

Ordinary least squares estimation has certain desirable features. First, it "fits" the regression equation to the data in such a way that the coefficient estimates are unbiased, and the sample correlation between the predicted and the actual values of the dependent variable is maximized. Second, when the errors are normally distributed with constant variance, the least squares estimates of the regression coefficients are maximum likelihood estimates, i.e., they maximize the probability of observing the given data.[4] Third, as the amount of data increases, the estimates become increasingly precise, the highly desirable property of consistency. Fourth, the least squares estimates are "efficient" linear estimates, meaning that they vary less in repeated sampling than other linear estimates; they therefore provide the most precise unbiased linear estimates of the "true" coefficients based on a given set of data.[5] For these reasons, least squares estimation has become the standard method for estimating linear multiple regression equations.[6]

Although least squares estimation has become standard and is amply justified on mathematical grounds, its use warrants a cautionary note. Since this method minimizes squared residual values, the largest residuals contribute disproportionately to the sum of squares (e.g., a residual that is twice as large as another contributes four times its weight in the sum). This means that data points creating the largest residuals, although few in number, can have a major

---

[4]For a discussion of maximum likelihood estimation in the binomial case see Section 5.6. Here the least squares estimates of the regression coefficients maximize the likelihood function based on the observed values of the dependent variable given the explanatory factors and normally distributed errors.

[5]The OLS estimate is "linear" because it in fact equals a weighted sum of the observed values of the dependent variable.

[6]But it is not the only method. For the weighted least squares technique, see Section 13.8 at p. 402. In other forms of regression (Section 14.7) the maximum likelihood estimates differ from the least squares estimates; this usually occurs when the errors are not normally distributed.

influence on the estimate of the coefficients. This may or may not be appropriate. Data points that lie more than two or three standard deviations from the mean, known as outliers, may seriously distort average values and regression estimates when they arise from mistakes in reporting, or from special circumstances not described in the model or likely to be repeated. It thus becomes important to check their accuracy and to investigate their effect on the regression estimates. In a similar vein, values of explanatory variables far removed from the main body of data exert a strong leverage on regression estimates, even without the occurrence of large residuals. Ways of testing the sensitivity of the regression model in this respect are discussed in Section 14.9. Regression "diagnostics" refers to various measures of the influence of particular data points with large residuals or high leverage. These are important clues to hidden defects. See R. Cook and S. Weisberg, *Residuals and Inference in Regression* (1982) for a discussion of regression diagnostics.

13.2.1 Western Union's cost of equity

In a 1970 ratemaking proceeding involving Western Union, the Federal Communications Commission sought to determine Western Union's cost of equity as a step in determining its total cost of capital. Although cost of equity traditionally is determined by the earnings-price ratio of a company's common stock, an expert for Western Union argued that its earnings had been abnormally depressed, so that the earnings-price ratio yielded manifestly absurd results. The expert proposed to estimate cost of equity for Western Union by examining cost of equity for other industry groups and adjusting for differences in risk. The risk adjustment was made by using an index of variability of earnings. Variability was defined as the standard deviation around the 1957–1960 trend in earnings (to avoid treating simple growth in earnings as variability) divided by the mean earnings per share (to eliminate the factor of absolute size of earnings per share). Research produced the statistics of cost of equity and earnings variability shown in Table 13.2.1.

TABLE 13.2.1. Cost of equity and variability of earnings for selected regulated industries

Utility group	Cost of equity	Variability of rate earned on book	Log of variability
AT&T	9.9%	3.6%	1.28
Electrics	10.4%	7.1%	1.96
Independent Telephones	10.4%	10.6%	2.36
Gas Pipelines	11.4%	11.9%	2.48
Gas Distributions	12.2%	12.9%	2.56
Water Utilities	10.0%	17.1%	2.84
Truckers	16.5%	44.0%	3.78
Airlines	18.9%	76.0%	4.33
Western Union	?	27.0%	3.30

Questions

1. Regress cost of equity on variability of rate earned on book to estimate a cost of equity for Western Union. (The log of variability is included here for Section 13.9.1.)

2. What objections would you envision?

Source

Michael O. Finkelstein, *Regression Models in Administrative Proceedings,* 86 Harv. L. Rev. 1442, 1445-1448 (1973), reprinted in *Quantitative Methods in Law* 215–219.

13.2.2   Tariffs for North Slope oil

The Trans-Alaska Pipeline System is a common carrier pipeline extending approximately 800 miles from Prudhoe Bay on Alaska's North Slope to Valdez on Alaska's south central coast. The pipeline carries a mixture of crude oils produced from various fields on the North Slope. A refinery near Fairbanks (ERCA) purchases oil from the pipeline and reinjects residual refinery oil into the system. The various oils shipped are of different qualities and, as a result of commingling in transportation, shippers in Valdez generally receive oil of a different quality from that tendered for transportation.

A so-called "Quality Bank" was established by the carriers to compensate or charge shippers for variations in quality between the oil tendered at Prudhoe Bay, the oil reinjected by ERCA, and the oil received at Valdez. For Quality Bank purposes, the quality of oil is measured in API Gravity (by degrees API)—the higher the degrees API, the higher the quality. When the carriers proposed a differential for the Quality Bank of 15 cents per barrel per degree, ERCA protested. ERCA, which received 26-27 degree API oil and reinjected 20 degree API oil, would be the principal payer into the Quality Bank; it recommended a 3.09 to 5.35 cent differential per degree.

Under Section 1(5) of the ICA (49 U.S.C. 1(5)), carriers are required to establish "just and reasonable" rates or tariffs. Hearings were held before an administrative law judge of the Federal Energy Regulatory Commission to determine whether a quality bank should be established and, if so, what the adjusting payment rate should be. The hearing examiner found that a quality bank concept was just and reasonable and then had to determine the amount of adjusting payment.

Because there were no posted prices for North Slope crude itself, prices for other oils were investigated.

The shippers' expert examined the prices of Mideast and certain domestic crudes. "[S]tudies showed good correlations between the prices of these crudes and either gravity, sulfur, or gravity and sulfur together, with one exception, namely the poor correlation between price and sulfur for domestic crudes. . . ."

TABLE 13.2.2. Persian Gulf crudes

°API	$/barrel
27.0	12.02
28.5	12.04
30.8	12.32
31.3	12.27
31.9	12.49
34.5	12.70
34.0	12.80
34.7	13.00
37.0	13.00
41.1	13.17
41.0	13.19
38.8	13.22
39.3	13.27

Table 13.2.2 shows the price per barrel and the degrees API for a sample of Persian Gulf crudes.

Questions

1. Plot the data and fit a linear regression model to the data (either by eye from a graph, or by the method of Section 13.2).

2. Apply regression analysis to these data. What conclusions do you draw?

3. As attorney for ERCA, what points would you make in response?

Source

Trans Alaska Pipeline System, 23 F.E.R.C. ¶63,048 (1983) (ALJ Initial Dec., dated May 3, 1983), and subsequent proceedings (rebuttal testimony of Jack B. Moshman).

13.2.3   *Ecological regression in vote-dilution cases*

The Voting Rights Act of 1965, as amended in 1982, forbids any practice that would diminish the opportunity for the members of any racial group to elect representatives of their choice. 42 U.S.C. §1973 (1988). In *Thornburg v. Gingles*, 478 U.S. 30 (1986), the Supreme Court held that, for a racial minority to make a case of vote dilution, plaintiffs must show that (i) the minority is geographically compact (i.e., there is a geographical district in which the minority would be a majority); (ii) the minority is politically cohesive (i.e., it votes substantially as a bloc for its desired candidate); and that (iii) the majority is also a voting bloc (i.e., it generally votes together to defeat the minority's candidate).

The Los Angeles County Board of Supervisors consisted of five members elected to serve 4-year terms in nonpartisan elections within supervisory districts. Before February 1990, no Hispanic had ever been elected as a supervisor despite the fact that, according to the 1980 Census, Hispanics made up 28% of the total population of Los Angeles County and 15% of the voting-age citizens.

In a suit to compel redistricting to create a majority-Hispanic district, plaintiffs presented an ecological regression model consisting of two regression equations to prove bloc voting by both Hispanics and non-Hispanics. In plaintiffs' model, the first regression equation was $Y_h = \alpha_h + \beta_h X_h + \epsilon_h$, where Y_h is the predicted turnout rate for the Hispanic candidate expressed as a percentage of registered voters in a precinct; α_h is the percentage of non-Hispanic registered voters who voted for the Hispanic candidate; β_h is the added vote for the Hispanic candidate per percentage point of Hispanic registered voters; X_h is the percentage of registered voters who are Hispanic; and ϵ_h is a random error term with 0 mean. Note that the model assumes that there is no census tract factor affecting turnout that correlates with X_h.

The above model purports to estimate the percentages of the Hispanic and non-Hispanic registered voters that turned out for the Hispanic candidate. To estimate the support rates (i.e., the proportion of those voting in each group that voted for the Hispanic candidate) it is necessary to estimate the voting rates for Hispanic and non-Hispanic registered voters. This was done by a second regression equation, $Y_t = \alpha_t + \beta_t X_h + \epsilon_t$, where Y_t is the total turnout expressed as a percentage of registered voters; α_t is the percentage of non-Hispanics voting for any candidate; β_t is the change in percentage voting associated with being Hispanic; X_h is, as before, the percentage of Hispanic registered voters in the precinct; and ε_t is a random error.

Using the results of a 1982 primary election for Sheriff of Los Angeles County, the investigators collected the following data for each of some 6,500 precincts: (i) total registered voters; (ii) Hispanic registered voters (based on Hispanic surnames, with certain adjustments); (iii) votes for Feliciano, the principal Hispanic candidate; and (iv) the total vote.

From these data the two regression models were estimated by OLS, as follows:

$$Y_h = 7.4\% + 0.11 X_h; \text{ and}$$
$$Y_t = 42.6\% - 0.048 X_h.$$

Figure 13.2.3 gives a scatter plot for the voting data. Each dot represents one precinct. The horizontal axis shows the percentage of registrants in the precinct who are Hispanic; the vertical axis shows the turnout rate for Feliciano. To make the figure more readable only every 10th district is plotted. The sloping line is the first equation, as estimated from the data.

In ruling for the plaintiffs, the district court accepted the ecological regression model, holding that "[e]cological regression [is] the standard method for inferring the behavior of population groups from data collected for aggregate

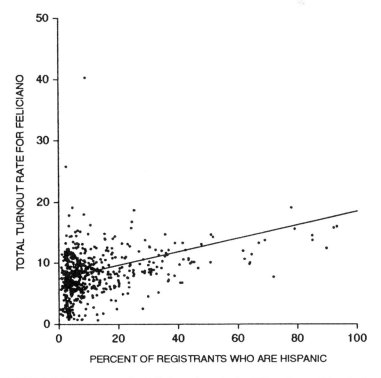

FIGURE 13.2.3. Turnout rates for Feliciano, the principal Hispanic candidate in the 1982 sheriff primary, Los Angeles County. Note: The unit of analysis is the precinct. The regression line is shown.

units." It further held that "[w]hile in theory there exists a possibility that ecological regression could overestimate polarization, experts for defendants have failed to demonstrate that there is in fact any substantial bias."

Questions

From the ecological regression model:

1. What are the average precinct rates and the difference in rates of Hispanic and non-Hispanic turnout for Feliciano?

2. What are the Hispanic and non-Hispanic average precinct voting rates?

3. Using these two estimates, what are the Hispanic and non-Hispanic average precinct support rates for Feliciano?

4. Assuming the ecological regression model is correct, what conclusions do you draw as to bloc voting?

5. What objections do you have to the ecological regression model?

6. What other ways are there to obtain the voting information purportedly furnished by the ecological regression model?

Source

Garza v. County of Los Angeles, 918 F.2d 763 (9th Cir.), *cert. denied*, 111 S. Ct. 681(1991). For an extensive discussion of the evidence and the issues, see Daniel L. Rubinfeld, ed., *Statistical and Demographic Issues Underlying Voting Rights Cases*, 15 Evaluation Rev. 659 (1991).

Notes

Plaintiffs interpreted the ecological regression model to mean that, on average across precincts, the Hispanic candidate received α_h% of the votes of the non-Hispanic registered voters and $(\alpha_h + \beta_h)$% of the votes of the Hispanic registered voters. The difference between the two—β_h percentage points—is the average difference in Hispanic and non-Hispanic support for the Hispanic candidate, and is thus a measure of polarized voting.

Defendants responded that plaintiffs' interpretation of the regression depended on their constancy assumption, and that one could interpret the regression quite differently. They argued that one could assume that within a precinct there was *no* bloc voting–i.e., that the same proportion of Hispanic and non-Hispanic registrants voted for the Hispanic candidate. Then β_h would be interpreted as the average percentage point increase for both groups in the vote for the Hispanic candidate for each percentage point increase in Hispanics among registered voters. In this interpretation, the percentage of Hispanic registered voters in a precinct is a proxy for economic and social conditions that affect the electoral choices of both Hispanic and non-Hispanic voters. Defendants' expert, David Freedman, called this a "neighborhood" model because it assumes that Hispanics and non-Hispanics in the same social and economic class tend to vote in the same way. For both groups, the percentage of registered voters who vote for the Hispanic candidate is $\alpha_h + \beta_h X_h$, with α_h and β_h as estimated from the regression model.

Although the neighborhood model posits no polarized voting within precincts, bloc voting may appear when the precincts are aggregated if disproportionate numbers of Hispanics live in districts with high proportions of Hispanic registered voters. If precincts are highly segregated, there will be little difference in estimates of polarized voting between the ecological regression model and the neighborhood model. But when Hispanics are a significant minority in many districts, as they are in Los Angeles County, the estimates of polarized voting will be higher under the ecological regression model than under the neighborhood model.

13.3 Measures of indeterminacy for the regression equation

A regression model allows for an explicit source of indeterminacy: the scatter of observations of the dependent variable about the regression mean (due to the error term). An additional source of indeterminacy arises when the regression equation is estimated from sample data.

Error of the regression

To illustrate the error of the regression we return to Galton's parents and sons data. For parents of a given midparent height, the heights of their sons vary. The scatter of the sons' heights around the regression mean of their height is represented by the error term in the regression model. The variance of this error is not affected by the number of observations. Nor is it affected by the conditioning factor of parents' height, as may be seen in Figure 13.3a: the scatter of points in the vertical direction is about the same for any value of midparent height. Finally, the distribution of sons' heights about the regression mean is approximately normal. The variance of the error term is usually denoted by σ^2 and its square root, the standard deviation of the error, by σ.

These properties are so typical of regression models that they have been canonized into four assumptions that define the standard regression model. These are: for any fixed values of the explanatory factors, (i) the errors of regression have zero mean; (ii) the errors are statistically independent; (iii) the errors have constant variance; (iv) the errors are normally distributed. The effect of one or more violations of these assumptions is discussed in Section 13.8.

In the usual case, we do not know either the "true" regression relation or the "true" errors; both must be estimated from sample data. Nature shows us the combined result, but hides the error. When the regression relation is estimated from sample data, the various regression means are uncertain, and the inherent uncertainty of a predicted value is enlarged by the sampling variability of the OLS estimates of the regression coefficients. For example, if we took repeated samples of parents of given heights, the average heights of their sons would vary from sample to sample. The estimated OLS regression equation would vary, and with it predictions of sons' heights.

Differences between values of the dependent variable and regression estimates are called residuals. They are to the estimated regression line what errors are to the "true" line. Since the errors are not known, their variance must be estimated from the residuals. In the standard model, the sum of the squared residuals is called the *residual sum of squares*. This is a basic measure of the unexplained variability in the model—unexplained because it primarily reflects the random variation of the dependent variable about the regression mean. The residual sum of squares divided by an adjusted sample size, called

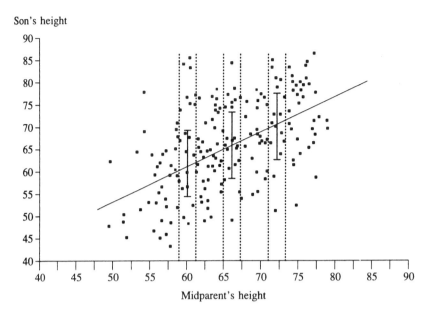

FIGURE 13.3a. Simulated parents-and-sons data illustrating constant error variance

the *error degrees of freedom* and denoted df$_e$, is an unbiased estimator of σ^2, the variance of the error. The estimate is called the *residual mean square*. The error degrees of freedom equals the sample size, n, less the number of explanatory variables in the equation (including the constant term).[7] The square root of the residual mean square, commonly (and confusingly) referred to as the *standard error of the regression*, is essential to the computation of statistical significance of the regression coefficients and confidence and prediction intervals for the dependent variable. See Sections 13.4 and 13.7. A convenient formula for its calculation is $\{[(n - 1)/(n - p)] \cdot s_y^2 \cdot (1 - R^2)\}^{1/2}$, where s_y^2 is the sample variance of the dependent variable, R is the multiple correlation coefficient (defined and discussed shortly), and p is the number of explanatory variables, including the constant.

A measure of the explained variability in the model—known as the *regression sum of squares*—is equal to the sum of the squared differences between the regression estimate of the dependent variable and its overall mean value in the sample. Variability is explained in the sense that the regression model accounts for variations in the regression mean as the explanatory factors vary. The regression sum of squares divided by the regression degrees of freedom (df$_r$) is called the *regression mean square*. The regression degrees of freedom

---

[7]The reduction in degrees of freedom is needed for an unbiased estimate. It reflects the fact that each explanatory factor added to the regression makes it possible to fit an equation more closely to a finite sample of data, and thereby to shrink the variance of the residuals compared with the variance of the errors.

is the number of explanatory variables in the equation (including the constant term) less 1.

The *total sum of squares* is simply the sum of squared deviations of the dependent variable about its overall mean. The total degrees of freedom (df) is the sample size less 1. It is a profound and useful mathematical fact that the total variability of the dependent variable can be resolved into two parts: variability explained by the regression model and variability due to the residuals. Specifically, the total sum of squares equals the regression sum of squares plus the residual sum of squares. This is an expression in sample data of the following theoretical property of random variables: For any two random variables X and Y,

$$Var(Y) = E[Var(Y|X)] + Var[E(Y|X)].$$

In words, the overall variance of Y is the sum of two components: the average of the conditional variances of Y for given values of X (the residual mean square), and the variance of the conditional means of Y given X (the regression mean square).

Measuring goodness of fit

Since the regression equation does not fit values of the dependent variable perfectly, it is important to measure how well the equation performs.

The extent to which a regression model fits the data is initially appraised by the standard error of regression and by the squared multiple correlation coefficient. The smaller the standard error, the better the fit in absolute terms, so that the standard error is often used as a criterion for selecting a regression model from among several candidates. This subject is taken up more fully in Section 13.5.

The multiple correlation coefficient, R, is the correlation between the regression estimates and the observed values of the dependent variable. The squared multiple correlation coefficient, R^2, is commonly used to describe goodness of fit because it can be shown to equal the ratio of the regression sum of squares to the total sum of squares. Thus, R^2 can be interpreted as the proportion of the total variability of the dependent variable that is explained or accounted for by the regression equation. R^2 ranges between 0 (no association) and 1 (a perfect fit). Figure 13.3b illustrates the explained and unexplained deviations on which R^2 is based.

Since the regression equation is optimally fitted to the data, R^2 is biased upward slightly as an estimate of the population value of the proportion of variability explained by the true regression. To compensate for this, R^2 is adjusted downward, by subtracting the quantity $(1 - R^2)(p - 1)/(n - p)$,

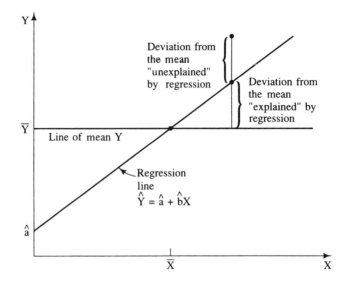

FIGURE 13.3b. Explained and unexplained deviations from the mean

where there are p predictors (including the constant term) and sample size n. The revised estimate is called *adjusted* R^2.[8]

One should remember that R^2 is a summary measure of fit, and a value near 1 does not necessarily imply that the model reflects causal patterns or even that it predicts outcomes very well. The problem of spurious correlation is most noticeable in time series data marked by strong trends; in those circumstances it is quite easy to obtain squared correlation coefficients exceeding 0.90, which seems a close fit. In fact, these coefficients may reflect concurrent trends rather than a causal relation, and in such cases the model fails to predict turning points in the trend. This is obvious in the sales data of Section 14.1.3, where Time and the Seasonal Indicator are not inevitable causes of growth in sales.

In addition, R^2 is affected by the span of the explanatory factors. A regression model based on data extending over a broad range of X values will almost always have a much larger R^2 than the same regression model evaluated on data extending over a narrow range of X values. Over a narrow range of X values, relatively less of the total variation in Y is explained by X, and relatively more is unexplained or error variation. See Figure 1.4.1.

Even when a high degree of correlation reflects a causal relation, there is no assurance that the equation reflects the true form. Models that are quite inconsistent with the true regression can "fit" the data equally well by the R^2 criterion. Each of the data sets shown in Figure 13.3c has the same correlation

---

[8]The effect is small if either R^2 is close to 1, or the number of explanatory factors is small compared with the number of observations of the dependent variable. For example, if $R^2 = 0.80$ based on 30 observations and 6 explanatory factors (including the constant term), adjusted R^2 would be 0.76.

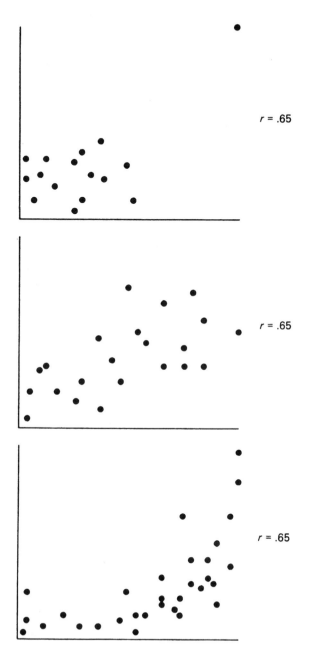

FIGURE 13.3c. Scatterplots with the same correlation coefficient

between the linear regression values and the actual data values. The patterns are quite different, however, and only in the middle case does the linear model appear to fit the data appropriately.

Consequently, it is essential to analyze the fit of the estimated equation in some detail, and not merely to rely on a summary measure.

13.3.1   Sex discrimination in academia

A class action on behalf of women was brought against the University of Houston alleging discrimination in faculty compensation. The court found plaintiffs' statistical proof of discrimination insufficient because it used comparisons between male and female salaries that did not simultaneously take account of all relevant factors.

The University presented two multiple regression studies including such factors, and the court considered plaintiffs' claim that these studies supplied the missing proof.

In the first study, salary was regressed on eight department, experience, and education variables; $R^2 = 52.4\%$. The second model was identical, except that sex was added as a factor; $R^2 = 53.2\%$. Plaintiffs claimed that their case was proved because the sex coefficient of \$694 per year (indicating that women on average earned that much less than equally qualified men) was statistically significant. The court rejected this argument, apparently adopting the position of the University's expert:

> According to the university's expert, despite the \$694 coefficient for sex in the second model, the multiple regression analysis [sic] do not indicate discrimination against women because the model with sex as a factor explained only 0.8 percent more of the total variation around the average salary than did the model without sex: 'It is not the values of the coefficients themselves which matter so much in a statistical interpretation of two models, it is the difference in proportion of variation that is explained.' Plaintiffs introduced no evidence that sex was not, in a statistical sense, independent of the other independent variables in the model (e.g., department, rank, experience, and degree); thus, there has been no showing that the incremental contribution of sex to the variation explained by the university's models is understated because part of its influence erroneously had been attributed to the other independent variables.[9] Further, plaintiffs did not introduce evidence that the value of the coefficient for the sex variable is significant notwithstanding the fact

---

[9] For example, if plaintiffs had proven that women were discriminated against with respect to promotion, the independent variable for experience as an assistant professor would be related to sex because women, due to the discrimination, would have more experience at that rank than would men. In that event, part of the variation in average salary that should have been attributed to sex would not have been because it would have been 'explained' by the experience-as-an-assistant-professor

that only 0.8 percent more of the total variation was explained when sex was added into the model. As we have no basis for rejecting the opinion of the university's expert, we decline to do so.

Questions

1. Does the 0.8 percentage point increase in R^2 when sex is added to the model imply an average 0.8 percentage point shortfall in female salaries relative to the salaries of equally qualified males? What does this small increase imply for the shortfall?

   After reading Section 13.4 answer the following:

2. What is the relation between the statistical significance of the change in R^2 when sex is added as an explanatory factor and the statistical significance of the sex coefficient?

Source

Wilkins v. *University of Houston*, 654 F.2d 388, 403–4 and n.19 (5th Cir. 1981), *reh. den.*, 662 F.2d 1156 (1981), *vacated*, 459 U.S. 809 (1982), *on remand*, 695 F.2d 134 (1983).

13.4   Statistical significance of the regression coefficients

Standard errors of the estimated coefficients

The variability of each of the estimated regression coefficients is measured in terms of quantities called the *standard errors of the coefficients* (not to be confused with the standard error of regression). These are the (estimated) standard deviations of the OLS estimators when viewed as random variables in repeated samples with fixed explanatory factor values. The size of the standard error of a coefficient depends on: 1) the standard error of regression; 2) the sample size; 3) the spread of the values of the explanatory factors; and 4) where there are multiple explanatory factors, the extent of linear correlation among them. In the simplest case, the linear model $Y = a + bX + e$ with $E[e|X] = 0$, if the variance of the error term is Var $(e|X) = \sigma^2$ then the variance of $\hat{b}$ in the

---

variable. As discussed above, plaintiffs did not prove discrimination with respect to promotion.

estimated regression equation $\hat{Y} = \hat{a} + \hat{b}X$ is

$$\text{Var } \hat{b} = \sigma^2 / \sum (X_i - \bar{X})^2,$$

and the variance of $\hat{a}$ for a sample size n is

$$\text{Var } \hat{a} = \sigma^2 \left(\frac{1}{n} + \bar{X}^2 / \sum (X_i - \bar{X})^2 \right).$$

These quantities are estimated by replacing σ^2 by $\hat{\sigma}^2$, the squared standard error of regression (i.e., the residual mean square). The standard error of $\hat{a}$ or $\hat{b}$ is then obtained as the square root of the corresponding estimated variance.

Consider first the fact that the variance of $\hat{a}$ and $\hat{b}$ depends on the variance of the error term and on the size of the sample from which the equation is estimated. If the error variance were zero, each sample of data with fixed values of the explanatory factors would have the same value of the dependent variable regardless of sample size. As the error variance increases, the influence of the random part of the equation causes the dependent variable to vary increasingly from sample to sample; as mentioned in Section 13.3 at p. 367, this produces different OLS equations.

On the other hand, for a given error variance, increasing the sample size diminishes the variability of the estimates due to sampling.

As for the spread of the explanatory factors, again consider the case of simple linear regression of Y on X. The OLS equation runs through the point defined by the means of Y and X. If the values of the explanatory factor X cluster together near the mean, while the Y values have a broad distribution, even small changes in the Y values can cause large swings in the estimated slope of the line (the $\hat{b}$ coefficient), as it responds like a see-saw. But if the X values are spread apart, the line is more stably located.

In more complex situations involving multiple explanatory variables, the fourth factor—the extent of linear correlation among the variables—has an important influence on the standard errors. Suppose that in a wage regression model almost all high-paid employees were men with special training and almost all low-paid employees were women without training. Sex and special training are then highly correlated. Data exhibiting such correlations are said to be collinear; where one factor is nearly a linear function of several other factors, the data are said to be multicollinear, a not uncommon occurrence in a multiple regression equation.

A high level of multicollinearity is an indication that the model is over-specified in the sense that more variables are included than are necessary or justified by the data. Multicollinearity does not bias regression estimates of the coefficients or of the dependent variable; its effect is simply to increase the standard errors of the coefficients, thus making significance tests less powerful and coefficient estimates less reliable.

The reason for the unreliability is not hard to see. In the example, if sex and special training were perfectly correlated, it would be impossible to distinguish their separate effects on wages because they would always move together. When explanatory factors are highly, but not perfectly, correlated, assessment of their separate effects depends on the few cases in which they do not move in tandem; the enlarged standard errors of the coefficients reflect the smallness of the effective sample size.

A simple (but not foolproof) test for multicollinearity involves looking for high correlations (e.g., in excess of 0.9) in pairs of explanatory variables (the correlation matrix printed by most multiple regression computer programs makes this easy). Other indications are large standard errors or the wrong sign for coefficients of important variables. Multicollinearity may also be suspected if, in the construction of a model, the addition or deletion of a variable causes a large change in the estimated coefficient for another variable, or if the addition or deletion of a data point causes a large change in the estimates. For a foolproof way to detect multicollinearity, see the definition of "tolerance" in Section 13.6 at p. 385.

While multicollinearity affects the precision of coefficient estimates, it is possible nevertheless to estimate regression means and predict values of the dependent variable with a small standard error, even though individual coefficients have large standard errors due to the multicollinearity. The situation is akin to locating a point on a map with highly oblique (non-orthogonal) grid lines. The position of the point on the map may be accurately fixed even though the coordinates of the point are quite uncertain, with many different coordinate pairs leading to approximately the same location on the map.

Tests for significance of regression coefficients

The statistical significance of a coefficient is usually determined by dividing the value of the coefficient by its (estimated) standard error. The resulting critical ratio is a t-statistic because, with normal errors, using an estimated value for the error variance gives the quotient Student's t-distribution (see Section 7.1). The attained significance level is obtained from tables, and depends on the error degrees of freedom, defined in Section 13.3 as the number of observations less the number of explanatory variables (including the constant).

The t-statistic described above tests significance of an individual estimated coefficient. It is also possible to test groups of coefficients, or all of them simultaneously, for joint significance. A test of all explanatory factors together (exclusive of the constant term) is, in effect, a test of the statistical significance of R^2. The effect of adding or deleting groups of explanatory factors is appraised by looking at F-statistics. See Section 7.2 for a discussion of the F distribution. In the present context, the F statistic is calculated as the ratio of (i) the absolute change in the residual sum of squares when explanatory factors are added to or subtracted from the equation, divided by the number of factors tested, i.e., the regression mean square for the added variables, to (ii) the sum of squared residuals with the factors included, divided by the error degrees of freedom, i.e., the error mean square. The numerator may also be expressed in terms of the absolute change in the regression sum of squares. Yet another expression for the F-statistic, often the most convenient, is the ratio of (i) the change in R^2 as variables are added or removed from the model, divided by the number of factors tested, to (ii) the value $1 - R^2$ from the larger model, divided by the df_e for the larger model. In symbols,

$$F = \frac{R_1^2 - R_0^2}{p} \bigg/ \frac{1 - R_1^2}{df_e},$$

where R_1^2 is from the larger model, R_0^2 is from the smaller model, the models differ by p parameters, and df_e is the error degrees of freedom from the larger model.

If the true values of the coefficients of all factors tested are 0 (i.e., the factors are not explanatory), the sampling distributions of the numerator and the denominator are independent chi-squared distributions divided by their respective degrees of freedom, and thus the ratio follows the F-distribution. If the reduction in residual variance is large when factors are added (or the increase is large when the factors are deleted), the numerator of the F-statistic becomes large, and the hypothesis that these factors should not be in the equation is rejected. For a single factor under test, the F-statistic reduces to the square of the t-statistic for that factor's coefficient, so that the t-test and the F-test are fully equivalent.

For a single explanatory variable, or for groups of variables, coefficients that are statistically significant usually justify rejection of the hypothesis associated with zero coefficients. The assertion that sex played no role in wages would be rejected if there were a statistically significant sex coefficient in a correct regression model. The converse situation is more open to interpretation. If there is an independent reason to believe that sex is a factor in wages, and the estimated coefficient is consistent with that hypothesis (e.g., it shows that equally qualified women earned less than men), but is not statistically significant, one would conclude that the data are consistent with the hypothesis, but that the observed differences between men and women may have been due to chance. The hypothesis of discrimination cannot be excluded, however, because the lack of significance may reflect lack of power (e.g., because of a small number of observations) rather than an absence of discrimination. On the other hand, if power were high or if the estimated coefficient were inconsistent with discrimination, e.g., it showed that equally qualified women earned more than men, a rejection of discrimination would be more justified. Our attitude toward a coefficient with an attained level of significance that fails to reach a critical level thus depends both on the substantive importance of the coefficient's value and on the power of the test to detect departures from the null hypothesis.

Confidence intervals for regression coefficients

A confidence interval for a true regression coefficient b is constructed by adding and subtracting from $\hat{b}$, the least squares estimate of b, a critical value times the standard error of $\hat{b}$. The critical value is obtained from tables of the t-distribution with df_e degrees of freedom. For example, a 95% confidence interval might take the form $\hat{b} \pm 2.09 \cdot se(\hat{b})$, where 2.09 is the 95% two-tailed critical value for the t-distribution on $df_e = 20$ degrees of freedom. Confidence intervals for the regression mean and prediction intervals for future observations are discussed in Section 13.7.

13.4.1 Race discrimination at Muscle Shoals

The Tennessee Valley Authority (TVA) has an Office of Agricultural and Chemical Development (OACD) at Muscle Shoals, Alabama. Jobs there are divided into schedules: Schedule A for administrative jobs; Schedule B for clerical; Schedule D for engineering or scientific; Schedule E for aides and technicians; Schedule F for custodial; and Schedule G for Public Safety. The OACD requires a BS degree in chemistry from an American Chemical Society (ACS) approved curriculum for initial hiring into a professional chemist (Schedule D) position. Individuals with non-ACS approved degrees are nevertheless considered for Schedule D positions provided their schools have met substantive requirements for ACS accreditation, or if the candidate has substantial related work experience. Recipients of non-ACS approved degrees are generally eligible for Schedule E (subprofessional) positions, from which they may be promoted to Schedule D jobs. Promotions are also made from Schedule B to Schedule A.

Plaintiffs brought a class action claiming discrimination against blacks in promotions. Each side performed multiple regression analyses. Plaintiffs used five explanatory factors (plus a constant term): technical degree, non-technical degree (no degree was a reference category), years of service at TVA, age, and race; annual salary was the dependent variable. The data were from Schedules A, B, D, E and F combined. There were only 7 blacks in Schedule E and 14 on Schedule D; there were a few blacks in Schedule B, but none in Schedule A. The estimated race effect was $1,685.90 against blacks, with a standard error of $643.15.

TVA's regressions differed in that they broke down degrees by area of major concentration: natural science, engineering science (other than chemical engineering), Code R (a doctoral degree in any field or a chemical engineering degree), and highest grade completed. The regressions were estimated separately for Schedules D and E. The choice of variables was the result of some trial and error by TVA's expert and a stepwise regression program (see Section 13.5).

The results were: for Schedule D the estimated race effect was $1,221.56 against blacks, with standard error $871.82; for Schedule E the estimated race effect was $2,254.28 against blacks, with standard error $1,158.40.

Questions

1. Compute the t-statistic for the race coefficient in each of the regression models and find its P-value (assume the df_e's are very large).

2. Why are the standard errors larger in defendant's regressions than in plaintiffs'?

3. Which model is preferable?

4. Should plaintiffs' model have included indicator variables for the different schedules? (See Section 13.5 at p. 380. If it had, what would have been the effect on the standard error of the regression coefficient for race?

5. Increase the power of TVA's model by combining the evidence from the two schedules using one of the methods of Section 8.1. What conclusions do you reach?

Source

Eastland v. *Tennessee Valley Authority*, 704 F.2d 613 (11th Cir. 1983).

13.5   Explanatory factors for a regression equation

Selection of factors

Ideally, the selection of explanatory factors is determined solely by strong, well-validated, substantive theory, without regard to the closeness of fit between the regression estimates and the actual data. Thus, a labor economist might use labor market theory to select certain types of experience and education as explanatory factors in a wage equation. Of course, if an employer has an explicit mechanism for awarding salary (such as wage increment guidelines), that mechanism should be reflected in an explanatory factor (unless it is tainted—see p. 382). This ideal is seldom met in practice, either because theory is not complete enough to dictate the choices, or because data for theoretically perfect factors are not available and surrogates must be used instead.

There are various techniques for selecting a set of explanatory factors from a group of candidates. One popular procedure is the so-called forward-selection method. In this method, the first variable to enter the equation is the one with the highest simple correlation with the dependent variable. If its regression coefficient is statistically significant, the second variable entered is the one with the highest partial correlation with the dependent variable. (The partial correlation is the correlation of the residuals of the dependent variable, after it has been regressed on the first variable, with the residuals of the second variable, after it has been regressed on the first variable.) In a so-called "step-wise" variant, when the regression is recomputed with two variables, the first is dropped if it proves no longer significant. This procedure of adding, retesting, and eliminating is continued until no new significant variables are found and all included variables remain significant.[10]

These techniques are widely practiced, but it should be remembered that the variables selected for the model may not reflect external validity. For example, a variable that increases R^2 too little might be excluded as not statistically significant even though its omission biases the remaining regression coefficients. Simulation studies employing the bootstrap technique have shown that the group of variables selected by stepwise regression is likely to vary widely depending on random factors in the data, and that the correct set of explanatory variables (i.e., the set assumed in the bootstrap experiment) is unlikely to be chosen. (See Section 14.9.) The final selection in a stepwise procedure should be viewed only as a candidate model with some degree of parsimony in explanatory variables and an adequate fit to the data given the set of variables on hand.

At least one court has rejected a regression equation, in part because the variables were selected by the forward-selection procedure. *Eastland* v. *Tennessee Valley Authority, supra.*

An alternate technique involves selecting the "best" from all possible subsets of explanatory factors. An index that assists in selecting the best set from among many candidates is Mallows' C_p,[11] defined as follows. Assume that there is a maximally inclusive model with a residual mean square that provides an unbiased estimate, say s^2, of the error variance. We wish to select a smaller model, say with p explanatory factors, including the constant, with no substantial degradation in the goodness of fit of the model to the data. The C_p index is defined as the ratio of the smaller model's residual sum of squares, RSS_p, to s^2, that ratio diminished by $(n - 2p)$, $C_p = (\text{RSS}_p/s^2) - (n - 2p)$. If there is adequate fit to the data, C_p will be approximately equal to p, whereas if there is a substantial lack of fit, C_p will exceed p. Thus when C_p is plotted against p for various models, those falling close to the line $C_p = p$ are preferred. C_p times σ^2 also estimates the total sum of squared deviations of the fitted model from the true but unknown model, combining error variance with squared bias. Therefore, the "best" models have the smallest values of C_p without departing far from the $C_p = p$ line. Plotting C_p vs. p for each competing model is recommended as a graphical aid to model selection.

The limitations of these methods—arising from the fact that the same data are used to fit the model and assess its goodness of fit—underscore the importance of validation for the model from external consistency, external theory, and replication with independent data.

---

[10]"Significance" here is merely synonomous with a sufficiently large F-statistic. Since the same data are used repeatedly to select a model, the actual Type I error rates are larger than nominal.

[11]Mallows, *Some comments on C_p*, 15 Technometrics 661 (1973); N. Draper and H. Smith, *Applied Regression Analysis*, ch. 15 (3rd ed. 1998).

Indicator or dummy variables

In many cases, one or more explanatory factors do not have "natural" values, but rather fall into categories. Examples are sex, race, academic rank, and so forth. Values may, of course, be assigned to each category, but assignment of successive integers to categories means that any shift between categories creates corresponding numerical changes in the dependent variable (the size of the change being measured by the coefficient of the coded variable). Often this kind of correspondence does not accurately reflect the phenomenon under study. Whether an employee finishes high school may affect his productivity very differently from whether he finishes college. To provide greater flexibility, the usual practice is to set up an indicator, or "dummy," variable for each category (save one reference category) which assumes the value 1 for observations in that category and 0 otherwise. Observations in the reference category are represented by zero indicator variables for all the other categories. With this arrangement, the differential effect of each category relative to the reference group is determined separately by the coefficient for the corresponding indicator variable. In the example given, separate dummy variables would be included for "finished high school but not college," and "finished college." The reference category would be "did not finish high school." The coefficients for the two dummy variables could differ, reflecting their different contributions to productivity.

Generally, to compare k means, one defines $k-1$ indicators, $X_1, \ldots, X_{k-1}$, and writes the model $E[Y|X_1, \ldots, X_{k-1}] = \mu + \beta_1 X_1 + \cdots + \beta_{k-1} X_{k-1}$. Then $\hat{\beta}_i$ is the difference between the ith sample mean and $\hat{\mu}$, the kth sample mean. The regression analysis is then entirely equivalent to the analysis of variance for comparing several means (see Section 7.2).

Consequences of misspecification

It is frequently the case (particularly in non-standard problems) that no detailed theory exists to dictate the choice of explanatory factors. Has the right set been chosen, and what if it hasn't? The lack of solid answers to these questions provides a fertile field for almost every challenge to a model. There are three important cases to consider: wrongly omitted variables, extraneous variables, and tainted variables.

Omitted variables

Omission of an *important* explanatory variable causes "misspecification" of the regression equation, and the resulting regression coefficient estimates are usually biased. Generally, the direction of the bias is to overstate the coefficients of included variables, as they "pick up" the correlated explanation that "belongs" to the omitted variable. If a wage regression model includes a sex variable, for example, the coefficient of that variable tends to be overstated if a significant productivity factor—which is positively correlated with sex after accounting for the included variables—is omitted.

On the other hand, failure to include an important explanatory variable does not bias the coefficient of a variable of interest if the two are uncorrelated for any fixed values of the other variables in the equation. Apart from this special case, however, omission of an important explanatory factor biases the coefficients of included variables and increases the residual mean square of the equation.

The effect on regression estimates of excluding an important explanatory factor also depends on the relation of the omitted factor to other explanatory factors in the equation. There are two cases. First, if the omitted variable is a linear function of the other variables in the equation plus an error term with mean zero, failure to include the omitted variable does not bias estimates of the dependent variable. That is, the estimated value of the dependent variable based on any set of fixed values of the included explanatory variables would equal the value of the estimate that would be obtained from the full model if the omitted factor had been evaluated at its estimated mean, given the included factors. As discussed above, however, the omission is likely to bias the coefficients of the included explanatory factors, and to increase the residual mean square of the equation. Second, if the omitted variable is not a linear function of other variables in the equation, the equation would be biased both with respect to estimates of the dependent variable and estimates of the coefficients; in addition, as before, the residual mean square of the equation would be increased.

Whether the omitted variable has a linear relationship with the variables included in the equation is commonly a matter of speculation; such variables usually are omitted because they cannot be observed. An indicator variable cannot have a substantial linear relation with a continuous variable over an infinite range, but might have an effectively linear relationship over an observed range. But a wispy specter of omitted variable bias should not *ipso facto* compel rejection of a model. Even if qualitatively plausible, a sensitivity analysis might show that misspecification produces only trivial bias.

The possibility of bias and the nature of omitted variables arose in New York State's and New York City's action against the Census Bureau to compel adjustment of the census to correct for the undercount (see Section 9.2.1). To estimate the undercount from post-enumeration survey data, New York proposed a regression equation in which the dependent variable was the undercount (estimated from the survey data) and the explanatory variables were percent minority, crime rate, and percent conventionally enumerated (face-to-face interview rather than mail-in, mail-back). Resisting this approach, Census Bureau experts argued that many other variables might be associated with the undercount (e.g., percentage of elderly people). Plaintiff's lead expert replied that omitted variables would not bias the regression estimate of the undercount. An expert for the Bureau rejoined that the estimate would be unbiased only if the omitted variables had a linear relationship with included variables and that some plausible omitted variables were unlikely to have that form (e.g., a dummy variable for "Central City"). The court did not resolve this issue. *Cuomo v. Baldridge*, 674 F. Supp. 1089 (S.D.N.Y. 1987).

Assume that a dummy variable like Central City should have been in the equation and (for all practical purposes) has a linear relation with the included variables. If the regression estimates were in fact unbiased, would that meet objections to the model based on the failure to include the Central City variable?

Extraneous variables

Including an unnecessary factor violates the desire for parsimony, but usually does not create misleading results. If the factor is uncorrelated with the dependent variable after accounting for the other variables in the model, the least-squares estimator of its coefficient would be near zero, and it would not affect either the regression estimates or the other regression coefficients, although it would reduce precision. [Do you see why?] A problem would arise only if there were substantial multicollinearity with other explanatory variables.

Tainted variables

The main reason for omitting a plausible variable is that it may be tainted. Academic rank is a good example of a potentially tainted variable in a wage regression involving university faculty. If the university is guilty of discrimination in awarding promotions, the inclusion of academic rank as an explanatory factor would effectively conceal wage discrimination consequent to promotion discrimination. On the other hand, if the university awards rank without discrimination, the omission of rank would deprive the wage regression of an important explanatory factor and might produce a spurious result indicating discrimination when none existed. The same problem could arise in a price equation used to measure the effect of a price-fixing conspiracy. It is common in such an equation to include cost of production as a factor. But that factor would be tainted if the conspiracy itself raised costs by reducing the incentive for cost reduction.

Some courts have dealt with the problem by looking to separate studies to see whether they indicate the presence of taint. In a sex discrimination case, rank was allowed as an explanatory factor in a wage regression because no statistically significant difference appeared between men and women in average waiting time for promotion.[12] An intermediate approach would be to attempt to remove the effect of any taint. Various strategies are possible. For example, in the case of academic rank, females might arbitrarily be coded with the next higher academic rank in cases identified by plaintiff as involving

---

[12] *Presseisen v. Swarthmore College*, 442 F. Supp 593 (E.D. Pa. 1977), *aff'd*, 582 F.2d 1275 (3rd Cir. 1978).

possible discrimination in the awarding of rank, but with salary and all other variables remaining at their actual values. The resulting measure of male-female disparity, adjusted for rank and other factors, might then be taken as an upper bound on the shortfall that would have been measured in the absence of taint.

Missing data

Almost all large data sets are beset by missing values. In sample surveys, data may be missing due to non-response to a survey item. In the context of controlled experiments, missing values occur due to unforseen mishaps, e.g., equipment malfunction. In multiple regression studies, data on a variable may be missing due to non-applicability of the variable to particular subjects. In all cases, data may be just lost or miscoded. What should be done about missing data in a multiple regression study?

The answer depends very much on the mechanisms that cause values to be unobserved. In the simplest case, data are missing at random, meaning that the probability that a unit to be sampled will be unobserved does not depend on the value to be observed in that unit. When data are missing at random, one can usually proceed with standard analyses on the observed data, on the theory that those data form a random subsample from the random sample, which of course is still a random sample. When this assumption is false, analyzing the observed data as if it were a random sample is unjustified and can cause severe bias. For example, in surveys designed to address sensitive areas, such as income or sexual behavior, it is far more likely to find missing values due to non-response when the true answer would be extraordinary than when the answer would be ordinary. In such cases the average income or rate of sexual deviance will be seriously underestimated from the observed data. Occasionally, the assumption of missing-at-random may be valid within strata or holding the value of some observed covariate fixed. In the income survey example, it may be valid to assume that, within a given age bracket, the probability of non-response is constant with respect to any person's income, although that probability may vary with age. In a regression analysis that adjusts for age, the observed data would be unbiased, since one is then estimating average income given age. The marginal distribution of income would still be biased, however, since the overall probability of non-response is not constant with age.[13]

Almost all remedies for missing values rely on an assumption of missing-at-random for their validity, although most often it is the exception rather than

---

[13]Technically, to ensure that all analyses remain unbiased in the observed data, one needs a stronger condition known as missing *completely at random*, which essentially postulates that the observable data are a random sample both marginally and conditionally with respect to covariates. The definition of missing at random therefore allows the probability that a unit to be sampled will be unobserved to depend on *observed* characteristics of the unit, but not unobserved characteristics. Missing completely at random allows no dependence at all.

the rule that one can verify the assumption. The simplest procedure analyzes cases with all relevant non-missing variables. In a regression context this is known as "listwise" deletion of missing values, because a case is dropped from the analysis whenever one or more variate values from the list of variables is missing for that case. This method is acceptable when there are only a few sporadic values missing, so that sample size is not substantially reduced, and the missing-at-random assumption is tenable. Sometimes, however, almost all cases have one or two missing values, in which case listwise deletion is inefficient. Another strategy utilizes "pairwise" deletion, the idea being to estimate means, variances, and covariances with as much data as possible. The correlation between two variables, for example, would be estimated from the set of data observed completely on the two variables, although this same data subset may not be used in estimating the correlation between a different pair of variables. While it makes use of more data than listwise deletion, this method has been shown to suffer from severe bias under certain circumstances with much missing data, and therefore should be avoided whenever possible.

Another class of methods relies on imputation of data to "fill in" the missing values, adjusting for the consequent inflation of apparent precision and significance. For example, in a multiple regression context, missing values of explanatory factors can be estimated using a subsidiary regression model that predicts the missing factors based on the present factors. The predicted values are then substituted as if they had been observed.

Yet another scheme attempts to model the probability of missingness as a function of observable covariates, and then uses the reciprocals of these probabilites as weights to "correct" for the unobserved portion of the sample based on the observed portion.

On the treatment of missing data by the courts, see the Notes, pp. 267–268.

Further Reading

N. Draper & H. Smith, *Applied Regression Analysis*, ch. 14 (dummy variables) (3d ed., 1998).

R. Little & D. Rubin, *Statistical Analysis of Missing Data* (1987).

13.6 Reading multiple-regression computer printout

The computer sheets that were exhibits in the *Bazemore* case (see Section 13.6.1) are output from a widely used computer program called *Statistical Package for the Social Sciences* (SPSS[x]).[14] Listed below are brief descriptions

---

[14]Many computer programs offer multiple regression routines. Among the better known of these are SAS, SPSS, and S-Plus.

of the terms appearing in the printout, most of which were defined in Sections 13.3 and 13.4.

The output begins with summary statistics for the dependent variable and each explanatory factor: the sample mean, standard deviation, sample size, and correlation matrix showing the simple correlation of each variable with every other.

Under the heading "Multiple Regression" are the names of the dependent variable and the variable entered in the current step of the forward-selection procedure. Beneath this are the values of the multiple correlation coefficient, squared multiple correlation ("R Squared"), adjusted R^2, and standard error of regression ("Standard Error").

Under the heading "Analysis of Variance" are the source of variability ("Regression" or "Residual" [i.e., error]), the respective degrees of freedom ("DF"), and the corresponding sum of squares and mean square. At the extreme right is the F-statistic, the joint test of overall significance of the explanatory factors.

Directly below this, under the heading "Variables in the Equation," is the estimated model, listing the variable name, the estimated regression coefficient (in the column headed "B"), the standardized coefficient (in the column headed "Beta," see below), the standard error of the regression coefficient, and the F-statistic for each coefficient (squared t-statistic).

Beta is the standardized form of coefficient obtained by multiplying B by the standard deviation of the explanatory variable and dividing by the standard deviation of the dependent variable. Equivalently, Beta (path) coefficients are OLS estimates that result when all the variables are first standardized, i.e., mean-centered and scaled by their sample standard deviations. Because the magnitude of the coefficient for an explanatory variable is affected by the relative sizes of the standard deviations of the explanatory and dependent variables, this transformation makes it possible to compare the relative contributions of explanatory factors expressed in different units (as changes in s.d. units of the dependent variable per s.d. change in the explanatory factor).

To the right of the regression model is a section labeled "Variables Not in the Equation." Here "Beta-In" refers to the value of Beta that would be calculated if that variable were entered next. "Partial" refers to the partial correlation between the dependent and the explanatory variable. (Note that the partial correlation squared is the proportionate reduction in $1 - R^2$ due to the addition of the new variable.) "Tolerance" is a measure of a variable's multicollinearity with other variables in the model. It is defined as the proportion of that variable's variance not explained by the other variables in the model. A tolerance close to zero (say, less than 0.01 in practice) indicates severe multicollinearity. "F" for a variable is the test statistic that would result from entering that variable.

A summary table at the end of the printout gives the multiple-R, R-squared, and the change in R^2 between adjacent steps, as variables are added to the model, and, for the final model, the simple correlation of each explanatory variable with the dependent variable, its regression coefficient, and standardized coefficient.

13.6.1   Pay discrimination in an agricultural extension service

The purpose of the North Carolina Agricultural Extension Service is to disseminate "useful and practical information on subjects relating to agriculture and home economics." Programs are carried out through local agents, who are divided into three ranks: full agent, associate agent, and assistant agent. "While the three ranks of agents perform essentially the same tasks, when an agent is promoted his responsibilities increase and a higher level of performance is expected of him."

The salaries of agents are determined jointly by the Extension Service and the county board of commissioners where the agents work. Federal, state, and county governments all contribute to the salaries.

Before August 1, 1965, the Extension Service had a separate "Negro branch," composed entirely of black personnel, that served only black farmers, homemakers, and youth. The white branch employed no blacks, but did on occasion serve blacks. On August 1, 1965, in response to the Civil Rights Act of 1964, the State merged the two branches. However, disparities in pay were not immediately eliminated. There were substantial county-to-county variations in contribution, but the counties contributing the least were not those with the greatest number of blacks.

In a suit by black agents claiming discrimination in pay (joined by the United States), plaintiffs relied heavily on multiple-regression analysis in which salary was regressed on four independent variables—race, education, tenure, and job title. These variables were selected on the basis of deposition testimony by an Extension Service official, who stated that four factors determined salary: education, tenure, job title, and job performance. Set forth below are the definitions of the variables used in the computer printout.

MS	= 1 if person has master's degree; 0 otherwise
PHD	= 1 if person has Ph.D.; 0 otherwise
TENURE	= Years with the service as of 1975
CHM	= Chairman, 1 or 0
AGENT	= Agent, 1 or 0
ASSOCIATE	= Associate Agent, 1 or 0
ASST	= Assistant Agent, 1 or 0

The printout of one stepwise OLS regression analysis introduced in the case is shown in Table 13.6.1.

Evidently, there was a preselection of variables because the stepwise regression computer printout from the record in the case shows only the variables that were ultimately selected for the model, and we know that others were excluded. The stepwise program without this preselection would be similar in operation except that there would be more variables in the table under the heading "Variables Not in the Equation" at each step, and possibly a different final model. Full disclosure of the method employed should normally include all of the variables from which selections were made.

TABLE 13.6.1. Multiple-regression computer output from *Bazemore v. Friday*

```
VARIABLE        MEAN      STAND. DEV.   CASES
SALARY      12524.0316    2487.0518      569
WHITE           0.8190    0.3854         569
MS              0.1634    0.3701         569
PHD             0.0       0.0            569
TENURE         10.6626    8.9259         569
CHM             0.1634    0.3701         569
AGENT           0.4763    0.4999         569
ASSOC           0.1459    0.3533         569
ASST            0.2144    0.4108         569
```

CORRELATION COEFFICIENTS

```
          SALARY    WHITE      MS PHD    TENURE      CHM      AGENT    ASSOC ASST
SALARY       1
WHITE    0.17511       1
MS       0.29322  0.12140        1
PRO         ---      ---       ---    1
TENURE   0.68522 -0.05157  0.12438  ---      1
CHM      0.67609  0.19023  0.19023  ---   0.42123      1
AGENT    0.17086 -0.10002 -0.02183  ---   0.35648 -0.42152      1
ASSOC    0.27590  0.03911 -0.08841  ---  -0.30483 -0.18267 -0.39409      1
ASST     0.57977 -0.08803 -0.06879  ---  -0.55116 -0.23092 -0.49820 -0.21590      1
```

Questions

1. From the Extension Service's point of view, what potentially explanatory variables were omitted from the model? Are the omissions justified?

2. From the black agents' point of view, what potentially objectionable variables were included in the model? Are the inclusions justified?

3. Interpret the coefficient for race in the final model. Is it statistically significant? Is the value of the coefficient affected by the fact that race was the last variable to be included in the model?

4. The larger pool of variables from which the variables in the model were selected included interaction variables for White Chairman, White Agent, and White Associate Agent. What was the reason for considering these variables for inclusion in the model? Why were they not included? What inference would you draw from their exclusion? If they had been included, how would you have used the model to measure the pay differential between the races?

Source

Bazemore v. Friday, 478 U.S. 385 (1986).

```
                    MULTIPLE REGRESSION RESULTS


DEPENDENT VARIABLE                      SALARY SALARY 1975
VARIABLE(S) ENTERED ON STEP NUMBER 1..  TENURE TENURE 1975


MULTIPLE R              0.68522
R SQUARE                0.46953
ADJUSTED R SQUARE       0.46859
STANDARD ERROR       1813.00693


ANALYSIS OF VARIANCE     DF     SUM OF SQUARES      MEAN SQUARE         F
REGRESSION                1.   1649596681.82313   1649596681.82313   501.85568
RESIDUAL                567.   1863725679.60747     3286994.14393


            VARIABLES IN THE EQUATION
VARIABLE        B       BETA     STD ERROR B      F
TENURE     190.9256   0.68522     8.52265     501.856
(CONSTANT)  10488.27


        VARIABLES NOT IN THE EQUATION
VARIABLE   BETA IN    PARTIAL    TOLERANCE     F
WHITE      0.21101    0.28933    0.99734     51.710
MS         0.21126    0.28780    0.98453     51.116
PHD          ---        ---        ---        ---
CHM        0.47104    0.58655    0.82256    296.862
AGENT     -0.08410   -0.10788    0.87292      6.665
ASSOC     -0.07389   -0.09663    0.90708      5.334
ASST      -0.29029   -0.33257    0.69622     70.384
```

```
VARIABLE(S) ENTERED ON STEP NUMBER 2..  CHM
INDICATOR VARIABLE                      I:CHAIRMAN 1975


MULTIPLE R              0.80749
R SQUARE                0.65203
ADJUSTED R SQUARE       0.65080
STANDARD ERROR       1469.67154


ANALYSIS OF VARIANCE     DF     SUM OF SQUARES      MEAN SQUARE         F
REGRESSION                2.   2290799475.06036   1145399737.53018   530.29375
RESIDUAL                566.   1222522886.37025     2159934.42822


            VARIABLES IN THE EQUATION
VARIABLE        B       BETA     STD ERROR B      F
TENURE     135.6400   0.48680     7.61748     317.069
CHM       3165.370    0.47104   183.71606     296.862
(CONSTANT)  10560.40


        VARIABLES NOT IN THE EQUATION
VARIABLE   BETA IN    PARTIAL    TOLERANCE     F
WHITE      0.11505    0.18910    0.94000     20.952
MS         0.14880    0.25734    0.96243     36.819
PHD          ---        ---      0.0          ---
AGENT      0.41184    0.48148    0.47560    170.510
ASSOC     -0.04590   -0.07396    0.90350      3.107
ASST      -0.29114   -0.41181    0.69622    115.386
```

```
DEPENDENT VARIABLE                       SALARY SALARY 1975
VARIABLE(S) ENTERED ON STEP NUMBER 3..   AGENT
INDICATOR VARIABLE                       I:AGENT 1975
MULTIPLE R                0.85598
R SQUARE                  0.73270
ADJUSTED R SQUARE         0.73128
STANDARD ERROR         1289.24145
```

ANALYSIS OF VARIANCE	DF	SUM OF SQUARES	MEAN SQUARE	F
REGRESSION	3.	2574221270.00844	858070423.33615	516.24328
RESIDUAL	565.	939111091.42217	1662143.52464	

VARIABLES IN THE EQUATION

VARIABLE	B	BETA	STD ERROR B	F
TENURE	61.13800	0.21942	8.78668	48.414
CHM	5088.825	0.75726	218.33633	543.229
AGENT	2049.039	0.41184	156.91900	170.510
(CONSTANT)	10064.50			

VARIABLES NOT IN THE EQUATION

VARIABLE	BETA IN	PARTIAL	TOLERANCE	F
WHITE	0.08560	0.15966	0.92989	14.754
MS	0.013637	0.25839	0.0	40.348
PHD	---	---	0.0	---
ASSOC	0.13395	0.21426	0.68395	27.138
ASST	-0.15574	-0.21426	0.50591	27.138

```
VARIABLE(S) ENTERED ON STEP NUMBER 4..   MS
INDICATOR VARIABLE                       I:MS DEGREE 1975
MULTIPLE R                0.86634
R SQUARE                  0.75055
ADJUSTED R SQUARE         0.74878
STANDARD ERROR         1246.56459
```

ANALYSIS OF VARIANCE	DF	SUM OF SQUARES	MEAN SQUARE	F
REGRESSION	4.	2636909630.52409	659227407.63102	424.23420
RESIDUAL	564	876412730.90652	1553923.28175	

VARIABLES IN THE EQUATION

VARIABLE	B	BETA	STD ERROR B	F
TENURE	60.63075	0.21760	8.49619	50.926
CHM	4895.584	0.72851	213.28961	526.829
AGENT	2006.770	0.40334	151.87048	174.602
MS	916.4253	0.13637	144.27257	40.348
(CONSTANT)	9971.838			

VARIABLES NOT IN THE EQUATION

VARIABLE	BETA IN	PARTIAL	TOLERANCE	F
WHITE	0.07345	0.14122	0.92199	11.456
PHD	---	---	0.0	---
ASSOC	0.13828	0.22889	0.68348	31.128
ASST	-0.16078	-0.22889	0.50556	31.128

```
DEPENDENT VARIABLE                          SALARY SALARY 1975
VARIABLE(S) ENTERED ON STEP NUMBER 5..      ASSOC
INDICATOR VARIABLE                          I:ASSOCIATE AGENT 1975

MULTIPLE R                0.87385
R SQUARE                  0.76362
ADJUSTED R SQUARE         0.76152
STANDARD ERROR            1214.54711

ANALYSIS OF VARIANCE    DF     SUM OF SQUARES        MEAN SQUARE        F
REGRESSION               5.    2682827168.60466    536565433.72093   363.74243
RESIDUAL               563.     830495192.82595      1475124.67642

              VARIABLES IN THE EQUATION
VARIABLE       B         BETA      STD ERROR B       F
TENURE      54.29896   0.19487       8.35543      42.231
CHM         5391.334   0.80228     226.01101     569.027
AGENT       2473.263   0.49710     169.95906     211.764
MS          937.0254   0.13944     140.61547      44.405
ASSOC       973.4755   0.13828     174.48178      31.128
(CONSTANT)  9590.787

              VARIABLES NOT IN THE EQUATION
VARIABLE    BETA IN    PARTIAL    TOLERANCE     F
WHITE       0.06117    0.12009     0.91098     8.224
PHD          ---        ---        0.0          ---
ASST         ---        ---        0.0          ---
```

```
VARIABLE(S) ENTERED ON STEP NUMBER 6      WHITE
INDICATOR VARIABLE                        I:WHITE

MULTIPLE R                0.87580
R SQUARE                  0.76702
ADJUSTED R SQUARE         0.76454
STANDARD ERROR            1206.82941

ANALYSIS OF VARIANCE    DF     SUM OF SQUARES        MEAN SQUARE        F
REGRESSION               6.    2694804647.82284    449134107.97047   308.37863
RESIDUAL               562.     818517713.60776      1456437.21283

              VARIABLES IN THE EQUATION
VARIABLE       B         BETA      STD ERROR B       F
TENURE      59.06173   0.21197       8.46689      48.659
CHM         5221.188   0.77696     232.28020     505.259
AGENT       2404.438   0.48327     170.57593     198.697
MS          898.5532   0.13371     140.36453      40.980
ASSOC       918.8206   0.13052     174.41745      27.751
WHITE       394.7963   0.06117     137.66904       8.224
(CONSTANT)  9291.513

              VARIABLES NOT IN THE EQUATION
VARIABLE    BETA IN    PARTIAL    TOLERANCE     F
PHD          ---        ---        0.0          ---
ASST         ---        ---        0.0          ---
```

F-LEVEL OR TOLERANCE-LEVEL INSUFFICIENT FOR FURTHER COMPUTATION

SUMMARY TABLE

VARIABLE	MULTIPLE-R	R-SQUARE	RSQ CHANGE	SIMPLE R	B	BETA
TENTURE	0.68522	0.46953	0.46953	0.68522	59.06273	0.21197
CHM	0.80749	0.65203	0.18251	0.67609	5221.188	0.77696
AGENT	0.85598	0.73270	0.08067	0.17086	2404.438	0.48327
MS	0.86634	0.75055	0.01785	0.29322	898.5532	0.13371
ASSOC	0.87385	0.76362	0.01307	0.27590	918.8206	0.15052
WHITE	0.87580	0.76702	0.00341	0.17511	394.7963	0.06117
(CONSTANT)					9291.513	

Notes

In the *Bazemore* case, the Supreme Court, in reversing the court of appeals' decision rejecting the multiple regression model, articulated a sensible approach to the question of omitted variables when "the major factors" are represented in the model:

> The Court of Appeals erred in stating that petitioners' regression analyses were "unacceptable as evidence of discrimination," because they did not include "all measurable variables thought to have an effect on salary level." The court's view of the evidentiary value of the regression analysis was plainly incorrect. While the omission of variables from a regression analysis may render the analysis less probative than it otherwise might be, it can hardly be said, absent some other infirmity, that an analysis which accounts for the major factors "must be considered unacceptable as evidence of discrimination." Ibid. Normally, failure to include variables will affect the analysis' probativeness, not its admissibility.

Id., 478 U.S. at 400. In a footnote the Court added that "there may, of course, be some regressions so incomplete as to be inadmissible as irrelevant; but such was clearly not the case here." *Id.* n.15.

The Court also indicated in a footnote that it is not enough merely to object to a regression on the ground that some factor had been omitted, and suggested that the party challenging a regression must demonstrate that the omitted factors would cause the disparity to vanish:

> Respondents' strategy at trial was to declare simply that many factors go into making up an individual employee's salary; they made no attempt that we are aware of—statistical or otherwise—to demonstrate that when these factors were properly organized and accounted for there was no significant disparity between the salaries of blacks and whites.

Id., 478 U.S. at 404, n. 14. The Court's prescription for rebuttal was applied by the Second Circuit in *Sobel v. Yeshiva University*, 839 F. 2d 18 (2d Cir. 1988); see Section 14.5.1.

13.6.2   *Public school financing in the State of Washington*

Public schools in Washington are financed by a combination of state funds and special levies voted and assessed on a district-wide basis. The state constitution provides that "it is the paramount duty of the state to make ample provision for the education of all children residing within its borders." In 1978, the Washington Supreme Court held, in a suit by the Seattle School District and others, that this provision required the legislature to define and fully fund a program of basic education. *Seattle School District v. State of Washington*, 90 Wash. 2d 476 (1978). The effect of this decision was to increase state funding and to reduce special levies by about two-thirds. Nevertheless, in 1981 the Seattle School District and other plaintiffs returned to court to complain that the state was spending too little on basic education. The school districts argued for increases in the scope of services, in staff salaries, in teacher-pupil ratios, and in budgets for equipment and supplies.

Student outcomes—as measured by scores on standardized tests—tend to be higher in more affluent districts that exact special levies and spend more on education. Is this relationship a direct result of the amount spent on education, or is it due to other correlated factors?

To answer that question, the state introduced an education production function study for Washington similar to studies conducted in many other states. The usual method is to regress a measure of performance (such as standardized test scores) on three groups of explanatory factors: *background variables* (such as parents' educational level, occupational level, and family income); *peer group variables* (such as percentage of minority students, social class composition of the school, ability composition of classmates); and *school resource variables* (such as class size, teacher postgraduate education, teacher experience, teacher salaries, and spending per pupil).

In the Washington study, the data were collected by district in the 88 districts that had an enrollment of at least 2,000 students (accounting for 83% of Washington's total public school enrollment).

The results of the regression study are shown in Table 13.6.2.

Questions

1. If the model correctly measured the factors causing differences in district reading scores, what conclusions relevant to the case would you draw from the regression study?

2. The model included prior test score as an explanatory variable. What is the relative importance of that variable compared with others in the equation?

3. What criticisms of the model would you have based on the inclusion of the prior test score variable?

TABLE 13.6.2. Regression for 1980 4th-grade language scores

	Estimated coefficient	t	Variable mean	Variable standard deviation
Dependent Variable				
4th-grade language	—	—	60.14	8.63
Background Variables				
Occupational index (%)	0.458	3.08	20.91	6.67
Median income ($000's)	−0.141	−0.59	2.07	3.76
Peer Group Variables				
Title I enrollment (%)*	−58.47	−4.02	0.09	0.06
Hispanic enrollment				
(1 if over 5%)	-3.668	−1.73	0.17	0.38
Logarithm of 4th-grade				
enrollment	−2.942	−2.34	5.67	0.53
Prior test score				
(district average,				
same grade, 1976)	0.200	3.60	53.40	12.31
School Variables				
Administrator-teacher ratio	17.93	0.42	0.09	0.01
Pupil-teacher ratio	0.689	0.42	28.35	3.63
Certificated staff-pupil ratio	−0.403	−0.27	0.45	0.50
Teacher experience:				
Level A (%) (less				
than three years)	−0.0614	−1.30	69.03	16.61
Level B (%) (more				
than six years)	−0.0382	−0.58	14.64	12.71
Constant	70.67	6.08	—	—
Standard error of regression:	5.57			
R^2	0.64			

* Refers to Title I of the Elementary and Secondary Education Act of 1965, 20 U.S.C. §§2701 (1982), which provided for financial assistance to school districts with concentrations of children from ages 5 to 17 in families below the poverty level as defined in the 1970 census.

Source

John Pincus & John E. Rolph, *How Much is Enough? Applying Regression Analysis to a School Finance Case*, in *Statistics and the Law* 257 (M. DeGroot, S. Fienberg & J. Kadane, eds., 1986).

Notes

In the U.S. Office of Education's groundbreaking study, *Equality of Educational Opportunity* (1966), mandated by the Civil Rights Act of 1964, J.S.

Coleman and others measured the relation between school factors and achievement on various types of tests. They found that (i) despite the wide diversity of school facilities, curricula, and teachers, and despite the wide diversity among student bodies in different schools, over 70 percent of the variation in achievement for each racial group was variation within the same student body; (ii) school-to-school variations in achievement at the beginning of grade 1 and compared with later years indicated that the component due to school factors was far smaller than that attributable to family background differences; and (iii) schools had an important effect on the achievement of minority group students.

The Coleman report has been cited to support the proposition that variations within a range of pupil-teacher ratios and higher teacher salary schedules do not affect the quality of education, *San Antonio School District* v. *Rodriguez*, 411 U.S. at 46, n.101 (see Section 1.4.2); that while expenditure by itself may not "matter," social composition of the student body does matter, *Commonwealth of Virginia* v. *United States*, 386 F. Supp. 1319, 1326 (D.C. 1974), *aff'd*, 420 U.S. 901 (1975); and that because attributes of other students account for far more variation in achievement by minority group children than do attributes of school facilities and slightly more than do attributes of staff (Coleman Rept. at 9), "achievement of Negro students was increased with the increasing degree of integration." *Johnson* v. *San Francisco Unified School District*, 339 F. Supp. 1315, 1331 (N.D. Calif. 1971).

Since the Coleman report, the relation between expenditure for education and student achievement has been much studied and debated. For meta-analyses of the conflicting studies, see Gary Burtless, ed., *Does Money Matter? The Effect of School Resources on Student Achievement and Adult Success* (1996).

In 1989, as part of the settlement of a desegregation case, 15 schools in Austin, Texas, serving low-income minority-group children, received $300,000 each, above normal school spending, for a period of five years. At the end of four years, achievement in 13 of the schools remained extremely low, as measured by state-wide tests, but the improvement in two schools was extraordinary. In those two schools, attendance had risen to the highest in the city, and test scores had risen to the city average, even though median family income for the families with children in those schools remained about $12,000. The difference was that, in those two schools, the principals made a package of changes–including reducing class size, changing instructional techniques, investing in raising student attendance, and increasing parental involvement– whereas in the other 13 schools the only change was reduction in class size. The authors pointed out that, in the statistical studies, data to capture these inputs were not available, and even if they had been, the complex interaction effects would not have been explored in the simple models commonly in use. Richard J. Murnane and Frank Levy, *Evidence from fifteen schools in Austin, Texas*, in Gary Burtless, ed., *Does Money Matter?* 93 (1996).

13.6.3   Public school financing in Pennsylvania

The Pennsylvania Association of Rural and Small Schools (PARSS) brought suit against the Commonwealth of Pennsylvania, attacking the system of local school district funding. The system relies substantially (although not wholly) on the local property tax on residential and nonresidential property. Plaintiff argued that districts with smaller property bases and low personal incomes are disadvantaged in a way that violates the Pennsylvania constitution's "education" clause, which places responsibility on the state "for the maintenance and support of a thorough and efficient system of public education."

There are substantial inter-district differences in amounts raised for education by the property tax. State and federal aid to poorer districts to some extent "flatten" the differences in spending per pupil that would exist without such help, but differences remain, and the lawsuit raised the question whether the remaining differences had educational significance. To help answer that question, the state commissioned a study by a statistician, Dr. William Fairley, of student performance on standardized tests. Using district data, Dr. Fairley performed a multiple regression study, in which student achievement in standardized tests was regressed on various district factors and two school factors, which he selected. The data and results for one model are in Table 13.6.3.

Pennsylvania has 502 school districts. Data for a few districts had missing values, so the regression model was estimated from the remaining 496 districts. The dependent and independent variables are district medians. The dependent variable is the average score in the district for the 11th grade PSSA test in 1995.

The eight independent variables are:

Parental Income = personal income per capita in the district in 1995;

Bachelor's Degree % = percent of persons in the district over age 18 having a bachelor's degree or higher in 1992–93;

Poverty % = percent of population in the district below poverty level income in 1994;

Instructional Expense = actual instructional expense in 1995 per weighted pupil (the weights reflect the differential costs of elementary and secondary education);

Score 89 = median reading and math scores in 5th grade in 1989 (presumably largely the same students as the 11th graders taking the test in 1995);

Plaintiff = 1 if district is a plaintiff and 0 if not;

Rural % = percent of district population living in rural areas as of 1994; and

Size (log) = size of district (measured as the log of a weighted per pupil figure in 1995).

TABLE 13.6.3. 1995 PSSA test model for 496 Pennsylvania school districts

Col #	1	2	3	4	5
Variable	Coeff.	P-value	Q1	Q3	Columns: $(4 - 3) \times 1$
Parental income ($)	0.0014	0.139	$9,319	$14,890	7.7994
Bachelor's degree %	5.0862	0.000	8.3	16.72	42.8258
Poverty %	−1.0474	0.000	12.62	79.1	−69.6312
Instructional expense	−0.021	0.000	$3,709	$4,598	−18, 669
Score 89	2.845	0.000	77.36	87.35	28.42155
Plaintiff	1.3554	0.839	0	1	1.3554
Rural %	0.248	0.007	8.4	100	22.7168
Size (log)	0.0928	0.984	7.46	8.32	0.079808

The regression results are shown in Table 13.6.3. Columns 3 and 4 are the first and third quartiles for the input variables and Column 5 is the effect over the interquartile range of the input variable on the dependent variable.

Questions

1. Do the model results help defendant's case?

2. How does the inclusion of the Score 89 variable affect the interpretation of the model?

Source

William B. Fairley, *Spending and economic base in Pennsylvania school districts*, Report to the Pennsylvania Department of Education and Office of Attorney General (November 3, 1995) in *Pennsylvania Association of Rural and Small Schools (PARSS) v. Ridge*, No. 11 M.D. 1991.

13.7   Confidence and prediction intervals

Important applications of the regression model are to *estimate* the value of the regression mean for a given X, or to *predict* the value of a future observation at a given X. We discuss these applications for simple linear regression.

Errors in estimation of the regression mean arise from two sources: error in estimation of the overall mean by the sample mean $\bar{Y}$, and error in the estimation of the slope coefficient by the OLS estimate $\hat{b} = r \cdot s_y/s_x$. It can be shown that these two sources of error are uncorrelated, so that the variance of the estimate $\hat{E}[Y|X] = \bar{Y} + \hat{b}(X - \bar{X})$ about its mean $E[Y|X]$ is equal to the sum of the variance of $\bar{Y}$ about EY, which is σ^2/n, and the variance of $\hat{b}(X - \bar{X})$ about its mean $b(X - \bar{X})$, which is $\sigma^2(X - \bar{X})^2/\sum(X_i - \bar{X})^2$. Thus,

$$\text{Var}\,\hat{E}[Y|X] = \sigma^2 \left(\frac{1}{n} + (X - \bar{X})^2 / \sum (X_i - \bar{X})^2 \right).$$

The variance of the errors, σ^2, may be estimated by replacing σ^2 by the residual mean square.

If the errors are normally distributed, a 95% confidence interval for $E[Y|X]$ is given by

$$\hat{E}[Y|X] \pm t_{\nu,.05} \cdot (\text{ estimated Var}\,\hat{E}[Y|X])^{1/2},$$

where $t_{\nu,.05}$ is the two-sided, 5% critical value of Student's t-distribution with $\nu = n - 2 = \text{df}_e$ degrees of freedom.

$\hat{a} + \hat{b}X$ is also a predictor of a future observation of Y at X; we will denote it by $\hat{Y}(X)$ to distinguish it from $\hat{E}[Y|X]$. The estimate $\hat{Y}(X)$ has minimum mean squared error $E(Y - \hat{Y})^2$ among all predictors $\hat{Y}$ that are linear functions of X. Prediction differs from estimation in that, in addition to the sources of sampling variability previously identified corresponding to uncertainty regarding the location of the regression mean, there is also the inherent variance σ^2 in any single observation about its regression mean even if $E[Y|X]$ were known exactly. Thus, for a single future observation at X, the mean squared error of prediction becomes

$$\text{Var}\,\hat{Y}(X) = \sigma^2 \left(1 + \frac{1}{n} + (X - \bar{X})^2 / \sum (X_i - \bar{X})^2 \right).$$

To predict the mean of m future observations at X, the leading term 1 in this formula is replaced by $1/m$. As before, the mean squared error of prediction may be estimated by replacing σ^2 by the residual mean square. A 95% prediction interval for the future value (or mean of the future values) at X is obtained as above, replacing the variance estimate of $\hat{Y}(X)$ by the appropriate estimated mean squared error of prediction. Note that increasing the sample size reduces the variability of the estimate of the regression mean, but the 95% prediction interval for a new observation will always be at least as wide as two standard errors of regression to account for inherent variability of the dependent variable.[15]

The relationship between confidence intervals and prediction intervals for the regression mean is illustrated by a case involving cash and futures wheat prices on the Chicago Board of Trade. When those prices rose dramatically during the last five days of trading in May 1963, the Department of Agriculture accused a firm of market manipulation. The firm argued that wheat prices had been artificially depressed and that the dramatic price rise was attributable to normal supply and demand factors. The staff of the Department of Agriculture presented a regression of the cash price for wheat on supply and demand factors; the study estimated a $2.13 price for May with a confidence interval for the regression mean of $2.10–$2.17. Because the actual average price for May was $2.136—very close to the regression mean and within the confidence interval—the study was successfully used to rebut the claim that wheat prices had been abnormally depressed. Alternatively, the consistency of these prices in the last few days of trading with normal supply and demand factors might have been tested by comparing those wheat prices with a prediction interval for each day or for the five-day averages.

See Michael O. Finkelstein, *Regression Models in Administrative Proceedings*, 86 Harv. L. Rev. 1442, 1453–1455 (1973), reprinted in *Quantitative Methods in Law* 211.

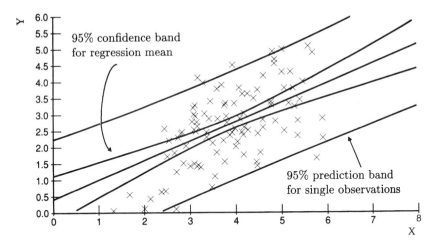

FIGURE 13.7. Confidence and prediction bands for the regression mean

---

[15]For either estimation or prediction, the farther away X is from the mean of the already observed X_i, the less precise the estimate. The most accurate predictions are for values near the mean of the X_i. The danger of extrapolation far from the body of data, and the increase in uncertainty of estimation of $E[Y|X]$ as X moves away from the center of the data, are evident in Figure 13.7 which shows a 95% confidence band for the regression mean of the form $\hat{E}[Y|X] \pm t_{v,.05} \cdot (\text{estimated Var} \hat{E}[Y|X])^{1/2}$, for a range of X values.

13.7.1   *Projecting fuel costs*

The Puerto Rico Maritime Shipping Authority, a carrier, filed for a rate increase with the Federal Maritime Commission. The Authority's forecast of rising fuel prices for 1981 was the single most significant item in its overall cost-based presentation, accounting for over half of its requested increase. (This was the pro forma test year for which the FMC required projections of revenues and costs; new rates would presumably be in place for a far longer period.) The carrier's projection relied on a straight line derived from least squares regression based on 1980 quarterly data. The quarterly fuel prices in dollars per barrel for 1980 were 19.26, 20.60, 21.39, and 28.03.

Questions

1. Using OLS, what are the projected quarterly fuel costs for 1981?

2. Assuming the regression model is correct, is the slope coefficient statistically significant?

3. Assuming the regression model is correct, do the regression estimates rule out a substantial flattening of fuel price increases in 1981?

4. What factors would lead you to doubt the model?

Source

Puerto Rico Maritime, etc. v. Federal Maritime Com'n, 678 F.2d 327, 337–42 (D.C. Cir. 1982)

13.7.2   *Severance pay dispute*

Able Company bought Baker Company under an agreement by which Able was to provide severance and related benefits for Baker employees involuntarily terminated by Able, and to do so within a specified period after the acquisition. These severance benefits were to be equivalent to the benefits provided under Baker's personnel policies. A Baker corporate policy paper issued before the acquisition stated that severance pay was based on three factors: (1) age, (2) length of service (using pension years of service), and (3) salary grade/position level.

When Able sold certain of the Baker assets, it laid off 200 employees. They received an average of 14.73 weeks of severance pay. One of these employees, claiming four to six weeks of severance pay for each year of service, filed a class action on behalf of all terminated employees. At the time of severance, plaintiff was age 45, had a salary grade of 20, seniority of 9 years, and received 18 weeks of severance pay.

To determine whether or not to settle and to prepare for a motion for class certification, Baker asked its statistician to determine whether plaintiff's weeks

TABLE 13.7.2. A regression model for weeks of severance pay

Variable	Estimate	t
Constant	14.221	14.771
Grade—12	0.526	5.227
Age—40	0.372	2.886
Seniority—5	0.944	4.410
R^2		0.426
St. err. reg. (root MSE)		9.202
Overall F		24.010
Model df		3
Error df		97
Sample size		101

	Weeks	Grade	Age	Seniority
Mean	15.859	15.000	40.286	4.949
St Dev.	11.964	9.302	7.852	4.734
Correlations				
Weeks of severance	1.000	0.395	0.432	0.426
Grade		1.000	0.102	−0.105
Age			1.000	0.392
Seniority				1.000

The dependent variable is number of weeks of severance pay.
Grade—12 is salary grade approximately centered by subtracting 12.
Age—40 is age in years approximately centered by subtracting 40.
Seniority—5 is seniority in years approximately centered by subtracting 5.

of severance pay followed the pattern of severance payments before the acquisition. Responding to this request, the statistician regressed weeks of severance pay on grade, age, and seniority with results shown in Table 13.7.2.

Questions

Compute confidence and prediction intervals at the means of the explanatory factors and at plaintiff's values of those factors, and answer the following questions.

1. Was the plaintiff's severance consistent with the earlier pattern? [Point of information: At the plaintiff's levels of the explanatory factors, the standard error of the regression estimate is $(0.0214 \cdot MSE)^{1/2}$.]

2. Is the overall level of severance consistent with the earlier overall level?

3. What relevance does the regression result have for the class certification issue?

TABLE 13.7.3. Machine votes and absentee ballots in 21 Pennsylvania senatorial elections

Election Y/Dist	Machine votes Democrat	Machine votes Republican	Machine votes difference (D-R)	Absentee votes Democrat	Absentee votes Republican	Absentee votes difference (D-R)
82/2	47767	21340	26427	551	205	346
82/4	44437	28533	15904	594	312	282
82/8	55662	13214	42448	338	115	223
84/1	58327	38883	19444	1357	764	593
84/3	78270	6473	71797	716	144	572
84/5	54812	55829	−1017	1207	1436	−229
84/7	77136	13730	63406	929	258	671
86/2	39034	23363	15671	609	316	293
86/4	52817	16541	36276	666	306	360
86/8	48315	11605	36710	477	171	306
88/1	56362	34514	21848	1101	700	401
88/3	69801	3939	65862	448	70	378
88/5	43527	562721	−13194	781	1610	−829
88/7	68702	12602	56100	644	250	394
90/2	27543	26843	700	660	509	151
90/4	39193	27664	11529	482	831	−349
90/8	34598	8551	26047	308	148	160
92/1	65943	21518	44425	1923	594	1329
92/3	58480	12968	45512	695	327	368
92/5	41267	46967	−5700	841	1275	−434
92/7	65516	14310	51206	814	423	391

13.7.3   Challenged absentee ballots

In a 1993 senatorial election held in District 2 in Philadelphia, Pennsylvania, there were allegations of fraud in the count of the absentee ballots. In the machine-counted ballots, the Democrat received 19,127 votes and the Republican received 19,691 votes, the latter candidate therefore winning the machine ballots. But when the absentee ballots were counted, the Democrat received 1,396 votes while the Republican received only 371 votes; thus, the Democrat won the election. An action was brought to set aside the election on the ground of fraud in the absentee ballots. In the preceding 21 senatorial elections in Pennsylvania on even years the data on machine votes and absentee ballots were as shown in Table 13.7.3.

Questions

1. Plot the difference between the Democrats and Republicans in the absentee ballots against the same difference in machine-counted ballots for the 21 elections listed above. Is there a pattern? Plot the point for the 1993 challenged election. Does it appear to be part of the pattern?

2. Regress the difference between the Democrats and the Republicans for the absentee ballots on the difference for the machine ballots. Compute a 95% prediction interval for the predicted value of the absentee ballots given the value for difference in the machine ballots in the 1993 election. Is the actual value of the difference in the absentee ballots in the 1993 election within that interval? What conclusion do you reach?

3. Are the computations of the prediction interval thrown into doubt by heteroscedasticity (see p. 396) in the data?

Source

Marks v. *Stinson*, 1994 U.S. Dist. Lexis 5273 (E.D. Pa. 1994) (Report of court-appointed expert O. Ashenfelter).

13.8   Assumptions of the regression model

Least squares calculations for the standard multiple regression model rest on four key assumptions about the error terms. If these assumptions are unwarranted, the OLS equation may be biased or inconsistent, or the usual computations of sampling variability may seriously misstate true variability and consequently misstate the statistical significance and reliability of the results. Let us consider these critical assumptions and the consequences of their failure in some detail.

First assumption: The errors have zero mean

The first assumption is that the expected value of the error term is zero for each set of values of the explanatory factors. In that case, regression estimates are unbiased; that is, if repeated samples are taken from the model and regression equations estimated, the overall average of the estimated coefficients will be their true values. When this assumption is not true, the estimates are biased, and if bias does not approach zero as sample size increases, the estimates are inconsistent. The assumption is violated whenever a correlation exists between the error term and an explanatory factor. This usually occurs because (i) some important explanatory factor has been omitted and that factor is correlated with factors in the equation; (ii) the equation is based on time-series data, a lagged dependent variable is included as a regressor, and there is serial correlation in the error term; or (iii) the equation belongs in a system of equations where one of the explanatory factors is determined within the system. We deal with the first of these situations here; the second is described below; the third situation is described in Section 14.6.

Omitted variable bias has been discussed in Section 13.5; a simple example here will illustrate the point. Suppose an employer gives employees yearly bonuses in varying amounts averaging $500 for each year of service. A regression of bonus on years of service would have a coefficient of $500 for the years-of-service variable. Suppose, further, that managers receive a $1,000 bonus in addition to the year-of-service amount. The increase in bonus per year of service is still $500, but the model is now misspecified because the important factor of managerial status has been omitted.

For a cohort of employees, the expected value of the error term in the earlier years (when no one is a manager) is zero, but becomes positive in the later years as the model predictions become systematically too low. If the regression of bonus on years-of-service is estimated from data, the coefficient of the years-of-service variable exceeds $500 because the least-squares criterion adjusts the estimate upward to explain as closely as possible the rise in bonuses due both to increase in years of service and the managerial bonus. Since the correct value for the years-of-service coefficient is still $500, the estimate is biased. A correctly specified model has a $500 coefficient for the years-of-service variable and a $1,000 coefficient for the manager status variable (coded 1 for manager, 0 for non-manager). Correlation between an explanatory factor and the error term arising from the omission of a necessary variable leads, therefore, to a biased estimate of a regression coefficient.

Second assumption: The errors are independent

The second assumption, that the errors are statistically independent across observations, is important for gauging the variability of sampling error. Independence means that knowing the error in one observation tells nothing about errors in other observations, i.e., the distribution of the error for one observation is unaffected by the errors for other observations. Errors that are not independent are said to be correlated. If the data have a natural sequence and correlations appear between errors in the sequence, they are said to be *autocorrelated* or *serially correlated*.

The problem of correlation of errors is most noticeable when the data have a natural ordering, as in a time series, but is by no means confined to that situation. It exists whenever the regression estimates tend to be systematically too high or too low for groups of related values of explanatory variables. For example, as discussed in Section 13.5 at p. 378, the plaintiffs in *Cuomo v. Baldridge* proposed to make estimates of the census undercount of the population in subnational areas using a regression equation estimated from a post-enumeration sample. The dependent variable was the undercount and the explanatory variables were percent minority, crime rate, and percent conventionally enumerated by face-to-face interviews. A statistician appearing for the Census Bureau pointed out that, because there were undoubtedly omitted explanatory variables, the equation was likely to have correlated errors for types of communities—such as inner cities—even though there was no natural ordering of the observations.

Third assumption: The errors have constant variance

The third assumption is that the errors have constant variance about the regression value, irrespective of the values of the explanatory factors. If this condition is met, the data are homoscedastic; if not, they are heteroscedastic (from the Greek word *skedastikos*, "able to scatter").

There are two consequences of heteroscedasticity. One is that the OLS estimates of the regression coefficients are no longer most efficient. If some observations have larger errors than others, those observations with large random effects contain less reliable information than observations with small effects and should receive less weight in the estimation. OLS regression, which gives the same weight to each observation, fails to take heterogeneity into account and, as a result, may have a larger variance for its estimates than other methods. One way to correct for this defect is "weighted" least-squares which, like OLS, estimates coefficients by minimizing a sum of squared deviations of observed values from fitted values, but where each deviation is weighted inversely to the variance of the errors. Other cures, involving transformations, are discussed in Section 13.9 at p. 409.

A second consequence of heteroscedasticity is that the usual measures of precision are no longer reliable. When there is fluctuation in the variance of error terms, the usual calculation of precision uses, in effect, an average variance based on the particular data of the construction sample. If another sample of data concentrated its predictors in a domain of greater variance in the error term, the estimated precision from the first sample would be overstated. This difficulty may seem esoteric, but it is by no means uncommon, particularly when variables in the regression equation are so structured that the entire scale of the equation (and hence the residuals) increases systematically. This often occurs with salary data, and is one reason for use of the logarithmic transformation.

An example of heteroscedasticity arose when the New York Stock Exchange sought to defend its minimum commission structure before the Securities and Exchange Commission. Using cost and transaction data for a large number of firms, the Exchange regressed total expenses on number of transactions and number of transactions squared. The coefficient of the quadratic term (transactions squared) was negative and statistically significant, indicating that the cost per transaction declined as the number of transactions increased. These economies of scale, argued the Exchange, made fixed minimum commissions necessary to prevent cut-throat competition in which larger firms, with their cost advantage, would drive out smaller firms.

The Department of Justice's reply pointed out that, because the Exchange used total firm costs as the dependent variable and total commissions as the explanatory variable, the entire scale of the equation would grow with increasing firm size, thereby increasing the size of the regression errors. This heteroscedasticity cast doubt on the statistical significance of the quadratic term. That doubt was reinforced when average cost per transaction was substituted as the dependent variable, which kept the scale of the equation unchanged as firm size increased. In this transformed equation, the coefficient of the squared term was positive and not statistically significant. Fixed commissions were abolished by the Exchange on May 1, 1975.

Fourth assumption: The errors are normally distributed

Because of mathematical tractability, it is frequently assumed that data are normally distributed when in fact there are departures from the normal. Although arbitrary, the assumption that regression errors are normally distributed in repeated sampling for any given values of the explanatory factors is not as dubious as one might think. The key is the correctness of the model. If all important influences on the dependent variable are accounted for by the model, what is left are many small influences whose sum is reflected in the error term. The central limit theorem then assures us that, to a good approximation, the error term is normally distributed. In addition, moderate departures from normality do not seriously affect the accuracy of the estimates of statistical significance. Nevertheless, it is appropriate to check the residuals for non-normality. Marked skewness of residuals often indicates that a transformation of the dependent variable is in order. See Section 13.9.

Validating the model

Whenever the reliability of the estimates is important, it is essential to test for, or at least carefully consider, the correctness of the assumptions of the standard model we have discussed. Since the assumptions concern the errors, and these are unknown, one set of tests involves looking at the residuals.

The OLS protocol forces residuals to have the characteristics of the expected values of errors in the true model. One characteristic is that the positive and negative residuals balance out, so that their algebraic sum is always exactly zero. Thus, if a dependent variable is regressed on a single explanatory factor, the estimated regression line will pass through the point of means of the data, $(\bar{X}, \bar{Y})$, with observations of the dependent variable falling above and below the regression estimate. In the true model, the average sum of errors over all different samples of data (the expected value) is zero, but, because of sampling variation, the sum of the errors for the particular sample used to estimate the equation will not necessarily be zero. Thus, the sum of residuals from the least squares equation is not necessarily identical to the sum of the errors with respect to the data used to estimate the equation.

A second characteristic is that residuals have exactly zero correlation with explanatory factors. (If there were a nonzero correlation between the residuals and an explanatory factor, the fit of the equation could be improved simply by adjusting the coefficient for that factor.) In the true model, errors have zero correlation with the explanatory factors. As before, though, this represents an expectation over the theoretical universe of data; because of sampling variation, some sample correlation between the true errors and explanatory factors can exist within the data from which the equation is estimated. In addition, while the residuals have zero correlation with explanatory factors, so that the best fitting *straight* line through a scatterplot of residuals vs. explanatories would have zero slope, nothing prevents the residuals from having a strong *nonlinear* dependence on the explanatory factor, as discussed below.

A third characteristic is that, because residuals are constrained to sum to zero, they must be negatively correlated with each other in any sample of data. (Selecting a positive residual increases the probability that the next selection will be negative, because positive and negative residuals must balance in the sample.) In the true model, the infinite universe of errors must also average to zero, but the errors are uncorrelated with each other in any finite sample of data.

Since the residuals must have these characteristics irrespective of the model, their existence is not evidence that the model is correct. The fact that residuals sum to zero does not show that the model is unbiased; absence of correlation between residuals and the explanatory factors does not mean that the model is correctly specified. However, we can look to residuals to indicate outliers, non-linearity, autocorrelation, and violations of the assumptions of homoscedasticity and normality. We discuss these subjects below.

Plots of residuals

The general method for analyzing residuals is to inspect computer-generated plots of residuals against other quantities to detect violations of the assumptions and to compute certain statistics. Residuals are generally plotted in standardized form, i.e., according to the number of standard deviations from the mean, which is zero. Usually, values of the standardized predicted dependent variable or any explanatory factor are plotted along the horizontal axis, and standardized residuals along the vertical axis. There are four types of residual plots.

First, residuals are plotted against explanatory factors not in the regression equation—such as time, or a variable that is a product of two variables already in the equation, or the square of a variable already in the equation. The investigator examines such plots to see if there is a trend in the residuals with respect to the new variable; if so, that variable is a candidate for inclusion in the model.

Second, the residuals are plotted against each explanatory factor in the equation. Although the least squares equation ensures a zero linear correlation between the residuals and each explanatory factor, there may be a systematic nonlinear relationship between them, which would indicate that the model was misspecified by having the wrong functional form for some explanatory factors. For example, if the true relation between Y and X is that $Y = X^2 + \varepsilon$, but one does not know this and simply regresses Y on X (but not X^2) for values of X symmetrically located about 0, the result is a regression line with a zero slope because X and X^2 are uncorrelated. The residuals from the fitted line have zero correlation with X, but a parabolic (U-shaped) dependence on X. Inspection of the residual plot would indicate that a quadratic term in X was missing from the model.

Third, residuals are plotted against the predicted dependent variable to see whether the data are homoscedastic. A symptom of heteroscedasticity is a residual plot that steadily fans out or narrows with increasing predicted values. If the error variance is constant, the cloud of points should look like a horizontal football: the spread should be greatest at the center of the plot and smallest at the ends. (The reason is that, in order to avoid large residuals that would prevent minimization of the residual sum of squares, the regression tends to stick more closely to those values of the dependent variable that correspond to influential extreme values of the explanatory factors than to the less influential middle values. The variance of the corresponding residuals is thus smaller than that of centrally located residuals.)

Fourth, residuals are examined for outliers, that is, points that depart markedly from the pattern of the rest of the data. Such points may have a large effect on the regression equation. The solution to the problem of outliers is not simply to eliminate them automatically, but to subject them to detailed examination, possibly leading to their elimination if their accuracy is questionable.

An example of a residual plot revealing moderate heteroscedasticity is shown in Figure 13.8.

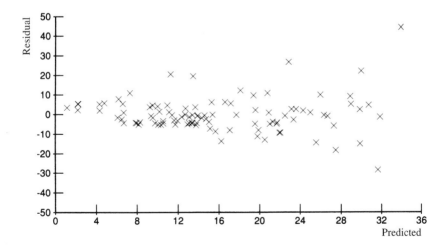

FIGURE 13.8. A plot of residuals

Tests of residuals

It is not always obvious from visual inspection of residual plots that one or more of the regression assumptions have been violated. Statistical tests provide an objective appraisal of residuals; the best known of these tests is the Durbin-Watson test for autocorrelation of the residuals.

The Durbin-Watson test is applied when there is a natural ordering of the data, as in a time series. The test assumes that the error term in the model for a given observation is equal to the preceding error times a correlation coefficient between 0 and 1, plus a random component. Of course, actual autocorrelations are likely to be more complex, with error depending not only on the immediately preceding error, but also, to a lesser extent, on more distant errors. Methods exist for analyzing such situations, but the Durbin-Watson test is simple and frequently used because the more complex autocorrelation structures often cause the residuals to fail the Durbin-Watson test as well.

The statistic used in the Durbin-Watson test is the sum of the squared differences between adjacent residuals divided by the sum of the squared residuals. If the correlation between successive errors is zero, the expected value of this statistic based on the residuals will be near two; as autocorrelation increases, the statistic approaches zero.[16]

The computed value of the observed test statistic is compared with tabulated critical values of the Durbin-Watson statistic, usually at the 1% or 5% levels of significance, under the hypothesis of no correlation in the errors. The results fall into one of three categories: if the statistic is close to 0, the hypothesis of zero correlation in the errors is rejected; if close to 2, the hypothesis is not rejected; if in the middle range, the test is indeterminate. See Appendix II, Table D.

Further Reading

N. Draper and H. Smith, *Applied Regression Analysis*, chs. 2 and 8 (3d ed., 1998).

---

[16] Assuming that the average absolute value of the errors remains the same in successive observations, the sum of squared differences between successive errors can be imagined as if all errors were of the same absolute magnitude, the only variation being whether they are positive or negative (i.e., above or below the regression estimate). Under the hypothesis of no correlation, half the differences would be zero (when successive errors are on the same side of the estimate) and half would be twice the error (when they are on opposite sides). The sum of the squared differences would therefore be four times the sum of the squared errors in half the cases, or two times the sum of the squared errors. Dividing this by the sum of the squared errors gives an expected value of two for the Durbin-Watson statistic under the hypothesis of no correlation. On the other hand, if there is high correlation between successive errors, positive errors would tend to follow positive errors, and negative errors to follow negative errors, so that the sum of the squared differences would tend toward zero. The Durbin-Watson statistic is approximately equal to twice the value of 1 minus the squared correlation coefficient between successive values of the errors.

David G. Kleinbaum & Lawrence L. Kupper, et al., *Applied Regression Analysis and Other Multivariable Methods*, 115–117 (1998).

13.9 Transformations of variables

Transformations are basic mathematical tools for re-expressing relations between variables in equivalent, but possibly more illuminating, forms. We focus on one-to-one transformations of data, which are such that, after applying the transformation, we could undo the transformation with its inverse, thereby regaining the original data. No information is lost by such transformations, nor is the purpose to distort the facts. We describe below several common transformations, the reasons for them, and the interpretation of the model coefficients in light of them.

Quadratic terms

The additive model with linear explanatory factors assumes that the effect on the dependent variable of a one-unit change in an explanatory factor is the same, regardless of the initial level of the explanatory factor. In many situations, however, this is unlikely to be true. For example, in an employment regression, salary increases tend to be greater per year of service for new employees than for those nearing retirement. On the other hand, profit per dollar of sales may rise with the number of sales, as fixed costs are spread over a greater number of transactions. To account for these kinds of changes, a regression model might include squared terms of the explanatory variables in addition to simple linear terms, e.g., $a+bx+cx^2$. If the coefficient of the squared term is negative, the change in the dependent variable per unit increase in the explanatory factor diminishes as the explanatory factor increases; if positive, the reverse is true. Thus, one would expect a positive coefficient for the years-of-service variable and a negative coefficient for a years-of-service-squared variable. This is typically the case. Quadratic terms also allow for reversals in trend in time-series regressions (see Section 14.1).

   Although the explanatory factors may be non-linear functions, if the regression model combines these terms in the usual way (multiplying by coefficients and adding), the model is still a linear regression model, which can be estimated in using OLS; the term "linear" refers to the dependence of the model on its parameters, not on the form of the explanatory factors. At other times, the relation cannot be expressed adequately or reasonably by a polynomial of fixed degree as, for example, when $y = a + bx^c$. That case requires non-linear regression analysis. Often, however, a relation that appears non-linear at first may be reduced to a linear relation by an appropriate transformation. This is true for multiplicative models, for which the log transform (discussed below) creates a linear relation. For another example, $y = x/(ax + b)$ can be re-expressed as $1/y = a + b(1/x)$, so that $1/x$ and $1/y$ are linearly related. The statistics of linear regression are simpler than that of non-linear regression, so it is preferable to use a linearizing transformation whenever possible. Thus, future events may be forecast by linear extrapolation, and the forecast reconverted to the original units by the inverse transform.

When a transformation is applied to a variable in a statistical model, the parameters of the model are generally transformed as well. For example, if a dependent variable Y is a count taking large positive values with mean μ, the square root transformation produces a variable $Y^{1/2}$ whose mean parameter is approximately $\mu^{1/2}$. Statistical hypotheses concerning parameters in the original problem can be reformulated as equivalent hypotheses in the new scale. Confidence intervals constructed in the transformed scale and then reconverted to the original scale are often more valid than those constructed from an inferior approximation in the original scale. Transformation may also aid in detecting subtleties of a model such as additivity of effects: it is easy to determine whether two lines are parallel, but the relation may not be obvious for the original curves.

Logarithmic terms

One of the most common transformations is to take the natural logarithm (denoted by log, sometimes by ln) of positive data; the regression equation is then estimated in linear form by OLS. A model in logarithmic form might be

$$\log Y = \log \alpha + \beta_1 \cdot \log X_1 + \cdots + \beta_k \log X_k + \text{error},$$

which is equivalent to a multiplicative model for the original data of the form

$$Y = \alpha \cdot X_1^{\beta_1} \cdots X_k^{\beta_k} \cdot \varepsilon,$$

where $\varepsilon = \exp(\text{error})$, the anti-log of the error.

As an example of a logarithmic transform, consider again salary and seniority. Since the qualitative relation is a rapid increase at the low end and slower at the high end, an alternative to the addition of a squared term for seniority is the form $\log \text{salary} = \alpha + \beta \log(1 + \text{seniority})$. (The unit is added to seniority before taking logs so that zero seniority before transformation becomes zero seniority after transformation.) This form produces a slower increase at the high end than by using a squared term, and also avoids the eventual downturn in predicted salary implied by a negative coefficient for the quadratic term.

Income distributions are another example. Typical income distributions are characterized by marked skewness and heavy tails at the high-income range. Summarizing such distributions is generally impossible without tabulations of percentiles, and statistical inferences based on ordinary standard deviation analysis are grossly inaccurate. If, however, the income distribution is normally distributed in the logarithmic scale (a "log-normal" distribution), then a complete summarization of the data is possible with the mean and standard deviation of log-income, and standard deviation analysis is then appropriate in the transformed scale. As an example, suppose we wish to find a 95% prediction interval for an individual income. The geometric mean of all incomes corresponds to the arithmetic mean μ of log incomes by the log transformation. In the transformed scale, the distribution is normal with mean μ and variance σ^2, say, and thus a 95% prediction interval for log-income is constructed as $(\mu - 1.96\sigma,\ \mu + 1.96\sigma)$. Applying anti-logs, $(e^{\mu - 1.96\sigma},\ e^{\mu + 1.96\sigma})$ is a 95% prediction interval for income. Note its asymmetry, reflecting the original asymmetry of the income distribution.

Linear models that assume normally distributed errors become applicable in non-standard cases when a transformation of the data can produce error terms that are approximately normal. Thus, for a model postulating $Y = \alpha e^X \varepsilon$, where the error term ε has a log-normal distribution with mean $\exp(\sigma^2/2)$, standard linear regression is applicable to the transformed model $\log y = \log \alpha + \beta X + e$, where the error term $e = \log \varepsilon$ has a normal distribution with zero mean and variance σ^2.

A closely related transformation appropriate specifically to percentages or proportions (in either dependent or explanatory variables) is the log-odds transform (sometimes called a folded log or "flog"). If x is a proportion strictly between 0 and 1, the transform is $y = \log[x/(1 - x)]$. For small x, the flog of x is approximately the same as $\log x$, while for x near 1, the flog is near $-\log(1 - x)$. This transform is particularly useful for variables that take on proportion values close to 0 and 1 because it makes their distribution more symmetrical. The same transformation is used in logistic regression (see Section 14.7).

In certain kinds of models, a comparison of two proportions is given by the odds ratio, $\Omega = [P_1/(1 - P_1)]/[P_2/(1 - P_2)]$ (see Section 1.5 at 36). The odds ratio is a number between zero and infinity, with the value 1 corresponding to $P_1 = P_2$. If $P_1 < P_2$, values of the odds ratio are constrained to lie between 0 and 1; if $P_1 > P_2$, the odds ratio can assume all values greater than 1. In large samples from two different populations we may estimate Ω as the odds ratio of sample proportions, say ω. Although the sampling distribution of ω is markedly asymmetrical, the sampling distribution of $\log(w)$ is nearly symmetrical and, in fact, approximately normal, with mean $\log \Omega$. With this transformation, approximate confidence intervals can be constructed. See Section 5.3 at 166.

Logarithmic or semi-logarithmic forms frequently are used in estimating the elasticity of response in econometric models, e.g., the effect of changes in price on sales. See Section 14.1.6 for an example.

When a logarithmic transformation is made, the meaning of the coefficients changes. Whereas in the original model a one-unit change in an explanatory variable X_i yields an average change of β_i units in the dependent variable, where β_i is the regression coefficient of X_i, in the logarithmic model a 1% change in the explanatory variable is interpreted as an approximate β_i percentage change in the dependent variable. This may be seen as follows. A 1% increase in a particular explanatory factor X_i is effected by multiplying X_i by 1.01; in a multiplicative model this multiplies the dependent vari-

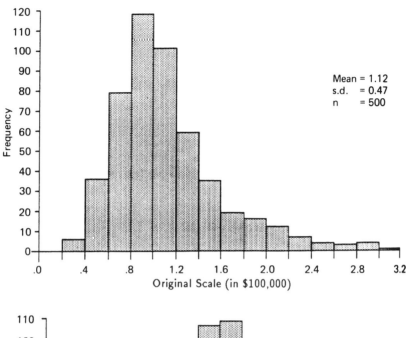

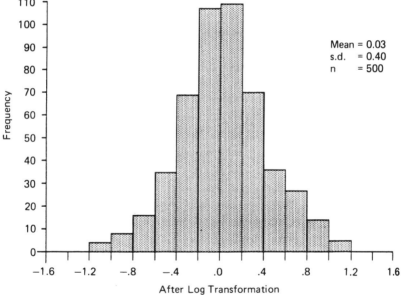

FIGURE 13.9. Sale prices for houses made more symmetrical by transformation into logs

able Y by 1.01^{β_i}. The relative change in Y (denote it "ΔY") is therefore $\Delta Y = [Y(1.01^{\beta_i}) - Y]/Y = 1.01^{\beta_i} - 1$. Hence $\Delta Y + 1 = 1.01^{\beta_i}$. Taking logarithms of both sides, we have $\log(1 + \Delta Y) = \beta_i \cdot \log(1.01)$. Since $\log(1+x) \approx x$ when x is close to 0, the foregoing expression becomes $\Delta Y \approx \beta_i \cdot (0.01)$. Expressing this in percentage points, we have $\Delta Y\% \approx \beta_i \cdot (0.01) \cdot 100\% = \beta_i\%$. Thus, we may say that in a multiplicative model (or its equivalent, a linear logarithmic model) a change of 1% in the explanatory variable X is accompanied by approximately a $\beta_i\%$ change in the dependent variable, when β_i is small.

Semi-logarithmic models are also useful in some circumstances. In one form of such models, the dependent variable is in logarithmic form, but the explanatory variable of interest is not. Such a model is appropriate when the explanatory factor of interest is itself a percentage. In that case, a unit (e.g., a percentage point) change leads to a *percentage* change in the dependent variable of $(e^{\beta_i} - 1) \cdot 100\%$ or approximately β_i 100% for small β_i. For example, a coefficient of $\beta_i = 0.02$ implies a $(e^{0.02} - 1) \cdot 100\% \approx 2\%$ change in the dependent variable per percentage point change in X_i.

When the dependent variable is a percentage, the reverse transformation may be more appropriate: the dependent variable remains in the original scale while the explanatory variables are transformed. In that case, each 1% change in the explanatory variable X_i corresponds to an additive change of $0.01\beta_i$ in the dependent variable.

Variance stabilizing transformations

The standard OLS estimates of regression coefficients are fully efficient when the error term has constant variance (homoscedasticity). It is not uncommon, however, for error terms to have greater variability for larger predicted values or at larger values of the independent variable. For example, counts often follow a Poisson distribution, in which the variance equals the mean. If such counts are the dependent variable in a regression model, the model is heteroscedastic. Although OLS estimates are unbiased in this case, they are not as precise as other available estimators. There are two approaches to this problem. The first transforms the dependent variable to achieve a constant variance in the transformed scale. For a Poisson distribution, the square root transformation $y^{1/2}$ achieves this goal, since the variance of the square root of a Poisson random variable with large mean is approximately 1/4, independent of the mean. Thus, we may entertain models of the form $Y^{1/2} = \alpha + \beta X + \varepsilon_1$, where the error term has constant variance. Transformations that achieve constant variance are called *variance-stabilizing* transformations. For binomial proportions, p, the variance-stabilizing transformation is $y = \arcsin \sqrt{p}$, with variance approximately equal to $1/(4n)$, independent of p, where n is the binomial sample size.

The second approach involves re-expression of the variables in the model to remove a parametric relationship in the variance of the error term. Suppose that examination of the residuals from a linear regression of y on x reveals that the variance of the residuals increases as the square of x, indicating a specific kind of heteroscedasticity. In this case the variance of the errors divided by x will be constant. This relation suggests dividing the original equation $Y = \alpha + \beta X + \varepsilon$ by X to obtain the new model $Y' = \beta + \alpha X' + \varepsilon'$, where

$Y' = Y/X$ and $X' = 1/X$, in which the error term ε' satisfies the assumption of homoscedasticity. This method is equivalent to estimation of the regression coefficients by the method of weighted least squares, where the weights are inversely proportional to the variances of the error term at each x. This method produces unbiased and efficient estimates of the regression coefficients if the assumed error structure (i.e., set of weights) is properly specified. Residual analysis is crucial in determining the success of any weighting system. The method of weighted least squares is applicable to a wide variety of regression problems, including logit and probit analysis (see Section 14.7).

On the problem of autocorrelation and the remedy of transformation to difference equations, see Section 14.1.

Further Reading

N. Draper & H. Smith, *Applied Regression Analysis*, ch. 13 (3d ed. 1998).

13.9.1   Western Union's cost of equity revisited

Questions

1. Calculate R^2 for the cost-of-equity data of Section 13.2.1 with variability-of-earnings both in logarithmic form and in the original scale.

2. What is the reason for the difference in R^2 between the models?

3. If goodness of fit were the only consideration, which model would be more appropriate for estimating Western Union's cost of equity?

13.9.2   Sex- and race-coefficient models for Republic National Bank

Blacks and women brought a class action against Republic National Bank claiming discrimination in hiring, promotion, and pay. To establish a prima facie case of pay discrimination, plaintiffs presented multiple regression studies in which the logarithm of salary was regressed on various sets of explanatory factors. Here was one set of factors used by the plaintiffs:

- Personal Characteristic Variables (Set D)
  - education (highest grade completed)
  - age (in years) [as a proxy for general labor market experience]
  - age squared
  - bank experience (in years) [using Republic National Bank hire dates]
  - bank experience squared

TABLE 13.9.2a. Plaintiffs' multiple-regression models for Republic National Bank salary levels in 1978

	#Employees	Personal characteristic variables	Job variables	Females	Blacks
Regression (2)	2,012	Set D	Not included	−0.3216**	−0.1985**
Regression (3)	1,806	Set D	Included	−0.0705**	−0.0948**

** Significant at the one-tailed 1%, level.

- Job Variables

    - bank officer (1 for bank officers, 0 otherwise)

    - other exempt (1 for employees in other exempt categories, 0 otherwise)

    - Hay points [A proprietary method for ranking jobs by content, which can be used in calculating their relative worth].

The regressions using these factors were computed separately for each year. The results for 1978 are shown in Table 13.9.2a.

To rebut plaintiffs' regressions, the bank's labor economist presented multiple regression studies based on data obtained from questionnaires submitted to a random sample of employees. The dependent variable was the logarithm of year-end salary. The 1978 results for white men and women are shown in Table 13.9.2b.

Questions

1. Interpret the results of plaintiffs' and defendant's regressions.

2. As attorney for defendant, what objections would you have to the explanatory factors used in plaintiffs' regressions?

3. As attorney for plaintiffs, what objections would you have to the study design and explanatory factors used in defendant's regressions?

Source

Vuyanich v. Republic Natl. Bank, 505 F. Supp. 224 (N.D. Texas 1980), *vacated*, 723 F.2d 1195 (5th Cir. 1984), *reh. denied*, 736 F.2d 160, *cert. denied*, 469 U.S. 1073 (1984). The district court opinion is a massive treatise on multiple regression that deals with many issues in addition to those discussed here.

TABLE 13.9.2b. Defendant's multiple-regression models for Republic National Bank 1978 year-end salary

Stratified by status at year-end 1978		
Status/variables Controlled Cumulatively	Female/male pay disparity (percent)	t
Status: Professional		
Sex (1=female), unadjusted	−25.75	−5.03**
Sex (1=female), adjusted cumulatively for:		
Service at RNB	−18.20	−3.42**
Highest educational level	−13.65	−2.71**
Banking major	−11.10	−2.12**
Length and type of prior experience	−6.63	−1.56
Non-productive time out of labor force	−6.00	−1.40
Investment in RNB career[1]	−4.22	−0.75
Career motivation[2]	−2.53	−0.46
Sample size 135		
Number of females 57		
Status: Non-professional		
Sex (1=female), unadjusted	−2.96	−0.89
Sex (1=female), adjusted cumulatively for:		
Service at RNB	−2.08	−0.73
Highest educational level	−2.13	−0.75
Banking major	−1.69	−0.60
Length and type of prior experience	−1.57	−0.57
Non-productive time out of labor force	−1.87	−0.67
Investment in RNB career[1]	−1.26	−0.44
Avg. ann. absence occasions from work	−0.69	−0.23
Career motivation[2]	−0.00	−0.00
Sample size 131		
Number of females 95		

** Statistically significant at the 1 percent level on a one-tailed test.

[1] Training programs and overtime worked after hire.

[2] Various subjective measures of interest in career at the bank.

CHAPTER 14

More Complex Regression Models

14.1   Time series

When observations of the dependent variable form a series over time, special problems may be encountered. Perhaps the most significant difference from the models discussed thus far is the use as an explanatory variable of the value of the dependent variable itself for the preceding period. A rationale for lagged dependent variables is that they account for excluded explanatory factors. Lagged dependent and independent variables may also be used to correct for "stickiness" in the response of the dependent variable to changes in explanatory factors. For example, in a price equation based on monthly prices, changes in cost or demand factors might affect price only after several months, so that regression estimates reflecting changes immediately would be too high or too low for a few months; the error term would be autocorrelated. A lagged value of the dependent variable might be used to correct for this. Note, however, that inclusion of a lagged dependent variable makes the regression essentially predict change in the dependent variable because the preceding period value is regarded as fixed; this may affect interpretation of the equation's coefficients. For previous examples, see Sections 13.6.2 and 13.6.3.

Regression estimates are unbiased in the presence of autocorrelation as long as the errors, autocorrelated or not, have an expected value of zero. However, when there is positive autocorrelation, the usual calculation seriously overestimates precision because the variability of residuals underestimates the true error variance.

To illustrate, suppose that a true regression model of price on time has expected values that, in the presence of autocorrelation, are either first consistently below and then consistently above the observed data, or vice versa; in repeated samples the overestimates and underestimates balance so that the

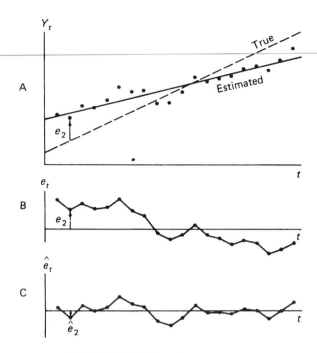

FIGURE 14.1. Effect of autocorrelation

equation is unbiased. Although the error is substantial, the OLS regression line is fitted to the points in a way that makes the residuals smaller than the error. In Figure 14.1 above, the OLS line passes through, rather than below and then above, the data points. An investigator unaware of the serial correlation would conclude that the regression model was quite precise, a false conclusion because, in another sample, the observations might first lie below the regression line, and then above it, reversing the prior pattern. Predictions would not be precise, and the new data would produce a very different estimate of the equation.

When the equation includes a lagged dependent variable, the existence of serial correlation in the error term may even result in biased estimates. The error at time t is of course correlated with the dependent variable at time t. If there is serial correlation in the error term, the error at t is correlated with the error at $t - 1$ and, since the error at $t - 1$ is correlated with the dependent variable at $t - 1$, it follows that the error at t is correlated with the dependent variable at $t - 1$. If the lagged dependent variable is included as an explanatory factor in the regression model, the error no longer has zero mean, and the correlation between regressor and error biases the estimated coefficients.

A remedy for serial correlation is based on the idea that the correlation can sometimes be eliminated if the regression is estimated from a *difference* equation: differences between successive observations of the dependent variable are regressed on differences between successive observations of the independent variables. A regression model based on single differences is called a *first difference* equation. In the so-called *generalized difference* equation, lagged values of both the dependent and explanatory factors are multiplied by an estimate, made from the residuals, of the correlation between successive error terms. The coefficients are usually estimated by maximum likelihood methods using iteration. This approach is sometimes referred to as the Cochrane-Orcutt method.

Further Reading

N. Draper & H. Smith, *Applied Regression Analysis* 179–203 (3d ed. 1998).

14.1.1   Corrugated container price-fixing

In an antitrust price-fixing case under the Sherman Act, damages are based on the difference between the prices paid by the plaintiff purchasers and the prices they would have paid in the absence of defendants' collusion. This difference is then trebled.

To estimate the "but for" prices in a case involving corrugated containers and containerboard, data during the conspiracy period (1965-1975) were used to estimate a regression relation between monthly average price of corrugated containerboard and the following explanatory variables: (1) price the previous month; (2) change in the cost of production from the previous month; (3) an index of the level of manufacturing output; and (4) wholesale price index for all commodities. The regression estimated for the conspiracy period (reflecting the effects of the conspiracy) was used to estimate prices in the post-conspiracy period (1976-79). These projected prices were higher than actual prices, with an average difference, as a percentage of the projected conspiracy price, of 7.8%. From this projection plaintiffs argued that the overcharge during the conspiracy period was at least 7.8%.

The specifications of the regression equation are given in Table 14.1.1. A plot of the actual and projected prices for containerboard is shown in Figure 14.1.1.

Question

As an expert for defendants, what procedures would you use to test the correctness of the regression model?

Source

In re Corrugated Container Antitrust Litigation, 756 F.2d 411 (5th Cir. 1985); Michael O. Finkelstein & Hans Levenbach, *Regression Estimates of Damages In Price-Fixing Cases*, Law & Contemp. Probs., Autumn 1983 at 145, reprinted in *Statistics and the Law* 93–95 (M. DeGroot, S. Fienberg, and J. Kadane, eds., 1986).

TABLE 14.1.1. Price of corrugated containerboard

Explanatory variable	Coefficient	Std. error	t
Intercept	0.732770	0.54660	1.34
Price of containerboard lagged by one-month	0.733244	0.05045	14.53
Change in cost of production from previous month	0.453225	0.10700	4.24
Level of manuf. output (1974 = 100)	0.012204	0.00659	1.85
Wholesale price index	0.245491	0.04726	5.20

The dependent variable is an index for the price of containerboard with price in 1974 = 100.
Monthly (1963:1 to 1975:12) observations ($n = 156$).

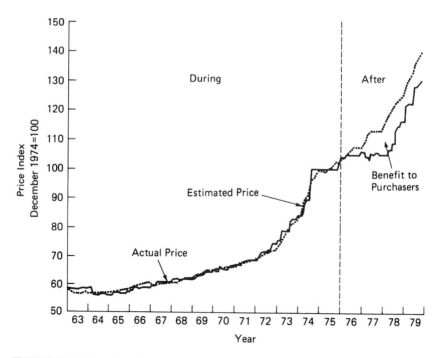

FIGURE 14.1.1. Actual and projected prices for corrugated containerboard during and after the conspiracy period

14.1.2 *Puppy Chow versus Chewy Morsels*

Alpo Petfoods and Ralston Purina are competing makers of puppy food. Beginning in December 1985 and continuing through October 1986, Ralston ran advertising in which it claimed that Purina Puppy Chow and Purina Chewy

Morsels had been reformulated to help lessen the severity of canine hip displasia ("CHD"). Contending these claims were false, Alpo sued for lost sales and profits. On the issue of damages, both sides introduced multiple regression models.

In one of Alpo's models, the dependent variable was Alpo's percentage share of 19 East Coast markets (where it sold most of its puppy food). In the period covered by the data, which was divided into 33 28-day periods between February 1, 1985, and July 17, 1987, this ranged between 16.1088% and 24.1604%. The independent variables were as follows:

- Total Alpo puppy food advertising expenditures by 28-day period.

- Total Alpo puppy food coupon redemption expenditures by 28-day period (a measure of consumer promotion expenditure).

- A dummy variable equal to 1 beginning in April 1985 when Purina Puppy Chow and Purina Chewy Morsels entered the market, and 0 prior to that.

- CHD ... A dummy variable equal to 1 during the 28-day periods between December 6, 1985, through October 10, 1986 (corresponding to the period of Ralston Purina's CHD advertising), and 0 outside that period.

- A dummy variable equal to 1 beginning in August 1986, the month in which a competing brand, Gaines Gravy Train Puppy Food, entered the market, and 0 before that.

- The price of ALPO Dry Puppy Food divided by the price of Purina Puppy Chow (excluding Chewy Morsels) based on the East Coast sales.

- Total Purina Puppy Chow and Chewy Morsels expenditures on advertising and promotion during the 28-day periods.

Using these variables and data for 33 28-day periods between February 1, 1985, and July 17, 1987, Alpo estimated a regression equation and reported a coefficient of -1.471 for CHD with a t-value of -2.238 and an associated P-value of 0.0344. The model had an R^2 of 0.697 and an adjusted R^2 of 0.612.

In one of Ralston Purina's models, the dependent variable was the same as in Alpo's, and the independent variables were similar, except that Ralston Purina added the following:

- A time trend variable (the number of months since February 1985) to capture underlying trends not captured by other explanatory factors.

- A dummy variable equal to 1 beginning in March 1986 to represent Purina's second reformulation of Puppy Chow and Chewy Morsels, and 0 prior to that.

- The number of puppy food brands on the market from the six leading competitors.

With these added variables, Purina reported a CHD coefficient of 1.832 with a standard error of 1.469.

Ralston Purina argued that Alpo's data and model were defective because: (1) Alpo counted as CHD advertising any advertisement that mentioned CHD no matter how fleeting the reference; (2) Alpo's model assumed that a given expenditure on advertising would yield the same dollar increase in sales regardless of the level of advertising, contrary to the economic law of diminishing returns for inputs; and (3) the model implicitly assumes that Ralston would not have run any advertising if it had not run the CHD advertisements.

Questions

1. Interpret the CHD coefficient in the two models.

2. What probably accounts for the difference in CHD coefficients in the two models? How would you test your surmise?

3. How would you resolve Ralston Purina's criticisms of Alpo's model?

Source

Alpo Petfoods, Inc. v. Ralston Purina Company, 997 F.2d 949, 955 (D.C. Cir. 1993) (expert reports).

14.1.3   Losses from infringing sales

Firm A sells a familiar consumer product and patents a new variation of that product. Sales of the variation are well established by 1972.

Beginning in 1981, Firm B begins selling a competing product that Firm A says infringes its patent. A sues B, claiming that its sales would have been much larger were it not for the infringer. B replies that a projection of A's sales from the pre-infringement period shows no lost sales. Sales data are shown in Table 14.1.3. In this table, "Firm A unit sales," expressed in millions, are total revenues divided by the Firm A price per unit. The seasonal indicator variable takes the value +1 for observations in the period May–October, and −1 in the period November–April of each 12-month period.

Questions

1. Estimate by OLS the regression coefficients of the linear regression of Firm A unit sales on Time and Seasonal indicator for the pre-infringement period. Use the estimate to project Firm A's 1981 sales based on the pre-infringement pattern.

2. Should other explanatory factors be included in the model?

TABLE 14.1.3. Sales data

Year	Bimonthly period	Firm A unit sales	Time	Seasonal indicator
1978	Jan/Feb	1.560	1	−1
	Mar/Apr	1.821	2	−1
	May/Jun	1.901	3	1
	Jul/Aug	2.106	4	1
	Sep/Oct	1.965	5	1
	Nov/Dec	1.875	6	−1
1979	Jan/Feb	2.005	7	−1
	Mar/Apr	2.057	8	−1
	May/Jun	2.139	9	1
	Jul/Aug	2.485	10	1
	Sep/Oct	2.603	11	1
	Nov/Dec	2.144	12	−1
1980	Jan/Feb	2.048	13	−1
	Mar/Apr	2.145	14	−1
	May/Jun	2.237	15	1
	Jul/Aug	2.440	16	1
	Sep/Oct	2.311	17	1
	Nov/Dec	2.284	18	−1
1981	Jan/Feb	2.150	19	−1
	Mar/Apr	2.400	20	−1
	May/Jun	2.947	21	1
	Jul/Aug	3.326	22	1
	Sep/Oct	2.789	23	1
	Nov/Dec	2.557	24	−1

	Firm A unit sales	Time	Seasonal indicator
1978–80 Means:	2.1181	9.5	0
1978–80 S.D.s:	0.2550	5.3385	1.0290
1978–80 Correlations:			
Firm A unit sales	1.0000	0.7264	0.5040
Time	0.7264	1.0000	0.0964
Seasonal indicator	0.5040	0.0964	1.0000

3. Does the model lead to an appropriate measure of A's lost sales?

14.1.4   OTC market manipulation

Between December 1975 and March 14, 1977, an account executive at Loeb Rhoades, Hornblower & Co., a large brokerage firm, manipulated the market for four over-the-counter stocks. Among other things, he bought and sold such stocks between accounts that he controlled. The SEC alleged that these transactions were made at artificial prices for the purpose of raising the prices of the securities, sustaining the prices once established, and absorbing shares

TABLE 14.1.4a. Regression of Olympic Brewing Company stock return on Daily Value Weighted Return (DVWR)

Source	df	Sum of squares	Mean square	F-value	Prob > F
Model	2	0.095021	0.047510	4.269	0.0186
Error	59	0.656682	0.011130		
Total*	61	0.751703			
Root MSE		0.105500	R-Square	0.1264	
Dep. Mean		0.014313	Adj R-Sq	0.1116	
C.V.		737.1111			

Variable	df	Parameter estimate	Standard error	t for H_0: Parameter = 0	Prob > $\|t\|$
Days	1	0.0004383	0.0006491	0.675	0.5022
DVWR	1	0.6968	0.2583	2.698	0.0091

* The intercept term is constrained to be 0. As a result, there is no reduction in total df for estimation of an intercept, and the total and model sums of squares are sums of squares but not squared deviations from the mean. R^2 is still the ratio of the model to total sums of squares, but is no longer the squared multiple correlation coefficient between dependent and explanatory variables. In this context, the constraint is reasonable because the assumption that the dependent variable is zero when both explanatory variables are zero seems correct.

Dependent variable is the return on Olympic Stock.

The standard error of regression (without centering on the mean) is 0.111.

of the securities being sold in the market. At times, his trades constituted a large fraction of total trades for these securities in the OTC market.

When the account executive was exposed on March 14, 1977, customers sued Loeb Rhoades, claiming that they had been injured by the manipulation. In response, the brokerage house prepared a regression study for each security involved in the manipulation, in which the price change of the security was regressed on the price change for the OTC market as a whole. This equation was estimated from data for a five-year period prior to the manipulation. The actual prices during the manipulation period were then compared with the regression estimates of those prices. Statistically significant differences were examined for special events, such as declaration of a dividend.

The regression equation for one security—Olympic Brewing Co.—is shown in Table 14.1.4. The dependent variable is the return on Olympic stock, defined as the difference between closing bid prices on the first and last days of the period for which the return is computed, plus any dividend paid during the period, divided by the closing bid price on the first day of the period. The explanatory variable, Daily Value Weighted Return (DVWR), is similarly defined on a value-weighted basis for all stocks in the OTC market. The other dependent variable, Days, is the number of trading days in the period over

TABLE 14.1.4b. Daily return data on Olympic stock and the market for selected days in January and March, 1977

Date	Price	Olympic return	Market return
03 Jan 77	46.00	0.00000	−0.00266
04 Jan 77	47.00	0.02174	−0.00998
05 Jan 77	47.50	0.01064	−0.00771
06 Jan 77	47.75	0.00526	0.00187
07 Jan 77	50.50	0.05759	0.00085
09 Mar 77	51.50	0.04040	−0.00697
10 Mar 77	48.50	−0.05825	0.00486
11 Mar 77	31.75	−0.34536	0.00074
14 Mar 77	28.375	−0.10630	0.00687

which the return is computed. A monthly period was used in the estimation of the equation; thus, Days was either 30 or 31 (except for February).

The account executive was particularly active in trading in December 1976 and January 1977. Shown in Table 14.1.4 are daily return data (defined for successive days in the same way as monthly data) for the first few days of January, and the last few days of March before he was discovered.

Questions

1. Using the regression equation, do you find a statistically significant difference between actual returns and the regression estimates in either period? (In determining the prediction interval for the regression estimate, assume that the coefficients of the regression equation are estimated without error. Also assume that the daily returns are statistically independent.)

2. It appears that the model fitted was of the form $Y = \beta_1 \cdot \text{Days} + \beta_2 \cdot \text{DVWR} + e$, where e is the error term with 0 mean given Days and DVWR. Is this model likely to be homoscedastic? What would be the effect of heteroscedasticity?

3. Is the model likely to have serial correlation in the error term? How would you test for such correlation? What would be its effect?

4. Does the model have sufficient power to justify an inference favorable to defendant? What conclusions relevant to the case would you draw?

Source

Loeb Rhoades, Hornblower & Co., Sec. Exch. Act Rel. No. 16823 (1980) (related SEC administrative proceeding).

Notes

A similar model was accepted without serious challenge in *Leigh v. Engle*, 669 F. Supp. 1390 (N.D. Ill. 1987) (testimony of Daniel Fischel). For other discussions of the model, see Franklin Easterbrook and Daniel Fischel, *Optimal Damages in Securities Cases*, 52 U. of Chi. L. Rev. 611 (1985); Daniel Fischel, *Use of Modern Finance Theory in Securities Fraud Cases*, 38 Bus. Law. 1 (1982); *Note, The Measure of Damages in Rule 10b–5 Cases Involving Actively Traded Securities*, 26 Stan. L. Rev. 371 (1974); see also *Basic Incorporated v. Levinson*, 99 L. Ed. 2d 194 (1988).

14.1.5   Fraud-on-the-market damages

Geriatric & Medical Centers, Inc. (Geri-med) operated nursing homes in Pennsylvania and New Jersey. Its stock was traded on NASDAQ. In 1992, purchasers of its stock brought a class action against the company alleging that it had perpetrated a "fraud on the market" by failing to disclose certain adverse developments. The jury rejected most of the charges, but found that Geri-med had failed to disclose the existence of a state grand jury investigation of patient deaths at two company-managed nursing homes. The period of non-disclosure found by the jury was from March 1, 1992, to September 1, 1992, when indictments were returned against company officials. During the class period, the stock trended down from around $4.00 at the beginning of March to $2.50 on August 31. After the disclosure on September 1, the stock dropped to under $2.00. Stockholders most clearly injured were those who bought during this "class period" and held through September 1, to see the value of their holdings decline. In order to quantify damages, an expert for Geri-med, Louis Guth, calculated by statistical projection what the price of Geri-med's stock would have been if disclosure of the investigation, which was known to the company, had been made at the beginning of the class period. He compared these "but for" prices with the actual prices for each day during the class period to come up with the excess amount paid by purchasers on each day. This amount varied from over $1 at the end of March to 0 on many days during the summer.

Guth's calculations used a regression equation of the form

$$\ln (P_t / P_{t-1}) = a + b \ln (x_t) + e_t,$$

where P_t, and P_{t-1}, are the Geri-med stock closing prices on trading days t and $t - 1$, respectively; a is the constant term; x_t is the ratio of closing prices on days t and $t - 1$ of an index of nursing home industry stock prices; and e_t is the error term on day t. This regression was estimated from data for a control period prior to the class period. The nursing home index declined from around $212 at the beginning of March to around $169 at the end of August.

The summary statistics are as shown in Table 14.1.5.

Since a was small and not statistically significant, it was ignored in subsequent calculations.

TABLE 14.1.5. Regression of Geri-med closing stock prices on nursing home company stock index

Number of observations	401
Mean of the dependent variable	0.720951E-03
S.D. of dependent variable	0.073265
Sum of squared residuals	2.1204
Variance of residuals	0.5314E-02
Std. error of regression	0.0728
R-squared	0.0124
Durbin-Watson statistic	2.5513

Parameter	Estimate	Standard Error	t-statistic
a	−0.4188E-03	0.3676E-02	−0.1139
b	0.53049	0.2368	2.2406

In models of this type, calculation of expected prices generally is made by "backcasting." In backcasting, one starts with the stock price and value for the index after full disclosure has ended the assumed distorting effects of the nondisclosure. One begins at the end rather than at the beginning because usually it is clearer when the problem ended than when it began. Then, working backwards day by day, the regression equation is used to compute expected stock returns to the beginning of the class period. For each day, the difference between actual and expected prices, if positive, is the first element in measuring damages for a person who bought shares on that day. If the shares were sold during the class period, the excess return received would be deducted from the excess return paid to arrive at a damage figure. If sold shortly after the class period, the excess return upon sale is presumed to be 0. In computing damages, actual or expected prices may be adjusted for the distorting effects of special events unrelated to the nondisclosures, such as earnings announcements.

To illustrate backcasting, we begin with the calculated expected closing price on April 2, 1992, and calculate the expected closing price on April 1. The expected price for April 2 is $3.233 from the backcasting calculation up to that point. The nursing home index number on April 2 is 193.796 and on April 1 is 198.733. The ratio of the two is 193.796/198.733 = 0.975158. The $\log_e$ of the ratio is −0.02516. From Table 14.1.5 the coefficient b of the log index variable is 0.53049. The coefficient times the log ratio is $0.53049 \times -0.02516 = -0.0252$. This is the logged return associated with the change in the index. Taking the anti-log, we have exp [−0.0252] = 0.9867 as the ratio of the stock prices on April 2/April 1. The expected price on April 1 is equal to the expected price on April 2 divided by the April 2/April 1 ratio, or 3.233/0.9867 = 3.276. Because the actual closing price on April 1 was $4.40, a purchaser who bought stock on that day and held to the end of the period would by this calculation have overpaid $1.124 per share.

Questions

1. In *In Re Control Data Corporation Securities Litigation*, 933 F.2d 616 (8th Cir. 1991), a similar fraud-on-the-market case, plaintiff's expert, John Torkelson, did not use regression to project the value of the stock during the class period. Instead, he assumed that the stock price had the same percentage changes day by day as an index (in that case, of technology stocks). If that method were applied to the April 1 calculation, based on the April 2 price, what would be the result? Which method is superior?

2. What are the arguments for and against using an upper end of a 95% prediction interval for the expected stock price? What effect would that have on the calculation of damages for April 1? (In determining the interval assume, as in the last problem, that the coefficients of the regression are estimated without error and that the daily returns are statistically independent.)

Source

Pearl v. Geriatric & Medical Centers, Inc., 1996 U.S. Dist. LEXIS 1559 (E.D. Pa. 1996)(testimony of Dr. Louis Guth). See also references cited in Section 14.1.4.

Notes

If an aggregate damage estimate is needed, either for settlement or litigation, this requires not only a measure of stock price inflation but also the number of shares bought on each day during the class period, and the day on which such shares are sold, if the sale occurs within the class period. Since such data are not available, various assumptions are made to model the process, which introduces another level of complexity. For a discussion of these models see Jon Koslow, *Estimating Aggregate Damages in Class-Action Litigation under Rule 10b-5 for Purposes of Settlement*, 59 Fordham L. Rev. 811, 826–842 (1991); Marcia Kramer Mayer, *Best-Fit Estimation of Damaged Volume in Shareholder Class Actions: The Multi-Sector, Multi-Trader Model of Investor Behavior* (National Economic Research Associates 1996).

Calculations of damages in securities actions must now take account of the cap on such damages imposed by Congress in the Private Securities Litigation Reform Act of 1995. See 15 U.S.C. §78u-4(e)(1997).

14.1.6   Effects of capital gains tax reductions

Capital gains are taxed only when the gain is realized. In any year, realizations of capital gains are only a small fraction of capital gains. (For example, net annual realized gains averaged only slightly over 3% of the stock of accrued gains between 1947 and 1980.) If the tax rate is sufficiently high, taxpayers

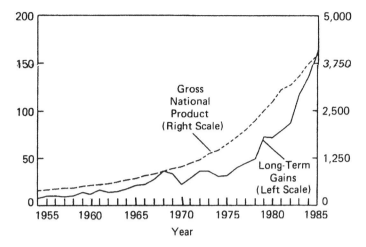

FIGURE 14.1.6a. Realized net long-term gains and Gross National Product, 1954–1985 (in billions of dollars)

have a strong incentive to delay realizations or to avoid the tax altogether by leaving the assets to heirs (at which time the tax basis of the assets is stepped up to market value, permanently avoiding the capital gains tax). Capital gains are concentrated among the wealthy. For example, in 1984, taxpayers with adjusted gross income (AGI) of $100,000 or more (the top 1% of returns) accounted for 54% of gains but only about 9% of other AGI.

The Tax Reform Act of 1986 abolished the preferential tax treatment of capital gains, effectively raising the maximum tax rate for wealthy individuals on such gains from 20% to 28%. A campaign for reinstatement of the preferential tax rate was based on the argument that lowering tax rates would unlock sufficient gains to more than offset the reduction in rates. Among the possible effects of a reduction in rates were: (i) a one-time unlocking of realizations, and (ii) a longer-term increase in realizations due to increased velocity of sales but, more significantly, a reduction in retention of assets until death. What could not be determined was whether these effects would be large enough to offset the effect of reduction in rates, other factors being held constant. For this a number of time-series regression studies were made.

The theory of the time-series studies was that capital gains realizations would grow in proportion to the level of total accrued gains. Total accrued gains cannot be measured directly, but are likely to follow overall growth in the economy and the value of corporate equities. In general, realized gains have moved upward with the growth of GNP and the value of corporate equities held by households. See Figures 14.1.6a and 14.1.6b. In the 1970s, realizations fluctuated without trend; they declined in years of recessions (e.g., 1970 and 1974–75) and increased in years of expansion (e.g., 1971–72 and 1976).

The data indicate that changes in tax rates have also influenced realizations. The Revenue Act of 1978 reduced the maximum marginal rate on

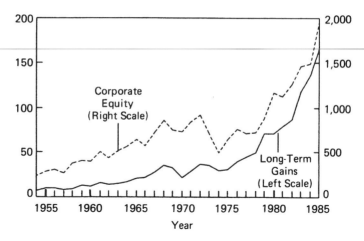

FIGURE 14.1.6b. Realized net long-term gains and corporate equity of households, 1954–1985 (in billions of dollars)

long-term gains from 39.875% (for most taxpayers) to 28%; the weighted-average marginal rates for wealthy taxpayers (defined as the top 1% in AGI) dropped from about 35.1% in 1978 to 25.9% in 1979.[1] This reduction evidently contributed to a dramatic 45% increase in realizations in 1979, which was concentrated in the high income classes that benefited most from the reductions. The increase, however, also occurred during a rising stock market and high growth of GNP, suggesting that part of the increase may have been due to these factors.

The 1981 Act further reduced the maximum tax rate on long-term capital gains to 20% for gains realized after June 20, 1981. The result was to drop the weighted average marginal rate for wealthy taxpayers from 26.1% in 1980 to 23.4% in 1981 to 20% in 1982. Capital gains realizations, however, increased by only about 5% in 1981 over 1980.

This weak response may have been due to the beginning of the 1981–82 recession in July 1981. Realizations rose only about 10% overall in 1982, perhaps reflecting the fact that the business cycle did not reach its nadir until late in the year (although the stock market revived in August). In 1983, both the stock market and realizations increased substantially.

To assess the separate effect of tax rate changes, Treasury analysts and others used various multiple linear regression models. Some of the data are shown in

---

[1] The average marginal rate is not the same as the maximum rate because not all taxpayers are in the highest bracket. Average marginal rates are computed as weighted averages of last-dollar marginal tax rates on capital gains faced by taxpayers with the average amounts of capital gains and taxable income in each income group. Changes in the income groups or in ordinary income rates can change average marginal rates on capital gains even when the maximum rate on capital gains does not change.

TABLE 14.1.6a. Capital gains realizations time series data

Year	Net long-term gains	Marginal tax rate	NYSE composite index	Real GNP	Corporate equities
1975	30.7	30.2	45.73	2695.0	637.4
1976	39.2	33.8	54.46	2826.7	752.0
1977	44.4	34.4	53.69	2958.6	706.6
1978	48.9	35.1	53.70	3115.2	703.2
1979	71.3	25.9	58.32	3192.4	857.4
1980	70.8	26.1	68.10	3187.1	1163.9
1981	78.3	23.4	74.02	3248.8	1102.0
1982	87.1	20.0	68.93	3166.0	1241.7
1983	117.3	19.7	92.63	3279.1	1422.5
1984	135.9	19.4	92.46	3501.4	1438.3
1985	165.5	19.5	108.09	3607.5	1890.1

Net long-term gains are net long-term capital gains in excess of net short-term losses in billions of dollars for top 1% AGI taxpayers.

Marginal tax rate refers to the average marginal tax rate on net long-term gains for top 1% AGI taxpayers.

NYSE composite index is an index of stock prices on the New York Stock Exchange.

Real GNP refers to the U.S. gross national product, adjusted for inflation.

Corporate equities refers to the value of household corporate stock holdings in billions of dollars.

Table 14.1.6a and the results of multiple regression analyses in Tables 14.1.6b and 14.1.6c.

During the 1988 presidential election campaign, one set of Treasury analysts, Darby, Gillingham, and Greenlees, argued, upon reanalysis of an earlier Treasury study, adding data for 1983–85 and refining some variables, that a reduction in rates would so increase realizations that revenue would be increased on a permanent basis. Their equation is Equation 1 in Table 14.1.6b. Another analyst, Joseph Minarik of the Urban Institute, disputed their conclusions and calculated Equation 2 in Table 14.1.6b.

The Congressional Budget Office (CBO) also created a regression model in logarithmic form using average marginal tax rate and the logarithms of GNP and corporate equities (in constant dollars) as explanatory variables, and the logarithm of net long-term realizations as the dependent variable. For the results, see Table 14.1.6c.

Questions

1. What is the difference in the estimate of realizations elasticity between the two models in Table 14.1.6b?

TABLE 14.1.6b. Time series linear regression models for capital gains realizations

Variable	Equation number			
	1		2	
	Estimate	t	Estimate	t
CONSTANT	−1186.4		−1568.78	
CRGNP	96.8	7.37	66.8	4.48
CIGNP	2.5	0.22	12.8	1.06
CSTK	36.2	6.03	35.2	3.65
CTX	−1795.0	−4.13	−1441.3	−3.75
CTX(-1)	50.9	0.11	818.1	1.84
CVALUE	Not Used	Not Used	24.5	3.29
R^2	0.845		0.875	
Durbin-Watson	1.917		1.411	
St. err. of reg.	not reported		3.712	
Sample period	1954-1985		1954-1985	

The dependent variable is the change in net long-term gains (in millions of dollars) as defined in Table 14.1.6a.

CRGNP and CIGNP refer to real and inflationary components of GNP change (in billions of dollars).

CSTK represents the change in corporate equities (in billions of dollars) as defined in Table 14.1.6a.

CTX and CTX(-l) refer to the current and lagged changes in the marginal tax rate as defined in Table 14.1.6a.

CVALUE refers to the change in NYSE composite Index as defined in Table 14.1.6a.

2. What is the reason for this difference? [Point of information: The original Treasury study, using variables substantially similar to those in Equation 1 on data for the period 1954-1982, obtained results similar to Equation 2. When that original Treasury equation was recalculated for data from 1954-1983, the results were similar to Equation 1. Hence, adding 1983 to the model without the CVALUE variable seemed to make the difference.]

3. Which model is more plausible?

4. Interpret the elasticity of net long-term gains for MTR in the CBO semilogarithmic model in Table 14.1.6c.

5. Is this form of model more plausible than the linear model?

6. Do the elasticities of realization estimated by the CBO model indicate that a reduction of average marginal tax rate from 20% to 15% would generate enough new realizations to offset the reduction in rate (assuming both rates would be flat rates, i.e., that all realizations would be taxed at that rate)?

7. What collateral effects might vitiate the revenue projections based on the model?

TABLE 14.1.6c. A semi-logarithmic regression model for capital gains realizations

Variable	Estimate	t
Constant	−10.900	4.93
Log-Price	0.901	3.13
Log-RCE	0.848	7.30
Log-RY	1.839	4.08
MTR	− 0.032	5.81
R^2	0.984	
Durbin-Watson	1.796	
St. err. of reg.	0.118	
Sample period	1954–1985	

The dependent variable is the logarithm of net long-term gains.

Log-Price refers to the logarithm of the GNP deflator.

Log-RCE refers to the logarithm of corporate equities (in constant dollars) held by individuals multiplied by the share of dividends received by top 1% AGI taxpayers.

Log-RY refers to the logarithm of the GNP (in constant dollars) multiplied by share of AGI received by top 1% AGI taxpayers.

MTR refers to the average marginal tax rate on net long-term gains for top 1% AGI taxpayers.

Source

Darby, Gillingham, and Greenlees, *The Direct Revenue Effects of Capital Gains Taxation*, Res. Paper No. 8801 (U.S. Dep't of Treasury 1988); Minarik, *The New Treasury Capital Gains Study: What is in the Black Box?* (The Urban Inst. 1988); Congressional Budget Office, *How Capital Gains Tax Rates Affect Revenues: The Historical Evidence* (March, 1988).

14.2   Interactive models

An *additive* linear model assumes that the impact of a change in any explanatory factor on the dependent variable is the same regardless of the levels of other factors. This is often not true. An employer's discrimination against women may well differ—either in absolute dollars or in percentage terms—at different levels in the work force. In this case, the coefficient of the sex factor represents a (complicated) average level of discrimination. A more realistic representation of the salary model must include one or more *interaction* or *cross-product* terms that combine sex and other explanatory factors. For example, to test whether the shortfall in salary increases with length of service, one might include, in addition to years-of-service, an interaction term, the product of the sex indicator and years-of-service. This cross-product would equal the number of years of service for men and zero for women (assuming the sex indicator was coded 1 for men and 0 for women). The coefficient of the simple years-of-service variable would reflect the return per year of service for women; the coefficient of the interaction term would reflect the added advantage given to men per year of service. The interactive model signals a different mode of discrimination than does the additive sex coefficient model—different returns per year of service, as opposed to a flat dollar shortfall. When interaction terms are added, the sex coefficient alone no longer represents the sole indicium of discrimination, or even the more important one; the interpretation is more complicated, but inclusion of interaction terms may pinpoint the sources of discrimination. In the interactive model of this example, the simple sex-coefficient describes the effect of sex for employees with zero years of service, while the coefficient of the interaction term describes how this effect increases or decreases per additional year of service.

Interaction terms are required when the effect of an explanatory factor, say X_1, depends on the level of another (or more than one other) variable, say X_2. Variable X_2 may be regarded as a modifier of the effect of X_1, and, symmetrically, X_1 may be regarded as an effect modifier of X_2. For example, consider the simplest interactive model,

$$Y = a + b{\cdot}X_1 + c{\cdot}X_2 + d{\cdot}X_1X_2 + \text{error}.$$

Suppose that X_1 is a dichotomous variable taking values 1 and 0, and X_2 is either continuous or dichotomous. When $X_1 = 0$, the model specifies the expected value of Y, given X_1 and X_2, as

$$E[Y|X_1 = 0, X_2] = a + c{\cdot}X_2,$$

since the interaction term X_1X_2 is zero. Thus, the regression coefficient c describes the effect of a unit change in X_2 *for the reference group* with $X_1 = 0$. (The constant term measures the mean of Y when X_1 and X_2 are both zero.) On the other hand, when $X_1 = 1$, the model specifies

$$E[Y|X_1 = 1, X_2] = (a + b) + (c + d){\cdot}X_2.$$

The change in Y per unit change in X_2 is now $c + d$; the presence of X_1 has modified the effect of X_2 from c to $c + d$. The coefficient of the interaction term, d, thus measures the *change* in the effect of X_2 as X_1 goes from 0 to 1. A symmetrical description is that the coefficient b of X_1 gives the difference in means of Y (when $X_1 = 1$ vs. 0) *for the reference subgroup* with $X_2 = 0$. That difference is modified by non-zero values of X_2 to $(b + d{\cdot}X_2)$, as can be seen by subtracting the model expression for Y when $X_1 = 0$ from that when $X_1 = 1$, holding X_2 constant.

When $d = 0$ the model is additive, in which case coefficients b and c may be interpreted as *main effects* (of X_1 and X_2, respectively). This terminology is unambiguous because a unit change in X_1 (or X_2) has the same effect, *viz.*, an average change in Y of b (or c) units, irrespective of the value of the other variable. When d is not zero, b and c are no longer usefully regarded as main effects, since that interpretation now applies only to the reference subgroups (when X_2 or $X_1 = 0$).[2] Interest usually shifts in this case from b or c to d. For example, in the employment discrimination model discussed in Section 13.5, the "sex-coefficient" b is the focus of attention in an additive model where X_1 is the sex indicator (0 for female, 1 for male). It describes the mean salary difference between males and females, adjusted for other factors in the model, say $X_2 =$ years of service. A single coefficient suffices because in the additive model the mean salary difference is a constant at all levels of experience. In the interactive model including $X_1 X_2$, the "sex-coefficient" b now measures the dollar shortfall only for employees with zero years of service, while d measures the difference between men and women in dollar return paid by the employer per year of service. The latter coefficient is likely to be a more important index of disparate treatment than the former. The coefficient b might even be 0, while if d were large, one would not say there was no discrimination.

Interactive models are also useful to describe changes in relations over time. If X_2 represents time measured from an initial moment, the interaction term allows for an effect of X_1 on Y with time dependent coefficients:

$$E[Y|X_1, X_2 = t] = (a + ct) + (b + dt) \cdot X_1.$$

The coefficients a and b describe the baseline relation when $t = 0$.

Sometimes categorical variables are coded with values other than 0 and 1, in which case the coefficients have a different meaning than described above. Suppose, for example, X_1 and X_2 are each dichotomous variables taking values $+1$ and -1. In this case neither of the "main effect" coefficients, b or c, describes a simple difference in means. For example, when $X_1 = 1$ the difference in means (for $X_2 = 1$ vs. -1) is $2(c + d)$ and when $X_1 = -1$ the difference in means (for $X_2 = 1$ vs. -1) is $2(c - d)$. It follows that c equals one-half the average of these differences and d equals one fourth the difference of these differences.

The above coefficients are familiar to statisticians from the analysis of variance, but are not as easily interpreted as the corresponding coefficients when $0 - 1$ coding is used. For this reason $0 - 1$ coding should be used whenever possible to avoid confusion.

In order to test for significance of an interaction coefficient estimated by OLS, the usual t and F tests may be applied (see Section 13.4) if the term is an explicit parameter in the model. Note that here an infelicitous choice of parameterization can lead to trouble. For example, another common coding scheme when studying four subgroups, as in the case of the double dichotomy above, is to use $0 - 1$ indicators X_1, X_2, and X_3 for three of the groups, leaving the fourth group as reference, which is then identified by its coding $X_1 = X_2 = X_3 = 0$. In the model

$$Y = a + b_1 \cdot X_1 + b_2 \cdot X_2 + b_3 \cdot X_3 + \cdots + \text{error},$$

---

[2]In general, depending on the coding, X_1 or $X_2 = 0$ may not even denote a realizable value, e.g., when $X_2 =$ age.

the coefficient b_i represents the difference between means for group i vs. the reference group ($i = 1, 2, 3$). An interaction coefficient, i.e., a difference between differences, would be expressed as $d = (\bar{Y}_1 - \bar{Y}_2) - (\bar{Y}_3 - \bar{Y}_4)$, where $\bar{Y}_1$, is the mean Y for observations in group 1 with $X_1 = 1$ and $X_2 = X_3 = 0$, and similarly for $\bar{Y}_2$ and $\bar{Y}_3$. $\bar{Y}_4$ is the mean of Y for observations in the reference group, for which $X_1 = X_2 = X_3 = 0$. Thus, $d = b_1 - (b_2 + b_3)$. If the interaction coefficient is the object of interest in the study, to assess its significance we require the standard error of the linear combination $\hat{b}_1 - \hat{b}_2 - \hat{b}_3$. The standard error here equals the square root of the sum of the variance of each coefficient, minus twice the sums of covariances between $\hat{b}_1$ and $\hat{b}_2$, $\hat{b}_1$ and $\hat{b}_3$, and $\hat{b}_2$ and $\hat{b}_3$. While the variance terms are usually available in the computer output, the covariance terms are often unavailable unless specifically requested. The better procedure is to express interaction terms explicitly.

The preceding discussion considered only "first order" interactive models, comprising equations with only pairwise products of variables. Higher-order interactions are occasionally required, comprising products of three or more variables. The higher-order interaction coefficients generally describe how lower-order interaction effects vary with other variables.

Two cautionary notes are in order. The first is that computer programs sometimes override the intended coding of an indicator variable, changing say from $0 - 1$ coding to ± 1 coding, even reversing the ordering of categories and, thus, the sign of the intended regression coefficient. The unwary user of such programs may be seriously misled by the printout.

The second caution is statistical: interaction effects usually have larger standard errors than their main-effect counterparts, reducing the power of hypothesis tests and widening confidence intervals. The reason for this may be seen readily in the case of the double-dichotomy with $0 - 1$ coding. Suppose, for simplicity, the four groups are of equal size n, and that the data in each group are distributed about the respective group mean with variance σ^2. In the interactive model, the interaction coefficient is estimated as $\hat{d} = \bar{y}_{11} - \bar{y}_{10} - \bar{y}_{01} + \bar{y}_{00}$, where $\bar{y}_{ij}$ are the sample means corresponding to $X_1 = i$, $X_2 = j$ ($i, j = 1$ or 0). The standard error for this coefficient is $(4\sigma^2/n)^{\frac{1}{2}} = 2\cdot\sigma/\sqrt{n}$. By contrast, the "main effect" coefficient for X_1 is estimated as $\hat{b} = \bar{y}_{10} - \bar{y}_{00}$, with standard error $\sqrt{2}\cdot\sigma/\sqrt{n}$, with a similar result for $\hat{c} = \bar{y}_{01} - \bar{y}_{00}$. In an additive model with $d = 0$, the main effect coefficient for X_1 is estimated as

$$\hat{b} = \left[\frac{1}{2}(\bar{y}_{10} + \bar{y}_{11}) - \frac{1}{2}(\bar{y}_{01} + \bar{y}_{00})\right] = \frac{1}{2}[(\bar{y}_{10} - \bar{y}_{00}) + (\bar{y}_{11} - \bar{y}_{01})],$$

in which the main effect of X_1 is averaged across the two levels of X_2, as is appropriate when there is no interaction. The standard error of $\hat{b}$ is thus the smallest, $\frac{1}{2}(4\sigma^2/n)^{\frac{1}{2}} = \sigma/\sqrt{n}$.

Further reading

David G. Kleinbaum, et al., *Applied Regression Analysis and Other Multivariable Methods* 188–93 (1998).

14.2.1   House values in the shadow of a uranium plant

The Feed Materials Production Center was located on a site of approximately 1,000 acres in a rural-residential area some 16 miles north and west of Cincinnati, Ohio. The plant was owned by the United States Government and was operated by a private contractor for the U.S. Department of Energy; it refined uranium trioxide powder into uranium metal and machined uranium metal for use in nuclear reactors, and in the manufacture of plutonium at other locations.

On December 10, 1984, the contractor (NLO, Inc.) announced that during the previous three months nearly 400 pounds of uranium trioxide powder had been released accidentally into the atmosphere. Approximately six weeks after that announcement, a class action suit was filed in federal district court alleging substantial personal injury and property damage to residents living in the area around the plant, as a result of the release.

The Rosen model

To support the plaintiffs' claim for loss of single-family home values, Dr. Harvey Rosen prepared a multiple regression study with home value as the dependent variable and explanatory variables to account for (i) differences in housing characteristics that could affect sale prices; (ii) differences in economic conditions over the period that could have affected sale prices, and (iii) location and time variables to reflect the effect of proximity to the plant both before and after the accident.

The data used to estimate the model were sales of single-family residential homes in three school districts adjacent to the plant, in the period 1983 through 1986. There were approximately 2,000 transactions, but due to incomplete data there were only about 1,000 sales that could be included in the analysis.

In the principal regression model, Rosen included the following indicator variables:

Pre-1985 In: coded 1 for a sale of a residence within five-miles of the plant perimeter prior to 1985, otherwise 0;

Pre-1985 Out: coded 1 for a sale of a residence beyond five-miles prior to 1985, otherwise 0; and

Post-1984 In: coded 1 for a sale of a residence within five-miles after 1984, otherwise 0.

The reference group consisted of the Post-1984 Out sales, coded 0 on the three preceding indicators.

TABLE 14.2.1a. Rosen model results

Variable	Estimate	t
Pre-1985 In	−1,621.11	−0.367
Pre-1985 Out	550.38	0.162
Post-1984 In	−6,094.45	−2.432

The regression included other factors—such as lot size and square-footage of house—designed to adjust for differences in house values. The coefficients of the indicator variables of interest were as shown in Table 14.2.1a.

From these results, Rosen concluded that the average loss in value to the homeowners within the five-mile radius was about $6,000 and the difference was statistically significant. He multiplied this figure by 5,500, the number of homes within the five-mile radius, to come up with a damage figure of $33 million.

Questions on the Rosen model

1. Assuming the Rosen model is correct, how would you interpret the −6,094 coefficient?

2. What objections would you have to use of this figure as a measure of damage?

3. What is the value of an interaction coefficient that would be more relevant for assessing damages? Is it statistically significant? (In computing significance, you may assume that the estimates of the indicator variable coefficients are statistically independent; in fact, they are correlated.)

The Gartside Model

Another expert for the plaintiffs, Peter Gartside, used data from a study by the defendants and calculated regressions using other indicator variables to code for distance and the pre-1985 and post-1984 years. He categorized distance ("RC") into 0–1 miles; 1–2 miles; and 2–4 miles. The reference category was 4–6 miles, for which observations were coded 0. A separate indicator was set up for each year between 1984 and 1986 ("PC"), with 1983 being the reference category. He then calculated a regression that included year, distance, and the interactions PC * RC, which are products of the year and distance indicators. The significance of each group of variables and the estimates are shown in Table 14.2.1b below. The OLS regression means were then shown in a 4 × 4 table (in dollar amounts and in deviation form, setting 1984 as the zero point). These results are shown in Tables 14.2.1c and 14.2.1d below.

From the fact that houses near the plant showed a decline or only modest rise in prices after 1984, while houses further away showed a substantial increase, Gartside concluded that the emissions had caused a diminution in value for houses nearer the plant.

TABLE 14.2.1b. Gartside model results

Variable	DF	SS	F Value	PR > F
PC	3	914.2	1.03	0.3780
RC	3	5,581.3	6.30	0.0003
PC * RC	9	3,644.9	1.37	0.1972

Variable	Estimate	t	PR > t
PC			
(1984)	−0.0921	−0.04	0.963
(1985)	2.1245	0.93	0.351
(1986)	8.8978	4.20	0.000
RC			
(0-1)	−4.5682	−0.26	0.796
(1-2)	0.8662	0.26	0.796
(2-4)	1.9371	0.56	0.574
PC * RC			
1984/0-1	11.047	0.57	0.573
1984/1-2	2.470	0.49	0.625
1984/2-4	12.414	2.50	0.013
1985/0-1	2.701	0.14	0.887
1985/1-2	− 2.675	−0.52	0.605
1985/2-4	3.510	0.80	0.425
1986/0-1	− 7.440	−0, 41	0.681
1986/1-2	− 5.790	−1.28	0.202
1986/2-4	3.327	0.79	0.429
Intercept	62.2		

TABLE 14.2.1c. Gartside regression model average price ($ 000's)

Distance	83	84	85	86
0-1	57.7	68.6	62.5	59.1
1-2	63.1	65.5	62.5	66.2
2-4	64.2	76.5	69.8	76.4
4-6	62.2	62.1	64.4	71.1

TABLE 14.2.1d. Gartside regression model—average price changes from 1984

Distance	83	84	85	86
0-1	−10.9	0	−6.1	−9.5
1-2	− 2.4	0	−3.0	0.7
2-4	−12.3	0	−6.7	−0.1
4-6	0.1	0	2.3	9.0

Questions on the Gartside Model

1. Interpret the PC, RC, and PC*RC coefficients, and describe the relevance of the PC*RC coefficients to the damage issue. Working from the 62.2 regression estimate for 1983 value of houses in the 4–6 mile band (these are the reference categories), do you see how Tables 14.2.1c and 14.2.1d were derived from Table 14.2.1b?

2. What objections would you have to Gartside's conclusions?

Source

In Re Fernald Litigation, Master File No. C-1-85-0149 (SAS) (H. Rosen and J. Burke, *Preliminary Report on the Property Value Effects of the Feed Materials Production Center at Fernald, Ohio on Single Family Residences Only* (*November 30, 1987*); P. Gartside, *Review of "Independent Analysis of Patterns of Real Estate Market Prices Around the Feed Materials Production Center, Fernald, Ohio"*).

Notes

A similar statistical study found that the accident at Three Mile Island caused neither an absolute decline in prices, nor a slower appreciation rate for houses within four miles of the plant in a seven-month period after the accident in 1979. Nelson, *Three Mile Island and Residential Property Values: Empirical Analysis and Policy Implications*, 57 Land Economics 363 (1981).

14.3   Alternative models in employment discrimination cases

The form of regression analysis used in many employment discrimination cases involves a model for salary (or log salary) as the dependent variable which is regressed on productivity factors and indicator variables for the various protected groups (usually women, minorities those over or age 40). If the coefficients for these groups are statistically significant, a prima facie case of discrimination is made, subject to the caveat that some courts have required a showing of particular instances of discrimination in addition to an overall pattern.

Courts often specify one or more failings when rejecting regression analysis: (1) certain explanatory factors are potentially tainted by discrimination; (2) omission of important wage-explaining variables; (3) incorrect aggregation of different groups into the same regression; (4) inclusion of observations from a time when discrimination was not illegal; (5) speculative assumptions that the independent variables have in fact determined decisions to raise pay or secure promotion.

The two-equation model

Campbell, Finkelstein, and others have argued that instead of a single equation with dummy variables designating protected groups, the coefficients of explanatory variables from two or more separate equations should be compared. Thus, determination of sex discrimination would involve a comparison of two equations, one for men and one for women. A larger coefficient for the experience variable in the men's equation, for example, would indicate that the employer was paying men more than women for each additional year of experience. The two-equation model is equivalent to a single-equation interactive model containing cross-product terms between the sex indicator and each of the other explanatory factors. The OLS coefficients of the cross-product terms equal the differences between corresponding OLS coefficients in the two separate equations.

The two-equation approach, arguably, is preferable for several reasons. If the null hypothesis of no discrimination is to be rejected, the single-equation model pre-specifies the form in which discrimination was practiced; the two-equation model does not. Using the two-equation model, the fact-finder may compare the coefficients of the various explanatory factors and pinpoint the source of any discrimination, thereby satisfying judicial insistence that the particular discriminatory practice be identified. For example, a markedly lower coefficient for years of experience in the women's equation suggests that the employer has discriminated with respect to that factor. Finally, back pay due to each individual can be estimated from the regression equations because the particular subgroup (e.g., experienced women) adversely affected by discrimination can be identified more accurately. See Thomas Campbell, *Regression Analysis in Title VII Cases: Minimum Standards, Comparable Worth, and Other Issues Where Law and Statistics Meet*, 36 Stan. L. Rev. 1299 (1984); see also Michael O. Finkelstein, *The Judicial Reception of Multiple Regression Studies in Race and Sex Discrimination Cases*, 80 Colum. L. Rev. 737, 739, n.12 (1980).

The trade-off implicit in the two-equation model is loss of power: comparison of several coefficients of explanatory factors requires higher levels of significance to compensate for the higher rate of Type I error when making multiple comparisons. See Sections 6.2 and 7.2. By contrast, the sex coefficient model tests only a single coefficient for significance, under the assumption that there are no interaction terms, i.e., that the return for each explanatory factor is the same for men and women.

The urn model

Levin and Robbins, focusing on the issue of statistical significance of unexplained disparities in mean wage between men and women, propose a different method to test that issue. Their approach, called an "urn model," regresses salary for all employees on explanatory variables, *excluding* indicator variables for group status, thereby obtaining a sex/race-neutral value for the expected salary of each employee. This adjusted mean salary represents the portion of an employee's salary explained by deviations of the explanatory factors from their mean. The role of OLS regression in this context is solely to produce an adjustment formula to define the "explained portion." The difference between the employee's actual salary and the explained portion is the residual or "unexplained portion" of salary. The difference between the mean residual for men and women is then tested for statistical significance.

In classical regression analysis, randomness is modeled through the assumption that replicate samples of employees would produce residuals with randomly different realizations. Levin and Robbins object that this view is highly artificial, if not untenable, because it posits a hypothetical population of employees (potential or future) from which the actual set is drawn by random sampling. The classical analysis also makes the strong assumptions that salaries are determined by a model formula plus error, that those errors are independent, and that they are normally distributed (see Section 13.8).

The urn model approach makes no assumption of hypothetical employees or random sampling errors, or any model-plus-error salary, or independence, or normality of errors. Instead, the urn model treats residuals like chips in an urn that are withdrawn at random without replacement. The chips withdrawn simulate the (e.g.) male residuals, while the remaining chips simulate the female residuals. The urn model treats the numbers of men and women and the values of their residuals as fixed, once given the adjustment formula. Its key assumption is that the residuals are *exchangeable*, i.e., that each possible assignment of sexes to the salary residuals is equally likely. This randomization device simulates situations that might appear under the null hypothesis of no influence of sex/race on salary.

If the average residual for men minus the average residual for women is so large that in simulations the difference would occur with small probability, say less than 0.05, then significance at the 5% level is declared. If the residual difference occurs fairly often in simulations, then the urn model shows that a sex-neutral random mechanism could have produced the observed disparity, which is then declared not significant. In practice it is not necessary to physically carry out the simulations, as simple formulas are available to approximate tail-area probabilities. See Bruce Levin and Herbert Robbins, *Urn Models for Regression Analysis, with Applications to Employment Discrimination Studies*, Law & Contemporary Problems, Autumn 1983 at 247. The urn model was accepted by the district court in *Sobel* v. *Yeshiva University*, 566 F. Supp. 1166 at 1169-70 (S.D.N.Y. 1983), but was rejected by the court of appeals on the ground that the model merely showed that the disparities could have occurred by chance, not that they *did*.[3] *Id.*, 839 F.2d 18 at 35–36. The urn model was accepted by the administrative law judge in *U.S. Dep't of Labor* v. *Harris Trust and Savings Bank*, No. 78-OFCCP-2 (ALJ Decision Dec. 22. 1986) as demonstrating statistically significant disparities against women and

---

[3]The objection by the court of appeals is misinformed, since no statistical test of significance does more than this and yet it is well established that a nonsignificant difference is evidence of nondiscrimination, if there is adequate power.

minorities. *Id.* at 63. *See* Section 14.4.1. It should be noted that the urn model is concerned only with testing the null hypothesis, and so does not allow the kind of detailed estimation of effects offered by the two-equation approach.

The test statistic in the urn model is described in terms of $\Delta \bar{r}$, the difference in mean residual between men and women, interpreted as the unexplained portion of the mean wage disparity between sexes. (When the wage variable is log dollars, $\Delta \bar{r}$ is approximately the fraction of the geometric mean wage ratio for the two groups that is unexplained by the adjustment procedure.) This difference is not intended to be an estimate of the sex coefficient in a single equation model; the principal intended use of the unexplained portion of the mean wage disparity is as a score by which one assesses statistical significance of unexplained disparities. In fact, the difference in mean residuals is systematically smaller than the sex coefficient by an amount that depends on the squared multiple correlation ($R^2_{Z(X)}$) between sex (Z) and the other predictors in the model (X). Specifically, the difference in mean residuals is equal to the sex coefficient reduced by the factor $1 - R^2_{Z(X)}$. If sex is uncorrelated with other explanatory factors, there is no shrinkage; if it is highly correlated, the shrinkage is nearly complete and the difference in mean residuals approaches zero.

The reason for this shrinkage is that the sex coefficient reflects the correlation of sex with salary after the effect of all other variables has been taken into account.[4] Each coefficient is computed in similar fashion, with magnitude depending on its factor's correlation with the dependent variable, the correlation with other explanatory factors, and the correlation of other explanatory factors with the dependent variable.[5]

The urn model differs in that the coefficients of the productivity factors are computed without regard to sex. To the extent that sex is a factor in determining salary, and is correlated with the productivity factors, its effect is assigned to those factors. Thus, the productivity coefficients are unbiased as estimators of productivity only under the null hypothesis of no discrimination. The urn model gives priority to the productivity factors since it works under the null hypothesis, asking whether the data are consistent with a nondiscriminatory explanation.

To calculate the significance of $\Delta \bar{r}$, one uses the statistic $z = \Delta \bar{r} / s \lfloor n/(mw) \rfloor^{1/2}$ where s^2, the variance of the residuals, is equal to $\sum r_i^2/(n-1)$; m is the number of men; w is the number of women; and n equals $m + w$. In large samples, z is approximately a standard normal variable. Generally, z is smaller in mag-

---

[4]Technically, the sex coefficient represents the simple linear regression coefficient of the residuals of salary (after salary is regressed on explanatory factors) on the residuals of sex (after sex is regressed on explanatory factors).

[5]The relationship in the case of sex and a single other explanatory factor may be seen from the following formula: $\hat{b}_Z = ((r_{YZ} - r_{YX} \cdot r_{ZX})/(1 - r_{ZX}^2)) \cdot (S_Y/S_Z)$ where Y denotes salary, Z is the sex indicator, X is the explanatory factor, the r's are correlation coefficients, the s's are standard deviations, and $\hat{b}$ is the sex coefficient. See Section 13.2.

nitude than t, the test statistic for the sex coefficient.[6] When the explanatory factors are uncorrelated with sex, the two statistics are asymptotically equivalent under the null hypothesis. If one is willing to assume the validity of the sex-coefficient model, the urn model represents a loss of efficiency, but such a loss is unimportant in large samples under the alternative hypothesis.

The urn model is most effective in stratified analyses where the assumption of exchangeability of residuals within homogeneous strata is most tenable. The evidence is combined across strata in a manner analogous to the Mantel-Haenszel statistic. See Levin and Robbins, *supra* at 412. In fact, the z-score for a stratified urn model, with dichotomous outcomes and no regression adjustments within strata, reduces exactly to the Mantel-Haenszel z-score.

Figure 14.3 illustrates the ideas behind the alternative forms of regression analysis discussed above.

14.4   Locally weighted regression

A single salary regression equation that is fitted globally to all employees may not accurately model the employer's salary process because it fails to represent different patterns of remuneration at different levels of qualifications, experience, responsibilities, or productivity. To extend the model to allow for different returns on these factors by adding interaction terms is a daunting task to accomplish properly in all but the simplest cases. Instead, one can fit a model to the data *locally* by using only those data points that fall within a neighborhood of the explanatory factors for a given employee, then repeating this procedure once for each employee. Restricting attention to local data in this way assures us that employees with very different profiles from a given employee will have no influence on the model that pertains to the given employee. Thus, the technique, known as *locally weighted regression*, explains each employee's salary in terms of other similarly situated employees.

The method requires a *distance* function to define how close a set of values of the explanatory factors is to the values for the referenced employee, and a *cutoff* criterion to define the extent of the local neighborhood. Within the neighborhood, weights may be used to give preferential emphasis in the fitting to those points closest to the employee's reference point, thereby diminishing the influence of data at the edges of the neighborhood. This refinement requires a *weighting* function. Locally weighted regression is a generalization of a smoothing algorithm for scatterplots. Practical details and elaborations in that

---

[6]The relation between them is given by the expression

$$z = \{t/[1 + t^2/(n - p - 1)]^{1/2}\} \cdot (1 - R^2_{Z(X)})^{1/2} \cdot [(n - 1)/(n - p - 1)]^{1/2},$$

where p is the number of explanatory factors in the adjusting equation, including the constant.

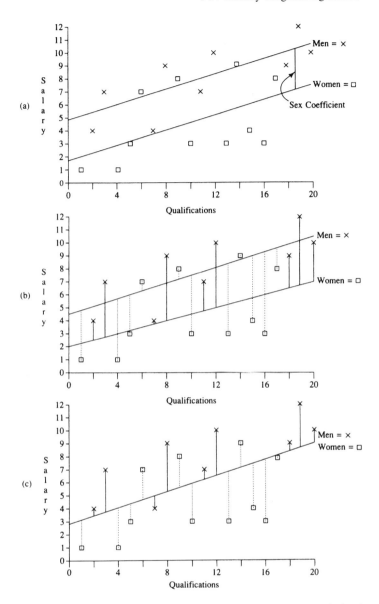

FIGURE 14.3. Alternative forms of regression analysis in employment discrimination cases. (a) The sex coefficient model—The model posits a constant dollar shortfall for women. (b) The two-equation model—predicted salary for men based upon the women's equation is lower than actual; predicted salary for women based upon the men's equation is higher than actual. (c) The urn model—male residuals tend to be positive while female residuals tend to be negative.

context are given in W. Cleveland, *Robust Locally Weighted Regression and Smoothing Scatterplots*, 74 J. Am. Stat. Assoc. 829 (1979).

As a rule, locally weighted regression produces fitted values with a substantially smaller residual sum of squares than that of global models, even when global models include higher order interaction terms not used in local fitting. This benefit counteracts the loss of a global model's convenient summary description through a single set of regression coefficients. Locally weighted regression is particularly well suited to the urn model method of analysis (see Section 14.3) because all that is required in that method is the unexplained portion of an employee's salary (which may be obtained as the residual from the local fit) and not necessarily any regression coefficients.

14.4.1   Urn models for Harris Bank

The Office of Federal Contract Compliance of the U.S. Department of Labor charged Harris Bank with discrimination against blacks and women in initial salary and advancement. Both sides prepared multiple regression studies. To avoid what defendant's expert called "nonhomogeneous job bias," employees were divided into cohorts, such as professional employees hired in the period 1973–1977. In one set of regression studies for this group the dependent variable was the log of initial salary; explanatory variables were age and age squared, indicators for prior professional job and prior job in finance, years of work experience and work experience squared, a series of indicator variables for higher education, time out of the work force, and business education.

Plaintiff's expert prepared two urn model studies for this cohort of 53 minority and 256 white employees. The first study included the above factors in a global OLS regression model to adjust log salary. The difference between the average residual for whites and average residual for minorities was 0.1070; the standard deviation of the residuals was 0.2171. The second study used locally weighted regression based on approximately 100 of the "nearest neighbors" for each employee. In this model, the difference in average residuals was 0.0519 and the standard deviation of the residuals was 0.1404.

Questions

1. Interpret the results of the global and locally weighted regressions.

2. Are the differences statistically significant? What does statistical significance mean in this context?

3. Does the difference between the global and locally weighted regression results cast doubt on the correctness of the global model?

4. If defendant had used the urn model and it had not shown statistical significance, what objections might plaintiffs have made?

Source

U.S. Dep't of Labor v. *Harris Trust and Savings Bank*, No. 78-OFCCP-2 (Dec. 22, 1986) (ALJ Decision).

14.5   Underadjustment bias in employment regressions

The problem of underadjustment bias in multiple regression models has become the focus of judicial attention and considerable academic debate in anti-discrimination class actions. The problem of underadjustment arises because the true productivity of employees cannot be observed directly, so that regression analysis, in adjusting for productivity differences, has to make use of proxies for productivity. These usually include education, experience, and other factors. The use of proxies, however, may be misleading because of their imperfect correlation with productivity.

To explore this problem we begin by looking more closely at the question of bias in the proxies. A proxy may be "unbiased" in that it applies to men and women equally, but this vague notion encompasses two assumptions that are in fact antithetical. The first assumption is that equally productive men and women will have, on average, the same value of the proxy. This assumption can also be expressed as "the regression of proxy on productivity is the same function of productivity for men as for women." The simplest example of this model occurs when proxy equals productivity plus an uncorrelated random error with zero mean for both sexes. The second assumption reverses the conditioning: men and women with equal values of the proxy are, on average, equally productive, i.e, productivity regressed on proxy is the same function of proxy for men and women. The simplest example of this assumption occurs when productivity equals proxy plus an uncorrelated random error with zero mean for both sexes. Note that either assumption is compatible with a difference between men and women in terms of both average proxy and productivity.

Because there are two regression relationships (see Section 13.1, at p. 350), if the proxy is an imperfect surrogate for productivity, the regression of proxy on productivity will not be the same as the regression of productivity on proxy. In fact, if men have different average proxy values than women, the truth of one assumption generally implies the falsity of the other. For example, if it is true that equally productive men and women have, on average, the same proxy value (first assumption), then it will not be true that men and women who have the same proxy are equally productive on average (second assumption); rather, assuming men have higher average scores on the proxy, women with the same score as men will tend to be less productive, on average. It follows that a fair employer who awards salary solely on the basis of productivity would pay men more than women with the same proxy, but the difference would falsely be attributed to sex discrimination in a multiple regression salary equation using that proxy for productivity. We call this "underadjustment" bias.

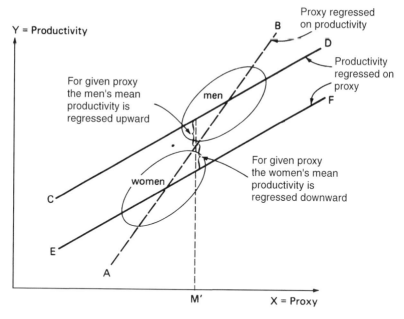

FIGURE 14.5. The mechanism of underadjustment bias

The reason for the seemingly paradoxical difference between male and female regressions of productivity on proxy can be understood by looking at
Figure 14.5, which illustrates the first assumption. Suppose the proxy is an evaluation score. If the regression of evaluation score on productivity is the same
for the two groups (see line AB on the figure, portraying $E[\text{eval}|\text{prod}] = \text{prod}$),
then the regression of productivity on evaluation score will be split into
separate regression lines for men (line CD) and women (line EF). This splitting
is a regression-to-the-mean effect, which is seen most clearly for an evaluation
score that lies between the mean scores for men and women (line at M'). Because the correlation between evaluation score and productivity is imperfect,
the average productivity for men at score M' regresses *upward* from the value
given by regression line AB toward the overall mean productivity of men,
while the average productivity for women at score M' regresses *downward*
from regression line AB toward the overall mean productivity of women. [7]

Under the second assumption, however, no underadjustment bias occurs.
The reason is that now men and women with equal proxy values are equally

---

[7]The same effect is exhibited if the conditioning proxy value is above or below the
mean values for both men and women. If above, the conditioning value represents a smaller
number of standard deviations above the mean for men than for women. As a result, the
average productivity for men with that level of proxy regresses less toward the mean overall
productivity of men than the average productivity of women at that value regresses toward
the mean overall productivity of women.

productive, on average, and should be paid equally, on average, by a nondiscriminatory employer. A regression based on that proxy would not be expected to show a sex effect, i.e., the sex-coefficient is unbiased.[8] The fact that men would have a higher proxy on average than equally productive women, though paradoxical, is irrelevant to whether equally productive employees are paid the same average salary.

Thus, whether a regression equation is threatened by underadjustment bias depends on the relation between the proxies and productivity. Whether the first or second assumption holds (if either) is an empirical question that seldom can be answered solely by the data. However, the nature of the proxies may make certain positions more or less tenable in the absence of specific evidence bearing on the issue. If a proxy measures *a cause of* productivity—such as years of education or work experience—then, absent evidence to the contrary, it would be reasonable to assume that the proxy plus a random error equals productivity because that is consistent with the causal model. From this it follows that there is no underadjustment bias. An employer who objects to the second assumption for these causal variables would have to introduce some evidence showing that there are omitted factors that correlate with sex and productivity (given the other included variables in the model) that would cause the second assumption to be violated.

If the causal direction is reversed, however, so that the proxy *reflects* or is *a consequence* of productivity, then, absent evidence to the contrary, it seems reasonable to make the first assumption, since the assumption that productivity plus a random error equals the proxy is consistent with the causal direction. In such circumstances, if men score higher on the proxy than women, on average, then the use of such a proxy in the regression would create underadjustment bias. An evaluation score is an example of a reflective proxy for which the first assumption is plausible for a nondiscriminatory employer. Plaintiffs who object to the first assumption for a particular reflective proxy would have to introduce evidence either of taint or of bias in the proxy. Defendants who object to regression results from models that include such a proxy will have an argument that the statistically significant sex-coefficient in the model is only a consequence of underadjustment.[9]

The foregoing indicates that judicial analysis of regression models depends on the type of proxy. The question in every case is whether, on average, men and women with the same values of the proxies would be equally productive. We have now identified two circumstances in which this question may be answered in the negative: (1) the inclusion of even unbiased reflective proxies with different means in the two groups; or (2) the existence of omitted identi-

---

[8] If the employer does award men a given amount more than equally productive women, the expected sex-coefficient will equal that amount.

[9] The line between causal and reflective proxies is not always as clear as this description may suggest. For example, rank at an academic institution, while primarily a reflective proxy, does have some causal aspects as well. It probably should be considered reflective.

fiable causal factors that are correlated with sex, after accounting for included variables.

The extent of underadjustment bias turns on the unobservable (to the analyst) correlation between the reflective proxies and productivity. For a nondiscriminatory employer, i.e., one who awards salary solely on the basis of productivity, the difference in the regression estimates of salary for men and women for a given level of the explanatory factors is equal to the unadjusted average salary difference for men and women times one minus the squared multiple correlation coefficient between the proxies and true productivity.[10]

The extent of correlation between the proxy and true productivity is therefore both crucial and generally unknowable. There are some guideposts, however.

Extensive studies of correlations between scores on tests designed to predict performance and actual academic performance have generally found values less than 0.60. See, e.g., College Entrance Examination Board, *ATP Guide for High School and Colleges* 28–29 (1989). If the correlation in the occupational context were no better than this, a salary difference equal to at least 64% of the unadjusted difference would be produced by underadjustment. For lower-level jobs in particular, validated proficiency tests or carefully structured and objective evaluation scores may have higher correlations, with the highest values attained for multiple reflective proxies or for proxies that are themselves elements of productivity. While hard data are lacking on this subject, we presume that correlations above 70% would be unusual. Thus, one might say that the lower limit of underadjustment bias would be at least 50% of the unadjusted salary difference, or of the difference that remains after adjustment for causal factors.

Another element to consider is the degree of correlation between the proxy and salary for men or for women alone. This correlation would set a lower bound on the correlation between productivity and proxy because errors in the assignment of salary reduce the correlation below the correlation between productivity and proxy. However, having a lower bound on the correlation gives one an upper bound on the salary difference that could be attributed to underadjustment bias, and if the residual salary difference exceeded that amount the employer could not fairly argue that the difference was attributable solely to underadjustment.[11]

---

[10]See Herbert Robbins & Bruce Levin, *A Note on the 'Underadjustment Phenomenon,'* 1 Statistics and Probability Letters 137–139 (1983). If the regression includes both reflective and causal proxies, and the employer is nondiscriminatory, the difference in regression estimates of salary for men and women for a given level of explanatory factors is equal to the difference between the regression estimates of salary for men and women using only the causal proxies, times one minus the partial squared multiple correlation coefficient between the reflective proxies and true productivity, given the causal proxies.

[11]This assumes that there is no direct effect of proxy on salary, only an indirect effect mediated through productivity. If there is a direct effect—salary awarded on the basis of, e.g., evaluation score—the correlation between productivity and proxy could be less than the correlation between salary and proxy.

As the foregoing analysis suggests, the threat of underadjustment bias will not exist in every case. In fact, the dynamics of the adversarial process may limit its scope. Plaintiffs tend to exclude reflective proxies from their regression studies because they are potentially tainted. Defendant employers are more likely to use them, but are generally reluctant to present regression studies unless the sex-coefficient is not statistically significant. Thus, regression models with reflective proxies and statistically significant sex-coefficients may be relatively few—at least in the first round. Models that raise the issue are more likely to appear in the second round, when plaintiffs may introduce one with reflective proxies to counter an employer's assertion that some essential variable was omitted in the first-round model; or the defendant employer may introduce one to show that some test or measure of productivity sharply narrows the gap between men and women, even if it is not closed entirely and remains statistically significant.

In such cases, the threat of underadjustment prevents us from finding discrimination simply from a statistically significant sex-coefficient. Instead the court must look at the degree to which reflective proxies are likely to correlate with productivity. When there is reason to think that the correlation between productivity and reflective proxies is high, and men score higher than women on the proxies, the potential for bias should be a reason for caution, but not the sole ground for rejecting the regression results. The weight given a statistically significant sex-coefficient should be diminished when the proxies are believed to be less accurate reflectors of productivity. The formulas for the degree of bias given in this section can be a starting point for the appraisal of such regression results.

14.5.1   Underadjustment in medical school

An issue of bias was raised by experts for the defense in *Sobel v. Yeshiva University*, 839 F.2d 18 (2d Cir. 1988), in which the Albert Einstein College of Medicine (AECOM) of Yeshiva University was accused of discrimination against women M.D.s with respect to pay and promotion. The difference in mean salaries for men and women without adjustment for differences in productivity was approximately $3,500. However, men had greater experience, education, and tenure at the institution, as well as higher rank.

Plaintiffs' experts initially presented a multiple-regression model in which salary was the dependent variable and the explanatory variables were indicators for education, work experience, tenure at the school, and year. After a critique of this model by defendant's experts, plaintiffs revised their model to estimate salary for each year separately, and added as variables license to practice medicine in New York State, rate of publications, and detailed departmental classifications. The sex coefficients were negative in every year, with the largest negative value being $-3,145$ ($t = -2.18$) for 1978. Since the indicator variable for sex was coded 1 for women and 0 for men, the coefficient indicated that women faculty members on average earned $3,145 less than equally qualified men.

Defendant also claimed that rank, time in rank, assignment of administrative responsibility, and assignment of research or clinical emphasis (since clinical departments were more highly paid than research departments) were variables that should have been included in the model. Defendant argued that rank in particular was needed to account for aspects of productivity not reflected in the other explanatory variables. Protesting that rank was not a cause of productivity but a result of it, and that the conferring of rank was tainted by the school's practices, plaintiffs nevertheless presented a revised regression adding defendant's variables. The resulting sex coefficient was $-2,204$ ($t = 1.96$), again indicating that on average women earned $2,204 less than equally qualified men. The effect of adding this last group of variables was to increase R^2 from 0.7297 to 0.8173.

The district court found for the defendant, holding that the difference in salary shown by the regressions might have been due to underadjustment bias:

> The evidence was clear that certain types of productivity, such as the procurement and management of large research grants, clinical expertise, the generation of revenue through private practice and affiliation practice, and a significant clinical work load, had a direct effect upon compensation and should have been accounted for as significant sources of productivity differences. In addition, it was necessary to take account of the less tangible but highly important factor of the quality of performance in the areas of research, teaching, and clinical practice. Although these factors were not easily quantifiable, we cannot conclude that their omission had no effect on the results of the multiple linear regression study. On the contrary, the fact that male faculty members scored higher on sixteen of the plaintiffs' twenty proxies strongly suggests that the omission of the less tangible factors distorted the results in favor of the plaintiffs. The Court agrees, therefore, with the defendant's conclusion that the failure to adequately account for productivity resulted in an underadjustment bias and plaintiffs' overstatement of the sex-coefficients.

Id., 566 F. Supp. at 1180.

The court of appeals reversed on issues not relevant here, and on remanding instructed that faculty rank should be used in the regression and that the underadjustment defense should be rejected. On the latter point, the court of appeals reasoned:

> Yeshiva's experts concluded that Sobel's regressions contained an "underadjustment bias" simply because men scored higher on the included variables. Insofar as Yeshiva argued that simply because men "scored" higher the imperfection of the included variables itself proved that the regression underadjusted for productivity, the argument is unpersuasive.

Men might have scored even higher had the variables perfectly reflected productivity, and this would have explained even more of the apparent gender disparity. But it is equally possible that the imperfection had the opposite effect; that women would have scored higher if the proxies were more accurate. On the present record, there is no way to tell which gender was disadvantaged by the imperfections.

Put another way, all that is known about the proxies used by plaintiffs in their regression is that they are not perfect measures of productivity, and that insofar as they do measure productivity they show that men on the AECOM faculty possess the attributes tied to productivity (e.g., experience) in greater measure. What is not known is whether variables that exactly measure productivity would show the same advantage for men (and thus would explain the same portion of the raw gender disparity as the imperfect proxies), a lesser advantage for men (and therefore explain less of the gender discrepancy), or a larger advantage. In short, the simple fact of imperfection, without more, does not establish that plaintiffs' model suffers from underadjustment, even though men score higher on the proxies.

Id., 839 F. 2d at 34.

Questions

1. The preceding section identified two circumstances in which the regression of productivity on proxies would not be the same for men and women. See p. 449–450. Which of these was found to exist by the district court? Did the district court deal with the other circumstance?

2. Was the court of appeals correct when it concluded that "the simple fact of imperfection, without more, does not establish that plaintiffs' model suffers from underadjustment, even though men score higher on the proxies"? Assuming the statement is true, does it justify the court's rejection of defendant's claim that the sex coefficient in plaintiffs' regressions was not a reliable indicator of discrimination?

3. Is it likely that the sex coefficient of $-2,204$ in plaintiffs' revised regression is entirely due to underadjustment?

14.6   Systems of equations

We have mentioned that a correlation between a regressor and the error term arises when the regression equation is part of a system of equations in which the regressor is determined within the system, i.e., appears itself as a dependent variable in one equation in the system (Section 13.8 at p. 402). Explanatory variables determined within the system are called *endogenous* variables;

those determined outside the system are called *exogenous* variables. A specific example may clarify these ideas.

Suppose we posit a "supply" function for murders (Section 14.6.1) in which the rate of murder is a weighted sum of (i) the risk of execution for those convicted, (ii) an exogenous variable (such as an index of violent crime other than murder), and (iii) an error term. In addition to the posited negative causal relation between execution risk and murder rate, there may be a reciprocal relationship: a sharp increase in the murder rate could raise a public outcry for more executions. This reciprocal relationship is represented by a separate function in which the execution rate is a weighted sum of the murder rate (an endogenous explanatory variable) and exogenous explanatory variables (probably different from those in the murder equation).

The reciprocity of the feedback relation compels us to abandon OLS as a method for determining the murder rate equation. Because of the feedback, correlation exists between the risk of execution and the error term. For example, if in a given period the error is positive (i.e., the rate of murder is above the mean value predicted by the equation), that higher rate would in turn produce a greater rate of executions (ignoring the lag factor, which obviously would be important here), and thus positive errors would be associated with higher values of execution risk. The resulting correlation means that OLS estimates of the murder rate equation would be biased and inconsistent because the correlated part of the error term would wrongly be attributed to the execution risk.

Here is a second example of a reciprocal relationship. In an antitrust case involving corrugated cardboard containers, plaintiffs introduced a regression equation in which the price of containers sold by manufacturers was regressed on various factors, including the price of containerboard sheets from which containers were made (see Section 14.1.1). Plaintiffs conceded that, if prices for corrugated containers dropped significantly, prices of containerboard sheets would also decline. This reciprocal relation creates a correlation between containerboard price and the error term that would bias OLS estimates of the price equation.

The point of these examples is that a full model involves more than a single equation. Econometrics usually calls for separate supply and demand equations, called structural equations, with the same dependent variable in both. Price appears in both as an endogenous explanatory variable, but otherwise the explanatory factors are different. A technically correct method of estimating the coefficients for price is to solve the two equations using special methods (discussed below).

In many cases, econometricians do not estimate separate structural equations, but content themselves with *reduced form equations* that combine supply and demand factors in a single equation. Although in the absence of reciprocal relations this practice is not objectionable per se, it may be desirable to estimate supply and demand equations separately. Structural equations disclose relationships concealed by the reduced form, which is essentially a black box;

estimating them separately may show perverse results (e.g., demand rising with price) that would reveal defects in the model.

Various techniques have been developed to deal with the problem of estimation when there are reciprocal relations in a system of equations. Perhaps the most widely used is the two-stage least-squares technique. In the first stage, the explanatory variable carrying a reciprocal relation is estimated in a separate equation in which it is the dependent variable and other exogenous explanatory variables in the supply and demand equations are the regressors. Since none of the exogenous explanatory variables is correlated with the error, neither their linear combination nor the regression values of the endogenous explanatory variable are correlated with the error in the original supply and demand equations. In effect, the endogenous variable has been purged of its correlation with the error. In the second stage, separate supply and demand equations are estimated by OLS using regression values for the endogenous variable (from the first stage) rather than its original values. The result is consistent (although not unbiased) estimates of the parameters in the supply and demand equations. Where there is more than one such endogenous variable, each must be estimated in a separate equation.

Two-stage least squares is an example of a "limited-information" technique, since it involves estimating only one equation at a time; it does not require estimation of all variables in all equations of the system. There are more complicated "full-information" techniques for simultaneously estimating all equations which require observations on all variables in the system. Full-information techniques include such methods as three-stage least squares and maximum likelihood. The advantage of such techniques is that they are more efficient than two-stage least squares.

Further Reading

R. Wonnacott & T. Wonnacott, *Econometrics*, chs. 7–9 (2d ed. 1979).
H. Theil, *Principles of Econometrics*, ch. 9 (1971).

14.6.1   Death penalty: Does it deter murder?

During the 1960s, attacks on capital punishment as "cruel and unusual," in violation of the Eighth and Fourteenth Amendments to the U.S. Constitution, focused attention on whether the death penalty deterred potential murderers. Contending that it might, the Solicitor General, appearing as amicus in *Gregg v. Georgia*, 428 U.S. 153 (1976), cited a then-unpublished study by Isaac Ehrlich of the University of Chicago. Ehrlich had used regression analysis and historical data to reach the tentative conclusion that every execution prevented seven to eight murders. This dramatic and publicized finding contradicted much previous research on the subject.

Ehrlich's data consisted of aggregate crime statistics for the United States from 1933 to 1969. The model treated estimates of the murder rate and the conditional probabilities of apprehension, conviction, and execution as jointly determined by a series of simultaneous regression equations. The key equation was a murder supply function in which the murder rate (murders and non-negligent homicides per thousand) was regressed on the product of endogenous and exogenous variables. The endogenous variables were probability of arrest, probability of conviction given arrest, and probability of execution given conviction. The probability of arrest was estimated from the ratio of arrests to murders. The execution risk was defined in several ways; the definition principally relied on by Ehrlich had one of the largest negative coefficients, and had the largest t-value. It was defined for year t as an estimate of executions in year $t + 1$ based on a weighted average of executions in the five preceding years (presumably to mimic the estimate that would be made by a person coldly calculating the risks of murder) divided by the number of convictions in year t.

The exogenous variables were: (1) fraction of the civilian population in the labor force; (2) unemployment rate; (3) fraction of residential population in the age group 14–24; (4) permanent income per capita; (5) chronological time; (6) civilian population in 1,000s; (7) per capita real expenditures of all governments (excluding national defense) in millions dollars; and (8) per capita (real) expenditures on police in dollars lagged one year.

The model was in the multiplicative form

$$Q/N = C \cdot X_1^{a_1} \cdot X_2^{a_2} \cdot X_3^{a_3} \cdot e^v,$$

where Q/N is the murder and non-negligent manslaughter rate; C is a constant; X_1 is the risk of execution given conviction; X_2 is a vector of the other endogenous variables, each with its own coefficient; X_3 is a vector of the exogenous variables, each with its own coefficient, and v is the error term, which is assumed to be correlated with its immediately preceding value because of the feedback relationship between the murder rate and the execution rate. The equation was estimated using a three-stage, least-squares procedure.

The simple correlation coefficient between the murder rate and execution risk was 0.096. But when the regression model was estimated using a modified first-difference linear equation in natural logarithms, the elasticity of execution risk became -0.065, with upper and lower 95 percent confidence limits of -0.01 and -0.10.

In Ehrlich's data, the average annual number of executions was 75, and the average annual number of murders and non-negligent homicides was 8,965.

Ehrlich conceded that his result was heavily influenced by the later years in his data. The national homicide rate, as reported by the FBI, rose precipitously in the middle and late 1960's; between 1962 and 1969 it increased almost 60% to a level exceeded only by the rate for 1933. At the same time, executions declined. The average numbers of executions for the five-year periods between 1930 and 1960 were 155, 178, 129, 128, 83, and 61. There was an average of 36 executions in 1961-1964; there were seven in 1965, one in 1966, two in 1967, and none after that (for a while).

TABLE 14.6.1. Estimated effect of execution risk on criminal homicide rate

Ending date of effective period	Log model		Natural values model	
	Estimate	(*t*-value)	Estimate	(*t*-value)
1969	−0.065	(−3.45)	.00085	(0.43)
1968	−0.069	(−4.09)	.00085	(0.41)
1967	−0.068	(−4.55)	.00039	(0.18)
1966	−0.056	(−3.40)	.00046	(0.22)
1965	−0.037	(−1.53)	.00087	(0.52)
1964	−0.013	(−0.40)	.00123	(0.78)
1963	0.048	(1.00)	.00189	(1.10)
1962	0.021	(0.35)	.00120	(0.80)
1961	0.050	(1.02)	.00216	(2.00)
1960	0.067	(1.36)	.00235	(2.17)

Excerpted from Tables IV and V of Bowers & Pierce, *The Illusion of Deterrence in Isaac Ehrlich's Research on Capital Punishment*, 85 Yale L.J. 187 (1975).

Questions

1. How did Ehrlich compute that each execution saved 7 to 8 lives?

2. What does Table 14.6.1, which estimates Ehrlich's model for sub-periods, suggest about Ehrlich's results?

3. What is the effect of using the logarithmic transformation to estimate this model?

4. What factors other than deterrence may account for the partial elasticity of the murder rate with respect to the execution risk in Ehrlich's model?

Source

Ehrlich's original study and the studies of his critics, together with other literature on the subject, are cited in *Gregg* v. *Georgia*, 428 U.S. 153, 184 n.31 (1976).

Notes

The classic work, by Thorsten Sellin, compared the homicide rates in six clusters of abolitionist and retentionist states and found no deterrent effect. T. Sellin, *The Death Penalty* (1959). Sellin's work was criticized by the Solicitor General in *Gregg* because it relied only on the statutory authorization for the death penalty and not on its actual use. Subsequent work, by Peter Passell, used Ehrlich's form of murder supply equation, but estimated it for cross-sectional data (on a state basis). The coefficient for the execution risk was not statistically significant. Passell, *The Deterrent Effect of the Death Penalty: A Statistical Test*, 28 Stan. L. Rev. 61 (1975). Bowers and Pierce reported that, in the 1960s, states that decreased executions were more likely to have an increase in homicide rate that was below the national average than states that increased executions. Bowers & Pierce, *supra*, 85 Yale L.J. at 203, n.42 (Table VIII).

14.7   Logit and probit regression

In many situations the outcome of interest in an analysis is simply whether a case falls into one category or another. When there are only two outcome categories, the parameter of primary interest is the probability that the outcome falls in one category (e.g., a "positive" response). This probability is represented as a function of explanatory factors. For example, the U.S. Parole Board (now abolished) used multiple regression to estimate the probability that an inmate would violate parole if released, based on type of offense, prior history, and behavior while incarcerated. In such studies the expected value of the dependent variable is a probability, and the accuracy of the estimate depends on the rate of violations for various categories of offenders. Inmates with a higher predicted probability of parole violation were denied parole or targeted for closer supervision when paroled.

The usual linear models should be avoided in this context because they allow the predicted outcome probability to assume a value outside the range 0 to 1. In addition, the ordinary least-squares model is inefficient because the errors do not satisfy the assumption of homoscedasticity: their variance depends on the values of the explanatory factors.[12]

A standard approach in this situation is to express the outcome odds in a logarithmic unit or "logit," which is the logarithm of the odds on a positive response, i.e., the logarithm of the ratio of the probability of a positive response to the probability of a negative response. This may seem arcane, but in fact "logit regression" (more often called "logistic regression") has several advantages.

First, when comparing two probabilities, the most useful statistic is usually a ratio of the probabilities, or a ratio of the odds, rather than a difference in probabilities or a difference in odds (see Section 1.5). When the outcome parameter in a regression equation is the log odds, the coefficient of the indicator variable for group status (in previous examples, the sex or race coefficient) is simply equal to the difference between the log odds for the two groups. This difference is the log of the ratio of the odds for the two groups. The coefficient is therefore referred to as the log odds ratio (l.o.r.) and its anti-log as the odds ratio or odds multiplier. For example, in a logistic regression involving success

---

[12]The regression estimates the probability of a positive (or negative) response of a binary variable, and the variance of that estimate is smallest when the probability is near 0 or 1 and greatest when it is near 0.5.

or failure on a test, since the anti-log of -0.693 is 0.5, a coefficient of -0.693 for a protected group implies that the odds on passing for a protected group member are one-half the odds on passing for a favored group member. If the coefficient is β, the percentage change in the odds is equal to $e^{\beta} - 1$; for small β, this is approximately equal to β. For example, when $\beta = 0.06$, the percentage change is 6.18%. While the coefficient of an explanatory factor in a linear model is the mean arithmetic change in the dependent variable associated with a unit change in that factor (other factors being held constant), the coefficient in a logit model is the (approximate) percentage change in the odds.

Second, the logarithm of the odds is appropriate because it can take on any positive or negative value without restriction, while probabilities and odds must be positive. If the probability of a positive response is less than 0.5, the odds are less than 1 and the logarithm is negative; if the probability is greater than 0.5, the odds are greater than 1 and the logarithm is positive. There is no restriction on the range of the logarithm in either direction. The regression thus cannot generate "impossible" results for the dependent variable probabilities. If L denotes a log-odds, $L = \log[p/(l - p)]$, then $p = 1/(1 + e^{-L})$, which is always between 0 and 1.

Third, unless the underlying probability is equal to 0.5, the sampling distribution of the sample proportion is skewed, with a longer tail on the side with the greater distance from 0 or 1, as the case may be. The sampling distribution of the odds is also skewed because, if the probability is less than 0.5, the odds are constrained to lie between 0 and 1; if greater than 0.5, the odds can range between 1 and infinity. The sampling distribution of the log odds tends to be more nearly symmetrical and closer to normal, so that standard normal distribution theory becomes applicable.

Fourth, in many data sets the transformation from probabilities to log odds yields a more linear relationship with explanatory factors than in other forms. In epidemiology, disease incidence models are often expressed in terms of log odds.

Fifth, logit models are useful for the classification problem in which one must assign subjects to one of two (or more) groups based on observed explanatory factors, such as in the parole board example above. It can be shown that, among all possible classification procedures, the one that minimizes the probability of misclassification is the rule that assigns subjects to group A vs. B according to whether the log-odds on membership in group A given the explanatory factor exceeds a certain criterion level. The log odds can be shown by Bayes's Theorem to be a linear function of the explanatory factors for a wide class of distributions governing those factors.

In addition to the utility of the logistic regression model, there is an entirely different genesis for logit and logit-like models when the outcome variable is a dichotomization of an underlying (and perhaps unobservable) continuous variable, T. Suppose the observable dichotomous outcome Y takes the value 1 if T falls below a given cut-off value c, and 0 if not. For example, T might represent a test score, with $Y = 1$ denoting failure to score above c. Suppose further that T is normally distributed in a population with mean value that depends linearly on explanatory factor(s) X. Then the probability that $Y = 1$ is not linearly related to X.[13] One strategy to retrieve linearity is to transform the conditional probabilities into values of the standard normal variate (z-scores) corresponding to the points on the normal curve at which such probabilities would occur. Thus, if the probability of being on one side of the cutoff is 0.20, a table of the standard normal curve shows that this is associated with a z-score of 0.842, and it is this value that would be linearly related to X. This approach is known as probit analysis.

Another example arises in bioassay problems, where T represents a hypothesized tolerance to a drug, and $Y = 1$ denotes a toxic response to the drug if tolerance T is below dosage level c. This time the mean and variance of the normally distributed T are fixed, but dosage levels c are chosen by the experimenter. The normal variate z-score corresponding to the probability that $Y = 1$ is again a linear function, of dose this time ($X = c$).

If the underlying variable T has what is known as a logistic distribution (which is similar to the normal distribution except that it has fatter tails) and there is a linear relation between that variable and independent variables X, then the logarithm of the odds that T would be on one side of the cutoff is a linear function of the independent variables.[14] This follows because the cumulative distribution function of the standard logistic distribution is $F(t) = 1/(1 + e^{-t})$; solving algebraically for the exponent yields $t = \ln[F(t)/(1 - F(t))]$. Thus, $P[Y = 1] = P[T < c] = F(t)$, where $t = \frac{c-\mu}{\sigma}$ is a linear function of X, so that $\ln(P[Y = 1]/P[Y = 0])$ is a linear function of X. We usually write $\log \frac{P[Y=1|X]}{P[Y=0|X]} = \alpha + \beta X$.

As mentioned above, ordinary least-squares techniques should not be used to estimate a logistic regression because the equation is not linear. However, at least two other techniques are available for estimating the equation: maximum likelihood and weighted least squares. We prefer the maximum likelihood method, which calculates the set of coefficients that maximizes the likelihood of the observed data (see Section 5.6). Maximum likelihood estimates can generally be calculated even when there is only a single binary outcome per vector of explanatory factors. By contrast, weighted least squares requires

---

[13]The reason lies in the nonlinear nature of the normal cumulative distribution. If T is approximately normally distributed and linearly related to explanatory variables, then the means of the conditional distributions of T (conditional, that is, on the values of the explanatory variables) increase linearly with increases in the explanatory variables. In effect the bell-shaped distribution of the dependent variable moves one notch higher above the dichotomizing point for each unit increase in the explanatory variables. But the one-notch increase in the dependent variable does not produce a constant increase in the *probability* that T is above the cutoff point. As the bell-shaped curve is shifted further to the right of the cutoff point, the change in probability from a unit shift diminishes; when the cutoff point is in the tail of the distribution, the change in probability becomes negligible.

[14]The word "logistic" in logistic regression thus refers both to "logits," and to a logistic random variable that has been dichotomized.

grouping of data to estimate probabilities. In large samples the two methods yield close answers, but not always in smaller samples.

In modern computer software for estimating logistic multiple regression models, the program may stop without convergence to a set of coefficient values. This does not necessarily mean that the technique cannot or should not be used. When small samples are involved, the program may stop because an explanatory factor is always associated in the data with a particular outcome, regardless of the presence or absence of other factors. In that case, the log odds multiplier associated with that factor is infinite and after a certain number of iterations the program shuts down. For the same reason, a log odds multiplier may be very large for such a factor, representing the point at which the program has stopped on its march to infinity, but this does not mean that the large value is wrong. One way to deal with this problem is to acknowledge that the factor is determinative or nearly so whenever it is present, regardless of other factors, and to reanalyze the data without cases in which that factor was present. By doing so, the program can reach closure with respect to the effect of other, nondeterminative factors.

In logistic regression, the standard errors of the coefficients of the explanatory factors have the same interpretation as they do in ordinary least squares regression, except that for testing the normal distribution is used instead of the t-distribution, because the estimates and inferential theory are valid only for large samples. Standard errors must generally be computed at the same time maximum likelihood estimates of the coefficients are calculated. For a simple fourfold table, the standard error of the log odds ratio is the square root of the sum of the reciprocals of the cell frequencies.

The correctness of the regression estimate of the odds (or probability) is usually measured by the correctness of classifications based on the estimate. The convention adopted is to predict the outcome that is most probable on the basis of the model (assuming the cost of either kind of misclassification is equal). The accuracy of the model is then assessed in terms of the rates of false positive errors and false negative errors (see Section 3.4). Section 14.9.1 gives a refinement.

With replicate observations per vector of explanatory factors, e.g., when the factors are discrete as in a contingency table, a measure for testing goodness of fit is a chi-squared type statistic known as the log likelihood ratio. See Section 5.6 at p. 195. This is defined as twice the sum (over both categories of the dependent variable in all cells of the table) of the observed frequency times the log of the ratio of the observed frequency to the estimated expected frequency. This statistic can test the goodness of fit of a logit model when the sample size in each of the categories is sufficiently large (more than 5). If sample sizes within cells are smaller than five, we may still use the difference between two log likelihood ratio statistics to assess the significance of particular variables included in one model, but excluded from the other.

Logistic regression models have been used in employment discrimination cases where the issue is dichotomous, such as hiring or promotion. See *Coser*

v. *Moore*, 587 F. Supp. 572 (E.D.N.Y. 1983) (model used by both sides to test whether women faculty members at the State University of New York were initially assigned lower faculty ranks than comparably qualified men); *Craik v. Minnesota State University Bd.*, 731 F.2d 465 (8th Cir. 1984) (same use of logistic regression by defendant St. Cloud State University).

14.7.1   Mortgage lending discrimination

Decatur Federal Savings and Loan in Atlanta, Georgia, was investigated by the U.S. Department of Justice for discrimination against blacks in making home mortgage loans. The investigators found that, although the bank had written guidelines, there was wide latitude for an underwriter's subjective assessment of the application, and the decision whether or not to make a loan might be influenced by personal, subjective criteria in addition to the financial data.

A statistical study of application files showed that, over a two-year period, after eliminating cases involving special circumstances, there were 896 applications for conventional, fixed-rate mortgages from whites, of which 86.9% were accepted, and 97 such applications from blacks, of which 55.7% were accepted. To adjust for possible differences in the ability of whites and blacks to meet the bank's underwriting guidelines, the investigators performed a multiple logistic regression analysis. The dependent variable was the log odds on being accepted. Explanatory variables were grouped in five categories: (1) the amount of money at risk, measured by various forms of the loan-to-value ratio of the house; (2) the borrower's ability to pay, measured, e.g., by the borrower's income in relation to his or her obligations; (3) the borrower's willingness to pay, measured by past credit history; (4) serious credit problems, such as bankruptcy or foreclosure; and (5) other. In the "other" category the investigators put a dummy variable, which took the value 1 if the borrower or co-borrower were black, and 0 otherwise. The model predicted 96% of the decisions correctly.

Questions

In the model, the coefficient for black was -0.8032, which was 2.52 standard deviations from its expected value under the null hypothesis.

1. Interpret this coefficient in terms of the probabilities of acceptance for whites and blacks.

2. What is the expected value of the coefficient under the null hypothesis and what is its approximate P-value?

Source

Bernard Siskin & Leonard Cupingood, *Use of Statistical Models to Provide Statistical Evidence of Discrimination in the Treatment of Mortgage Loan*

Applicants: A Study of One Lending Institution in *Mortgage Lending, Racial Discrimination, and Federal Policy*, ch. 16 at 451 (Urban Institute Press, John Goering and Ron Wienk, eds., 1996). The seminal work of this type was a market-wide study by the Federal Reserve Bank of Boston; see Alicia H. Munnell, et al., *Mortgage Lending in Boston: Interpreting HMDA Data*, 86 Am. Econ. Rev. 25 (1996).

14.7.2   Death penalty in Georgia

In *McCleskey* v. *Zant*, 481 U.S. 279 (1987), the petitioner, a black man who had been sentenced to death for the murder of a white policeman, argued that Georgia's death penalty was unconstitutional because it was applied in a racially discriminatory manner. To prove his claim, petitioner relied on a study by Professor David Baldus, which assessed the effect of race in the decision to impose the death penalty on those convicted of homicides in Georgia. Baldus conducted two separate studies, the smaller of which, the Procedural Reform Study, consisted of "594 defendants arrested in the state of Georgia between March 28, 1973, and June 30, 1978, who were convicted after trial of murder and sentenced either to life imprisonment or to death, or who received death sentences after pleading guilty to a charge of murder."[15]

The data collection effort was massive. Data for each case consisted of information about the defendant, the victim, the crime, aggravating and mitigating circumstances, the evidence, and the race of killer and victim. In all, over 250 variables were collected in each case. Information was gathered by examining records of the Georgia Parole Board and, in some cases, Georgia Supreme Court opinions, by examining records of the Bureau of Vital Statistics, and by interviewing lawyers involved in the trial.

Table 14.7.2a lists eight explanatory factors for a sample of 100 cases from the "middle range" (in terms of aggravation of the crime) in the Procedural Reform Study. Cases are ranked by the predicted probability of a death sentence based on the multiple logistic regression model shown in Table 14.7.2b, which was estimated from the data of Table 14.7.2c.

Questions

1. Sift through the data and compile two fourfold tables, each table to classify death sentence outcome for white victim in the first row and "other" victim in the second row. The left-hand column should be for death sentences and the right-hand column for other sentences. The first table should be for cases involving stranger victims and the second table for victims not strangers.

---

[15]The data presented in this section come from Baldus's Procedural Reform Study. Most of the analysis and conclusions of the courts dealt with the other study, the Charging and Sentencing Study. The methods of the two studies were substantially the same, except that in the latter, a more comprehensive model included a larger number of variables.

TABLE 14.7.2a. Variable codes for 100 death-eligible cases

Death:	1 = Death sentence
	0 = Life sentence
BD:	1 = Black defendant
	0 = White defendant
WV:	1 = One or more white victims
	0 = No white victims
AC:	Number of statutory aggravating circumstances in the case
FV:	1 = Female victim
	0 = Male victim
VS:	1 = Victim was a stranger
	0 = Victim was not a stranger
2V:	1 = Two or more victims
	0 = One victim
MS:	1 = Multiple stabs
	0 = No multiple stabs
YV:	1 = Victim 12 years of age or younger
	0 = Victim over 12 years of age

TABLE 14.7.2b. Logistic regression results for the Procedural Reform Study data

Variable	Coefficient	MLE	S.E.
Constant	β_0	−3.5675	1.1243
BD	β_1	−0.5308	0.5439
WV	β_2	1.5563	0.6161
AC	β_3	0.3730	0,1963
FV	β_4	0.3707	0.5405
VS	β_5	1.7911	0.5386
2V	β_6	0.1999	0.7450
MS	β_7	1.4429	0.7938
YV	β_8	0.1232	0.9526

2. Compute the odds ratio and log odds ratio relating sentence to race of victim for each table.

3. Estimate the common log odds ratio by taking a weighted average of the two log odds ratios, with weights inversely proportional to the variances of the individual log odds ratios. Take anti-logs to arrive at an estimated common odds ratio. What conclusions do you draw?

4. Referring to Tables 14.7.2a and 14.7.2b, the model fitted was

$$\log \frac{P[\text{ death }|X]}{P[\text{ other }|X]} = \beta_0 + \beta_1 X_1 + \cdots + \beta_k X_k.$$

The maximum likelihood estimates of the coefficients are shown in Table 14.7.2b. In the above equation, each of the coefficients is a "main effect" for the corresponding variable. Using this equation, how would you describe the main effect for the race-of-victim and race-of-defendant variables? How

TABLE 14.7.2c. 100 death-eligible cases from the middle range

#	Death	Pred	BD	WV	#AC	FV	VS	2V	MS	YV
1	0	0.0337	1	0	1	1	0	0	0	0
2	0	0.0338	1	0	2	0	0	0	0	0
3	0	0.0461	1	0	2	0	0	1	0	1
4	1	0.0584	1	0	3	0	0	1	0	0
5	0	0.0655	1	0	2	1	0	1	0	1
6	0	0.0687	1	0	4	0	0	0	0	0
7	0	0.0769	1	0	3	1	0	0	0	1
8	0	0.1155	1	0	4	1	0	1	0	0
9	0	0.1288	1	0	1	1	0	0	1	0
10	1	0.1344	1	0	5	1	0	0	0	0
11	0	0.1423	1	1	2	0	0	0	0	0
12	0	0.1627	0	1	1	0	0	0	0	0
13	0	0.1735	1	0	2	0	1	0	0	0
14	1	0.1735	1	0	2	0	1	0	0	0
15	0	0.1770	1	0	3	0	0	0	1	0
16	0	0.2197	0	1	1	1	0	0	0	0
17	0	0.2201	0	1	2	0	0	0	0	0
18	0	0.2201	0	1	2	0	0	0	0	0
19	0	0.2336	1	0	3	0	1	0	0	0
20	0	0.2336	1	0	3	0	1	0	0	0
21	0	0.2380	1	0	4	0	0	0	1	0
22	1	0.2558	0	1	1	1	0	1	0	0
23	0	0.2588	1	1	3	1	0	0	0	0
24	0	0.2902	0	1	2	1	0	0	0	0
25	0	0.2902	0	1	2	1	0	0	0	0
26	1	0.3063	1	0	3	1	1	0	0	0
27	0	0.3068	1	0	4	0	1	0	0	0
28	0	0.3068	1	0	4	0	1	0	0	0
29	0	0.3167	0	1	3	0	0	0	0	1
30	0	0.3335	0	1	3	0	0	1	0	0
31	1	0.3369	1	1	5	0	0	0	0	0
32	1	0.3369	1	1	5	0	0	0	0	0
33	1	0.3730	0	1	4	0	0	0	0	0
34	0	0.3730	0	1	4	0	0	0	0	0
35	0	0.3730	0	1	4	0	0	0	0	0
36	1	0.3730	0	1	4	0	0	0	0	0
37	0	0.3730	0	1	4	0	0	0	0	0
38	0	0.3906	1	0	4	1	1	0	0	0
39	1	0.3912	1	0	5	0	1	0	MS	0
40	0	0.3912	1	1	5	0	1	0	0	0
41	1	0.4208	0	1	4	0	0	1	0	0
42	0	0.4208	0	1	4	0	0	1	0	0
43	0	0.4245	1	0	6	0	0	0	0	0
44	0	0.4294	0	1	4	0	1	0	0	0
45	1	0.4511	0	1	4	0	0	1	0	1
46	1	0.4629	0	1	4	1	0	0	0	0
47	0	0.4635	0	1	5	0	0	0	0	0
48	1	0.4635	0	1	5	0	0	0	0	0
49	0	0.4821	1	0	5	1	1	0	0	0
50	0	0.4827	1	0	6	0	1	0	0	0

TABLE 14.7.2c. (continued)

#	Death	Pred	BD	WV	#AC	FV	VS	2V	MS	YV
51	0	0.4936	0	1	4	1	0	0	0	1
52	0	0.5050	1	1	3	0	0	0	1	0
53	0	0.5172	1	1	7	0	0	0	0	0
54	0	0.5172	1	1	7	0	0	0	0	0
55	1	0.5221	0	0	5	0	1	0	0	0
56	1	0.5564	0	1	6	0	0	0	0	0
57	0	0.5904	1	1	2	1	1	0	0	0
58	0	0.5910	1	1	3	0	1	0	0	0
59	1	0.5910	1	1	3	0	1	0	0	0
60	1	0.5910	1	1	3	0	1	0	0	0
61	0	0.6285	0	1	2	0	1	0	0	0
62	1	0.6343	0	1	3	0	0	0	1	0
63	1	0.6343	0	1	3	0	0	0	1	0
64	0	0.6383	1	1	3	0	1	1	0	0
65	0	0.6451	0	1	6	1	0	0	0	0
66	1	0.6767	1	1	3	1	1	0	0	0
67	1	0.6772	1	1	4	0	1	0	0	0
68	0	0.6772	1	1	4	0	1	0	0	0
69	1	0.6772	1	1	4	0	1	0	0	0
70	1	0.6772	1	1	4	0	1	0	0	0
71	1	0.6772	1	1	4	0	1	0	0	0
72	0	0.6788	0	1	2	1	0	1	1	0
73	1	0.7102	0	1	2	1	1	0	0	0
74	1	0.7102	0	1	2	1	1	0	0	0
75	0	0.7102	0	1	2	1	1	0	0	0
76	1	0.7107	0	1	3	0	1	0	0	0
77	1	0.7107	0	1	3	0	1	0	0	0
78	1	0.7153	0	1	3	1	0	0	1	0
79	1	0.7500	0	1	3	0	1	1	0	0
80	1	0.7524	1	1	4	1	1	0	0	0
81	1	0.7529	1	1	5	0	1	0	0	0
82	0	0.7529	1	1	5	0	1	0	0	0
83	1	0.7529	1	1	5	0	1	0	0	0
84	1	0.7529	1	1	5	0	1	0	0	0
85	1	0.7751	1	1	5	0	1	0	0	1
86	1	0.7751	1	1	5	0	1	0	0	1
87	1	0.7807	0	1	3	1	1	0	0	0
88	1	0.7807	0	1	3	1	1	0	0	0
89	1	0.7811	0	1	4	0	1	0	0	0
90	1	0.7811	0	1	4	0	1	0	0	0
91	0	0.7811	0	1	4	0	1	0	0	0
92	1	0.7811	0	1	4	0	1	0	0	0
93	1	0.7849	0	1	4	1	0	0	1	0
94	1	0.7849	0	1	4	1	0	0	1	0
95	1	0.8153	1	1	5	1	1	0	0	0
96	1	0.8592	1	1	2	1	1	0	1	0
97	1	0.8595	1	1	3	0	1	0	1	0
98	0	0.8635	0	1	5	0	1	1	0	0
99	1	0.8827	0	1	6	0	1	0	0	0
100	1	0.9018	0	1	6	0	1	1	0	0

would you describe the joint race-of-victim and race-of-defendant effect against the base category (i.e., $BD = WV = 0$)?

5. Calculate the odds ratio on death sentence vs. other, comparing white victim cases with non-white victim cases. How does this result compare with the answer to question 3?

6. Show that the probability of a death sentence is $P[\text{death} \,|\, X] = e^L/(1 + e^L)$, where $L = \beta_0 + \beta_1 X_1 + \cdots + \beta_k X_k$.

7. Assume that McCleskey shot a white male police officer who was a stranger, and that there were three aggravating factors in the case. Using the result in question 6, calculate the predicted probability of a death sentence. Repeat with the same factors assuming, however, that McCleskey's victim was not white. Do the data show that it is more likely than not that McCleskey would not have received the death penalty if his victim had not been white?

8. Using the rule that a death sentence is predicted when the estimated probability of such a sentence is more than 0.50, calculate the proportions of correct predictions in Table 14.7.2c.

Notes

Baldus's method

In Baldus's data, blacks who killed whites received the death penalty in a far greater proportion of cases than any other racial combination of defendant and victim. It was, however, argued that such cases usually involve strangers and more aggravating factors. To isolate the racial factor, Baldus began by examining unadjusted measures of the impact of case characteristics on death sentencing rates. For example, he compared the death sentencing rate for cases in which assailant and victim were strangers ($74/449 = 0.16$) with the rate for cases in which the parties knew each other ($54/2035 = 0.03$). To assess the impact of such differences, he used four measures: (i) the difference between death sentencing rates for the two groups; (ii) the ratio of death sentencing rates for the two groups; (iii) the OLS regression coefficient for the group indicator; and (iv) the odds-ratio multiplier (by how much an offender's odds on receiving the death sentence increase or decrease in the presence of the relevant characteristic).

Next, Baldus looked at racial characteristics while adjusting, one at a time, for background factors that correlated with death sentence to see whether the apparent effect of racial characteristics could be attributed to the nonracial factor. He examined, for example, whether the fact that the defendant was black and the victim white was an important factor in its own right or whether it merely pointed to a higher probability that the murder was between strangers and involved a contemporaneous felony.

Baldus used two methods to control for background characteristics: cross-tabulation and multiple linear regression. In the cross-tabulation method, the association between death sentence and the racial characteristic of interest is compared in two tables at different levels of the background factor; a persistence of the association in both tables is evidence that the effect of the characteristic of interest is not explained by background. In the example given above, two 2×2 tables would be constructed for black defendant/white victim cases versus other cases, one involving contemporaneous felonies, and the other not. In the multiple regression method, a variable for the background factor is added to the regression equation and the resulting change in the coefficient for racial characteristic observed. If the coefficient becomes statistically insignificant, then the change indicates that the apparent racial effect was merely a mask for the effect of the background factor.

Baldus then expanded his analysis to correct for more than one background factor at a time. While cross tabulation becomes ineffective with more than a few characteristics, because the numbers in each cell become too small, multiple regression is not so limited. Baldus ran multiple regressions on several different sets of background characteristics, the largest of which contained over 230 nonracial background variables. In the largest model he found significant regression coefficients of 0.06 for both the variable representing race of the victim and the variable representing race of the defendant.

In addition to weighted least squares, Baldus also used a logistic regression model. He described the result this way: "The overall race-of-victim disparity estimated with a logistic coefficient measure was 1.31 (with a death odds multiplier of 3.7)."

The district court opinion

The Federal District Court for the Northern District of Georgia rejected the validity of Baldus's study, citing three specific defects. *McCleskey* v. *Zant*, 580 F. Supp. 338 (N.D. Ga. 1984).

1. "[T]he data base has substantial flaws." The court criticized the failure of the data to capture all nuances of a case, the unavailability of certain information, the miscoding of information, and the application of a coding convention in which unknown data were coded as a U and then recoded to show the factor's absence (for 39 characteristics, over 10 percent of the cases were coded U).

2. "[N]one of the models utilized by the petitioner's experts are sufficiently predictive to support an inference of discrimination." Specifically, the court rejected any model with fewer than the 230 variables Baldus expected to explain the sentence; it found the R^2 (approximately 0.48) in the 230 variable model too low; and it took issue with (what it asserted was) the assumption that information available in retrospect had also been available to the fact finders.

3. "The presence of multi-colinearity [sic] substantially diminishes the weight to be accorded to the circumstantial statistical evidence of racial disparity." The court found white victim cases more aggravated than black victim cases and suggested that, due to multicollinearity, the race of the victim might be a surrogate for unaccounted-for aggravation otherwise not accounted for.

4. The stepwise regression technique used by Baldus was invalid since it searched for correlation without considering causality.

5. The study did not prove that race was a factor in any given case.

Baldus also selected 32 cases from Fulton County (where the murder took place), dividing them into three groups: more aggravated than *McCleskey*, equal in degree of aggravation, and less aggravated. Baldus concluded that these cases, analyzed individually, demonstrated discrimination. The district court analyzed several of the "equal to McCleskey" cases and found differences that it felt justified not imposing the death penalty.

The court of appeals opinions

The Court of Appeals for the Eleventh Circuit chose not to deal with the asserted invalidity of the Baldus study. It accepted *arguendo* its validity, but found its results insufficient to justify upsetting the system.

The court noted that the simple unadjusted figures showed that death sentences were imposed in 11% of the white victim cases and 1% of the black victim cases. After adjusting for various explanatory factors (the 230 variable model), the difference was described by the court as a 0.06 race-of-victim effect, signifying that, on average, a white-victim homicide was 6% more likely to result in a death sentence than a black-victim homicide. The court also cited Baldus's tables in which, in intermediate aggravation cases, according to the court, "white victim crimes were shown to be 20% more likely to result in the death penalty than equally aggravated black victim crimes." *McCleskey* v. *Kemp*, 753 F.2d 8771 896 (11th Cir. 1985). Is the court's interpretation of the 6% figure correct?

The court found that "the 6% bottom line" was insufficient to overcome the presumption that the statute was constitutional and insufficient to prove that racial factors played "a role in the outcome sufficient to render the system as a whole arbitrary and capricious," or that they operated in any given case. *Id.* at 897. As for the 20% figure, the court objected that a valid challenge could not be made using only a middle range of cases, that Baldus had not explained the rationale of his definitions, and that statistics alone could not demonstrate that McCleskey's sentence was determined by the race of his victim.

One dissenting opinion, however, addressed the validity of the Baldus study. Finding the district court clearly erroneous in holding the study invalid, Judge Johnson raised the following points.

1. The flaws in the data base were not significant. Many of the mismatches (inconsistent coding between the two studies) were the result of improvements in coding technique, and the remainder were not sufficiently important to invalidate the result. The amount of missing data was minor and perfection was unnecessary.

2. Multicollinearity was not a relevant concern since it either reduced the statistical significance of the coefficient of interest or distorted the interrelationship among coefficients. In neither case, Judge Johnson asserted, did it operate to increase the race-of-victim coefficient (in the former it depressed statistical significance and in the latter it dampened the race-of-victim coefficient).[16]

3. Bigger is not always better in models, which must balance the risk of omitting a significant variable with the risk of multicollinearity. A parsimonious model may be better than a more inclusive model.

4. An R^2 of approximately 0.48 is not too low. The explanatory power of a given model must be viewed in its context. Because with death sentences the random effects are likely to be great, the model might be sufficient. The district court's statement that a model with an R^2 less than 0.5 "does not predict the outcome in half of the cases" was incorrect.

The Supreme Court opinions

A narrowly divided Supreme Court affirmed the court of appeals (5-4). Writing for the majority, Justice Powell repeated some of the criticisms of the district court, but assumed that the model was statistically valid. However, given the necessity for discretion in the criminal process and the social interest in laws against murder, McCleskey had to prove by exceptionally clear evidence that the decision makers had acted with intent to discriminate in his case; the statistical pattern was not sufficient for that purpose. The dissenters pointed out that McCleskey's case was in the middle range of aggravation, for which the statistics showed that 20 out of every 34 defendants convicted of killing whites and sentenced to death would not have received the death sentence had their victims been black. It was therefore more likely than not that McCleskey would not have received the death sentence had his victim been black. His sentence thus depended on an arbitrary racial factor in violation of the Eighth and Fourteenth Amendments.

After his retirement from the Court, Justice Powell told an interviewer that the action he most regretted while on the bench was his vote in *McCleskey*.

---

[16]We note that it is not true that multicollinearity always distorts coefficients in the direction of zero. Severe multicollinearity causes unpredictable perturbations that reflect round-off and minor errors.

Notes

A description of the *McCleskey* studies may be found in David C. Baldus, George G. Woodworth & Charles A. Pulaski, *Equal Justice and the Death Penalty: A Legal and Empirical Analysis* (1990). Subsequent studies, led by Professor Baldus, showed race-of-defendant and race-of-victim effects in New Jersey and Philadelphia. For New Jersey, see David C. Baldus, Special Master, *Death Penalty Proportionality Review Project Final Report to the New Jersey Supreme Court*, Tables 18 and 18A (1991) (finding a 19 percentage point race-of-defendant disparity and at least a 14 percentage point race-of-victim disparity). For Philadelphia, see David C. Baldus, et al., *Racial Discrimination and the Death Penalty in the* Post-Furman *Era: An Empirical and Legal Overview, with Recent Findings from Philadelphia*, 83 Cornell L. Rev. 1638 (1998) (strong race-of-defendant and race-of-victim disparities found, but results sensitive to the particular models and data used).

14.7.3   Deterring teenage smoking

In debates over proposed tax increases for cigarettes as a way of discouraging teenage smoking, the Treasury Department argued, based on studies, that every 10% increase in price would decrease the prevalence of teenage smoking by about 7%. In particular, the Treasury asserted, adding about $1.00 to the price of cigarettes (an increase of about 50%) would decrease teenage smoking by more than 30%. This estimate was derived from cross-sectional studies that regressed smoking among teenagers on a variety of student and parental control factors, including state cigarette taxes. This conclusion was disputed by a group at the Department of Policy Analysis & Management at Cornell University. They used data for a cohort of about 10,000 students who were in 8th grade in 1988 and 12th grade in 1992. First, they looked at cross sections of all students in the data set in 1988 and 1992, and regressed a probit of smoking on student and parental control factors (including state cigarette taxes) to model the probability that an individual student would smoke in 8th and 12th grades. The results are shown in Table 14.7.3 below. They then excluded students who had smoked in 8th grade, and created an onset-of-smoking model. In this model they regressed a probit for smoking in the 12th grade on essentially the same control factors, including the level of state cigarette taxes and the increase in such taxes between 1988 and 1992. The results of this model are also shown in Table 14.7.3.

Questions

1. For the two cross-sectional models and the onset model, use the coefficients to compute the estimated percentage decrease in demand associated with a 10% increase in cigarette price (about $0.20) due to a tax increase in that amount.

TABLE 14.7.3. Tax coefficients in probit regression of teenage smoking

Tax regressors	1988 Cross section Coefficient (t-ratio)	1992 Cross section Coefficient (t-ratio)	Onset model 88-92 Coefficient (t-ratio)
Cigarette tax 1988 (cents/pack)	−0.0059 (−2.35)	−0.0031 (−2.57)	−0.0015 (−0.83)
Maximum	38	48	38
Minimum	2	2.5	2
Mean	18.71	26.27	18.71
Increase in cigarette tax 1988-1992 (cents/pack)	n/a	n/a	−0.00042 (−0.22)
Maximum			25
Minimum			0
Mean			7.56
Sample size	12866	12036	11271
Overall rate of smoking	5.5%	24.4%	21.7%

2. What might account for the difference in estimates between the cross-section and onset models?

3. What do these models tell us about the probable effect of an additional $1.00 tax?

Source

Philip DeCicca, Donald Kenkel & Alan Mathios, *Putting Out the Fires: Will Higher Taxes Reduce Youth Smoking?* (Department of Policy Analysis & Management, Cornell University 1998; unpublished ms.)

14.8   Poisson regression

Poisson regression is a close relative of logistic regression. The dependent variable Y is a count, taking values $0, 1, 2\ldots$, which follows a Poisson distribution. See Section 4.7. The mean of the distribution at given values of explanatory factors $X_1, \ldots, X_k$, which we denote $\mu(X_1 \ldots, X_k)$ or $\mu(\underline{X})$ or $E[Y|\underline{X}]$, varies with the levels of the explanatory factors. Because the mean is a positive quantity, it is natural to model the logarithm of the mean as a linear function of the explanatory factors. Thus, $\ln\{\mu|\underline{X}\} = \beta_0 + \beta_1 X_1 + \ldots + \beta_k X_k$. In this model, $\exp[\beta_j]$ gives the multiplicative increase in the expected value of the Poisson count per unit increase in the explanatory factor X_j, holding other factors fixed. As is general in log transformations (see Section 13.9 at p. 409), if β is not too large $100 \times \beta_j$ is the approximate percentage increase in μ per unit increase in X_j.

Typically, the regression coefficients are estimated by maximum likelihood. Standard computer packages routinely produce these estimates along with standard errors. Weighted least squares is a less desirable alternative method. Ordinary least squares estimation should be avoided because the variance of a Poisson distribution is equal to its mean; as the regression mean changes, the variance also changes, contrary to the basic OLS assumption of homoscedasticity.

In the Poisson model the sum of squared differences between observed and expected counts divided by the expected count has a chi-squared distribution with degrees of freedom $n - k - 1$, where n is the number of observations of the dependent variable and k is the number of explanatory factors in the regression, excluding the constant term. The chi-squared statistic can thus be used to test the goodness-of-fit of the model.

Although in theory the variance of a Poisson random variable is equal to its mean, it is common to find in data sets—especially when there are replicate observations for given values of the explanatory factors—that the actual dispersion exceeds the mean. This is a consequence of unexplained heterogeneity—there are unspecified factors that affect μ beyond what is accounted for by the explanatory factors $\underline{X}$ and this heterogeneity causes overdispersion in the data. That is, the conditional distribution of Y given $\underline{X}$ is not a pure Poisson random variable with mean $\mu(\underline{X})$, but has a distribution of a *mixture* of Poisson variables with mean $\mu(\underline{X})$ and variance $\mathrm{var}(Y|N) > \mu(\underline{X})$.

Overdispersion generally does not affect the predicted mean values from a model, but does influence the significance of the regression coefficients and the precision of predicted counts. Levels of significance reported in computer output should therefore be viewed with caution. A simple adjustment that is often made in cases of overdispersion is to multiply the standard errors reported under the Poisson assumption by the overdispersion factor—which is equal to the square root of the ratio of the chi-squared goodness-of-fit statistic to its degrees of freedom.

Poisson regression is commonly used in survival analysis where the counts are failures or deaths and the explanatory factors are etiologic.

14.8.1   Challenger disaster

On January 26, 1986, Flight 51-L of the space shuttle *Challenger* ended in a spectacular disaster. Within a minute after launch, an explosion destroyed the shuttle and killed the seven astronauts on board. A presidential commission appointed to investigate the catastrophe concluded that the explosion was caused by the failure of an O-ring seal between two lower segments of the right solid rocket motor. The O-rings were designed to prevent hot gases from leaking through the joint during the propellant burn of the rocket motor. It appears that the escaping gases penetrated an external fuel tank, leading to the explosion. The leak occurred because the O-ring did not adapt itself to changes in the gap between the tang and the clevis of the joint, which varied as the joint worked itself during take-off.

Although other factors may have played a role, the most prominent reason for the loss of flexibility in the O-ring was the low ambient temperature at the time of the launch. The commission noted, for example, that at 75 degrees Fahrenheit the O-ring returned to its original shape after being compressed about five times faster than at 30 degrees. In a dramatic demonstration at the commission's public hearings, physics professor Richard Feynman dipped a section of O-ring into ice water and then squeezed it with pliers to demonstrate its loss of resilience. On the day of the Challenger launch, the ambient temperature was 36 degrees Fahrenheit (about 15 degrees colder than the next coldest launch day) and the side of the solid rocket motor away from the sun was 28 degrees plus or minus 5 degrees. (Even the side toward the sun was only 50°.) Citing some of the data from earlier launches set out below, the commission concluded that, "a careful analysis would have revealed a correlation of O-ring damage and low temperature. Neither NASA nor Morton Thiokol, Inc. [the maker of the O-rings] carried out such an analysis; consequently, they were unprepared to properly evaluate the risks of launching the 51-L mission in conditions more extreme than they had encountered before."

As Table 14.8.1 shows, the lowest ambient temperature at a launch was 53° Fahrenheit. To project what the risks of O-ring damage would have been at some lower temperature, say 36°, a Poisson regression model can be fitted to the data. In such a model the number of damaged O-rings is a Poisson count with the mean varying with the temperature. To estimate the model, it is natural to use the logarithmic transformation. The logarithm of the mean number of O-ring failures at any given temperature is a linear function consisting of a constant plus a slope coefficient times the ambient temperature. In our calculations we expressed the temperature variable in degrees Fahrenheit minus 36°. (We subtracted 36° to center temperature on the particular value of interest; this centering does not affect the results of the model.) The equation is $ln\mu(T) = a + b(T - 36)$.

The maximum likelihood results are as follows: The intercept is 2.8983, with standard error of the estimate 1.1317; the slope coefficient is -0.1239, with standard error of the estimate 0.0407.

TABLE 14.8.1. Number of damaged O-rings (Row B) by ambient temperature in degrees Fahrenheit (Row A) at launch for 24 shuttle flights prior to Challenger flight 51-L.

A	53	57	58	63	66	67	67	67	68	69	70	70
B	3	1	1	1	0	0	0	0	0	0	1	1
A	70	70	72	73	75	75	76	76	78	79	80	81
B	0	0	0	0	2	0	0	0	0	0	0	0

Questions

1. Plot the observed data and model estimates on the same graph, with temperature minus 36° on the X-axis and the number of damaged O-rings on the Y-axis.

2. What does the model estimate for the average number of damaged O-rings at 36° compared with the risk at 53°, the lowest temperature previously encountered?

3. What does the model indicate is the approximate percentage increase in the average number of damaged O-rings for each degree reduction in temperature?

4. Compute an approximate 95% confidence interval for the expected value of the dependent variable when the explanatory temperature variable is at 36°. Does the width of the confidence interval affect your view of the increase in risk at that temperature vs. the risk at 53°?

5. Are the data consistent with the Poisson model? The chi-squared statistic is 32.3 for these data.

6. What is the degree of overdispersion in the data? What is the reason for the overdispersion and what correction would you make?

Source

The Presidential Commission on the Space Shuttle Challenger Accident Report, June 6, 1986, at 148; *cf. Boisjoly v. Morton Thiokol, Inc.*, 706 F. Supp. 795 (D. Utah 1988).

14.9 Jackknife, cross-validation, and bootstrap

In theory, the standard errors of regression coefficients reflect sampling variation, but in fact it is not uncommon to find that, in repeated samples, actual variation from sample to sample exceeds the theoretical prediction.

One important source of sensitivity is not reflected in standard errors. When a regression model is selected and estimated from the same sample of data (the construction sample), there is usually deterioration in the goodness of fit of the equation when it is applied to other data. The greatest deterioration tends to occur if the regression has been constructed using a stepwise procedure or some other data-dredging technique. In the context of logistic regression, the difference between the apparent misclassification rate in the construction sample and the true error rate for independent samples is called overoptimism.

Computer technology has opened up ways of making direct assessment of the stability of regression coefficients and the degree of overoptimism. Given enough data, one could assess both by dividing the data into groups and recomputing the regression for each group to determine how the coefficients vary in new samples. With respect to overoptimism, the parallel technique is to divide the data into construction and validation samples, estimate the regression from the construction sample, and test its accuracy with the validation sample. In many cases, however, there is not sufficient data to proceed in this way, and the statistician turns to devices known as the *jackknife*, the *bootstrap*, and *cross-validation* to create replicate regressions for testing stability of the coefficients and the accuracy of regression classifications.

In the jackknife technique (so named for its versatility), the regression equation is computed many times, each time omitting a different case from the data base. Computer programs do this efficiently by adjusting the estimates of the coefficients for the addition of one point and the deletion of another. The variation of the coefficients in these replications forms the basis for a direct estimate of their standard errors.[17] The technique is more robust than the classic method in that standard errors can be estimated without any assumption as to the correctness of the model. Confidence intervals for the coefficients may be based on the jackknifed estimate of standard errors using the t-distribution, as in the classical method. Direct assessment of variability is a useful check on theoretical calculations, particularly in cases in which the significance of coefficients is at issue.

The resampling idea can also be used to estimate the degree of overoptimism by using the model fitted with a case deleted to predict the outcome for the deleted data point. The average accuracy of those predictions is a more accurate estimate of the true error rate than that obtained from the construction sample. This technique is called cross-validation.

The other technique for replicating regression equations is called the bootstrap. Its essential idea is to create replications, not by deleting data points, but by treating existing data points as a population from which simulated samples equal in size to the actual sample may be selected. Variation is possible because after each point is selected it is replaced in the population. Thus, if there are 30 points of data, a bootstrap resample consists of 30 selections at random with replacement from this group. Most such samples involve multiple selections of some individuals and the omission of others. The number of possible different

---

[17] The calculation is as follows. Let b denote a coefficient based on all the data, and let $b_{(i)}$ denote the not-i coefficient based on deleting observation i, one at a time, for $i = 1, \ldots, n$. Define the ith *pseudovalue* $b_i^* = nb - (n-1)b_{(i)}$. Then the jackknifed estimate of the s.e. of b is the s.d. of the n pseudovalues,

$$\left[\frac{1}{n-1} \sum (b_i^* - b^*)^2 \right]^{1/2},$$

where $b^* = (1/n) \sum b_i^*$ is the jackknifed estimate of the coefficient. The jackknife estimator was originally introduced to convert a biased estimator b into a largely unbiased estimator b^*.

replications is much larger than in the jackknife. For 30 points, the number of jackknife resamples is 30; the number of different bootstrap resamples is 30^{30}. Usually, however, a few hundred bootstrap samples give results close to much larger samples.

In the multiple linear regression context, there are two ways to make selections. If the regression model is correct, for each given set of values of the explanatory variables, the value of the dependent variable is computed by taking the regression estimate and adding a residual selected at random from the population of residuals. The selected residual is "replaced" in the population and the procedure repeated for the next set of values of the explanatory factors. In this way a simulated set of values for the dependent variable is generated, and a new regression is computed based on the new values. From these replicate regressions the variability of the regression coefficients is assessed.

This approach is valid only if the linear regression model is correct. A more robust procedure, which does not depend on the correctness of the model, involves treating each observation vector as a single data point, and randomly selecting (with replacement) a number of observations equal in size to the sample. If there are 30 data points in the sample, repeated samples of size 30 are selected from the population. As before, replicate regressions are derived from the data. In logistic regression, the extent of overoptimism of a fixed prediction equation can be assessed using the average rate of error of that equation in the bootstrap samples.

Further Reading

Bradley Efron & Robert J. Tibshirani, *An Introduction to the Bootstrap* (1993).

14.9.1   Georgia death penalty revisited

Table 14.9.1a shows the predicted probabilities of a death sentence based on a regression computed from a "construction sample" of cases from Table 14.7.2c. The equation was then applied to the remaining cases (the "test sample"), with the results shown in Table 14.9.1b.

Questions

1. Using the rule that a death sentence is predicted when the estimated probability of such a sentence is greater than 0.50, calculate the proportion of correct predictions in the test and construction samples. Do the data show overoptimism?

2. Refine your calculations by computing the positive predictive value (i.e., those who received the death penalty as a proportion of those predicted to receive it) and the negative predictive value (i.e., those who did not receive the death penalty as a proportion of those predicted not to receive it).

TABLE 14.9.1a. Death sentences and predicted probability of death—construction sample

#	Death	Pred	BD	WV	#AC	FV	VS	?V	MS	YV
1	0	0.0066	1	0	1	1	0	0	0	0
2	0	0.0105	1	0	4	0	0	0	0	0
3	0	0.0204	1	0	1	1	0	0	1	0
4	0	0.0211	1	0	3	0	0	0	1	0
5	0	0.0324	1	0	4	0	0	0	1	0
6	0	0.0504	1	0	2	1	0	1	0	1
7	0	0.0505	1	1	2	0	0	0	0	0
8	0	0.0634	1	0	3	0	1	0	0	0
9	0	0.0806	0	1	2	0	0	0	0	0
10	0	0.0951	1	0	4	0	1	0	0	0
11	0	0.1325	0	1	4	0	0	1	0	0
12	1	0.1362	1	0	3	1	1	0	0	0
13	0	0.1404	1	0	5	0	1	0	0	0
14	0	0.1477	0	0	4	0	1	0	0	0
15	0	0.1615	1	1	3	1	0	0	0	0
16	1	0.1663	1	1	5	0	0	0	0	0
17	0	0.1697	0	1	2	1	0	0	0	0
18	0	0.1747	0	1	4	0	0	0	0	0
19	1	0.1747	0	1	4	0	0	0	0	0
20	0	0.2065	1	1	3	0	0	0	1	0
21	0	0.2366	1	1	6	0	0	0	0	0
22	0	0.2475	0	1	5	0	0	0	0	0
23	1	0.2475	0	1	5	0	0	0	0	0
24	0	0.2757	1	0	5	1	1	0	0	0
25	0	0.3171	0	1	2	1	0	1	1	0
26	0	0.3250	1	1	7	0	0	0	0	0
27	0	0.4493	1	1	3	0	1	0	0	0
28	0	0.4640	0	1	2	0	1	0	0	0
29	1	0.4925	0	1	3	0	1	1	0	0
30	0	0.4937	0	1	3	0	0	0	0	1
31	1	0.5222	0	1	4	0	0	1	0	1
32	0	0.5436	0	1	6	1	0	0	0	0
33	0	0.5590	1	1	4	0	1	0	0	0
34	1	0.5590	1	1	4	0	1	0	0	0
35	1	0.5590	1	1	4	0	1	0	0	0
36	1	0.5736	0	1	3	0	1	0	0	0
37	1	0.6084	0	1	4	1	0	0	1	0
38	0	0.6632	1	1	5	0	1	0	0	0
39	1	0.6632	1	1	5	0	1	0	0	0
40	1	0.6686	0	1	2	1	1	0	0	0
41	0	0.6686	0	1	2	1	1	0	0	0
42	1	0.6764	0	1	4	0	1	0	0	0
43	1	0.6764	0	1	4	0	1	0	0	0
44	0	0.6764	0	1	4	0	1	0	0	0
45	0	0.7008	0	1	5	0	1	1	0	0
46	1	0.7582	0	1	3	1	1	0	0	0
47	1	0.7845	0	1	6	0	1	1	0	0
48	1	0.7940	1	1	2	1	1	0	1	0
49	1	0.8211	1	1	5	1	1	0	0	0
50	1	0.9337	1	1	5	0	1	0	0	1

TABLE 14.9.1b. Death sentences and predicted probability of death—test sample

#	Death	Pred	BD	WV	#AC	FV	VS	2V	MS	YV
1	0	0.0044	1	0	2	0	0	0	0	0
2	1	0.0049	1	0	3	0	0	1	0	0
3	0	0.0176	1	0	4	1	0	1	0	0
4	0	0.0223	1	0	2	0	0	1	0	1
5	1	0.0371	1	0	5	1	0	0	0	0
6	0	0.0417	1	0	2	0	1	0	0	0
7	1	0.0417	1	0	2	0	1	0	0	0
8	0	0.0534	0	1	1	0	0	0	0	0
9	0	0.0634	1	0	3	0	1	0	0	0
10	0	0.0806	0	1	2	0	0	0	0	0
11	1	0.0867	0	1	1	1	0	1	0	0
12	0	0.0895	0	1	3	0	0	1	0	0
13	0	0.0951	1	0	4	0	1	0	0	0
14	0	0.1026	1	0	3	1	0	0	0	1
15	0	0.1162	0	1	1	1	0	0	0	0
16	1	0.1325	0	1	4	0	0	1	0	0
17	1	0.1404	1	0	5	0	1	0	0	0
18	1	0.1663	1	1	5	0	0	0	0	0
19	0	0.1697	0	1	2	1	0	0	0	0
20	1	0.1747	0	1	4	0	0	0	0	0
21	0	0.1747	0	1	4	0	0	0	0	0
22	0	0.1747	0	1	4	0	0	0	0	0
23	0	0.1968	1	0	4	1	1	0	0	0
24	0	0.2024	1	0	6	0	1	0	0	0
25	1	0.2122	0	0	5	0	1	0	0	0
26	1	0.3003	0	1	3	0	0	0	0	0
27	1	0.3003	0	1	3	0	0	0	1	0
28	0	0.3250	1	1	7	0	0	0	1	0
29	1	0.3304	0	1	4	1	0	0	0	0
30	1	0.3382	0	1	6	0	0	0	0	0
31	0	0.3705	1	1	3	0	1	1	0	0
32	1	0.4493	1	1	3	0	1	0	0	0
33	1	0.4493	1	1	3	0	1	0	0	0
34	1	0.5001	0	1	3	1	0	0	1	0
35	0	0.5503	1	1	2	1	1	0	0	0
36	1	0.5590	1	1	4	0	1	0	0	0
37	1	0.5590	1	1	4	0	1	0	0	0
38	1	0.5736	0	1	3	0	1	0	0	0
39	1	0.6084	0	1	4	1	0	0	1	0
40	1	0.6553	1	1	3	1	1	0	0	0
41	1	0.6632	1	1	5	0	1	0	0	0
42	1	0.6632	1	1	5	0	1	0	0	0
43	1	0.6686	0	1	2	1	1	0	0	0
44	1	0.6764	0	1	4	0	1	0	0	0
45	1	0.7198	1	1	3	0	1	0	1	0
46	1	0.7471	1	1	4	1	1	0	0	0
47	1	0.7582	0	1	3	1	1	0	0	0
48	0	0.7793	0	1	4	1	0	0	0	1
49	1	0.8346	0	1	6	0	1	0	0	0
50	1	0.9337	1	1	5	0	1	0	0	1

Appendix I:
Calculations and
Comments on the Cases

In this appendix we furnish our calculations of statistical results for most of the problems requiring some calculation. We do not suggest that these are the only "correct" solutions; there are usually a variety of analyses appropriate for a statistical question. The methods we present are designed to illustrate the topics discussed in the text. There are also comments on some of the statistical issues, and occasionally some supplementary technical discussion for those inclined to dig further. For the most part, we have left discussion of the legal issues to the reader.

Chapter 1. Descriptive Statistics

Section 1.2.1. Parking meter heist

1. Average total revenue/month for the ten months of the Brink's contract prior to April 1980 was $1,707,000; the comparable figure for CDC the next year was $1,800,000. The city argued that the difference of about $100,000/month was due to theft.

   An alternative calculation can be made based on the questionable assumption that the $4,500 recovered when the employees were arrested represented an average theft per collection. Since there were about 200 collections in the preceding ten months, the theft in that period was $200 \times 4,500 = 900,000$, close to the figure obtained from the time-series analysis.

2-3. Brink's argued that the difference was due to an underlying trend, pointing to the figures in Area 1-A, where no theft occurred. Average monthly revenue for Area 1-A during the last ten months of the Brink's period was $7,099; its average during the CDC period was $7,328. The difference of $229 is a percentage increase of 3.2%. A 3.2% increase applied to $1,707,000 comes to $1,762,000, which leaves only $38,000/month, or $380,000 for ten months, unaccounted for. At the trial, Brink's adduced the evidence from Area 1-A, but did not stress the comparison. Its chief defense was that confounding factors had contributed to an overall trend that biased the city's estimate. The jury found for the city and awarded $1,000,000 compensatory damages and punitive damages.

Section 1.2.2. Taxing railroad property

1. The language of the statute, which refers to the assessed value of "other commercial and industrial property in the same assessment jurisdiction," suggests using the ratio of the aggregate assessed value to the aggregate market value; that would be a weighted mean. In its generic sense of "not unusual," the word "average" is not informative on the choice between the mean and the median. However, in its specific sense the "average" is the mean, and "weighted average" could only be the weighted mean. The issue is made murky by the comment of the railroad spokesman, which suggests that the mean was not intended to be used in applying the statute.

2. Because the median minimizes the sum of absolute deviations, it might be said to minimize discrimination with respect to the railroads.

3. Total assessment is calculated as the mean assessment rate times the number of properties; use of the mean would therefore produce no change in revenue.

4. The median is the middle measurement for all properties. Its treatment of all properties as equal can be regarded as most equitable to the railroads.

5. The weighted mean assessment/sales ratio is a weighted harmonic mean because it is the reciprocal of the weighted mean of the reciprocals of the assessment/sales ratios in each stratum, with the weights being the proportions that the assessments in each stratum bear to total assessments.

Section 1.2.4. Hydroelectric fish kill

The sample arithmetic mean is equal to the sum of the observations in the sample divided by the sample size, n. From this definition, it is obvious that, if one is given the mean of a sample, the sum of the sample observations can be retrieved by multiplying the mean by n. The sample can then be "blown up" to the population by multiplying it by the reciprocal of the proportion that the sample size bears to the whole. In short, the arithmetic mean fish kill per day times the number of days in the year estimates the annual fish kill. However, for the geometric mean the first step cannot be taken because the geometric mean is the nth root of the product of the n observations; multiplying that by n does not yield the sum of the sample observations. Hence, the geometric mean fish kill per day times the number of days in the year does not estimate the annual kill. Notice also that if on any day in the sample of days there were no fish killed, the geometric mean would equal zero, clearly an underestimate. So the geometric mean procedure is inappropriate.

Section 1.2.5. Pricey lettuce

1. The ratio of new to old lettuce prices, using the arithmetic mean index with constant dollar shares (R_A), can be expressed algebraically as $R_A = \sum S_{0,i}(P_{t,i}/P_{0,i})$, where $S_{0,i}$ is the dollar share of the ith product in the base period; $P_{t,i}$ is the price of the ith product at time t; $P_{0,i}$ is the price of the ith product in the base period; and the summation is across products being averaged. Thus the arithmetic average ratio of new to old lettuce prices is $(1/2)(1/1) + (1/2)(1.5/1) = 1.25$, indicating a 25% increase in price of lettuce. The new quantities implied by this increase are 1.25 lbs of iceberg and 0.8333 lbs of romaine lettuce. (These are the quantities that keep the expenditure shares for each kind of lettuce equal, given that the total amount spent on lettuce increases by 25%.) Since the reduction in amount of romaine is less than the increase in quantity of iceberg, the assumption of equal satisfaction implies that consumer satisfaction is greater per unit of romaine than of iceberg.

2. Using the geometric mean, index, the ratio of new to old prices can be expressed as $R_G = \prod(P_{t,i}/P_{0,i})^{S_{0,i}}$, where $\prod$ indicates the product of the relative price changes across items. The geometric average ratio of new to old lettuce prices is $(1/1)^{1/2} \times (1.5/1)^{1/2} = 1.225$, for a 22.5% average increase. The new quantities implied by this increase are 1.225 lbs of iceberg and 0.816 lbs of romaine lettuce. Again, consumer satisfaction is deemed greater per unit of romaine than of iceberg, although the difference is smaller than under the arithmetic mean computation.

3. The arithmetic mean ratio if romaine drops back from \$1.5 to \$1 is $(1/2)(1/1) + (1/2)(1/1.5) = 0.8333$, which is greater than 0.8, the reciprocal of 1.25. The geometric mean ratio is $(1/1)^{1/2} \times (1/1.5)^{1/2} = 0.816$, which is the reciprocal of 1.225. The geometric mean ratio would appear superior to the arithmetic mean ratio from this point of view because it matches reciprocal price movements with reciprocal changes in the index.

Section 1.2.6. Super-drowsy drug

1. Baker's expert used in effect an average percentage reduction in sleep latency weighted by the baseline latency. His weighted average was greater than the FTC simple or unweighted average because, in the data, larger percentage reductions are associated with longer baseline latencies. If there

were a constant mean percentage reduction of sleep latency for all baselines, either the weighted or unweighted average would be an unbiased estimator of it. Our choice would then depend on which was more precise.

2. If the variability of reduction percentages were the same for all baseline latencies, the unweighted sample average (the FTC estimate) would be the preferred estimator because it would have the smallest sampling variance. But if, as Figure 1.2.6 indicates, variability in the reduction percentage decreases with increasing baseline latency, the more precise estimator is the weighted average (the Baker estimate), which gives greater weight to larger baseline latencies. Regression methods (see Chapter 13) can be applied to study the percentage reduction as a function of baseline latency.

Able's exclusion of data in the baseline period is clearly unacceptable, as the week 2 latency average would appear shorter even with exactly the same values as in week 1. Baker's entry criteria and non-exclusion of data are more reasonable, defining a target population of insomniacs as those with longer than 30 minute latencies on at least 4 out of 7 nights. Such entry criteria ordinarily lead to some apparent reduction in sleep latency, even without the sleep aid, by an effect known as the regression phenomenon (see Section 13.1 at p. 350). The total effect is a sum of drug effect, placebo effect, and regression effect.

Section 1.3.1. Texas reapportionment

1. a. The range is 7,379. The range in percentage terms is from $+5.8\%$ to -4.1%, or 9.9 percentage points.

   b. The mean absolute deviation is 1,148 (mean absolute percentage deviation = 1.54%).

   c. Using the formula

   $$S = \left\{ \left(\frac{n}{n-1} \right) \left[\left(\frac{1}{n} \sum x_i^2 \right) - \left(\frac{1}{n} \sum x_i \right)^2 \right] \right\}^{1/2},$$

   the standard deviation of district size is 1454.92 (if the exact mean 74,644.78 is used), or 1443.51 (if 74,645 is used). This formula is sensitive to the least significant digits in the mean. A better way to calculate S is from the mean squared deviation

   $$S = \left\{ \left(\frac{n}{n-1} \right) \left[\frac{1}{n} \sum (x_i - \bar{x})^2 \right] \right\}^{1/2},$$

   which does not suffer from this instability. The sd of percentage deviation is 1.95%. We give the above formula because it is the usual one used for estimating the standard deviation from samples; in this context it could well be argued that the data are the population, in which case the factor $n/(n-1)$ should be deleted.

d. The interquartile range is $75,191 - 73,740 = 1,451$. In percentage terms it is $0.7 - (-1.2) = 1.9$.

e. The Banzhaf measure of voting power (see Section 2.1.2, Weighted voting) argues for the use of the square root of the population as a measure of voting power. Transforming the data by taking square roots, the range becomes $280.97 - 267.51 = 13.46$. For percentage deviation from the average root size of 273.2, the range is $+2.84\%$ to -2.08%, or 4.92 percentage points.

2. The choice among measures depends initially on whether one is most influenced by the most discrepant districts or by the general level of discrepancy. The former points to the range; the latter to one of the other measures. In any event, there would seem to be no reason to use squared differences, i.e., the variance or standard deviation, as a measure.

Section 1.3.2. Damages for pain and suffering

1. Treating the data as the population ($ in 000's), the variance is $913,241 - 558,562 = 354,679$. The standard deviation is $\sqrt{354,679} = 595.55$. The mean is 747.37. Hence, the maximum award under the court's two-standard-deviation rule would be $747.37 + 2 \times 595.55 = 1,938.47$. Probably it is more correct to treat the data as a sample from a larger population of "normative" awards. In that case, the variance would be estimated as $354,679 \times \frac{27}{26} = 368,320.5$. The standard deviation would be $\sqrt{368,320.5} = 606.89$, and the maximum award under the court's rule would be $747.37 + 2 \times 606.89 = 1,961.15$.

2. The court's use of a two-standard-deviation interval based on a group of "normative" cases to establish permissible variation in awards assumes that the pain and suffering in each case was expected to be the same, apart from sampling variability, assumed permissible, and was similar to that suffered in the Geressy case. This seems unlikely since, e.g., an award of $2,000,000 would probably have been deemed excessive if made to the case that had received $37,000. If, as seems likely, different levels of pain and suffering are involved over the range of cases, one would have to locate *Geressy* on that continuum, and then construct an interval for that level. Two standard deviations calculated from the normative cases would no longer be appropriate because that would reflect both variation at a particular level and variation across levels, the latter being irrelevant to the reasonableness of the award in the particular case.

These are linear regression concepts – see Section 13.3.

Section 1.3.3. Ancient trial of the Pyx

1. The remedy was 7,632 grains. If $\sigma = 0.42708$, then the standard deviation of the sum of $n = 8,935$ differences of a sovereign's actual weight from the standard is $\sigma \cdot \sqrt{n} = 40.4$ gr. Thus, the deviation represented by the remedy in standard deviation units is $7,632/40.4 = 188.9$ sd. The probability of finding an absolute sum of n differences greater than 188.9 standard deviations is, by Chebyshev's inequality, less than $1/188.9^2 = 2.8 \times 10^{-5}$, vanishingly small.

2. The remedy grows too large because it increases by the factor of n instead of $\sqrt{n}$. The remedy of $3\sigma\sqrt{n}$ should have amply protected the Master of the Mint without giving him excessive leeway; in the example, it would have been $3 \times 0.42708 \cdot \sqrt{8935} = 121.1$ gr. Whether Newton would have appreciated the flaw in the remedy formula has been the subject of some conjecture. See Stigler, *supra*, in the *Source*.

The square root law may be inadequate if the individual coins put into the Pyx were not statistically independent, e.g., because they were from the same mint run. Such dependence tends to reduce the effective sample size.

Section 1.4.1. Dangerous eggs

1. Pearson's correlation coefficient between mean egg consumption (EC) and mean ischemic heart disease (IHD) is 0.426.

2. The probative value of ecologic studies such as this is considered extremely limited. The correlational study of national means says little about the strength of correlation among individuals within the societies. The "ecologic fallacy" results from an uncritical assumption that the observed association applies at the individual level. In fact, it is theoretically possible for EC and IHD to be uncorrelated among individuals within a given society, while societal mean levels of EC and IHD co-vary with a confounding factor that itself varies across societies. For example, red meat consumption in societies with high-protein diets might be such a confounding factor, i.e., a factor correlated with both EC and IHD.

Counsel for the egg producers might argue that, in view of the ecologic fallacy, statistical comparisons should be limited to countries similar to the United States, at least in egg consumption. When that is done, however, the correlation between IHD and egg consumption virtually disappears.

3. Errors in the data would tend to reduce the correlation.

Section 1.4.2. Public school financing in Texas

1. The correlation in Texas between state and local revenues per pupil and median family income from 1960 for the five groups (unweighted) is 0.875. On a weighted basis, treating each district as separate but using median values for each group, the Texas correlation is 0.634. (The effect of grouping districts is to remove some of the variability in the data, thereby increasing the correlation.) Using the first figure one would say that $0.875^2 = 76.6\%$ of the variation in revenues per pupil is accounted for by variation in median family income; using the weighted correlation the figure is $0.6342^2 = 40.2\%$.

The overall correlation is reduced by the inverse relation in the center of the data. The inversion is not very important because the median income figures in the three districts are quite close to the overall median figure.

2. The possible objection that only a few extreme districts were involved cannot be resolved because, remarkably, it does not appear that Professor Berke stated how his 10% sample of districts had been selected.

Section 1.5.1. Proficiency test with a disparate impact

1. The computations are as follows:

	Pass rates:	Fail rates:
Blacks:	$448/770 = 0.582$	$322/770 = 0.418$
Whites:	$240/341 = 0.704$	$101/341 = 0.296$

Difference in rates:

Blacks to Whites:	$0.582 - 0.704 = -0.122$	$0.418 - 0.296 = 0.122$
Whites to Blacks:	$0.704 - 0.582 = 0.122$	$0.296 - 0.418 = -0.122$

Ratio of rates (relative risk):

Blacks to Whites:	$0.582/0.704 = 0.827$	$0.418/0.296 = 1.412$
Whites to Blacks:	$0.704/0.582 = 1.210$	$0.296/0.418 = 0.708$

Odds Ratio:

Blacks to Whites: $\dfrac{(0.582/0.418)}{(0.704/0.296)} = 0.585$ $\dfrac{(0.418/0.582)}{(0.296/0.704)} = 1.708$

Whites to Blacks: $\dfrac{(0.704/0.296)}{(0.582/0.418)} = 1.708$ $\dfrac{(0.296/0.704)}{(0.418/0.582)} = 0.585$

2. Under the express terms of the four-fifths rule, the data do not indicate adverse impact because the relative risk is 0.827, which exceeds 0.80. However, in light of the policy of the rule, there might be said to be adverse impact because the odds of a black passing are only 0.585 of the odds that a white would pass.

If the test purported to be one of minimum qualifications (e.g., absence of an arrest record), the principal focus would be on failure rates. From that perspective, the case for adverse impact is stronger on these data since the white failure rate is only 0.708 of the black failure rate. Notice that the odds ratio for black failure vs. white failure is the same (0.585) as the odds ration for white pass vs. black pass. If the EEOC had used odds ratios rather than relative risks in its rule, it would have avoided the problem of possible inconsistency in results between pass and fail rates.

Section 1.5.2. Bail and bench warrants

1. The rates of bench warrants previously issued among those currently issued (54/146) and not currently issued (23/147) are not dependent on the sample sizes for current issuance. However, these rates are only of indirect interest.

2. The statistics of direct interest are the rates of bench warrants currently issued among those previously issued (54/77) and not previously issued (92/216). But these rates are dependent on the sample sizes with respect to the currently issued warrants, which are arbitrary features of the study design.

3. Since the study was retrospective, to express the value of using prior issuance as a predictor, one must use the odds ratio, which is

$$\frac{54/92}{23/124} = \frac{54/23}{92/124} = \frac{54 \times 124}{23 \times 92} = 3.16.$$

In its prospective interpretation, this says that the odds on a current bench warrant issuing if one has been previously issued are three times the odds on a current bench warrant issuing if one has not been previously issued.

4. One cannot tell from these data how many non-appearances would be avoided if prior issuance of a warrant automatically led to denial of bail because that figure (known as the attributable risk, see Section 10.2) requires knowledge of the proportion of prior issuance and the relative risk of a warrant currently issuing given a prior issuance (or not) and those quantities cannot be determined from these data.

Section 1.5.3. Non-intoxicating beer

1. The relative risk of DWI arrest by gender is $2/0.18 = 11.11$. The attributable risk due to maleness, using formula (3) on p. 287, is

$$\frac{x}{1 + x} = \frac{5.06}{1 + 5.06} = 0.835,$$

where x = the proportion of males (0.5) times (R.R. $- 1$) = $(0.5)(11.11 - 1) = 5.06$. Thus, even though relatively few men are involved, the higher arrest rate of men accounts for over 80% of the arrests.

2. Justice Brennan's statement is confusing, but appears to address an issue beyond the large relative risk. His conclusion assumes that the low percentage of arrests reflects the extent of the drinking and driving problem among young people who drink 3.2% beer. In that case, the small numbers involved would make a gender-discriminatory statute less acceptable in terms of social necessity, even in the face of the large relative risk. A difficulty is that the data are not satisfactory for assessing the extent of the problem. On the one hand, the number of arrests clearly underestimates the extent of the drinking and driving problem. But on the other hand, arrests for all alcoholic beverages clearly overstate the problem for 3.2% beer. There is no reason to think that these biases would balance out. In addition, since the study was made after the statute was passed, it is not clear what direct bearing it has on assessing the conditions that led to and would justify its passage.

Chapter 2. How to Count

Section 2.1.1. DNA profiling

1. There are 20 distinguishable homozygous and $(20 \times 19)/2 = 190$ heterozygous genotypes possible at the locus (not distinguishing between chromosomes of the pair). Thus, there are 210 distinguishable homozygous and heterozygous pairs of alleles at a locus.

2. With four loci, there are $210^4 = 1.94 \times 10^9$ possible homozygous and heterozygous genotypes at the loci (again, not distinguishing between chromosomes of the genotype at a locus).

Section 2.1.2. Weighted voting

1. There are 182 ways the supervisors from the eight small towns can vote so that the supervisor from the larger town can cast the deciding vote, as shown in the third column of the table below.

Yes	No	# of Combinations
4	4	70
5	3	56
3	5	56
		182

2. There are 42 ways a small town supervisor can find himself casting a deciding vote:

Yes	No	# of Combinations
big town (3)+2	5	21
5	big town (3)+2	21
		42

3. Total number of decisive votes: $182 + 8 \cdot 42 = 518$. Proportion of decisive votes for the big town: $182/518 = 35.1\%$. Since it has only 3/11 or 27.3% of the total population, it may be said to be overrepresented, if the definition of voting power is accepted. The smaller towns are correspondingly underrepresented. They each have 42/518 of voting power, or 8.1%, while they have 1/11 or 9.1% of the total population.

4. If the theory is applied at the voter level, the supervisor from the larger town should have not three times the voting power, but $\sqrt{3}$ times the voting power, because a voter's probability of breaking a tie for election of a representative declines not in proportion to the increase in number of votes, but in proportion to the square root of their numbers. This result can be obtained in various ways, among them by applying Stirling's approximation (see Section 2.1 at p. 44) to the binomial coefficient $\binom{2N}{N}$.

Section 2.1.3. Was the bidding rigged?

The nine low bids can be distributed in the occupancy sequence 2, 2, 1, 1, 1, 1, 1 in

$$\frac{9!}{2!2!1!1!1!1!1!} = 90,720$$

ways. The total number of ways of distributing the occupancy sequence among the seven firms is $\binom{7}{2} = 21$. Thus, the total number of ways of obtaining the observed occupancy numbers without regard to sequence is $21 \times 90,720 = 1,905,120$. The total number of ways of distributing the 9 low bids among the 7 firms is $7^9 = 40,353,607$. Thus, the probability of observing the actual set of occupancy numbers is $1,905,120/40,353,607 = 0.047$, assuming equally likely outcomes.

The model of equal probability seems appropriate since the firms claim that they are similar and act independently. Under that model, the observed distribution is *not* in fact the most likely. (For example, the occupancy sequence 3, 2, 2, 1, 1, 0, 0 has probability 0.236.) Among all occupancy sequences, 2, 2, 1, 1, 1, 1, 1 has the smallest range (maximum minus minimum), which is another statistic suggesting collusive allocation.

Section 2.1.4. A cluster of leukemia

We seek the probability that, in a uniform multinomial distribution with 6 cells and sample size 12, the largest cell frequency would be 6 or more. While a computer algorithm is generally indispensable for calculating a quantity of this type, in the present case exhaustive enumeration of all cases involving a maximum frequency of 6 or more is feasible. There are 29 such sets of occupancy numbers (for which multinomial probabilities must be computed). The exact answer is 0.047544. In the bid-rigging problem (Section 2.1.3), the distribution was too uniform while in this problem it is too clumped. The point is that either extreme is improbable under the hypothesis of equal probability for all cells.

Bonferroni's inequality may also be applied with excellent accuracy; see Section 3.1 at p. 57.

Section 2.1.5. Measuring market concentration

1. The number of ways N customers can be divided among i firms with a_i customers of firm i is the multinomial coefficient $N!/\prod a_i!$

2. Setting $1/n_i = a_i/N$, we have $a_i = N/n_i$. Substituting in the multinomial coefficient we have $N!/\prod (N/n_i)!$

3. Using the crude form of Stirling's approximation for the factorials, we have

$$\frac{N^N e^{-N}}{\prod (N/n_i)^{N/n_i} e^{-N/n_i}}.$$

Since $\sum 1/n_i = 1$, we have $\prod e^{-N/n_i} = e^{-N}$ and $\prod N^{N/n_i} = N^N$, so that canceling e^{-N} and N^N from numerator and denominator and taking the Nth root, we have $\prod n_i^{1/n_i}$ as the entropy measure of concentration.

4. In the example, the value of the entropy index is obtained as follows. The value of n_i is 5 for the 20% firms, 50 for the 2% firms, and 100 for the 1% firms. When there are 10 small firms, the entropy measure is $5^{0.20 \times 4} \times 50^{0.02 \times 10} = 7.92$ equal firms; when there are 20 small firms the entropy measure is $5^{0.20 \times 4} \times 100^{0.01 \times 20} = 9.10$ equal firms. Thus, by the entropy measure, the number of small firms is relevant to competition. By the Herfindahl index (HHI), when there are 10 small firms, HHI is $20^2 \times 4 + 2^2 \times 10 = 1640$, or $10,000/1640 = 6.10$ equal firms; by a similar calculation, when there are 20 small firms, HHI is $10,000/1620 = 6.17$. By HHI, the number of small firms makes little difference.

These two indices form the basis for two classical tests of the null hypothesis of uniform distribution of customers in a multinomial model. The entropy measure gives rise to the likelihood ratio statistic (see Section 5.6), while HHI gives rise to the Pearson chi-squared statistic (see Section 6.1).

Section 2.2.1. Tracing funds for constructive trusts

1. Under the reflection principle of D. André, to find the number of paths that begin at $10, touch or cross the 0 ordinate, and end at 10, we count the number of paths that end at −$10. This is equal to $\binom{100}{40}$. The total number of paths that end at 10 is $\binom{100}{50}$. Hence, the probability that a path would touch or cross 0 and end at 10 is

$$\binom{100}{40} \Big/ \binom{100}{50} = \frac{50!^2}{40! \cdot 60!} = 0.136.$$

2. The expected maximum reduction in the trust is $0.627\sqrt{100} = 6.27$ dollars.

3. The probability of no $10 depletion is $1 - 0.136 = 0.864$. This number raised to the 10th power is 0.232, the probability of no $10 depletion on any of the 10 days. Therefore, the probability that a $10 depletion occurs at some point during the ten day period is $1 - 0.232 = 0.768$.

The fact that we apply probability theory here should not be construed as a recommendation for the rule of law that suggests the excursion into fluctuation theory, but which otherwise seems to have little to recommend it.

Chapter 3. Elements of Probability

Section 3.1.1. Interracial couple in yellow car

1-2. The identifying factors are almost certainly not independent. However, if probabilities are interpreted in the conditional sense, i.e., $P[e|f] = 1/10$; $P[d|e$ and $f] = 1/3$; $P[c|d$ and e and $f] = 1/10$, etc., the probability of the factors' joint occurrence would be equal to the product.

3. The calculation in the court's appendix assumes that the selection of couples who might have been at the scene from the larger universe was made *with replacement*. This assumption would produce a 0.41 probability of including a C-couple twice in the selection in any case in which a C-couple had been included once, even if they were the only such couple in the population. An analysis assuming sampling without replacement would have been more appropriate.

4. The prosecutor assumed, in essence, that the frequency of C-couples in the population was the probability of the Collinses' innocence. It is not that, but the probability that the guilty couple would have been a C-couple *if* the Collinses' were innocent. This inversion of the conditional is sometimes called the prosecutor's fallacy. The two probabilities would be of the same magnitude only if the other evidence in the case implied a 50% probability of the Collinses' guilt. See Sections 3.3 and 3.3.2.

5. The last clause of the last sentence of the court's appendix assumes that the probability of the Collinses' guilt is one over the number of C-couples. This is called the defendant's fallacy because it assumes that there is no other evidence in the case except the statistics, which is generally not true and was not true here. See the sections referred to above.

Section 3.1.2. Independence assumption in DNA profiles

1. The weighted average frequency of the homozygous genotypes consisting of allele 9 is $\left(\frac{130}{916}\right)^2 \times \frac{916}{2844} + \ldots + \left(\frac{52}{508}\right)^2 \times \frac{508}{2844} = 0.01537$. For heterozygous genotypes consisting of alleles 9 and 10, the weighted average calculation is

$$\left[\frac{130}{916} \times \frac{78}{916} \times 2\right] \frac{916}{2844} + \ldots + \left[\frac{52}{508} \times \frac{43}{508} \times 2\right] \frac{508}{2844} = 0.02247.$$

2. The frequency of a homozygous genotype consisting of allele 9 using the total population figure is $\left(\frac{350}{2844}\right)^2 = 0.01515$. For the heterozygous genotype consisting of alleles 9 and 10, the calculation based on total population figures is $\frac{350}{2844} \times \frac{261}{2844} \times 2 = 0.02259$.

3. In both the homozygous and heterozygous cases, the agreement with the weighted average calculation is very good, indicating that HW is justified.

4. A sufficient condition for HW in the total population is that the rates of the alleles in the subpopulations are the same (or, HW holds approximately, if the rates are not very different).

5. The weighted average for Canadians and non-Canadians is 0.3782 for a homozygous genotype of allele 9. The figure based on population totals is 0.25. HW doesn't hold because the rates of allele 9 in the two subpopulations are very different.

Section 3.1.4. Telltale hairs

1. The total number of pairs among the 861 hairs is $\binom{861}{2} = 370,230$. Of these, there were approximately 9 hairs per person or $\binom{9}{2} = 36$ intra-person pairs, yielding a total of 3.600 intra-person pairs. Of the net total of 366,630 inter-person pairs, 9 were indistinguishable, a rate of $9/366,630 = 1/40,737$; the rest of the derivation is in the Gaudette–Keeping quotation in the text.

Gaudette–Keeping assumes that the probability of one pair's distinguishability is independent of the distinguishability for other pairs. Although such independence is unlikely, violation of the assumption does not matter much because the probability of one or more matches is less than or equal to $9 \cdot 1/40,737$ (Bonferroni's inequality), or 1/4,526—a result close to the 1/4,500 figure given in the text. The substantial problems with the study are its lack of blindedness and the unbelievability of the number of asserted comparisons.

The probability reported is *not* the probability that the hair came from another individual but the probability that it would be indistinguishable if it came from another individual.

Section 3.2.1. L'affaire Dreyfus

1. The expert computed the probability of *exactly* four coincidences in the particular words in which they occurred.

2. The more relevant calculation is the probability of *at least* four coincidences in *any* of the thirteen words. This is 0.2527, on the assumption that the occurrence of a coincidence in one word is independent of coincidences in others.

Section 3.2.2. Searching DNA databases

1-2. Multiplying by the size of the database is a way of applying Bonferroni's adjustment for multiple comparisons. On the assumption that no one in the database left the incriminating DNA, and that each person's DNA had the same probability of matching, the approximate probability of one or more matches is given by the Bonferroni adjustment. This probability would seem to be only peripherally relevant to the identification issue.

3. In the fiber-matching case, the question was whether *any* fibers matched; in this case, the question is whether the DNA of a particular person matched by coincidence, not whether *any* person's DNA matched.

4. It should make no difference to the strength of the statistical evidence whether particular evidence leads to the DNA testing, or is discovered afterward. Because no adjustment for multiple comparisons would be made in the former case, none should be made in the latter. As for the argument that with a large database some matching would be likely because there are multiple trials, if a match occurs by coincidence there is unlikely to be particular evidence supporting guilt, and probably some exonerating evidence. This means that the prior probability of guilt would be no greater than 1 over the size of the suspect population and the posterior probability of guilt would be quite small.

5. Because the probative effect of a match does not depend on what occasioned the testing, we conclude that the adjustment recommended by the committee should not be made. The use of Bonferroni's adjustment may be seen as a frequentist attempt to give a result consistent with Bayesian analysis without the clear-cut Bayesian formulation.

Section 3.3.1. Rogue bus

4. The prior odds that it was company B's bus are $0.20/0.80 = 1/4$. The likelihood ratio associated with the testimony that it was company B's bus is $0.70/0.30 = 7/3$. The posterior odds that it was company B's bus, given the prior odds and the testimony, are $\frac{1}{4} \times \frac{7}{3} = 0.583$. The posterior probability that it was company B's bus is $0.583/(1 + 0.583) = 0.368$.

Section 3.3.2. Bayesian proof of paternity

1. The expert's calculation is wrong, but the arithmetic error is harmless. Given the prior odds of 1 and a genotype rate of 1%, the posterior odds are $1 \times 1/0.01 = 100$. This translates into a posterior probability of paternity of $100/101 = 0.991$. Since the expert used her own prior, which might have been irrelevant, it would seem that her calculation should not have been admitted.

The appellate court held it was improper for the expert to use a 50-50% prior, but indicated that the jurors might be given illustrative calculations based on a range of priors, as suggested in question 3.

Section 3.4.1. Airport screening device

The defendant will argue that among those persons who screen as "high risk," or "+," only the proportion

$$P[W|+] = \frac{P[+|W] \cdot P[W]}{P[+|W] \cdot P[W] + P[+|\bar{W}] \cdot P[\bar{W}]}$$

actually carry a concealed weapon. In the example $P[W] = 0.00004$, $P[+|W] = \text{sensitivity} = 0.9$ and $P[+|\bar{W}] = 1 - \text{specificity} = 0.0005$. Thus,

$$P[W|+] = \frac{0.9(0.00004)}{0.9(0.00004) + 0.0005(0.99996)} = 0.067,$$

or 6.7% of "high risk" individuals carry weapons. This rate is arguably too low for either probable cause or reasonable suspicion.

Assuming independence of test results for any person in the repeated test situation of the Notes, sensitivity falls to $P[+|D] = P[+_1|D] \cdot P[+_2|D, +_1] = 0.95 \cdot 0.98 = 0.9310$. Specificity increases such that $P[+|\bar{D}] = P[+_1|\bar{D}] \cdot P[+_2|\bar{D}, +_1] = 0.05 \cdot 0.02 = 0.001$. Then PPV $= (0.001 \cdot 0.931)/(0.001 \cdot 0.931 + 0.001 \cdot 0.999) = 0.4824$, which is still rather low as a basis for significant adverse action. Specificity would have to be much greater to have adequate PPV.

Section 3.4.2. Polygraph evidence

1. Using the Defense Department data, the PPV of the test is no greater than 22/176=12.5%.

2. A statistic to compute would be either NPV or $1 - $ NPV. The latter is the probability of guilt given an exonerating polygraph.

3. It seems reasonable to conclude that PPV $> 1 - $ NPV because the probability of guilt when there is an incriminating polygraph should be greater than when there is an exonerating polygraph. Thus, the probability of guilt given the exonerating test is less than 12.5%. The test seems quite accurate when used for exoneration even if not very persuasive when used as evidence of guilt. Because

$$
\begin{aligned}
\frac{\text{NPV}}{1 - \text{NPV}} &= P[\text{not guilty}|\text{exonerating exam}]/P[\text{Guilty}|\text{exon}] \\
&= \frac{P[\text{exon}|\text{not } G]P[\text{not } G]}{P[\text{exon }|G]P[G]} \\
&= \frac{\text{specificity}}{1 - \text{sensitivity}} \times \text{prior odds on not guilty,}
\end{aligned}
$$

the NPV will be high if the sensitivity of the test is high, or if prior probability of guilt is low.

4. The statistical point is that rates of error for a test that would seem acceptably small when the test is used as evidence in an individual case are unacceptably large in the screening context; the screening paradigm is more exacting than the individual case, not less. Because $7,616 - 176 = 7,440$ persons tested negative in the Defense Department screening program, if $1 - $ NPV were in fact as high as 12.5%, the number of guilty but undetected personnel would be $0.125 \times 7,440 = 930$. A screening test that passed so many questionable personnel would clearly be unacceptable. The fact that the Defense Department uses the test for screening is thus significant evidence that it is sufficiently reliable to be admitted, at least as exonerating evidence.

Section 3.5.2. Cheating on multiple-choice tests

Under the assumption of innocent coincidence, the likelihood of 3 matching wrong answers in dyads with 4 wrong answers is 37/10,000. Under the assumption of copying, the expert used 1 as the likelihood of 3 matching wrong answers out of 4. Thus, the likelihood ratio is $\frac{1}{37/10,000} = 270.27$. The interpretation is that one is 270 times more likely to observe these data under the assumption of copying than of innocent coincidence.

Chapter 4. Some Probability Distributions

Section 4.2.1. Discrimination in jury selection

1. The probability of no black selections at random out of 60 is $(0.95)^{60} = 0.046$. This is statistically significant under the usual 5% standard (using a one-tailed test, see Section 4.4), but only marginally so.

2. The Court computed the probability of observing exactly 7 blacks. The more relevant is the probability of observing 7 *or fewer* blacks.

3. From Table B, for $n = 25$, $p = 0.25$, the probability of 3 or more blacks is 0.9679; the probability of 2 or fewer is $1 - 0.9679 = 0.0321$.

Section 4.2.3. Small and nonunanimous juries in criminal cases

1. The probability of selecting one or more minority jurors, by the binomial model and Table B in Appendix II, is 0.931; two or more is 0.725; three or more is 0.442; and four or more is 0.205. Thus, the probability that one or more minority jurors must concur in a 12-person verdict is 0.931, and in a 9–3 verdict it is 0.205.

2. Using the Kalven and Zeisel overall proportion of guilty votes on first ballots, $1828/2700 = 0.677$, the binomial probability of 9 or more guilty votes is 0.423 and the probability of 5 unanimous guilty votes is 0.142. A prosecutor is therefore much more likely to get at least 9 votes for conviction than a unanimous 5, on the first ballot.

3. The Kalven and Zeisel data are inconsistent with the binomial model because the expected number of cases arrayed by first ballot votes does not correspond with the actual number, as the following table shows:

	First ballot guilty votes				
	0	1 – 5	6	7-11	12
Observed number of cases	26	41	10	105	43
Expected number of cases	0	12.7	22.7	187.5	2.09

See Section 6.1 for a formal test of goodness of fit. The data show many more unanimous verdicts than one would expect on the basis of the model, suggesting that the greater source of variation in guilty votes is the strength of the case, which affects all jurors, rather than variation in individual jurors.

Section 4.2.4. Cross-section requirement for federal jury lists

1. The probability of 0 or 1 blacks in a 100-person venire, with probability of picking a black, $p = 0.038$ (the functional wheel), is $\binom{100}{1}(0.038)^1(0.962)^{99} + (0.962)^{100} = 0.103$. With $p = 0.0634$, the pro-

portion of blacks in a representative wheel, the probability of so few blacks is 0.011. By this probability test, the functional wheel seems unrepresentative. The comparable probabilities for Hispanics are 0.485 and 0.035 for the functional and representative wheels, respectively.

2. As for absolute numbers: $6.34 - 3.8 = 2.54$ blacks and $5.07 - 1.72 = 3.35$ Hispanics would have to be added to the average venire chosen from the functional wheel to make it representative. These numbers might be regarded as too small to indicate that the wheel was unrepresentative.

3. One problem with the probabilities test is that the hypothetical number of minorities used to compare probabilities (here a maximum of 1) is fairly arbitrary. A problem with the absolute numbers test is that, when the percentages are small the numbers will be small, even if the underrepresentation, in proportionate terms, is large.

Section 4.3.1. Alexander: *Culling the jury list*

1. The result is approximately correct. The calculations are as follows. Given $n = 400$ and $p = 1015/7374$,

$$P[X \leq 27]$$
$$= P\left[z < \frac{27.5 - 400(1015/7374)}{\sqrt{400 \cdot (1015/7374) \cdot (6359/7374)}} = -4.00\right]$$
$$< 0.00003,$$

from Table A1 or A2 in Appendix II.

2. The largest proportion of blacks in the qualified pool, p^*, consistent with the data would satisfy the equation

$$\frac{27.5 - 400p^*}{\sqrt{400p^*q^*}} = -1.645.$$

Solving for p^* yields $p^* = 0.0926$. (This is a one-sided upper 95% confidence limit for the proportion; see Section 5.3.) To calculate the qualification rate ratio, note that of Q denotes qualification, B black, and W white,

$$\frac{P[Q|W]}{P[Q|B]} = \frac{P[W|Q]P[Q]/P[W]}{P[B|Q]P[Q]/P[B]} = \frac{P[W|Q]/P[B|Q]}{P[W]/P[B]}.$$

The odds on W vs. B among the qualified is at least $0.9074/0.0926 = 9.800$, while in the initial pool the odds on being white are $6359/1015 = 6.265$. The qualification rate ratio is thus at least $9.8/6.265 = 1.564$, i.e., whites qualify at a minimum of 1.5 times the rate that blacks qualify.

Section 4.4.1. Hiring teachers

1. Given $n = 405$ and $p = 0.154$,

$$P[X \le 15] = P\left[Z < \frac{15.5 - 405 \cdot 0.154}{\sqrt{405 \cdot 0.154 \cdot 0.846}} = -6.45\right] \approx 0.$$

given $p = 0.057$,

$$P[X \le 15] = P\left[Z < \frac{15.5 - 405 \cdot 0.057}{\sqrt{405 \cdot 0.057 \cdot 0.943}} = -1.626\right] \approx 0.052.$$

The exact result is 0.0456.

2. Justice Stevens's clerk used a one-tailed test; the *Castaneda* two- or three-standard-deviations "rule" reflects the general social science practice of using two-tailed tests.

Section 4.5.1. Were the accountants negligent?

1. $\binom{983}{100} \cdot \binom{17}{0} / \binom{1000}{100} = \frac{983 \cdot 982 \cdots 884}{1000 \cdot 999 \cdots 901} = 0.164.$

Because this computation is somewhat laborious, the normal approximation would commonly be used instead.

2. If the sample were twice as large, the probability of finding none would be

$$P[X = 0] = P[X < 0.5]$$

$$= P\left[Z < \frac{0.5 - 17(0.2)}{\sqrt{\frac{200 \cdot 800 \cdot 17 \cdot 983}{1000^2 \cdot 999}}} = -1.77\right] \approx 0.04.$$

3. The accountants should (and frequently do) stratify their sample, with separate strata and heavier sampling for the large invoices.

Section 4.5.2. Challenged election

1. Let a denote the number of votes for Candidate A and b the votes for Candidate B. Let m denote the number of invalid votes. If X denotes the unknown number of invalid votes cast for A then a reversal would occur upon removal of the invalid votes if $a - X \le b - (m - X)$, or $X \ge (a - b + m)/2$. On the facts, if 59 of the invalid votes were cast for the winner, the election would be tied by the removal of all the invalid votes.

2. $E(x) = 101 \cdot 1422/2827 = 50.8.$

3. Using the normal approximation for the hypergeometric, the probability that $X \geq 59$ is

$$P\left[Z \geq \frac{58.5 - 101 \cdot 1422/2827}{\sqrt{\frac{101 \cdot 2726 \cdot 1422 \cdot 1405}{2827^2 \cdot 2826}}} = 1.559\right] = 0.06.$$

When the election is close ($a \approx b$) the probability of reversal is approximately the normal tail area above

$$z = (a - b - 1) \cdot \sqrt{\frac{1}{m} - \frac{1}{a + b}}$$

(1 subtracted from the plurality reflects the continuity correction).

The incumbent, who controls the election machinery, may have greater opportunity to create fraudulent, as opposed to random, errors.

Section 4.5.3. Election 2000: Who won Florida?

1. From column (7), the expected total net for Gore is -811.9. Thus Bush would have won.

2. From column (8), the standard error of the expected total net for Gore is the square root of 9802.2, or 99.01. The z-score is therefore -8.20, so that we reject the null hypothesis that Gore won Florida, with an extremely significant P-value.

3. Considering Miami-Dade only, the expected net for Gore among the under-vote from column (6) is 145.0, with a standard error of square root of 2271.2, or 47.66. Letting the actual net for Gore from Miami-Dade be denoted by the random variable X, a reversal would have occurred if and only if X is greater than or equal to 195, the plurality in favor of Bush from column (4). The probability of this event is approximately (using continuity correction) $P[z > 1.0596] = 0.145$.

Section 4.6.1. Heights of French conscripts

1. Chi-squared is 446.51, which on 6 to 8 degrees of freedom, has a vanishingly small P-value, $P \approx 0.0000$. The null hypothesis that the data are normally distributed must be rejected.

2. There are several departures from normality, of which the largest is the excess number of men shorter than 1.570 meters. Quetelet hypothesized that since 1.570 meters was the cutoff for service in the army, the excess number of men below that height represented cheating to avoid military service.

3. If the first two rows are combined, chi-squared drops to 66.8, which is still highly significant. Under either calculation, the null hypothesis that the data are normally distributed must be rejected.

Section 4.6.2. Silver "butterfly" straddles

1. The expert has it backward: the correct statement is that, if the data were normally distributed, there is only a 0.01 chance of observing D that large or larger.

2. Again the expert has it backward: If the data were not normally distributed due to positive kurtosis, the distributions would have fatter tails than the normal and there would be greater opportunities for profit.

3. Since 4-week price changes are the sum of daily price changes, if the daily changes were independent, the central limit theorem tells us that their sum would tend to be normally distributed. Even if the daily changes are negatively autocorrelated, as they appear to be, there are central limit theorems implying normality of sums, so independence is not necessary for normality.

4. If daily price changes were independent, the sd of the 8-week price changes would be $\sqrt{2}$ times the sd of the 4-week price changes; the sd of the 4-week price changes would be $\sqrt{4} = 2$ times the sd of the 1-week price changes. The data show no such increase.

5. In order to break even after commissions, the price change must be \$126. For a 2-2 contract held 4 weeks, this is approximately $126/32.68 = 3.86$ sd's. The probability of a move at least that large is, by Chebyshev's inequality, not greater than $1/3.86^2 = 0.067$. Thus, there is at best a small probability of breaking even.

6. The expert's holding period statistic does not reflect whatever opportunities exist for profitable liquidation within the 4-week, etc., periods.

7. Instead of a probability model, the expert might simply have inspected the data to determine the number of profitable points for liquidation, given the holding period. Chebyshev's inequality indicates that at most these would have been few.

Section 4.7.1. Sulphur in the air

Given a Poisson random variable with mean 1, $P[X \geq 2] = 1 - P[X = 0 \text{ or } 1] = 1 - [e^{-1} + e^{-1}] = 0.264$, as the probability of two or more days with excess emissions in a year. This assumes a simple Poisson with mean 1. Because the mean will vary from day to day, the distribution of excess emissions will be a compound Poisson and will have a larger variance than a simple Poisson. In normal distributions, an increase in variance (with fixed mean) would imply an increased probability of a tail event, such as $X \geq 2$. In the case of the Poisson distribution, when μ is small (as it is in this case, where $\mu = 1$) an increase in variance would, as expected, increase the probability of a left-tail event, such as $X \leq 1$, but may *decrease* the probability of a right-tail event, such as $X \geq 2$. In such a case, the probability of two or more days of excess emissions in a year would be less than calculated, not more, as one might think by analogy to the normal distribution.

Section 4.7.2. Vaccinations

The expected number of cases of functional damage given the risk of functional damage that the plaintiffs accepted is $\mu = 300,533 \times \frac{1}{310,000} = 0.969$. For this mean, the probability of 4 or more cases is $1 - e^{-\mu} \sum_{x=0}^{3} \mu^x / x! = 1 - e^{-0.969}(1 + 0.969 + 0.969^2/2! + 0.969^3/3!) = 0.017$. The null hypothesis that $\mu = 0.969$ can be rejected at the 0.05 level.

Section 4.7.3. Is the cult dangerous?

The expected number of deaths per 4,000 drivers over a 5 year period is $5 \cdot 4,000 \cdot 13.2/100,000 = 2.64$, so that 10 deaths is $(10 - 2.64)/\sqrt{2.64} = 4.5$ standard deviations above expectation. The normal approximation cannot be used, but Chebyshev's inequality gives us $1/4.5^2 = 0.049$ as an upper bound on the tail area probability.

The exact tail area probability is given by

$$1 - \sum_{i=0}^{9} e^{-\mu}(\mu^i/i!),$$

which can be expressed for calculation purposes as nested quantities as follows:

$$1 - e^{-\mu} \cdot \left(1 + \mu \left(1 + \frac{\mu}{2} \left(1 + \frac{\mu}{3} \left(1 + \cdots + \frac{\mu}{8} \left(1 + \frac{\mu}{9}\right) \cdots\right)\right)\right)\right).$$

With $\mu = 2.64$, this gives the tail area probability as 0.000422. Note that the probability of 10 alone is 0.0003, so that the first term dominates the sum. Cult members have a higher rate of death from suicide or automobile accidents than the general population, but the difference may be due more to the type of person who joined the cult than the exhortations of the leader.

Section 4.7.4. Incentive for good drivers

Consider the group of drivers with a total of n accidents in the two time periods. Any driver contributes to the sought-after sum (i.e., the number of accidents in the second time period by those having zero accidents in the first) either (i) n accidents (if the driver has none in the first); or (ii) 0 (for those with one or more accidents in the first period). The expected contribution to the sum is

n times the probability that the driver had all n accidents in the second time period. Since by hypothesis an accident occurs with equal probability in the two time periods, the probability of all n accidents occurring in the second period is $(1/2)^n$. Thus, the expected contribution is $n(1/2)^n$.

On the other hand, the probability that a driver will have exactly one accident in the first period, given that he has a total of n accidents, is by the binomial formula $\binom{n}{1}(1/2)^1(1/2)^{n-1} = n(1/2)^n$. Because this is the same as the expected contribution of the $(0, n)$ driver, and holds for any n, we conclude that the observed number of drivers with one accident in the first period is an unbiased predictor of the number of accidents drivers with no accidents in the first period will have in the second period.

Robbins's expression for $E(\mu|X = i)$ furnishes a way to estimate the number of accidents in the second period by those with zero in the first. With $i = 0$, estimate $P[X = 1]$ by $1{,}231/7{,}842$ and $P[X = 0]$ by $6{,}305/7{,}842$. Then the expected number of accidents among the $6{,}305$ zero-accident drivers is $6{,}305(1{,}231/7{,}842) \div (6{,}305/7{,}842) = 1{,}231$, i.e., the number of drivers with one accident in the first period.

Let $m(j)$ denote the number of drivers who had j accidents in the first period. Our prediction is $m(1)$. It can be shown that a 95% prediction interval for the quantity of interest is

$$m(1) \pm 1.96(2[m(1) + m(2)])^{1/2}.$$

In the example, the prediction interval takes the value $1{,}231 \pm 109$. (The original problem did not separate those drivers with two from those with more than two accidents; substituting 306 for $m(2)$ provides a slightly conservative estimate for the width of the prediction interval.) Since the actual number of accidents for the zero-accident group was $1{,}420$, there is significant evidence of deterioration in performance. The data also support the hypothesis that drivers with one or more accidents improved in performance.

Section 4.7.5. Epidemic of cardiac arrests

1. Treat each of the $n = 34$ deaths as binary trial with respect to the presence or absence of Nurse i, with parameter p_i for Nurse i obtained from the proportion of evening shifts worked for that nurse. Calculate the z-score $z_i = (X_i - 34p_i)/(34p_i q_i)^{1/2}$ for each nurse. Nurse 32's z_i is 4.12, Nurse 60's z_i is 2.10, and the rest are all less than 1.1. The probability that among eight z-scores the largest would equal or exceed 4.12 is $P[\max z_i \geq 4.12] = P[\text{at least one } z_i \geq 4.12] \leq \sum_i P[z_i \geq 4.12]$ by Bonferroni's inequality, where the sum is over $i = 1, \ldots, 8$. Since each term is about 0.00002, $P[\max z_i \geq 4.12] \leq 8 \cdot 0.00002 = 0.00016$, i.e., a z- score as large as 4.12 is highly unlikely even when the worst nurse is not identified beforehand. If there had been other reasons for singling out Nurse 32, the Bonferroni inequality would be unnecessary.

2. The relative risk of death for shifts when Nurse 32 was on duty, opposed to that when she was not on duty, is (27 deaths/201 shifts) ÷ (7 deaths/253 shifts = (0.1343 deaths / shifts) ÷ (0.0277 deaths / shift) = 4.85.

Section 4.8.1. Marine transportation of liquefied natural gas

1. The exponential model is a good candidate since the risk of an accident arguably will not increase with time.

2. The death density function for failure time is $\theta e^{-\theta t}$. This is greatest in the first year.

3. Since the probability of survival for more than 10 years is $e^{-10\theta}$, the probability of failure in the first ten years is $1 - e^{-10\theta}$. If $\theta = 1/7000$ the above expression is approximately equal to $10/7000 = 0.0014$.

4. There is no bunching of probability around the 7000 year mark. The standard deviation of time to first accident is 7000.

Section 4.8.2. Network affiliation contracts

1. The intuitive estimate of the discrete hazard rate is $\hat{p} = n/S$, the number of terminations n, divided by the total number of affiliation years commenced, S. This can be shown to be the maximum likelihood estimate.

For the 1962 data, the mean life of a contract is estimated at $1767/88 = 20^+$ years, corresponding to a yearly discrete hazard of $\hat{p} = 0.05$.

2. The model was ultimately found inappropriate because the risk of termination declined as life continued. A Weibull model should have been used instead. See Section 11.1.

Section 4.8.3. Dr. Branion's case

1. No.

2. They would be independent unless Branion had planned to call by a certain time. In any event, the joint probability (that driving took less than 6 minutes and garroting less then 15 minutes) underestimates the probability that the sum took less than 21 (or 27) minutes, because the joint occurrence of the two events is only one of the many ways in which the sum of the two times could be less than 21 (or 27) minutes.

3. If the distribution of the total driving and garroting times were skewed, there would probably be a right tail that was fatter than the left tail, i.e., longer times would be more probable. That would help Branion's case. The exponential distribution probably would not be appropriate to model the distribution of driving and garroting times because at any intermediate point the elapsed time would be relevant to the time necessary to finish the work.

4. Use Chebyshev's inequality: 27 minutes is 4.5 minutes from the mean of $9 + 22.5 = 31.5$ minutes. The standard deviation of the sum would be, under the assumptions, $(1^2 + 2.5^2)^{1/2} = (7.25)^{1/2}$, so the deviation is $\frac{4.5}{\sqrt{7.25}}$ sd units. By Chebyshev's inequality, such deviations occur no more than with probability $1 \big/ \left(\frac{4.5}{\sqrt{7.25}}\right)^2 = \frac{7.25}{20.25} = 0.358$.

Chapter 5. Statistical Inference for Two Proportions

Section 5.1.1. Nursing examination

Using the hypergeometric distribution, we have:

$$P = \binom{9}{4} \cdot \binom{26}{26} \big/ \binom{35}{30} = 0.000388,$$

a highly significant result.

Section 5.2.1. Suspected specialists

The data may be summarized as follows:

	Unaccounted for	Accounted for	Total
Favorable	74	146	220
Unfavorable	8	123	131
Total	82	269	351

The formula

$$X^2 = \frac{N \cdot (|ad - bc| - N/2)^2}{m_1 \cdot m_2 \cdot n_1 \cdot n_2}$$

yields 33.24, which is highly significant. The data do not support the specialists' position.

Section 5.2.2. Reallocating commodity trades

		Profitable	Unprofitable	Total
Account	F	607	165	772
	G	98	15	113
Total		705	180	885

The corrected z-score $= -1.871$. The one-sided P-value is about 0.03 for these data. The data are not consistent with the broker's defense.

Section 5.2.3. Police examination

The data are

	Pass	Fail	Total
Hispanics	3	23	26
Others	14	50	64
Total	17	73	90

$$z = \frac{|(3.5/26) - (13.5/64)|}{\sqrt{(17/90) \cdot (73/90) \cdot (26^{-1} + 64^{-1})}} = 0.84, \qquad p \approx 0.20,$$

a non-significant result (one-tailed).

Section 5.2.4. Promotions at a bank

The aggregate data for Grade 4 are as follows:

	B	W	Total
Promotions	39	34	73
Non-Promotions	85	41	126
Total	124	75	199

1. This model requires the hypergeometric distribution, for which the normal approximation may be used, as follows:

$$P \left[z \leq \frac{39.5 - 73(124/199)}{\left(\frac{73 \cdot 126 \cdot 124 \cdot 75}{(199)^2 \cdot 198} \right)^{1/2}} = \frac{-5.9874}{3.3030} = -1.813 \right] \approx 0.035,$$

a result significant at the 5% level with a one-tailed test. If the balls are returned to the urn, we have a one-sample binomial model, as follows:

$$P \left[z \leq \frac{39.5 - 73(124/199)}{\left(\frac{73 \cdot 124 \cdot 75}{(199)^2} \right)^{1/2}} = \frac{-5.9874}{4.1405} = -1.446 \right] \approx 0.074.$$

2. If there are two urns, we must use a two-sample binomial, as follows:

$$P \left[z \leq \frac{39.5/124 - 33.5/75}{\left(\frac{73 \cdot 126}{(199)^2} \left[\frac{1}{124} + \frac{1}{75} \right] \right)^{1/2}} = \frac{-0.1281}{0.0705} = -1.817 \right] \approx 0.035,$$

a result that is in close agreement with the hypergeometric.

3. The statistical text cited by the court was referring to the fact that the t-distribution closely approximates the normal for samples of at least 30, so that the normal distribution may be used instead. This is not of relevance here. Despite the court's statement, the hypergeometric is the better model.

First, to justify sampling with replacement, the replacement would have to be of the same race as the promoted employee and have on average the same probability of promotion. Neither of these conditions is likely to be true. Second, the hypergeometric model is substantially equivalent to the two-sample binomial model in large samples, which would appear to be an appropriate alternative choice.

4. The expert was justified in using a one-tailed test, but not for the reasons given. The point is that, in appraising the level of Type I error, only discrimination against blacks should be considered because the court would take action only in that event.

5. There are serious objections to aggregating over years and grades. Aggregating over years assumes either a new employee group each year or that decisions not to promote in one year are independent of those in another year even for the same employees. Neither of these assumptions is likely to be true and, as a result, the effective sample size is smaller than indicated from the aggregate numbers (and the significance is less). Aggregation over grades is also risky (but perhaps less so) since the results can be misleading if promotion rates are different in the two grades. See Section 8.1.

Section 5.3.2. Paucity of Crossets

A reasonable upper bound is provided by a one-sided 95% upper confidence limit. Assuming a binomial model for the number of Crossets found, $X \sim$ Bin (n, P) where $n = 129,000,000$ with $X = 0$ observed, one solves the equation $0.05 = $ lower tail area probability $= (1 - P)^n$. This yields $P_u = 1 - 0.05^{1/n} = 23 \times 10^{-9}$. An alternative method is to use the Poisson approximation to the binomial distribution since n is large and P is small. If $X \sim$ Poisson with mean μ, and $X = 0$ is observed, the upper 95% confidence limit for μ is the solution of the equation $e^{-\mu} = 0.05$, or $\mu_u = - \ln 0.05 \approx 3$. The confidence limit for P is then obtained from $\mu_u = n \cdot P_u$, or $P_u = 3/129,000,000 = 23 \times 10^{-9}$.

Section 5.3.3. Purloined notices

1. The dispute concerns the proportion of Financial's bond redemption notices that were copied by Moody's, among the roughly 600 instances in which copying could not be ruled out. Consider the 600 notices as chips in an urn, some of which were copied, the others not. The existence of an error is tantamount to withdrawing a chip at random from the urn and being able to inspect it as to copied status. Thus, each error can be viewed as an independent binomial trial of copied status with the problem being to estimate the smallest p, the probability that a random chip would be copied, that is consistent at the given level of confidence with the number of copied chips among those withdrawn.

Looking just at the 1981 data, there were 8 errors and all of them were copied. Using a 99% one-sided confidence interval, we have $P_L^8 = 0.01$, or $P_L = 0.56$. Thus, the minimum proportion of copied notices, consistent with finding 8 out of 8, is about 56%.

When not all the errors have been copied, it is necessary to solve the formula for the cumulative binomial for p. This can be done by trial and error or by using tables of the F-distribution. Combining the two years data, there were 18 errors, of which 15 were copied. Asserting that $P_L = 0.54$, we have

$$\sum_{i=15}^{18} \binom{18}{i}(0.54)^i(0.46)^{18-i} = 0.01,$$

which confirms the assertion.

Tables of the F-distribution can be used as follows. Given n trials, suppose that c are copied. If $F_{a,b;\alpha}$ denotes the critical value of the F-distribution with $a = 2(n-c+1)$ and $b = 2c$ degrees of freedom that cuts off probability α in the upper tail, then

$$P_L = b/(b + a \cdot F_{a,b;\alpha})$$

is the solution to the expression for the cumulative binomial distribution when the upper tail is set equal to α. In our example, $n = 18$, $c = 15$, and $F_{8,30;01} = 3.17$; the above expression for $P_L = 0.54$.

It is of some computational interest to compare this exact solution with the approximate 99% confidence interval (± 2.326 standard errors). For both years pooled, $P_L = 0.63$, which is about 9 percentage points too high. Using the more precise approximation given by Fleiss (at p. 172), $P_L \approx 0.53$.

2. The lawyer for Moody's might point out that Financial's model assumes that whether an error was made is independent of whether the notice was copied. To this he might object that Moody's was more likely to copy the more obscure notices, and those notices were more likely to involve errors.

Section 5.3.4. Commodity exchange reports

Required: to find n such that

$$1.96(0.05 \cdot 0.95/n)^{1/2} \le 0.02.$$

Solving for n yields

$$n \ge \frac{1.96^2 \cdot 0.05 \cdot 0.95}{0.02^2} = 456.2.$$

Thus, a sample of 457 is required.

Errata: *Statistics for Lawyers,* Second Ed., by M. O. Finkelstein and B. Levin

Please note that the mailing address for Bruce Levin is:

The Joseph L. Mailman School of Public Health
Department of Biostatistics
Columbia University
New York, NY 10032
USA

Page 60, line 19: $P[A_i]$ should be $P[A_1]$

Page 104, line 7: $21 - i$ in exponent should be $24 - i$

Page 125, line 16: The first appearance of $\binom{60}{9}$ in the expression for $P[X = x]$ should be $\binom{60}{8}$: so it should read $\binom{40}{2}\binom{60}{8}/\binom{100}{10}$

Page 167, footnote 8, line 3: $a = 2(S + 2)$ should be $a = 2(S + 1)$

Page 218, line 2: at p. 157 should be at p. 162

Page 233, ANOVA table under Expected mean square: $\sigma^2 + \sum_i n_i(\mu_i - \bar{\mu})^2$ should be $\sigma^2 + \sum_i n_i(\mu_i - \bar{\mu})^2/3$

Page 234, line 10 from bottom, in formula for D: subscript in F should be $k - 1$: $D = [(k - 1) \cdot F_{k-1, N-k;\alpha}]^{1/2}$

Page 243, line 17, right side: it should be if

Page 254, line 10: $(w_i = \bar{w})^2$ should be $(w_i - \bar{w})^2$

Page 271, line 8 from bottom: NutraTaste should be NatraTaste

Page 318, lines 11 and 14: $\prod_t$ should be $\prod_j$

Page 364, line 9, 14, 22: $\in$ should be ε so $\in_h$ and $\in_t$ will read ε_h and ε_t

Page 459, line 17: $L = \log[p/(l - p)]$ should be $L = \log[p/(1 - p)]$

Page 473, line 22: $\mathrm{Var}(Y|N)$ should be $\mathrm{Var}(Y|\underline{X})$

Section 5.3.5. Discharge for dishonest acts

An approximate 95% confidence interval for the proportion of blacks among employees discharged for dishonest acts committed outside of employment is given by $(6/18) \pm 1.96 \cdot \sqrt{6 \cdot 12/18^3}$ or $(0.116, 0.551)$. The workforce proportion of blacks is well below the lower 95% confidence limit.

The small sample size is already taken into account in the standard error formula and is not proper grounds per se for rejecting the statistical argument. A valid objection to small sample statistics could be raised if the sample were not random, but of course that objection applies also to large samples. Sometimes a "small" sample means a sample of convenience, or worse, a biased sample; if so, that is where the objection should be raised. Sometimes there is variability in the data not represented by sampling variability, but which is difficult to measure in small samples. However, it is not valid to reject a statistically significant finding merely on grounds of sample size.

Section 5.3.6. Confidence interval for promotion test data

The ratio of black to white pass rates is 0.681. The approximate 95% confidence limits for the log relative risk are at $\log(p_1/p_2) \pm 1.96\sqrt{[q_1/(n_1 p_1)] + [q_2/(n_2 p_2)]} = -0.384 \pm 0.267 = (-0.652, -0.117)$. Exponentiating to obtain a confidence interval for the R.R. in original units gives

$$0.521 < R.R. < 0.890.$$

Section 5.3.7. Complications in vascular surgery

The 95% two-sided confidence interval is $(0.021, 0.167)$.

Section 5.3.8. Torture, disappearance, and summary execution in the Philippines

1. There can be no guarantee that the half-width of the 95% confidence interval based on $n = 137$ will not exceed plus or minus five percentage points because the width of the confidence interval depends on the results of the sample. For example, if the sample proportion were $69/137 = 0.504$, the approximate 95% confidence interval would be ± 0.084, thus exceeding the 5 percentage point range. A sample size of about $n = 400$ is required for the guarantee. The given sample size of 137 meets the $\pm 5\%$ goal only if the sample proportion is ≥ 0.92 or ≤ 0.08, which would occur roughly only half the time under the expert's assumption of $P = 0.90$.

2. For summary execution, a 95% confidence interval for the average award is $128,515 \pm \frac{1.96 \times 34,143}{\sqrt{50}} = 128,515 \pm 9,464$, or $(119,051, 137,979)$. The argument that defendant has no cause for complaint at use of the sample

average award to compute the total award for the class seems reasonable because the sample average is just as likely to be below the population average as above it, assuming the approximate normality of the distribution of the sample average based on $n = 50$ observations. (Notice, however, that this is not the same as saying that the population average is just as likely to be above the sample average as below it, which is what the argument assumes.) In any event, the point cannot be pressed too far because, if the sample is very small, the confidence interval would become very wide and defendant might then justly object that the results were too indeterminate. Here the coefficient of variation for the average summary execution award is $4{,}829/128{,}515 \approx 3.8\%$, which seems acceptable.

3. Given the broad variation in individual awards, the use of an average for all plaintiffs within a class seems more questionable as a division of the total award for that class. It may be justified in this case by the arbitrariness of the amounts given for pain and suffering and the administrative impossibility of processing the large number of claims.

Section 5.4.1. Death penalty for rape

In Table 5.4.1e, the black p for unauthorized entry was 0.419, and the white, 0.15. Assuming these are the population values, the power of the test to detect a difference as large as 0.269 in either direction is the probability that a standard normal variable would exceed

$$\frac{1.96[0.3137 \cdot 0.6863 \cdot (31^{-1} + 20^{-1})]^{1/2} - 0.269}{[(0.419 \cdot 0.581/31 + 0.15) \cdot (0.85/20)]^{1/2}} = -0.07,$$

which is slightly larger than one-half. Given such a difference, there is only a 50% chance that it would be found statistically significant.

The corresponding calculations for Table 5.4.1f yield probability of death given unauthorized entry $= 0.375$; probability of death given authorized entry $= 0.229$. The difference is 0.146. The power to detect a difference that large in absolute value corresponds to the probability that a standard normal variable would exceed 0.838, which is 20%.

Dr. Wolfgang's conclusion that the absence of significance implies an absence of association ignores the lack of power.

Section 5.4.2. Is Bendectin a teratogen?

1. This is a one-sample binomial problem. Let X denote the number of defects of any kind in a sample of n exposed women. To test the null hypothesis that $p = p_0 = 0.03$, where p denotes the true malformation rate, we reject H_0 when $X > np_0 + 1.645 \cdot \sqrt{np_0q_0}$. Under the alternative hypothesis $p = p_1 = 0.036$, the probability of rejection, i.e., power, is obtained from

$$P\left[X > np_0 + 1.645 \cdot \sqrt{np_0q_0}\right]$$

$$= P\left[\frac{X - np_1}{\sqrt{np_1q_1}} > \frac{\sqrt{n}(p_0 - p_1)}{\sqrt{p_1q_1}} + 1.645 \cdot \frac{\sqrt{p_0q_0}}{\sqrt{p_1q_1}}\right]$$

$$\sim \Phi\left(\sqrt{n} \cdot \frac{p_1 - p_0}{\sqrt{p_1q_1}} - 1.645 \cdot \frac{\sqrt{p_0q_0}}{\sqrt{p_1q_1}}\right),$$

where Φ is the standard normal cdf. With $n = 1000$, the power is approximately $\Phi(-0.488) = 0.31$.

2. In order for power to be at least 90%, the argument of the normal cdf must be at least equal to the upper 10th percentile value, $z_{0.10} = 1.282$. We may solve

$$\sqrt{n} \cdot \frac{p_1 - p_0}{\sqrt{p_1q_1}} - 1.645 \cdot \frac{\sqrt{p_0q_0}}{\sqrt{p_1q_1}} \geq 1.282$$

for n to find the requirement

$$n \geq \left(\frac{1.282 \cdot \sqrt{p_1q_1} + 1.645 \cdot \sqrt{p_0q_0}}{p_1 - p_0}\right)^2,$$

which in the present problem is $n \geq 7,495$.

3. Now $p_0 = 0.001$ and $p_1 = 0.002$, which yields $n \geq 11,940$. As p_0 approaches 0 with $p_1 = 2p_0$, the required sample size grows approximately as

$$\frac{(z_\alpha + \sqrt{2} \cdot z_\beta)^2}{p_0},$$

where z_α is the critical value used in the test of H_0 (1.645 in the example) and z_β is the upper percentile corresponding to the Type II error (1.282 in the example).

Section 5.4.3. Automobile emissions and the Clean Air Act

1. Let $n = 16$, and let X denote the number of cars in the sample of n cars that fail the emissions standard. Assume $X \sim \text{Bin}(n, P)$, where P is the fleet (i.e., population) proportion that would similarly fail. Using Table B of Appendix II for $n = 16$ we find that the rejection region $X \geq 2$ has power 0.9365 when $P = 0.25$, thus meeting the EPA regulation for 90% power. Under this test, Petrocoal fails.

2. The Type I error for this procedure is out of control. Using the table, if the fleet proportion were $P = 0.10$, a possibly allowable proportion, there would still be about a 50% chance of failing the test. The EPA regulation is silent on the maximum allowable proportion P and Type I error rate for that proportion.

3. To achieve a Type I error rate between, say 0.05 and 0.10, if the fleet proportion failing the emissions standard were $P = 0.05$, and meets the

90% power requirement at $P = 0.25$, we enter Table B for larger values of n. With a sample of $n = 20$ cars and rejection region $X \geq 3$, the test has Type I error rate 0.0755 at $P = 0.05$ and power 0.9087 at $P = 0.25$. With a sample of $n = 25$ cars and rejection region $X \geq 4$, the Type I error can be limited to 0.0341 when $P = 0.05$, with power 0.9038 at $P = 0.25$.

Section 5.5.1. Port Authority promotions

1. The pass rate for whites was $455/508 = 89.57\%$ and for blacks was $50/64 = 78.13\%$. The ratio of the black rate to the white rate is $78.13/89.57 = 0.8723$, which exceeds 80%. This is not a sufficient disparity for action under the four-fifths rule.

   The P-value for the ratio is obtained by taking logs: $\ln(0.8723) = -0.1367$. The s.e. of the log ratio (see p. 173) is $\left[\frac{1-0.8957}{0.8957 \times 508} + \frac{1-0.7813}{0.7813 \times 64}\right]^{1/2} = 0.0678$. Then $z = (-0.1367)/0.0678 = -2.02$. The ratio is statistically significant at the 5% level and meets the Supreme Court's *Castaneda* standard.

2. If two more blacks had passed the test, the black pass rate would rise to $52/64 = 0.8125$. The ratio would be 0.9071 and $\ln(0.9071) = -0.0975$. The s.e. of the log ratio becomes 0.0619 and $z = (-0.0975)/0.0619 = -1.58$, which is not significant. The court of appeals' conclusion as to significance is thus correct, but since calculations of significance are intended to take account of sampling error, it seems incorrect to vary the sample as if mimicking such error and then to make another allowance for randomness. This does not mean that change-one or change-two hypotheticals are not valid ways to appraise the depth of the disparities. As discussed in Section 5.6, a better indication of the *weight of evidence* is the likelihood ratio, which is the maximized likelihood under the alternative hypothesis (where $P_1 = 455/508$ and $P_2 = 50/64$) divided by the maximized likelihood under the null hypothesis (where $P_1 = P_2 = 505/574$). Thus the likelihood ratio is

$$\frac{\left(\frac{455}{508}\right)^{455} \left(\frac{53}{508}\right)^{53} \left(\frac{50}{64}\right)^{50} \left(\frac{14}{64}\right)^{14}}{\left(\frac{505}{572}\right)^{505} \left(\frac{67}{572}\right)^{67}} = 21.05,$$

which is fairly strong evidence against the null hypothesis in favor of the alternative. If two more blacks had passed the test, the likelihood ratio would have been

$$\frac{\left(\frac{455}{508}\right)^{455} \left(\frac{53}{508}\right)^{53} \left(\frac{52}{64}\right)^{52} \left(\frac{12}{64}\right)^{12}}{\left(\frac{505}{572}\right)^{505} \left(\frac{65}{572}\right)^{65}} = 5.52.$$

Thus the weight of evidence against the null hypothesis falls by a factor of 3.8 upon alteration of the data. The altered data would still be 5.5 times more likely under the disparate impact hypothesis than under the null hypothesis, but would be considered fairly weak evidence.

3. The ratio of black to white pass rates is $5/63 \div 70/501 = 56.80\%$, with $\ln(0.5680) = -0.5656$. The z-score is -1.28, which is not significant. The court of appeals' statement (i) is technically correct, but it does not follow that the lack of significance is irrelevant to proof of disparity when sample sizes are small. Lack of significance in a small sample, while not affirmative evidence in favor of the null hypothesis as it would be in a large sample, implies that the sample data are consistent with either the null or the alternative hypothesis; there is not sufficient information to choose between them. That is relevant when plaintiff asserts that the data are sufficient proof of disparity. The court's more telling point is (ii): statistical significance is irrelevant because the disparities were not caused by chance, i.e., the court decided not to accept the null hypothesis, but to accept the alternate hypothesis of disparate impact under which a Type II error occurred. This seems correct since only 42.2% of the blacks achieved a score (76) on the written examination sufficient to put them high enough on the list to be promoted, while 78.1% of the whites did so. The court's opinion is more penetrating than most in analyzing the reason for disparities that *could* have been caused by chance.

Section 5.6.1. Purloined notices revisited

The mle for the alternative hypothesis is $(0.50)^{15}(0.50)^3$, and for the null hypothesis is $(0.04)^{15}(0.96)^3$. The ratio of the two is 4×10^{15}. Moody's claim is not supported by the evidence.

Section 5.6.2. Do microwaves cause cancer?

1. The maximum likelihood estimates are $\hat{p}_1 = 18/100 = 0.18$ and $\hat{p}_2 = 5/100 = 0.05$. The mle of p_1/p_2 is $\hat{p}_1/\hat{p}_2 = 3.6$ by the invariance property.

2. The maximized likelihood under $H_1 : p_1 \neq p_2$ is

$$L_1 = (18/100)^{18}(82/100)^{82}(5/100)^5(95/100)^{95},$$

while under $H_0 : p_1 = p_2$ the maximized likelihood is

$$L_0 = (23/200)^{23}(177/200)^{177}.$$

The likelihood ratio is

$$L_1/L_0 = 79.66, \text{ against } H_0.$$

The data are approximately 80 times more likely under H_1 than they are under H_0.

To assess the significance of a likelihood ratio this large, we calculate the log-likelihood ratio statistic,

$$G^2(H_1 : H_0) = 2\log(L_1/L_0) = 8.76.$$

Alternatively, by the formula at p. 195,

$$G^2 = 2[18\log(18/11.5) + 5\log(5/11.5)$$
$$+ 82\log(82/88.5) + 95\log(95/88.5)]$$
$$= 8.76.$$

This exceeds the upper 1% critical value for χ^2 on 1 df.

3. Students should be wary of the multiple comparisons problem: with 155 measures considered, one or more significant results at the 1% level are to be expected, even under the null hypothesis. The case for a statistical artifact is strengthened by the fact that the rate of cancer in the cases was about as expected and was below expectation in the control and by the apparent absence of any biological mechanism.

Section 5.6.3. Peremptory challenges of prospective jurors

1. Without loss of generality, we may assume that the prosecution first designates its 7 strikes and the defense then designates its 11 strikes, without knowledge of the prosecution's choices. Letting X and Y designate the number of clear-choice jurors for the prosecution and the defense, respectively, the hypergeometric probability of one overstrike is:

$$\binom{7-X}{1}\binom{32-7-Y}{11-Y-1} \bigg/ \binom{32-X-Y}{11-Y}.$$

2. (i)   If there were all clear-choice jurors save one for the prosecution ($X = 6$) and one for the defense ($Y = 10$), the probability of one overstrike would be

$$\binom{7-6}{1}\binom{32-7-10}{11-10-1} \bigg/ \binom{32-6-10}{11-10} = 1/16 = 0.063.$$

(ii)   If there were no clear-choice jurors, the probability of one overstrike would be

$$\binom{7}{1}\binom{32-7}{11-1} \bigg/ \binom{32}{11} = 0.177.$$

(iii)   If the prosecution had three clear-choice jurors and the defense had five, the probability of one overstrike would be $\binom{7-3}{1}\binom{32-7-5}{11-5-1} \big/ \binom{32-3-5}{11-5} = 0.461$. This is the mle estimate. The same result follows if $X = 4$ and $Y = 5$.

Chapter 6. Comparing Multiple Proportions

Section 6.1.1. Death-qualified jurors

Under the null hypothesis that there is no difference in the verdict distributions of death-qualified and Witherspoon-excludable juries, the probabilities in the cells are best estimated from the pooled data in the margins. For the pre-deliberation data this leads to the following (cell expectations in parentheses):

	Death-qualified		Excludable		Total
First degree murder	20	(18.81)	1	(2.19)	21
Second degree murder	55	(55.54)	7	(6.46)	62
Manslaughter	126	(120.04)	8	(13.96)	134
Not guilty	57	(63.60)	14	(7.40)	71
Total	258		30		288

For these values, $X^2 = \sum(\text{observed} - \text{expected})^2/\text{expected} = 10.184$ on 3 df, which is significant ($p = 0.017$). (If exact cell expectations are used, without rounding, the answer is 10.192.)

The principal contributions to X^2 are the manslaughter and not-guilty categories, with the Witherspoon-excludables providing more not-guilty verdicts and fewer manslaughter verdicts than expected. The omission of the Witherspoon-excludables thus increases the probability of conviction, but not of first degree murder conviction.

For the post-deliberation verdicts, $X^2 = 9.478$ on 3 df, which is also significant ($p = 0.024$).

Section 6.1.2. Spock jurors

1. The data on observed and expected numbers of men and women panelists, assuming a constant $p = 86/597 = 0.144$ of selecting a woman panelist, are as follows:

	Observed		Expected	
	Women	Men	Women	Men
	8	42	7.203	42.797
	9	41	7.203	42.797
	7	43	7.203	42.797
	3	50	7.635	45.365
	9	41	7.203	42.797
	19	110	18.583	110.417
	11	59	10.084	59.916
	9	91	14.405	85.595
	11	34	6.482	38.518
Totals	86	511	86	511

$$X^2 = \sum_1^9 \frac{(\text{observed} - \text{expected})^2}{\text{expected}} = 10.60,$$

on 8 df (not significant). We do not reject the null hypothesis of binomial sampling with constant p. Since X^2 is also not too small (it exceeds its expected value of 8), there is no basis for inferring a non-binomial sampling distribution with too little variation, such as might be produced by a quota. See Section 6.1.3 for an example with too little binomial variation.

To illustrate the assessment of post-hoc findings, note that the fourth, eighth, and ninth juries contribute 9.335 to X^2, the remaining six juries only 1.266. If there were some a priori reason to separate the juries in this way, there would be grounds for rejecting the null hypothesis of constant rate. However, as things stand, this is a post-hoc comparison that is properly adjusted for by calculating the usual X^2 for the collapsed table,

		W	M	Total
Juries 4, 8	Low	12	141	153
Juries 1 – 3,5 – 7	Med	63	336	399
Jury 9	Hi	11	34	45
Total		86	511	597

but assessing significance still with respect to the critical value of X^2 on 8 df. Such assessment allows arbitrary regroupings of the original 9×2 table. For the regrouped 3×2 table, $X^2 = 9.642$ is not significant.

2. Assuming that there is a constant expected proportion of women for each jury of 0.144, the standard deviation is 0.014, based on the 597 persons seated on the panels. A 95% confidence interval for p is $0.116 \le p \le 0.172$. This result supports Spock's position, since the 29% average for other judges is far above the upper limit of this confidence interval (the variance of the 29% estimate is negligible under the binomial model).

On the other hand, the difference between the proportion of women on the panel in Spock's case (9/100) and in the other 8 panels for the same judge (77/497) is of borderline significance: $z = (9.5/100 - 76.5/497)/[(86/597) \cdot (511/597) \cdot (100^{-1} + 497^{-1})]^{1/2} = -1.53 (p = 0.063)$. Hence, the statistical evidence is weak for asserting that the selection practice for Spock's judge specially excluded women in Spock's case.

An odd point of the case is that, because the trial judges had no ostensible role in selecting the panels (which was done by the clerks from jury lists), no reason for the difference between the Spock judge and the other judges is apparent.

Section 6.1.3. Grand jury selection revisited

1. Using the calculation formula in Section 6.1 at p. 200, we have

$$\sum 12 \cdot (p_i - 0.1574)^2 / (0.1574)(0.8426) = 0.56,$$

on 8 df. This is far into the (unconventional) lower tail of chi-squared on 8 df. A calculation shows that $p = 0.000205$. This results indicates that one must reject the null hypothesis that black jurors were selected in binomial trials with constant p since the variation in number of blacks is too small to be consistent with the binomial model. Thus, the hypothesis of random selection must be rejected even if one accepts the state's position that the overall proportion of blacks appearing on juries reflected their proportion of the qualified and available pool. The test is unconventional in that the usual chi-squared test involves the upper tail of the chi-squared distribution, detecting departures from expectation that are too large due to heterogeneity in the proportions.

The accuracy of the chi-squared approximation might be questioned because of the small expected cell frequencies. An exact computation of $P[X^2 \leq 0.56]$, given fixed margins of $(12,\ldots,12)$ and $(17, 91)$ in the multiple hypergeometric model for the 2×9 table yields

$$P = 9 \cdot \binom{12}{2}^8 \cdot \binom{12}{1} / \binom{108}{17} = 0.00141.$$

2. Using the binomial distribution with $n = 12$ and $p = 17/108 = 0.1574$:

Binomial probability for outcome:		Expected #s	Observed #s
0 or 1	=0.4151	3.736	1
2	=0.2950	2.655	8
3 or more	=0.2899	2.609	0

$X^2 = 15.37$, which is highly significant at 2 df ($p < 0.0005$). The actual dfs are between 1 and 2 since p was estimated from the uncategorized data. See the answer to Section 6.1.4. Since the observed value of chi-squared exceeds even the critical value of chi-squared on 2 df, it is unnecessary to be more precise. Although the expected cell frequencies are less than five, an exact calculation shows that $P[X^2 \geq 15.37]$ is less than 0.001, a result that is consistent with an evaluation of chi-squared on 2 df.

Section 6.1.4. Howland Will contest

The value of X^2 for the 14-category outcome $(0, 1, 2, 3, \ldots, 12, 13 - 30)$ is 141.54, where we have used the value $p = 5325/25830$ for calculating the expected cell frequencies from the terms of the binomial distribution. Referring the test statistic to critical values of chi-squared with 13 df, one must reject the binomial model for lack of fit. Much of the discrepancy comes from the tail region, as we discuss below. There are two details to consider first, however, one minor, the other more serious.

Because the probability of coincidence per downstroke was estimated from the data, one would ordinarily lose one degree of freedom, leaving 12 dfs. But because the rate was estimated from averages in the original data (as opposed to an estimate based on the multinomial likelihood for the grouped frequencies), the correct critical value of the statistic lies between that of a chi-squared on 12 and 13 df. See Chernoff and Lehmann, *The use of maximum likelihood estimates in chi-squared tests for goodness of fit*, 25 Ann. Math. Stat. 579 (1954). In this case, of course, the difference is inconsequential.

Of greater concern in the questionable validity of the chi-squared analysis, because the 861 pairs are not independent data points. In addition to possible reasons for dependence between pairs of signatures related to how and when the signatures were made, there may be correlation between paired comparisons because the same signatures are used to form many pairs and some signatures may be more likely than others to have many (or few) coincidences.

Consider, for example, assessing the likelihood of observing as many as 20 coincidences of 13 or more downstrokes. The probability for thirteen or more coincidences in Peirce's binomial model is (using $p = 1/5$ for simplicity)

$$P[X \geq 13] = \sum_{i=13}^{30} \binom{30}{i} \left(\frac{1}{5}\right)^i \left(\frac{4}{5}\right)^{30-i} = 0.003111.$$

Students who use the normal approximation with continuity correction will find

$$P[X \geq 13] = P[X > 12.5]$$
$$= P\left[z > \frac{12.5 - 30 \cdot (1/5)}{[30 \cdot (1/5)(4/5)^{1/2}]} \approx 2.967\right]$$
$$\approx \Phi(-3) \approx 0.001.$$

The expected number of occurrences in 861 pairs is $861 \cdot 0.003111 = 2.68$. The deviation of observation from expectation is thus $20 - 2.68 = 17.32$. What is a correct standard error for this deviation? If the number of coincidences in all $N = 861$ pairs were truly independent, the variance of the deviation would be the binomial variance $861 \cdot 0.00311 \cdot (1 - 0.00311) = 2.67$, so that the deviation of 17.32 would be more than 10 standard errors (accounting for most of the value of the chi-squared statistic). Because the pairs were overlapping and since signatures probably would vary in the likelihood of coincidences, the variance is likely to be greater than that given by binomial theory. See E.L. Lehmann, *Nonparametrics: Statistical Methods Based on Ranks*, App. Sec. 5 at 362 (1975) (theory of "U-statistic"). Peirce's binomial assumption thus understates the frequency of highly coincident signature pairs.

Does this undermine Peirce's conclusion that 30 coincidences is exceedingly rare? Not really. In a nonparametric form of analysis, one might consider the set of signature pairs as a finite population, each equally likely to be the pair scrutinized under the hypothesis of authenticity of the disputed signature. In this probability space one observes as many as 30 coincidences only once; the probability of this event is thus one over the number of signature pairs examined. For example, if one includes the pair found in the Howland will with the 861 other pairs examined, the probability would be $1/862$. If one assumes that none of the other pairs were in as much agreement among the $\binom{44}{2} = 946$ pairs potentially examinable, the probability would be $1/946 \approx 0.001$. The assumption of equally likely paired outcomes might be challenged based on the basis of how close in time the signatures were made.

A test more closely tailored to the tracing hypothesis might restrict attention to those 43 signature pairs that included the authentic signature on the will from which the tracing was supposed to have been made.

Section 6.1.5. Imanishi-Kari's case

Chi-squared for the questioned data is 30.94; on $9 - 1 = 8$ df the P-value is 0.0001. For the control group, chi-squared is 12.17; P-value $= 0.14$.

Section 6.2.1. Wage additives and the four-fifths rule

The data are:

	+	−	Total
WM	20	123	143
BM	6	32	38
HM	1	6	7
AOM	0	8	8
WF	5	98	103
BF	4	55	59
HF	2	36	38
AOF	2	11	13
Total	40	369	409

$$X^2 = \sum n_i (p_i - \bar{p})^2 / (\bar{p} \cdot \bar{q}) = 10.226$$

on 7 df, which is not significant ($p > 0.10$). An alternative test procedure in this problem is to compare each group to the WM group, and reject as significant any z-score statistic at or beyond the nominal $\alpha/7 = 0.05/7 = 0.007$ level. Since the largest z-score without continuity correction is 2.33 (for WM vs. WF), which is significant only at the one-tailed 0.01 level, we would not reject the overall hypothesis, in agreement with the chi-squared procedure. On the other hand, if the EEOC would only find disparate impact within race and across gender, or within gender and across race, there are only four comparisons with white male, and the difference is statistically significant at the 5% level.

Section 6.2.2. Discretionary parole

1. Using the calculation formula for chi-squared in Section 6.1 at p. 200, the value of chi-squared is 17.67 on 3 df, which is significant at the 0.001 level. This is the test procedure appropriate for the null hypothesis of no association between discretionary parole and race or ethnicity.

2. The court used a binomial model when it should have used a hypergeometric model and the corresponding chi-squared statistic or z-score utilizing the hypergeometric variance.

3. The court's method of using four-fold tables, but adding the finite population correction factor, yields for Native Americans vs. whites

$$z = \frac{|24.5 - 34.73|}{\left[\frac{382 \cdot 267 \cdot 590 \cdot 59}{(649)^2 \cdot 648}\right]^{1/2}} = \frac{10.23}{3.607} = 2.84.$$

This is equivalent to $X^2 = 8.04$. As a test of the a priori hypothesis that Native American inmates received disproportionately few paroles, one can refer 8.04 to the usual table of chi-squared on a single degree of freedom, $P < 0.005$. Similarly for Mexican Americans: the z-score $= 2.56$ and $X^2 = 6.54$, $P \approx 0.01$.

4. In one sense, the selection of the Native Americans or Mexican Americans might be viewed as determined by the data. To adjust for such post hoc selection and to achieve a given level of overall Type I error when there are multiple comparisons, the method described in Section 6.1 should be applied. In the present case, we could assess significance of each 2×2 chi-squared by referring each value to the upper critical value indicated for the chi-squared distribution for the 3 df in the original 2×4 table. The 5% critical value for 3 df is 8.0, which would make the result for Mexican Americans non-significant using a two-tailed test.

It would probably be more appropriate to take account of only the three comparisons of interest: Native Americans, Mexican Americans, and blacks vs. whites. Using Bonferroni's method, assuming three comparisons, the attained significance level is 3 times the level based on a single df P-value of 0.005 for Native Americans and 0.01 for Mexican Americans. Both would remain significant at the 5% level.

Section 6.3.1. Preventive detention

Computation of τ_b requires the joint probabilities p_{ij} and the expected probabilities under no association. These are computed as follows.

	A or B	Neither	Total
Low	0.1098	0.5002	0.61
Med	0.0560	0.1440	0.20
High	0.0703	0.1197	0.19
Total	0.2361	0.7639	1.00

The table of expected probabilities under no association, $p_{i.} \cdot p_{.j}$, is

	A or B	Neither	Total
Low	0.1440	0.4660	0.61
Med	0.0472	0.1528	0.20
High	0.0449	0.1451	0.19
Total	0.2361	0.7639	1.00

The value of τ_B from these data is 0.03. Thus, there is only a 3% reduction in classification errors due to knowledge of risk group.

Chapter 7. Comparing Means

Section 7.1.1. Automobile emissions and the Clean Air Act revisited

1. A 90% (two-sided) confidence interval is

$$0.0841 \pm 1.753 \cdot 0.1672/\sqrt{16}$$

or (0.011, 0.157). Petrocoal fails this test.

2. The two-sample t-statistic is $t = 0.0841/[0.5967 \cdot (2/16)^{1/2}] = 0.399$. Petrocoal would pass this test. The difference in result is due to the much larger estimated s.e. of the difference when pairing is ignored.

3. The two-sample t-test requires that each sample be selected from a normal population; that the variances of the two populations be the same; and that the selections for each sample be independent of each other and of the other sample.

Assume that the ith car has an expected emission value θ_i for the base fuel, i.e., θ_i is the long run average if the ith car were tested repeatedly with the base fuel. Under the over-simplified assumption that θ_i is the same for all cars with respect to each fuel, so that the only variation is measurement error, assumed statistically independent, the above assumptions are met and the t-test would be valid. However, it is apparent from the data that car-to-car variation makes the assumption of constant θ_i's untenable. While the car-to-car variation may be combined with measurement error for the overall error term, thereby restoring the assumption of selection from two normal populations with given means, the two groups are no longer independent: a car that has a below-average rate of emissions for the base fuel tends to have below-average emissions for Petrocoal. Thus, the two-sample

t-test would not be valid. The paired t-test would remain valid because only the paired differences enter into the test statistic.

Section 7.1.2. Voir dire of prospective trial jurors

The best way of doing this problem is to regard the state and federal voir dire times for each judge as naturally paired and to look at the distribution of the paired differences. The mean of the differences (state minus federal) is 27.556 minutes and the estimated standard deviation of the differences is 60.232. Student's t is

$$27.556/(60.232/\sqrt{18}) = 1.94.$$

Since $t_{17, 0.05} = 2.11$ and $t_{17, 0.10} = 1.74$, the result is not significant at the two-tailed 5% level but is significant at the 10% level.

The reciprocal transformation yields a mean difference of -0.0026 with standard deviation 0.0114. Thus, $t = |-0.0026|/(0.0114/\sqrt{18}) = 0.968$, which is much less significant than for the untransformed data. The reason for this non-significant result is that the reciprocal transformation emphasizes differences for shorter impaneling times relative to the same differences for longer times; in these data judges who already had short impaneling times under the state method did not consistently further shorten their times by using the federal method. The non-significant result for the transformed data suggests that the result is sensitive to the outlier for Judge A in the original scale and calls into question the assumption of normality needed for the t-test in the original data. "Waiting time" data like these frequently have exponential distributions with much thicker tails than the normal distribution. See Section 12.21 for a nonparametric test of these data.

Section 7.1.3. Ballot position

The mean percentage point difference is $64.3 - 59.3 = 5.0$. The pooled variance estimate for a single election is

$$s_p^2 = \frac{148.19 \cdot (121 - 1) + 177.82 \cdot (31 - 1)}{152 - 2} = 154.116.$$

The variance of the difference between the averages over elections is estimated as:

$$s_{\bar{X}_1 - \bar{X}_2}^2 = 154.116 \cdot (121^{-1} + 31^{-1}) = 6.245.$$

The standard error of the difference in means is the positive square root, 2.50.

Thus, the difference is almost exactly two sd's. While the distribution of vote percentages cannot be truly normal, we may rely on the large sample size and robustness of the t-statistic for an approximate analysis. In fact, with 150 df, the t-distribution is approximately equal to the normal; the difference is statistically significant at about the two-tailed 5% level. Notice, that since incumbents appeared in the second position so infrequently, the assumption that position was determined by lot must be rejected (the observed proportion was more than 7 sd's from 50%). Incumbents who appeared second may have received a smaller proportion of the vote because they were politically less powerful.

Section 7.1.4. Student's t-test and the Castaneda rule

1. The court is wrong for the binomial distribution. The t-distribution is unnecessary because there is no need to estimate the standard deviation, which is $\sqrt{npq}$, once p is determined under the null hypothesis.

2. The sample size of 5 illustrates the fallacy of the court's suggestion. The most extreme result (0 blacks) is only 2.24 sd's from the mean of 2.5, yet the exact binomial probability of such a result is $(1/2)^5 = 0.03$, which is significant at the 5% level.

3. The need for an independent estimate of the variance introduces added variability to the standardized test statistic.

4. If the percentage of blacks in the voting rolls were constant, a simple binomial test on the aggregated data would suffice. It shows $z = -2.82$, $p < .0025$, a significant shortfall. If the percentages varied, each year would have to be treated separately and the methods of Section 8.1 used to combine the evidence.

Section 7.2.1. Fiddling debt collector

1. $SS_w = \sum_1^3 (39)(0.3)^2 = 3 \times 39 \times 0.09 = 10.53$

   $MS_w = SS_w/(N-3) = \dfrac{10.53}{120-3} = 0.09$

2. $SS_b = \sum_1^3 n_i(\bar{Y}_i - \bar{\bar{Y}})^2$

   $= 40(0.71 - 0.70)^2 + 40(0.70 - 0.70)^2 + 40(0.69 - 0.70)^2 = 0.008$

   $MS_b = SS_b/(k-1) = 0.008/2 = 0.004$

3. The F-statistic is $F = \dfrac{MS_b}{MS_w} = \dfrac{0.004}{0.09} = 0.044$ with 2 and 117 df.

4. The null hypothesis is that the samples were selected at random. The alternative hypothesis is that the process was manipulated to create results close to 0.70 in each sample. Given the alternative hypothesis, the lower tail of

the F-distribution should be used to see whether the means are too close to 0.70.

5. The P-value is $P[F_{2,117} \leq 0.044]$. We don't need special F-tables to evaluate this because of the relationship $P[F_{2,117} \leq 0.044] = P\left[F_{117,2} \geq \frac{1}{0.044} = 22.5\right] < 0.05$ from Table F in Appendix II. Since the P-value is less than 0.05, we reject the null hypothesis that the samples were selected at random from populations with an average collection rate of 70%. However, use of the F-distribution assumes that the percentages collected of individual invoices are normally distributed. This is unlikely since the collection percentage only varies between 0 and 100% and may well be U-shaped, reflecting the fact that invoices are either collected or not. If the distribution is J-shaped, with most invoices largely collected, simulations suggest that the above calculation may be conservative, i.e., the probability of such close adherence to 0.70 is even smaller than indicated.

Chapter 8. Combining Evidence Across Independent Strata

Section 8.1.1. Hiring lawyers

Strata 1970-72 may be disregarded as non-informative. The elements of the Mantel-Haenszel statistic are as follows (taking white hires as the reference category):

Observed whites hired (1973–82)	60
Expected number	55.197
Excess number whites (or shortfall for blacks)	4.8
Sum of the hypergeometric variances	7.2706

The continuity corrected z-score is $z_c = (4.8 - 0.5)/\sqrt{7.2706} = 1.59$ which is just short of significant at the one-tailed 0.05 level. Note that if the squared form of the Mantel-Haenszel statistic is used, the P-value corresponding to the upper tail of the chi-squared distribution is a two-sided test because both positive and negative deviations in the z-score contribute to the upper tail. The P-value should be divided by two for a one-sided test.

In the actual case, data from other positions were combined, the result was significant, and the magistrate accepted the Mantel-Haenszel statistic.

Fisher's exact test is appropriate here for each year and the corresponding hypergeometric model is in fact the basis for the Mantel-Haenszel test that we use. Fisher's method of combing the evidence applies to continuous distributions and is inappropriate here because of the small numbers and the resulting discreteness of the P-values. Pooling should be avoided because of the variations in outcome rate (hiring) and exposure rate (race) across years (the stratification variable).

Section 8.1.2. Age discrimination in employment terminations

1. In 10 out of 15 cases, the age of the terminated employee exceeded the average age of employees. There is a total of 89.67 years in excess of expectation on the hypothesis of no discrimination.

2. Following the logic of the Mantel-Haenszel test, for each termination we generate an expected value for age, which is the average age (or that times the number of terminations when there are more than 1). The differences between observed and expected ages are then summed over the termination events. The sum is divided by the square root of the sum of the variances of the aggregated ages at each termination in a sampling experiment without replacement for those at risk of termination. When only one person is terminated this variance is simply the variance of the ages of those at risk. When more than one is terminated, it is the number of terminations times the variance of the ages of those at risk times the finite population correction.

The result is a standard normal deviate,

$$z = \frac{653 - 563.4}{\sqrt{1276.84}} = 2.51,$$

which is significant at the one-tailed 1% level ($P = 0.006$).

Section 8.2.1. Bendectin revisited

1. With weights $w_i = 1/(\text{s.e. } \log RR)^2$, the estimated log relative risk is $\hat{\theta} = \sum w_i \theta_i / \sum w_i = -10.025/269.75 = -0.0372$. The s.e. of $\hat{\theta} = 1/\sqrt{269.75} = 0.0609$.

2. A 95% confidence interval is

$$-0.0372 \pm (1.96)(0.0609) = (-0.157, 0.082).$$

3. Exponentiating, the estimated relative risk is 0.96 with confidence interval (0.85, 1.09).

4. Suppose there were a true common log relative risk underlying each of the studies. In order for the log confidence interval found in Answer 2 to exclude the null value of 0, we must have the estimate exceed 1.96 standard errors. For this event to occur with 80% probability, we must have

$$\frac{\hat{\theta} - \theta}{\text{s.e. } \hat{\theta}} > \frac{1.96 \text{ s.e. } \hat{\theta} - \theta}{\text{s.e. } \hat{\theta}}.$$

Since the left-hand side of the above equation is approximately a standard normal variable, the probability of rejecting the null hypothesis will exceed 80% when the right-hand side is less than -0.84. This implies that $\theta >$

$(2.8)(0.0609) = 0.17$, corresponding to an RR of approximately 1.2. Power is sufficient to detect the relative risk posited by plaintiff's expert. (A one-tailed test would yield even greater power.) This result indicates that the lack of significance is not due to a lack of power, but rather to the inconsistency in direction of the study results.

5. Under the hypothesis of homogeneity, Q has a chi-squared distribution on nine df. Since the value $Q = 21.46$ has a tail probability of 0.011, we reject the hypothesis of homogeneity and conclude that the true $ln(RR)$'s vary from study to study.

6. $\tilde{\tau}^2 = 0.058$; $\tilde{\tau} = 0.241$. $\tilde{\tau}$ is the estimated standard deviation of the true $ln(RR)$'s in the hypothetical population. Using weights adjusted for this factor, the mean of the population of true $ln(RR)$'s is estimated as -0.136. The antilog is 0.873, with s.e. 0.1088.

Chapter 9. Sampling Issues

Section 9.1.1. Selective Service draft lotteries

2. Dichotomizing sequence numbers in those ≤ 183 or ≥ 184 yields the following 2×12 table:

	Jan	Feb	Mar	Apr	May	Jun	Jul	Aug	Sep	Oct	Nov	Dec	
≥ 184	19	17	21	19	17	16	17	12	13	18	9	5	183
≤ 183	12	12	10	11	14	14	14	19	17	13	21	26	183
	31	29	31	30	31	30	31	31	30	31	30	31	366

The chi-squared statistic is appropriate to test for random assortment in a multiple hypergeometric model with all margins fixed. The value of X^2 is 31.139 on 11 df ($p \sim 0.001$). There is a significantly decreasing trend in the proportion of high-end sequence numbers from beginning to end of year. The later months were more likely to receive lower sequence numbers because the chronological sequence in which the slips were put into the bowl put them on top and the stirring was inadequate (as it usually is).

The results from the 1971 lottery are much more uniform:

	Jan	Feb	Mar	Apr	May	Jun	Jul	Aug	Sep	Oct	Nov	Dec	
≥ 184	9	18	16	14	15	16	14	17	19	14	14	16	182
≤ 183	22	10	15	16	16	14	17	14	11	17	16	15	183
	31	28	31	30	31	30	31	31	30	31	30	31	365

For these data $X^2 = 11.236$ on 11 df ($p > 0.4$).

Section 9.1.2. Uninsured motor vehicles

1. Using Fleiss's formula at p. 172, a 95% confidence interval is (93.7%, 98.7%).

2. The sample is not too small; the confidence interval in theory reflects the uncertainty attributable to the given sample size.

3. The true rate of uninsured vehicles is probably greater than the confidence interval would indicate because the missing vehicles may well have a lower insurance rate. If all the missing vehicles in the sample of size 249 are assumed to be uninsured, the 95% confidence interval would be (86.3%, 93.9%).

Section 9.1.3. Mail order survey

The mail-order house argued that the respondents were "representative" of all purchasers with respect to the resale/personal-use issue because respondents and nonrespondents had similar distributions of order size and shipping address. In general, reference to such covariates is appropriate to justify treating a sample as random despite nonresponse. If the sample is random, the company's position is correct. If the sample is deemed not random, there are no clear benchmarks as to the right percentage, but the Board's position that all nonresponse should be deemed in the personal-use category would seem extreme. On the randomness issue, however, a chi-squared test of homogeneity in the distribution of orders by size for respondents and nonrespondents is 22.15, on 3 df ($P < 0.0001$), indicating a significant departure from the homogeneity hypothesis implicitly argued by the mail-order house. Chi-squared for the shipping address table is 2.67, on 2 df, which is not significant, a result that is consistent with the company's position.

Section 9.1.5. NatraTaste v. NutraSweet

1. The entry criterion was criticized by the court because respondents included users as well as buyers, and a user might well have not been a buyer, e.g., those who used the product in restaurants or used it at home, but did not buy it.

   The study method was criticized because (i) the respondents were not shown the NutraSweet box together with the other boxes, as they would be seen in a store; (ii) the question was leading; and (iii) there was no instruction against guessing.

2. The expert's analysis was flawed because the control boxes did not control for similarities in names and colors, which were not protected elements of trade dress. The court analyzed the reasons given by respondents for their choice of the NatraTaste box and found that in most cases these elements played a role in respondents' choices.

Section 9.1.6. Cocaine by the bag

1. Let p be the proportion of the 55 bags that contains cocaine. We seek the value of p for which the probability of picking three bags at random, all of which contain cocaine, is 0.05 (this value is the lower end of a one-sided 95% confidence interval). This is given by $p^3 = 0.05$; $p = (0.05)^{1/3} = 0.368$, suggesting that a minimum of about $0.368 \times 55 = 20$ of the 55 bags have cocaine. A one-sided interval is used because there is no p larger than that shown by the data ($p = 1$).

2. Using the slightly more accurate hypergeometric distribution, the probability of finding cocaine in all 3 bags selected at random, if there were 21 bags with cocaine, is given by

$$\frac{\binom{21}{3}\binom{34}{0}}{\binom{55}{3}} = \frac{1,330}{26,235} \cong 0.05.$$

Section 9.1.7. ASCAP sampling plan

1. ASCAP's expert's statements:

Randomness

a. Not too bad. The reference to chance should be "knowable" to reflect the purpose of random sampling.

b. Not "determined by statistical laws" in the sense that other days, times, etc., couldn't be selected. The statement is correct if understood to mean that the characteristics in repeated samples will vary in accordance with statistical laws.

c. True, in theory.

Sampling Precision

a. False. Sampling precision is the degree to which results will vary from sample to sample.

b. Not quite true. The less the sampling precision, the greater the average absolute or squared understatement or overstatement of the number of a member's playings.

c. Under commonly assumed distributions for counts, such as the Poisson or the binomial, this statement is true in the proportional sense, but not in the absolute sense. For a composition with many playings, the variance will be greater in terms of number of playings, but smaller in proportion to the mean than for a piece with few playings.

d. True, and also true for a piece with many playings.

2. Estimates of accuracy of the samples:

In a sampling plan as complex as this, the court should require some replicated, split-sample technique, such as the jackknife or the bootstrap, to
support estimates of precision. See Section 14.9.

Section 9.1.8. Current population survey

1-3. Stratified sampling is used in creating the strata composed of similar PSUs;
its purpose is to reduce sampling error by creating more homogeneous sampling units. Cluster sampling is used in creating clusters of four neighboring
housing units for the systematic sample. This increases efficiency, but also
increases sampling error compared with a random sample of the same size.
Ratio estimation is used in estimating the number of unemployed in each
subgroup in the sample; its use reduces sampling error.

4. One factor is size. Sampling ratios should be greater in small PSUs than in
large ones because it is primarily the size of the sample, not its ratio to the
population, that determines sampling error. Thus, PSUs in low population
states should have higher sampling ratios.

5. Recall that CV = sd/mean. Assuming the mean to be 6%, sd =
$(0.06)(0.019) = 0.00114$. Hence, a 95% c.i. is about 6.0 ± 0.2.

Section 9.2.1. Adjusting the Census

1. Putting aside the provision for erroneous enumerations in brackets (for
which there is no provision in the fish example), the DSE formula is equal
to the number of fish in the first sample (census) × the number of fish
in the second sample (N_p–the estimated population from the post-census
sample) divided by the number of tagged fish in the second sample (M–the
weighted-up number of matches).

2. The DSE underadjusts for black males and overadjusts for black females.
This suggests that black males who are hard to catch in the census are also
hard to catch in the post-census survey. [Do you see why?] The reason for
the overadjustment of black females is unknown.

3. That adjustment should be made for this reason was argued by the plaintiffs
in the first round of litigation (to compel adjustment). The Second Circuit
in effect agreed, but the Supreme Court unanimously reversed. *Wisconsin
v. City of New York*, 517 U.S. 1 (1996).

4. This remedy has been suggested by some statisticians. See the article by
L.D. Brown, et al., cited at p. 281.

Chapter 10. Epidemiology

Section 10.2.1. Atomic weapons tests

The relative risk for males and females combined and control groups combined in Table 10.2.1a is $4.42/2.18 = 2.03$. Let p_1 denote the disease rate among those exposed and p_0 the rate among those unexposed. Among those exposed, if p_0 are cases assumed to have arisen from other causes, while the remaining $p_1 - p_0$ are assumed to have been caused by the exposure, then $(p_1 - p_0)/p_1 = 1 - 1/RR$ is the proportion of exposed cases caused by the exposure. Thus, the probability that Orton's leukemia was caused by the exposure is $1 - 1/2.03 = 0.507$, or slightly more likely than not. On the other hand, when the earlier years are dropped, the relative risk falls below 2 (1.99 for Southern Utah in the Lyon study, and 1.60 in Southern Utah and 1.33 for Northern Utah in the Land study) and the data do not indicate causation is more likely than not. In addition, the data for at least Eastern Oregon also indicate an elevated risk (1.48), suggesting that events other than fallout may have been a contributing factor.

Section 10.3.1. Dalkon Shield

Because the study was retrospective, use the odds ratio. The odds ratios for the group of Dalkon Shield users vs. each group separately are as follows:

O.C.	(35/15)/(127/830)	= 15.2
Barrier	(35/15)/(60/439)	= 17.1
Other IUDs	(35/15)/(150/322)	= 5.0
No Method	(35/15)/(250/763)	= 7.1

The data support the suggestion of a much higher risk of PID with the Shield than with other methods or with no method.

Since the study was conducted after extensive and negative publicity about the Shield, the proportion of Shield wearers among the controls appears to be smaller than their proportion in the general population at the time plaintiff wore the Shield. Robins estimated that 6.5% of women in the 18–44 age group wore the Shield, while only $15/2,369 = 0.63\%$ of the controls wore the Shield. Such a change would cause the study to overstate the risk of the Shield. It also could be argued that those cases still wearing the Shield at the time of the study were more careless about their health and thus more likely to contract PID than women who wore the Shield when the plaintiff did. Such a change would also cause that study to overstate the effect of the Shield. On the other hand, it could be argued that many of those who would contract PID from the Shield had already done so before the study recruitment period, thus, rate of Shield use for those with PID was below that rate when plaintiff wore the Shield. Such a change would cause the study to understate the risk of the Shield.

Section 10.3.2. *Radioactive "cocktails" for pregnant women*

1. **(i)** Since the control group had 0 childhood cancer deaths, RR = ∞.

   (ii) Using population data and including the liver cancer, SMR = 4/0.65 = 6.15. Excluding the liver cancer, SMR = 4.62.

2. Since the exposed and control groups are about the same size, the probability that a cancer death would be in the exposed group, under the null hypothesis, is about 1/2; the probability that all four cases would be in the exposed group is $1/2^4 = 1/16 = 0.0625$ and the probability that all 4 cases would be in either group is $2 \times 0.0625 = 0.125$. With three cases, the probability of all being in either group is 0.25. In neither case are the data significant. Comparing the exposed group with the expected number from general population data, we treat X, the number of childhood cancer deaths, as a Poisson random variable with mean 0.65. Then, including the liver cancer,

$$P[X \geq 4] = 1 - P[X \leq 3]$$
$$= 1 - e^{-0.65}\{1 + 0.65 + (0.65)^2/2 + (0.65)^3/(3 \times 2)\}$$
$$= 0.0044$$

   for the upper-tail P-value. The point probability for $X = 4$ is 0.0039. Since the only value in the lower tail is $X = 0$, and the probability of that is $e^{-0.65} = 0.522$, there is no additional P-value for the lower tail by the point probability method (see p. 121). The two-tailed test is thus significant. Excluding the liver cancer, the two-tailed test is still significant.

3. An approximate 95% confidence interval for the log number of cancer deaths is $\ln 4 \pm 1.96(1/4)^{1/2} = (0.4063, 2.3663)$. Exponentiating, we have (1.50, 10.66) as the lower and upper confidence limits, respectively, for the number of deaths. Dividing by 0.65 gives (2.3, 16.4) as the limits for the SMR. The size of the interval is an indication of the evidence's weakness. If the liver cancer is excluded, the approximate interval is (0.968, 9.302), which includes 1. But an exact computation gives limits (1.258, 13.549) that are broader, shifted to the right, and as a result exclude 1.

4. Defense experts argued that the 0.65 figure was biased downward because the followup was for an average of 18.44 years, but childhood cancer deaths only through age 14 were included. Plaintiffs' experts argued in reply that to include later cancer deaths would introduce a biasing factor of adult cancer. One of the defense experts recalculated the expected number of deaths through 20 years of age and found 0.90 expected deaths in the exposed cohort. With this expectation, $P[X \geq 4] = 0.013$ and $P[X \geq 3] = 0.063$.

5. Plaintiff's experts argued that the liver cancer death should be included because the population figures did not exclude deaths from hereditary cancers. They also argued that the radiation "promoted" the liver cancer, causing

death at an earlier age. Defendant's experts replied that the first argument
did not justify including the liver cancer, but making an allowance for the
rate of hereditary cancer in computing the expected number. They rejected
the promotion theory as speculative.

Section 10.3.3. Preconception paternal irradiation and leukemia

Given that human spermatozoa do not live more than a few days, the data
for the previous six months would seem to be the most relevant. Within that
table, to compute an odds ratio we must decide what categories to use. On the
theory that only the highest exposures have any significant effect, it would be
most reasonable to compare the ≥ 10 and 0 dosage categories and to use local
controls as the closest to the cases. The result is:

Father's preconception dose (in mSv)	Leukemic children	Local controls	Totals
0	38	246	284
≥ 10	4	3	7
Totals	42	249	291

For this table $OR = \frac{246 \times 4}{3 \times 38} = 8.63$, which has an exact (one-sided) P-value of
0.0096. However, had this been one of many comparisons arrived at post hoc,
it would have been necessary to apply an adjustment for multiple comparisons
in computing significance. For example, if this table were considered merely
as one of nine possible 2×2 subtables obtained without summing from the
original 3×3 table, the Bonferroni-adjusted P-value is $0.0096 \times 9 = 0.086$,
which is not significant. The loss of significance that may occur when making
multiple comparisons underscores the importance of an a priori specification
of the primary analysis.

Section 10.3.4. Swine flu vaccine and Guillain-Barré Syndrome

1. If the government's experts were right, the decline in GBS cases among
   the vaccinated as time passed after the moratorium was due to the fact that
   any risk from the vaccine declined with time and largely, if not completely,
   disappeared by 13 weeks; the decline was not due to underreporting of GBS
   among the vaccinated, as plaintiff claimed. Since plaintiff's use of population rates of GBS among the *un*vaccinated in the post-moratorium period
   was admittedly biased downward, the relative risk was biased upward. It
   would seem probable that underreporting would be greater among unvaccinated than vaccinated persons, given the medical and media attention given
   to the GBS risk of the vaccine.

2. Denote GBS as G, vaccination as V, and illness as I. Then, using Bayes's theorem, for vaccinated persons,

$$P[G \mid I, V] = \frac{P[I \ and \ G \mid V]}{P[I \mid V]} = \frac{P[I \mid G, V]P[G \mid V]}{P[I \mid V]}.$$

Because this also holds for unvaccinated persons, the ratio of the two yields the relative risk equation in the text.

3. Although 33/62 may be a reasonable adjustment to reflect $RR[I \mid G]$ in the numerator of the relative risk equation in the text, a similar effect may obtain for $RR[I]$ in the denominator of the equation. To the extent that the two effects are similar, they would cancel. This highlights a difficulty with the data: the unconditional relative risk of illness for vaccinated vs. unvaccinated people was not estimated, and the absence of such an estimate makes problematic any adjustment for illness as a GBS risk factor.

Section 10.3.5. Silicone breast implants

1. The weighted average adjusted $\ln(OR) = 43.518/259.697 = 0.1676$; exponentiating, $OR = 1.18$.

The standard error of the weighted average is 1/(sum of weights)$^{1/2}$ or $1/(259.697)^{1/2} = 0.0621$. The approximate 95% confidence interval is $\ln(OR) = 0.1676 \pm 1.96(0.0621) = [0.046, 0.289]$; exponentiating, the limits are $[1.05, 1.34]$.

2. Using Table 10.3.5b and our earlier result, $12.17 - (259.697)(0.1676)^2 = 4.88$, which is non-significant for chi-squared on $k - 1$ (here 10) degrees of freedom ($P = 0.899$). Notice that there are two different null hypotheses for the studies, the first being that the log odds ratios for all the studies equal zero; the second being that the studies have the same log odds ratios, but not necessarily zero. The test for homogeneity used here is for the second null hypothesis.

Based on a variety of analyses, the panel concluded that there was no "consistent or meaningful" association between breast implants and connective tissue disease.

Chapter 11. Survival Analysis

Section 11.1.2. Defective house sidings

Since the slope coefficient is far from unity, the exponential model must be rejected in favor of the more general Weibull model. From the fitted equation $\log_{10} H(t) = \log_{10} \theta_0 + c \cdot \log_{10} t = -3.742 + 2.484 \log_{10} t$ and $\theta_0 = 10^{-3.742} = 0.0001811$. Then, the estimated Weibull survival function is $S(t) = \exp(-\theta_0 \cdot$

$t^c) = \exp(-0.0001811 \cdot t^{2.484})$. The probability of a siding lasting for at least five years ($t = 60$ months) is 0.0088.

The data shown in the problem are questionable because the 9-month hazard is 0.0246, which suggests that there were initially about 40 houses ($1/40 = 0.025$). In that case, however, the hazard should rise as the number of failures reduces the risk set, but it essentially does not. The relative constancy of the hazard would normally suggest the exponential distribution, but in these data the unequal spacing of failure events implies a non-exponential distribution and leads us to fit the more general Weibull distribution.

Section 11.1.3. "Lifing" deposit accounts

1. The hazard rates (one minus the proportions remaining) are decreasing with age. This suggests consideration of a Weibull distribution with $c < 1$.

2. A plot of the log cumulative hazard estimated by the summation of the d_j/n_j method gives slope coefficient $c \cong 1/2$ and intercept, $\ln\theta_0 \cong -1/2$, or $\theta_0 \cong e^{-1/2}$.

3. The median based on these parameters (with $1/c$ rounded to 2) is $\{(\log 2)/e^{-1/2}\}^2 = 1.3$. The mean is $\cong \Gamma(1+2)/(e^{-1/2})^2 = 2! \times e \cong 5.4$.

Section 11.2.1. Age discrimination in employment terminations revisited

The simplest proportional hazards model postulates that the hazard at time t for an employee who at time t is $X(t)$ years older than an arbitrary reference age is $\theta_0(t)e^{bX(t)}$, where $\theta_0(t)$ is the hazard at time t for the arbitrary reference age. Using the closed form (one-step) approximation for the log-proportional hazards constant referenced in Section 11.2 at p. 324–325, and the data of Table 8.1.2c at p. 249, we have

$$\tilde{\beta} = \frac{653 - 563.4}{1276.84} = 0.070,$$

implying that each year of age increases the risk of termination by about 7%.

Section 11.2.2. Contaminated wells in Woburn

1. The risk set for Case No. 1 is all children with the same birth year who are alive and without diagnosis of leukemia in Woburn just prior to the time Case No. 1 is diagnosed. The analysis is thus stratified by birth year, which is more refined than a risk set consisting of all children, without regard to birth year, who would be followed from the sames age as Case No. 1 at diagnosis.

2. The expected exposure under the null hypothesis is the average cumulative exposure for children in the risk set, for the period of residence in Woburn of the child who was diagnosed at the time of such diagnosis. For Case No. 1 this was 0.31 child-years for cumulative exposure; 33% of the children in the risk set had some exposure.

The variance is the finite population variance of these exposure scores in the risk set at each diagnosis. For the binary exposure variable, this is the product of the proportions exposed and unexposed at diagnosis, which sums to 3.493 in Table 11.2.2.

3. The excess number of cases with some exposure is $9 - 5.12 = 3.88$. For the continuous variate the excess of $21.06 - 10.55 = 10.61$ is in terms of child-years of exposure, which cannot be interpreted in terms of cases.

4. Using the one-step approximation given in Section 11.2 at p. 324–325, the log-proportional hazard constant is

$$\tilde{\beta} = \frac{21.06 - 10.55}{31.53} = 0.333,$$

which yields an estimated relative risk of $e^{\tilde{\beta}} = 1.40$ per child-year of cumulative exposure.

5. The Mantel-Haenszel estimate for the odds ratio on leukemia for exposed vs. unexposed children is given by

$$\Omega_{MH} = \frac{\left(\begin{smallmatrix}\text{sum of proportion unexposed among those}\\\text{at risk at the time of each exposed death}\end{smallmatrix}\right)}{\left(\begin{smallmatrix}\text{sum of proportion exposed among those at}\\\text{risk at the time of each unexposed death}\end{smallmatrix}\right)}$$

$$= \frac{(1 - 0.33) + (1 - 0.25) + (1 - 0.36) + (1 - 0.32) + \cdots}{0.26 + 0.29 + 0.38 + 0.25 + \cdots}$$

$$= \frac{6.21}{2.33} = 2.67.$$

6. The excess leukemia cases continued in children born after the wells were closed and that the cruder dichotomous measure of exposure shows a larger relative risk than the presumably more accurate continuous exposure metric.

Section 11.3.2. Ethylene oxide

1. Using the two-stage model extrapolated from male rats, with the scaling factor for humans given in footnote (a) of Table 11.3.2a ($S = 0.129898$) and the coefficients given in footnote (b) of Table 11.3.2b ($q[1] = 0.008905$ and $q[2] = 0.000184$), the excess lifetime risk of cancer per 10,000 workers is 634 for 50 ppm exposure, as follows:

$$P_{\text{excess}}(d) = 1 - \exp[-(0.008905)(50 \times 0.129898) + (0.000184)(50 \times 0.129898)^2]$$
$$= 0.0635,$$

or 635 in 10,000. The term $q[0]$ does not appear because it cancels out of the expression $\{P(d) - P(0)\}/\{1 - P(0)\}$ for $P[E]$. When ppm = 1 a similar calculation indicates that the excess lifetime risk is 12. The decrease is clearly significant under the Supreme Court's one-in-a-thousand standard in *American Petroleum Institute*.

2. Although the P-value ($P = 0.12$) for chi-squared indicates a non-significant departure from the linear model, a plot of the data indicates that the relationship is nonlinear, with the curve flattening at higher doses.

3. The Weibull model, extrapolated from the female rat data, for an exposure of 50 ppm, estimates $P_{\text{excess}}(d) = 1 - \exp[-0.0829 \times (50 \times 0.129898)^{0.4830}] = 0.1851$, or 1,851 excess cancers per 10,000. A similar calculation for an exposure of 1 ppm estimates that the excess lifetime risk is 305. The large ratios of female to male cancer rates (almost 3 to 1 at 50 ppm and over 25 to 1 at 1 ppm) in the two models, not replicated in the observed data, casts some doubt on the extrapolations. (The large ratios apparently are not due to the difference between the two-stage model used for males and the Weibull model used for females because OSHA calculated a Weibull model for males and got essentially the same results as the two-stage model.)

4. Chi-squared measures the fit of the model with the observed cancers at the higher doses. A good fit with the observed data is a necessary condition, but by no means ensures that the extrapolations to low doses will be valid.

Chapter 12. Nonparametric Methods

Section 12.1.1. Supervisory examinations

1. Evaluating the binomial probability using the normal approximation with continuity correction, we find $P[X \leq 7|n = 32, p = 1/2] = P[X < 7.5] = P[z < (7.5 - 16)/\sqrt{32 \cdot 1/2 \cdot 1/2} = -3.0] \approx 0.0013$.

The shortfall of tests in which blacks and Hispanics combined passed at a greater rate than whites is thus statistically significant.

2. Black and Hispanic test-takers exceeded their 10% proportion of the eligible population ($818/5, 910 = 13.8\%$), suggesting that they may have been less self-selective, and perhaps less qualified as a group, than whites.

Section 12.2.1. Voir dire of prospective trial jurors revisited

1. The sum of the ranks corresponding to impaneling times longer for the federal method than for the state method is $W = 14 + 8.5 + 13 + 6 = 41.5$, corresponding to Judges C, G, P, and R, where we have used the midrank 8.5 for the tied differences of Judges G and O. Table H1 of Appendix II

with $n = 18$ gives $P[W \le 41.5]$ just above 0.025 (or 0.05, two-tailed). Use of Table H1 ignores the zero difference ($d_0 = 1$) for Judge J and the two groups of tied rankings with two tied values in each ($d_1 = d_2 = 2$). With the formulas for mean and variance adjusted for zeros and ties given in the text, we find $EW = (18 \cdot 19 - 2)/4 = 85$ and $\text{Var}(W) = [(18 \cdot 19 \cdot 37 - 1 \cdot 2 \cdot 3)/24] - (2 \cdot 3 + 2 \cdot 3)/48 = 527 - 0.25 = 526.75$. Then $P[W \le 41.5] \approx P[z \le (41.5 - 85)/\sqrt{526.75} = -1.90] \approx 0.03$. The significance of the difference in impaneling times is about the same as found in Section 7.1.2.

2. The nonparametric test used here does not assume a normal distribution and removes the influence of outliers as an issue.

Section 12.3.1. Sex discrimination in time to promotion

1. The Wilcoxon rank-sum for females is $23 + 24 = 47$. In a random selection from an urn containing 25 ranks, there are $\binom{25}{2} = 300$ ways to choose 2 ranks. Of these, only 4 selections produce a rank sum ≥ 47: these are $23 + 24$, $23 + 25$, $24 + 25$, and $22 + 25$. Thus, the exact attained significance level is $4/300 \approx 0.0133$.

Defendant's calculation is also combinatorial. Ignoring the ties among waiting times, there are 24 possible ways to interpose the waiting time of either woman among the 23 times of the men, so that the 24 slots can be filled in $\binom{24}{2} + 24 = 300$ ways (the extra 24 for the event in which both waiting times fall between two times for men or at either end). Of these, $\binom{12}{2} + 12 = 78$ result in both women's waiting times exceeding the median time for men. The probability is $78/300 = 0.26$.

2. Plaintiff's test, by using ranks, does not depend on an assumed distribution.

3. Defendant's test has unacceptably low power because the null hypothesis would never be rejected.

A variation on defendant's approach, analogous to the sign test but still less powerful than the Wilcoxon test, would be to calculate a hypergeometric probability based on categorization of all 25 times as either $\le$ the combined median (13 such) or $>$ median (12 such). Treating the two times for females as a random sample without replacement under the null hypothesis, the probability that both exceed median is $\binom{12}{2} \cdot \binom{13}{0} / \binom{25}{2} = 0.22$. Again, power is too low.

Section 12.3.2. Selection for employment from a list

1. The rank-sum for minorities is $S = 19{,}785$. With $N = 738$ who passed the test and $m = 38$ minorities, S has null expectation $38 \cdot 1/2 \cdot (738 + 1) = 14{,}041$ and variance $38 \cdot 700 \cdot 739/12 = 1{,}638{,}116.67$. The upper tail area

probability is approximately

$$P[S \geq 19,785] = P[S > 19,784.5]$$

$$= P\left[z > \frac{19,784.5 - 14,041}{\sqrt{1,638,116.67}} = 4.487\right]$$

$$\approx 4 \times 10^{-6}.$$

The lower scores of minorities are statistically significant.

2. The t-test would not technically be valid since the sample size for minorities is small and the truncation skews the distribution of scores and makes it markedly nonnormal.

3. Using the relation between the Mann-Whitney U-statistic and Wilcoxon rank-sum statistic, $U = (38)(700) + \frac{1}{2}(38)(39) - 19,785 = 7,556$. Thus, the sought-after probability is $7,556/(38 \cdot 700) = 0.28$. Under the null hypothesis it should be 0.50.

If the minority group scores had a smaller standard deviation around their mean than did the majority group scores, even if those means were the same, the average rank among those passing would tend to be lower for the majority group. For the above result to prove a lower mean score for minorities, the standard deviations for the two groups must be assumed to be comparable. A careful analysis would therefore include a look at the passing data for non-minorities.

Section 12.3.3. Sentencing by federal judges

1. For a single judge among 50 who sentence equally, the rank for the judge on a single case has expectation 25.5 and variance $(50^2 - 1)/12$; for 13 independent cases, the average has expectation 25.5 and variance $(50^2 - 1)/(12 \cdot 13) = 16.02$.

2. The mean squared deviation is $1,822.47/50 = 36.45$. (The division is by 50 rather than 49 because the mean of the ranks is known (25.5) and is not estimated from the data.) The expected value under the null hypothesis is 16.02, as given in the answer to question 1. Thus, the mean squared deviation is 2.275 times larger than its expected value. This is a significant difference, as shown by X^2, which is $(50 - 1)(2.275) = 111.5$, a highly significant value by Table C. The data show significant dispersion in average rankings among the judges.

Section 12.4.1. Draft lottery revisited

1. Spearman's rank correlation coefficient is -0.226 for the 1970 draft lottery, indicating a tendency for lower sequence numbers to occur toward year's

end. The z-score $z = -0.226\sqrt{366 - 1} = -4.3$ is significantly different from 0.

2. The rank correlation for the 1971 draft lottery is 0.0142, with z score $= 0.27$, which is not significant.

Chapter 13. Regression Models

Section 13.1.1. Head Start programs

The data were simulated with zero treatment effect. The supposed effects are due entirely to regression to the mean.

1. Since the correlation coefficient is $1/2$, the mean value of the post-test score for a given pre-test score regresses one-half the distance toward the grand mean of the post-test scores.

   For the Experimental Group with pre-test score 55, for example, the average post-test score is 62.93, approximately halfway between the overall mean (60) and the five-point differential that would be naively predicted by the five-point spread in pre-test scores above the average (55–50). For the control group, the average post-test score is 70.29, roughly halfway between the group average of 80 and the 15 point shortfall naively predicted on the basis of the 15 point shortfall in pre-test scores below the average (55–70).

2. The above shows that the regression-to-the-mean effect would produce approximately the difference in post-test scores for those with the same pre-test scores *even if both groups (or neither) were given the same treatment*. The point is that, in the presence of the initial disparity in pre-test means, the post-test means move in opposite directions, although this has nothing to do with a true "treatment effect."

   The director's argument is similarly fallacious, also being based on the regression phenomenon, for pre-test scores regressed on post-test scores.

   Students should be alerted to the serious flaw in experimental design, namely the confounding between treatment and pre-test scores, for which regression analysis only partly adjusts. See Section 14.5.

3. Since the average difference is equal to the difference in averages, the post-test minus pre-test difference for both treatment and control groups is simply 20, indicating no treatment effect.

Section 13.2.1. Western Union's cost of equity

1. The estimated regression is cost of equity $= 9.46 + 0.13\times$ (variability of rate) + residual. The predicted cost of equity for Western Union is 13.0%.

Section 13.2.2. Tariffs for North Slope oil

1. The mean value of price is \$12.73/barrel; the mean value of °API is 34.61. The OLS regression line is

$$\hat{Y} = 12.73 + 0.095 \cdot (X - 34.61) = 9.442 + 0.095 \cdot X.$$

2. The estimated return per °API is 9.5¢. In this example, the correlation coefficient is 0.96.

3. The attorney for ERCA might question whether this relationship would hold for °API values below the range of the data on which the regression was computed.

Section 13.2.3. Ecological regression in vote-dilution cases

Using the two regression equations, plaintiffs made the following estimates of the turnout and support rates for Feliciano:

1. The average Hispanic turnout rate for Feliciano (i.e., percentage of registrants voting for Feliciano) was 18.4% and the average non-Hispanic turnout rate was 7.4%.

2. 42.6% of the non-Hispanics and 37.8% of the Hispanics voted in the election.

3. The non-Hispanic support rate for Feliciano (i.e., percentage of non-Hispanics in the district who voted for Feliciano) was $7.4/42.6 \approx 17\%$; the Hispanic support rate was $18.4/37.8 \approx 49\%$.

4. There is significant Hispanic and non-Hispanic bloc voting.

5. The model depends on the constancy assumption, i.e., that, subject to random error, the same proportion of Hispanics in all districts votes for the Hispanic candidate. This is probably not true since bloc voting is likely to be more intense in poor, Hispanic districts than in more affluent mixed districts. The "neighborhood" model, which assumes that Hispanics and non-Hispanics in all districts vote in the same proportions for the Hispanic candidate, is at the other extreme and is also unlikely to be completely true, since ethnicity is well known to affect voting.

6. An alternative to a regression model would be exit polling, which is commonly done to give early forecasts of election results.

Section 13.3.1. Sex discrimination in academia

The 0.8 percentage point increase in R^2 is not directly related to the percentage shortfall in salaries for women relative to men. The change in R^2 is more closely related to the significance of the sex-coefficient when added as an explanatory factor, but, like the t-statistic, that describes only the ratio of the coefficient's magnitude relative to its standard error, not the importance of the coefficient per se.

As a special case of the formula given in Section 13.4 at p. 376, when adding a single variable to a regression model (the sex-coefficient), the F-statistic = (change in R^2) ÷ $(1 - R_1^2)/\mathrm{df}_e$ where R_1^2 denotes the value in the larger model (0.532) and df_e denotes the error degrees of freedom in that model. It may also be shown that $F = t^2 = \hat{\beta}^2/se(\hat{\beta})^2$ where $\hat{\beta}$ = the sex-coefficient. Thus, the sex-coefficient is related to the change in R^2 by the formula

$$\hat{\beta}^2 = \frac{[se(\hat{\beta})]^2 \cdot \mathrm{df}_e}{1 - R_1^2} \cdot (\text{change in } R^2).$$

While the change of 0.008 may be small, it must be multiplied by a potentially huge factor before it can describe the salary shortfall.

Section 13.4.1. Race discrimination at Muscle Shoals

The t-statistic for the race coefficient in plaintiff's regression is $t = 1,685.90/643.15 = 2.62$ ($p = 0.0044$, one-sided). For defendant's regressions, the t-statistics are:

Schedule D : $t = 1,221.56/871.82 = 1.40$ ($p = 0.081$, one-sided)

Schedule E : $t = 2,254.28/1,158.40 = 1.95$ ($p = 0.026$, one-sided).

The standard errors are larger in defendant's model because the sample size for each regression is smaller. Defendant's model is preferable because it avoids the possibly confounding effect of different schedules. Plaintiffs should either have used indicator variables or separate regressions and combined their results to compensate for loss of power.

Fisher's test is given by $-2 \cdot (\ln 0.091 + \ln 0.026) = 12.093$, distributed as chi-squared on 4 df under the null hypothesis of no discrimination in either schedule, and therefore significant at the 0.025 level. Using weights $w_1 = 1/(871.82)^2$ and $w_2 = 1/(1,158.40)^2$, the weighted average of the two coefficients $(w_1 \cdot 1,221.56 + w_2 \cdot 2,254.28)/(w_1 + w_2) = \$1,594.99$. The s.e. is the reciprocal square root of the sum of w_1 and w_2, equal to 696.58, leading to a z-score of 2.29 ($p = 0.011$, one-tailed).

Section 13.6.1. Pay discrimination in an agricultural extension service

1. The Extension Service objected that: (i) the regression did not screen out the effects of pre-Title VII discrimination (the year 1972 was the cutoff for the governmental entity); (ii) county-to-county variations were not reflected; and (iii) performance on the job was not reflected. The Supreme

Court answered these objections by pointing out that (i) pre-Act discrimination could not be perpetuated; (ii) a regression need not account for all
factors and in any event studies showed that black employees were not located disproportionately in counties that contributed only a small amount
to Extension Service salaries; and (iii) a regression analysis by the Extension Service that included a performance variable showed an even greater
disparity for 1975.

2. The black agents might object that the regression included job title, an
explanatory factor that might be tainted by discrimination.

3. The 394.80 coefficient for WHITE in the final model indicates that whites
received on average $395 more than equally qualified blacks (as measured
by the explanatory factors in the model). The coefficient is statistically
significant: $F = 8.224$; $p \approx 0.002$. Because the equation is recomputed
each time a variable is entered, the fact that race was entered last does not
affect the value of the coefficient, given the selection of variables in the
model.

4. The reason for including White Chairman, White Agent, and White Associate Agent was to pinpoint the possibility of different degrees or extent
of discrimination at different levels of rank. These factors were probably
not included by the stepwise regression program because the numbers were
too small for statistical significance. If they had been included, the coefficient for WHITE would have applied only to assistant agents, the reference
category, for which no separate variable was included. The interaction coefficients would then give the differences between the effect of WHITE at the
other levels and the effect of WHITE at the assistant level. The subject of
such interaction terms is an important one that is taken up in Section 14.2.

Section 13.6.2. Public school financing in the State of Washington

The inclusion of the lagged variable "Prior Test Score" radically changes the
interpretation of the regression from one of absolute prediction to one of change
in language scores since 1976. While school variables may affect the absolute
level of language scores from district to district, unless they also changed in
the four-year period, their influence would be understated by the equation.

Section 13.7.1. Projecting fuel costs

The OLS predictor of fuel costs is $\hat{Y}(X) = 15.545 + 2.71 \cdot X$ where $X = 1, 2, 3,$ or 4 for the 1980 quarterly data. For 1981, with $X = 5, 6, 7,$ and
8, the projections are 29.10, 31.80, 34.52, and 37.22. The slope coefficient,
with standard error $\hat{\sigma}/\sqrt{\sum(X_i - 2.5)^2} = 2.130/\sqrt{5} = 0.952$ just fails to reach
significance at the two-sided 10% significance level ($t = 2.845, t_{2;0.10} = 2.920$),
because of the small sample size. The 95% prediction interval for first quarter

1981 is $29.10 \pm 4.303[1 + \frac{1}{4} + (5 - 2.5)^2/5]^{1/2} = [22.3, 35.9]$, while that for last quarter 1981 is $37.22 \pm 4.303[1 + \frac{1}{4} + (8 - 2.5)^2/5]^{1/2} = [25.6, 48.8]$.

Section 13.7.2. Severance pay dispute

The estimated regression mean of severance weeks at the mean value of the explanatory factors is simply the sample mean, 15.859, with a confidence interval of 15.859 plus or minus 2.0 times $(MSE/n)^{1/2} = 15.859 \pm 2.0 \cdot 9.202/\sqrt{101} = 15.859 \pm 1.831$, or from 14.0 to 17.7. The predicted value of an observation at the mean is also the sample mean, although a 95% prediction interval is this value plus or minus $2 \cdot 9.202 \cdot (1 + 1/101)^{1/2} = 18.495$, or from -2.6 (i.e., 0) to 34.4.

1. At plaintiff's values of the explanatory factors, the prediction is $24.07 = 14.221 + 0.526 \cdot (20 - 12) + 0.372 \cdot (45 - 40) + 0.944 \cdot (9 - 5)$. A 95% confidence interval for the estimated regression mean at plaintiff's values is 24.07 plus or minus $2.0 \cdot 9.202 \cdot (0.0214)^{1/2} = 24.07 \pm 2.7$, or from 21.4 to 26.8. The prediction interval is $24.07 \pm 2.0 \cdot 9.202 \cdot (1 + 0.0214)^{1/2} = 24.07 \pm 18.6$, or from 5.5 to 42.7.

2. At the mean value of the explanatory variables, the standard error of the difference between actual and predicted mean severance weeks is the root MSE times $(200^{-1} + 101^{-1})^{1/2} = 9.202 \cdot 0.122 = 1.123$. Since the difference is $14.73 - 15.859 = -1.13$, or -1 standard error, there is no significant difference between actual and predicted mean values.

Section 13.7.3. Challenged absentee ballots

2. The slope coefficient, b, of the regression line is equal to $0.6992 \times \frac{442.84}{24,372.3} = 0.0127$. The intercept is $256.05 - (0.0127 \times 30,066.7) = -125.90$. For D–R= -564, the prediction is $-125.90 + 0.0127(-564) = -133.1$.

A 95% prediction interval is obtained, first, by computing the standard error of regression; second, by using the standard error of regression to compute a standard error for the regression prediction at the value -564; and, third, by taking plus or minus an appropriate number of standard errors from the t-distribution for $n - 2$ degrees of freedom. In this case the standard error of regression is $\left[\frac{20 \times (442.84)^2 \times (1 - 0.6992^2)}{19}\right]^{1/2} = 324.8$. The standard error of the prediction is $324.8\left[1 + \frac{1}{21} + \frac{(-564 - 30,066.7)^2}{20 \times 24,372.3^2}\right]^{1/2} = 344.75$. The relevant t-value is $(t_{19,0.05} = 2.093)$. Finally, the 95% prediction interval is $-133.1 \pm (2.093 \times 344.75)$, or approximately $(-855, 588)$. Since the actual difference was $1,396 - 371 = 1,025$, it was well outside the 95% prediction interval.

3. A scatterplot of the data does not show them to be markedly heteroscedastic. However, because the data appear to be somewhat more widely dispersed

at the lower end of the machine D–R vote, where the challenged election lies, the width of the prediction interval for that value may be somewhat understated.

Section 13.9.1. Western Union's cost of equity revisited

R^2 for the untransformed data is 0.922. When variability of earnings is transformed into logs, $R^2 = 0.808$. The predicted value in the log scale is 14.35%, compared with 13% in the untransformed scale.

R^2 is reduced in the logarithmic model because of the foreshortening in the range of the explanatory factor. A graph of the data suggests that the relationship between cost of equity and variability of earnings is more nearly linear when variability is transformed into logs. Hence, in terms of fit, the logarithmic model should be preferred even though it has a smaller R^2.

Section 13.9.2. Sex- and race-coefficient models for Republic National Bank

The district court in *Vuyanich* held that in plaintiffs' regressions age was an inaccurate and biased proxy for the general work experience of women at the bank, but not of black men. The court reached this conclusion partly on the basis of the bank's data, which showed that the average male was out of the labor force and not in school for about four months, whereas the average female was out of the labor force and not in school for about 30 months. [The court apparently interpreted this statement about crude averages as a stronger conditional statement, namely, that for men and women of a given age, women were out of the labor force more than men.] Nor did the plaintiffs compensate for this defect by offering a regression that controlled for a job level by the use of both Hay points or job grades and actual general work experience. *Id.* at 314–316.

The district court rejected the bank's regression on the ground that there was no variable to account for "the job the individual is performing (such as would be done in a crude fashion through the use of Hay points)." *Id.* at 308. Plaintiffs might also have objected that some of the explanatory variables in the bank's model might have been tainted (e.g., career motivation), and that by separating employees into two groups and computing separate regressions, the bank substantially reduced the power of the model.

Chapter 14. More Complex Regression Models

Section 14.1.1. Corrugated container price-fixing

The dominance of the lagged price as an explanatory variable in the equation means that the model tends to project a smooth trend from wherever the fit period is ended. Franklin Fisher, the expert for the defendant in the Corrugated Container case, showed that stopping the fit period a little earlier led to projections of collusive prices that were actually below the competitive prices in the competitive period. Fisher also tested the model using other data by estimating it from the competitive period data and then using it to "backcast" competitive prices for the collusive period, beginning in 1963. The result was that estimated competitive prices in the collusive period were far above the actual collusive prices—a perverse result indicating that the explanatory factors were not good predictors of price movement.

Section 14.1.3. Losses from infringing sales

The OLS regression estimate of Firm A Unit Sales on Time and Seasonal Indicator is:

	Coefficient	Standard Error	t
Time	0.033	0.0066	4.96
Seasonal Ind.	0.109	0.034	3.18
Constant	1.808	0.071	25.40

$R^2 = 0.718$, $F = 19.08$ on 2 and 15 df, error mean square = 0.0208.

Applying the equation, the 1981 projections of Firm A Unit Sales are: 2.326, 2.359, 2.610, 2.643, 2.676, and 2.491 for a 1981 annual total of 15.105, compared with the actual total of 16.17.

Section 14.1.4. OTC market manipulation

A 95% prediction interval here is simply the regression estimate plus or minus 1.96 standard errors, which is $\pm 1.96 \cdot 0.1055 = 0.2068$.

	Actual Olympic Return	Regression Predictions	Difference
03 Jan 77	0.00000	−0.00142	0.00142
04 Jan 77	0.02174	−0.00652	0.02826
05 Jan 77	0.01064	−0.00493	0.01557
06 Jan 77	0.00526	0.00174	0.00352
07 Jan 77	0.05759	0.00103	0.05656
⋮			
09 Mar 77	0.04040	−0.00442	0.04482
10 Mar 77	−0.05825	0.00382	−0.06207
11 Mar 77	−0.34536	0.00095	−0.34631
14 Mar 77	−0.10630	0.00523	−0.11153

1. The only day on which the actual value is outside the prediction interval is March 11.

2. Under the model, the returns for Olympic stock increase with the DVWR returns; because the scale of the equation increases with increasing DVWR, the errors are likely also to increase, making the model heteroscedastic.

Apart from a loss of efficiency in estimating the coefficients (which is not relevant here since we assumed that the coefficients were estimated without error), the effect is to overstate the precision of the Olympic stock price regression prediction when DVWR is in the high range; as a result, the model is even less powerful in that range than it appears to be. However, since the prediction interval, as calculated, is so wide, the fact that it should in some contexts be even wider is of little importance.

3. The model would have serial correlation in the error term if there were periods extending over more than a month when special factors affected the price of Olympic stock vis-à-vis the market; this possible defect does not seem very important. If serial correlation is suspected, the correlation of successive residuals should be tested with the Durbin–Watson statistic. The effect of serial correlation would be to make the estimated 95% prediction interval appear more precise than the true interval, which, as noted, is of little practical importance in this context.

4. The prediction interval is so wide that the model has power to detect only the largest abnormal stock price movements. For example, on 07 Jan 77 the actual return on Olympic stock (0.05759) was more than 5.6 percentage points greater than the regression estimated return (0.00103), but nevertheless was within a 95% prediction interval for the estimate. In addition, the model has no power to detect as abnormal any manipulation that simply maintained Olympic prices in relation to the rest of the market.

Section 14.1.5. Fraud-on-the-market damages

1. Torkelson's method in effect unjustifiably assumes that $R^2 = 1$ in the regression model, which is far from the case. The regression method would seem to be more defensible. For a single backcast the difference is not large. Starting at 3.233 on April 2 (note that this is a calculated figure using the regression model; under Torkelson's method the starting point would be different), the April 1 price would be $3.233 \div 193.796/198.733 = 3.315$, compared with 3.276 using the regression model. However, the cumulative difference between the two methods could become large if the calculation is extended over many days.

2. Assuming the regression is estimated without error, a 95% prediction interval for a single backcast would be the regression estimate ± 1.96 standard errors of regression. Since the estimate for each day is built on the preceding estimate, the errors are additive in the sense that the standard error for a given day is equal to the square root of the sum of the squared standard errors for the preceding estimates. Thus, a regression estimate for April 1 (the outcome of 153 backcasts from September 1 to April 1) has a standard error of $[(0.0728)^2 \times 153]^{1/2} = 0.9005$. The 95% prediction interval for April 1 is therefore $3.276 \pm 1.96 \times 0.9005$, or from 1.51 to 5.04. The errors of

estimating the regression coefficients would make this interval still wider. But even as it stands, if the upper limit were used, there would be no damages for a purchase on that day. Note that purchasers closer to September 1 would have a better chance to recover damages because the prediction intervals for the regression estimates on their purchase days would be narrower, although it would seem as a matter of equity that they should be treated no differently than purchasers earlier in the class period. Finally, given the imperfection of the methods necessarily used, it would seem inappropriate to resolve uncertainties in measurement in favor of the defendant, who is, after all, a wrongdoer.

Section 14.1.6. Effects of capital gains tax reductions

The difference in the estimate of realizations elasticity between Equations 1 and 2 is that, in Equation 2, the larger value of the lagged term (CTX(−1)) implies that in the second year after the tax decrease the realizations would decline substantially (818). No significant decline is shown in Equation 1. The reason for the difference appears to be that since Equation 1 did not include the NYSE Composite Index, when the stock market surged in 1983, to explain the increase in realizations that occurred as a result required two changes: (i) a large negative increase in the coefficient of CTX, since the marginal tax rate decreased only slightly between 1982 and 1983, and (ii) a large reduction in the lagged coefficient CTX(−1) to offset the negative effect on realizations that would otherwise have been caused by the substantial decrease in the marginal tax rates that occurred between 1981 and 1982.

The interpretation of the coefficient of MTR in the CBO semilogarithmic model (Table 14.1.6c) is that, for each percentage point decrease in rates, realizations would increase by 3.2%. According to the model, reducing rates from 20% to 15% would increase realizations by $3.2 \times 5 = 16\%$. However, the 25% reduction in rate would more than overcome the increase in realizations, with the net result being a reduction of 13% in revenue.

Section 14.2.1. House values in the shadow of a uranium plant

The Rosen Model

1. The −6,094 coefficient means that after 1984 houses inside the five-mile radius sold on average for $6,094 less than similar houses outside the five-mile radius.

2. The objection to this figure as a measure of damage is that it does not measure the *change* in the negative premium for being inside the five-mile radius before and after the announcement. The change in the negative premium may be expressed as [(Post-1984 In) − (Post-1984 Out)] − [(Pre-

1985 In) − (Pre-1985 Out)]. This equals $(-6,094) - [(-1,621) - (550)] = -3,923$.

3. The interpretation of this interaction coefficient is that the negative premium for being inside the five-mile radius increased by $3,923 between 1983 and 1986.

Assuming independence of the component coefficients, the standard error of the interaction coefficient is the square root of the sum of the component variances:

$$\left[\left(\frac{6,094}{2.432} \right)^2 + \left(\frac{1,621}{0.367} \right)^2 + \left(\frac{550}{0.162} \right)^2 \right]^{1/2} = 6,109.$$

Thus,

$$t = \frac{-3,973}{6,109} = -0.65,$$

which is not significant. The covariances between estimated coefficients have been ignored.

The Gartside Model

1. The PC coefficient for each year represents the change in average house value over the base year, 1983, for houses in the 4–6 mile band (the base distance category). Similarly, the RC coefficient for a particular distance band represents the change in average house value over the base distance band, 4–6 miles, with respect to 1983 values (the base year category). The interaction coefficients represent differences of the differences. For example, since there was a negative premium of $4,568 for houses at 0–1 miles compared with 4–6 miles in 1983, the coefficient for 1986/0–1 of −7,440 implies that this negative premium increased by 7,440 between 1983 and 1986. This would seem to be a relevant measure of damage. If the interaction coefficients were 0, there would be no basis for damages because that would imply that the negative premium on being inside the five-mile radius had not changed after the announcement.

   The cells of Table 14.2.1d can be derived by beginning with the value 62.2 for 1983 values at 4–6 miles, adding the main effects to adjust for year and distance separately, and then adding the interaction term to adjust for both together.

2. One would object to Gartside's conclusions because the interaction coefficients are not statistically significant either as a group (the F value for $PC * RC = 1.37$, $P \sim 0.2$) or individually (with one irrelevant exception). There is also a question whether the data are normally distributed, or whether a few high sale prices are affecting the coefficents.

Section 14.4.1. Urn models for Harris Bank

1. In the global log model, for which $\Delta \bar{r} = 0.107$, the adjusting factors fail to account for about 10% of the ratio of the geometric mean wage for blacks to the geometric mean wage for whites. The figure is about 5% for the locally weighted regression.

2. For the global urn model, $z = 0.1070/\{0.2171 \cdot [309/(53 \cdot 256)]^{1/2}\} = 3.27$. For the locally weighted regression model, $z = 0.0519/\{0.1404 \cdot [309/(53 \cdot 256)]^{1/2}\} = 2.45$. Both are significant.

3. Since the z-score in the locally weighted regression is smaller than in the global model, apparently relatively more of the mean wage disparity is explained by the locally weighted regression, which suggests, but does not conclusively prove, that the larger z-score of the global model may be due to model misspecification rather than discrimination.

4. Plaintiffs might have objected that the equation was biased because the productivity factors picked up part of the salary difference that was correlated with both race and productivity, but should have been attributed to race alone.

Section 14.5.1. Underadjustment in medical school

1. The district court assumed that the regression of productivity on proxies was not the same for men and women, due to omitted factors on which men were likely to have higher values than women because they had higher values for the included factors. This extrapolation was rejected by the court of appeals. The district court did not deal with the effect of using reflective proxies on which men score higher than women.

2. The court of appeals' statement is correct, but its argument is irrelevant to AECOM's experts' point, which evidently was not understood. If the model is correct, whether or not perfect proxies would fully explain the disparity in pay between men and women turns on whether the employer was discriminatory. If AECOM was nondiscriminatory and the proxies were perfect, the sex-coefficient would become nonsignificant; if AECOM was discriminatory and the proxies perfect, the sex coefficient would remain significant. But AECOM's experts were not addressing whether AECOM was discriminatory, and not even whether the plaintiffs' regressions overstated or understated the degree of discrimination. Their point was that even if AECOM were nondiscriminatory, a regression of salary on proxies might nevertheless show a statistically significant sex coefficient because the proxies were imperfect and men scored higher on them. The statisticians passed judgment only on the question of whether the appearance of a statistically significant sex coefficient in the regression was unambiguous evidence of discrimination. They concluded it was not because, on the facts in *Sobel*,

a significant sex coefficient would be produced as a statistical artifact even for a nondiscriminatory employer. To this observation, the court's speculation about what would happen if the proxies were perfect—i.e., whether AECOM actually discriminated—was essentially irrelevant.

3. The sex-coefficient in the regression model without the reflective proxies (except for publication rate) is $-3,145$, and with all the reflective proxies is $-2,204$. Under the Robbins and Levin formula discussed in Section 14.5 at p. 450, n.10, if underadjustment were to account for the entire sex-coefficient of $-2,204$, the reflective proxies would have to account for no more than about 30% of the variation in productivity after accounting for the causal proxies:

$$(-3,145) \cdot (1 - R^2) = -2,204, \qquad R^2 = 0.30.$$

This would seem a substantial possibility because the squared partial multiple correlation between salary and the proxies (including the important reflective proxies for rank), argued to be necessary by Yeshiva and added by plaintiffs in their last round model, is $(0.8173 - 0.7297) \div (1 - 0.7297) = 0.324$, or 32.4%. (See note on partial correlation in Section 13.6 at p. 385.)

Section 14.6.1. Death penalty: does it deter murder?

1. Since Ehrlich's model is in logs, each 1% increase in the execution risk is associated with a 0.065% decrease in the murder rate. At the average values, the expectation of an additional execution in year $t + 1$ is equal to $1/75 = 1.333\%$ increase in executions, which the model equates to a $1.333 \times -0.065\% = -0.0867\%$ decrease in murders. Since the average number of murders in Ehrlich's data was 8,965, Ehrlich computed that the saving was $0.000867 \times 8,965 = 7.770$, or between 7 and 8 murders per execution. Ehrlich's statement that every execution saves between 7 and 8 murders misrepresented the results of his model, since the effect would only be achieved by increasing the average number of executions over the preceding five years.

2. When the period after 1964 is eliminated, the apparent deterrent effect of executions disappears as the coefficients of the execution risk become positive (although non-significant). This casts doubt on the model, since a finding of deterrence should not depend on the particular period selected.

3. The logarithmic transformation tends to emphasize variations at the lower range of a variable; at such values the change in the logarithm may be larger than the change in natural value of the variable. Because the execution risk declined to extremely low values in the middle and late 1960s, the log transform emphasizes this decline. And because the murder rate soared as the execution risk declined, this emphasis in the log model probably

accounts for the negative elasticity of its murder rate with respect to the execution risk, which does not appear in the natural values model.

4. The previous answers suggest that Ehrlich's conclusion depends on an assumed causal relation between the dramatic increase in murder and decline in executions that occurred in the late 1960s. However, the two trends may be unrelated: the increase in murder may be due primarily or entirely to factors that produced a general increase in violent crime in that period, without significant dependence on the decline in execution risk.

Section 14.7.2. Death penalty in Georgia

1. The two tables are as follows:

	Stranger victim			Non-Stranger victim		
	Death	Other	Total	Death	Other	Total
WV	28	9	37	15	22	37
Non WV	4	10	14	2	10	12
Total	32	19	51	17	32	49

2. The odds ratios are: 7.78 and 3.41
   The log odds ratios are: 2.05 and 1.23

3. The weights used for averaging the two log odds ratios are:

$$w_1 = 1/(28^{-1} + 9^{-1} + 4^{-1} + 10^{-1}) = 2.013$$
$$w_2 = 1/(15^{-1} + 22^{-1} + 2^{-1} + 10^{-1}) = 1.404.$$

Weighted average $= (2.05 \cdot w_1 + 1.23 \cdot w_2)/(w_1 + w_2) = 1.713$. Exponentiating, we have $e^{1.71} = 5.53$ as the corresponding odds ratio. The variance of the weighted average log odds ratio is the reciprocal of the sum of the weights, or $1/(2.013 + 1.404) = 0.293$. The log odds ratio is significantly different from 0, i.e., the odds ratio is significantly different from 1.

4. In cases involving a white victim, the odds that a defendant would be sentenced to death are $e^{1.5563} = 4.74$ times the odds on a death sentence if the victim were not white. In cases involving a black defendant, the odds on a death sentence are only $e^{-0.5308} = 0.588$ as great as when a non-black defendant is involved (the difference, however, is not statistically significant). In the white victim/black defendant cases, the joint effect is $e^{1.5663-0.5308} = 2.79$, indicating that the odds on a death sentence are almost three times as great for a black who kills a white then for the reference category (non-black defendant, non-white victim). Note that this is an additive model, so that there is no synergism between BD and WV effects. Also, the effect of WV is the same for any fixed values of other factors.

5. As above, the odds ratio on death sentence vs. other, comparing WV with non-WV, is $e^{1.5563} = 4.74$. This is somewhat less than the weighted average odds ratio of 5.53 computed in question 3, which indicated that a part of the odds ratio computed there was due to factors other than the factor that the victim was or was not a stranger.

6. The term e^L is the odds on a death sentence. Since in general odds equal $P/(1 - P)$, solving for P yields $P = \text{odds}/(1 + \text{odds}) = e^L/(1 + e^L)$.

7. The log-odds on death penalty at the values of McClesky's factors is

$$\text{log-odds} = (-3.5675) + (-0.5308 \cdot 1) + (1.5563 \cdot 1)$$
$$+ (0.3730 \cdot 3) + (0.3707 \cdot 0) + (1.7911 \cdot 1)$$
$$+ (0.1999 \cdot 0) + (1.4429 \cdot 0) + (0.1232 \cdot 0)$$
$$= 0.3681,$$

corresponding to a probability of death penalty of $e^{0.3681}/(1+e^{0.3681}) = 0.59$. If the victim were not white, the log-odds would be $0.3681 - 1.5563 = -1.1882$, corresponding to a probability of death penalty of $e^{-1.1882}/(1 + e^{-1.1882}) = 0.23$. Since the relative risk of a death penalty is $0.59/0.23 = 2.57 \geq 2$, it is more likely than not in the model that McClesky would not have received the death sentence if his victim had not been white.

8. There are 35 correct predictions of a death sentence and 37 correct predictions of no death sentence. The rate of correct predictions is $(35+37)/100 = 72\%$. See Section 14.9.1 for a more refined method of calculating accuracy.

Section 14.7.3. Deterring teenage smoking

1. Assuming that the overall prevalence of smoking (p_0) corresponds to the mean of cigarette taxes in 1988 ($\bar{X}$), probit (p_0) = constant + $\beta\bar{X}$, while probit (p_1) = constant + $\beta(\bar{X} + 20)$ where p_1 is the estimated prevalence of smoking corresponding to a tax 20 cents larger than $\bar{X}$. Subtracting, gives us probit (p_1) = probit(p_0) + $\beta \cdot 20$, or, equivalently, $z(p_1) = z(p_0) + 20\beta$, where $z(p)$ is the standard normal deviate cutting off probability p in the lower tail. For the 1988 cross-sectional model, $z(p_0) = (0.055) = -1.598$, so $z(p_1) = -1.598 + 20(-0.0059) = -1.716$, a normal deviate cutting off proportion $p_1 = 0.043$. The change is estimated to be $0.055 - 0.043 = 1.2$ percentage points, or a $1.2/5.5 = 21.8\%$ decrease (cross-sectionally, in 1988). For the 1992 cross-sectional model, the 20 cent increase in 1988 taxes is associated with a 7.4% reduction from the baseline rate of 24.4%. For the onset model, the estimated change in smoking incidence is from 24.4% at the baseline (corresponding to a 7.56¢ increase in taxes) to 21.49% for a 20¢ increase over that, or a 0.946% decrease. The change is not statistically significant.

2. The elasticity of response to tax increases in cross-sectional models may reflect the fact that cigarette taxes may be high in states with strong anti-smoking sentiment, so that the estimated effects of taxes also reflect the influence of that sentiment on teenage smoking decisions.

3. The effect of adding $1 to cigarette taxes cannot be estimated with any assurance because the variation in taxes did not approach that amount, the maximum increase being only 25¢.

Section 14.8.1. Challenger *disaster*

2. At 36°, the expected number of damaged O-rings is about 18.1; at 53° the expected number is about 2.2.

3. The percentage increase in expected number of damaged O-rings for each degree decrease in temperature is about 12.4%, or, more precisely, $\exp(0.124) - 1 = 13.2\%$.

4. An approximate 95% confidence interval for the expected number of damaged O-rings at 36° is given by exponentiating the confidence interval for the intercept term, $a = \ln \mu(36°)$: $\exp(2.8983 \pm 1.96 \cdot 1.1317)$, or from 2.0 to 167.

5. Pearson's chi-squared comparing expected and observed counts is 32.262 with 22 df; $P > 0.05$; the Poisson distribution cannot be rejected for these data.

6. The overdispersion factor for the data is $(32.262/22)^{1/2} = 1.21$. Hence, a 95% confidence interval for the expected number of damaged O-rings at 36° is $\exp(2.8983 \pm 1.96 \cdot 1.1317 \cdot 1.21)$, or from 1.2 to 266. The uncertainty is large due to extropolation down to 36° F. Evidently, temperature is not the only factor affecting the risk of damage to the rings.

Section 14.9.1. Georgia death penalty *revisited*

1. For the construction sample with a 50% classification rule, the outcome was as follows:

		Predicted sentence		
		D	$\bar{D}$	
Actual	D	14	5	19
sentence	$\bar{D}$	6	25	31
		20	30	50

The sensitivity of the classification is apparently $14/19 = 0.74$ and the specificity is apparently $25/31 = 0.81$. In the test sample, the outcome was

		Predicted sentence		
		D	$\bar{D}$	
Actual	D	15	15	30
sentence	$\bar{D}$	2	18	20
		17	33	50

Now, sensitivity is only $0.50 = 15/30$ while specificity has increased to $0.90 = 18/20$. The proportion of correct classifications falls from 78% in the construction sample to 66% in the test sample.

2. The PPV is $14/20 = 0.70$, and the NPV is $25/30 = 0.83$ in the construction sample. In the test sample, PPV $= 15/17 = 0.88$, but the NPV is $18/33 = 0.55$.

Appendix II: Tables

Table A1:   The Cumulative Normal Distribution

z	$P[Z < z]$	z	$P[Z < z]$	z	$P[Z < z]$	z	$P[Z < z]$	z	$P[Z < z]$
0.00	0.50000	0.80	0.78814	1.60	0.94520	2.40	0.99180	3.50	0.99977
0.02	0.50798	0.82	0.79389	1.62	0.94738	2.42	0.99224	3.55	0.99981
0.04	0.51595	0.84	0.79955	1.64	0.94950	2.44	0.99266	3.60	0.99984
0.06	0.52392	0.86	0.80511	1.66	0.95154	2.46	0.99305	3.65	0.99987
0.08	0.53188	0.88	0.81057	1.68	0.95352	2.48	0.99343	3.70	0.99989
0.10	0.53983	0.90	0.81594	1.70	0.95543	2.50	0.99379	3.75	0.99991
0.12	0.54776	0.92	0.82121	1.72	0.95728	2.52	0.99413	3.80	0.99993
0.14	0.55567	0.94	0.82639	1.74	0.95907	2.54	0.99446	3.85	0.99994
0.16	0.56356	0.96	0.83147	1.76	0.96080	2.56	0.99477	3.90	0.99995
0.18	0.57142	0.98	0.83646	1.78	0.96246	2.58	0.99506	3.95	0.99996
0.20	0.57926	1.00	0.84134	1.80	0.96407	2.60	0.99534	4.00	0.99997
0.22	0.58706	1.02	0.84614	1.82	0.96562	2.62	0.99560	4.05	0.99997
0.24	0.59483	1.04	0.85083	1.84	0.96712	2.64	0.99585	4.10	0.99998
0.26	0.60257	1.06	0.85543	1.86	0.96856	2.66	0.99609	4.15	0.99998
0.28	0.61026	1.08	0.85993	1.88	0.96995	2.68	0.99632	4.20	0.99999
0.30	0.61791	1.10	0.86433	1.90	0.97128	2.70	0.99653	4.25	0.99999
0.32	0.62552	1.12	0.86864	1.92	0.97257	2.72	0.99674	4.30	0.99999
0.34	0.63307	1.14	0.87286	1.94	0.97381	2.74	0.99693	4.35	0.99999
0.36	0.64058	1.16	0.87698	1.96	0.97500	2.76	0.99711	4.40	0.99999
0.38	0.64803	1.18	0.88100	1.98	0.97615	2.78	0.99728	4.45	1.00000
0.40	0.65542	1.20	0.88493	2.00	0.97725	2.80	0.99744	4.50	1.00000
0.42	0.66276	1.22	0.88877	2.02	0.97831	2.82	0.99760	4.55	1.00000
0.44	0.67003	1.24	0.89251	2.04	0.97932	2.84	0.99774	4.60	1.00000
0.46	0.67724	1.26	0.89617	2.06	0.98030	2.86	0.99788	4.65	1.00000
0.48	0.68439	1.28	0.89973	2.08	0.98124	2.88	0.99801	4.70	1.00000
0.50	0.69146	1.30	0.90320	2.10	0.98214	2.90	0.99813	4.75	1.00000
0.52	0.69847	1.32	0.90658	2.12	0.98300	2.92	0.99825	4.80	1.00000
0.54	0.70540	1.34	0.90988	2.14	0.98382	2.94	0.99836	4.85	1.00000
0.56	0.71226	1.36	0.91309	2.16	0.98461	2.96	0.99846	4.90	1.00000
0.58	0.71904	1.38	0.91621	2.18	0.98537	2.98	0.99856	4.95	1.00000
0.60	0.72575	1.40	0.91924	2.20	0.98610	3.00	0.99865	5.00	1.00000
0.62	0.73237	1.42	0.92220	2.22	0.98679	3.05	0.99886		
0.64	0.73891	1.44	0.92507	2.24	0.98745	3.10	0.99903		
0.66	0.74537	1.46	0.92786	2.26	0.98809	3.15	0.99918		
0.68	0.75175	1.48	0.93056	2.28	0.98870	3.20	0.99931		
0.70	0.75804	1.50	0.93319	2.30	0.98928	3.25	0.99942		
0.72	0.76424	1.52	0.93574	2.32	0.98983	3.30	0.99952		
0.74	0.77035	1.54	0.93822	2.34	0.99036	3.35	0.99960		
0.76	0.77637	1.56	0.94062	2.36	0.99086	3.40	0.99966		
0.78	0.78230	1.58	0.94295	2.38	0.99134	3.45	0.99972		

Table A2:   Critical Values of the Standard Normal Distribution

The table gives critical values z for selected values of P, and tail probabilities P for selected values of z.

z	$P[Z < z]$	$P[Z > z]$	$P[-z < Z < z]$	$P[|Z| > z]$
0.0	.50	.50	.00	1.0000
0.1	.5398	.4602	.0797	.9203
0.126	.55	.45	.10	.90
0.2	.5793	.4207	.1585	.8415
0.253	.60	.40	.20	.80
0.3	.6179	.3821	.2358	.7642
0.385	.65	.35	.30	.70
0.4	.6554	.3446	.3108	.6892
0.5	.6915	.3085	.3829	.6171
0.524	.70	.30	.40	.60
0.6	.7257	.2743	.4515	.5485
0.674	.75	.25	.50	.50
0.7	.7580	.2420	.5161	.4839
0.8	.7881	.2119	.5763	.4237
0.842	.80	.20	.60	.40
0.9	.8159	.1841	.6319	.3681
1.0	.8413	.1587	.6827	.3173
1.036	.85	.15	.70	.30
1.1	.8643	.1357	.7287	.2713
1.2	.8849	.1151	.7699	.2301
1.282	.90	.10	.80	.20
1.3	.9032	.0968	.8064	.1936
1.4	.9192	.0808	.8385	.1615
1.440	.925	.075	.85	.15
1.5	.9332	.0668	.8664	.1336
1.6	.9452	.0548	.8904	.1096
1.645	.95	.05	.90	.10
1.7	.9554	.0446	.9109	.0891
1.8	.9641	.0359	.9281	.0719
1.9	.9713	.0287	.9426	.0574
1.960	.975	.025	.95	.05
2.0	.9772	.0228	.9545	.0455
2.1	.9821	.0179	.9643	.0357
2.2	.9861	.0139	.9722	.0278
2.242	.9875	.0125	.975	.025
2.3	.9893	.0107	.9786	.0214
2.326	.99	.01	.98	.02
2.4	.9918	.0082	.9836	.0164
2.5	.9938	.0062	.9876	.0124
2.576	.995	.005	.99	.01
2.6	.9953	.0047	.9907	.0093
2.7	.9965	.0035	.9931	.0069
2.8	.9974	.0026	.9949	.0051
2.813	.9975	.0025	.995	.005
2.9	.9981	.0019	.9963	.0037
3.0	.9987	.0013	.9973	.0027
3.090	.999	.001	.998	.002
3.1	.9990	.0010	.9981	.0019
3.2	.9993	.0007	.9986	.0014
3.291	.9995	.0005	.999	.001
3.3	.9995	.0005	.9990	.0010

Table A2   (Cont'd)

z	$P[Z < z]$	$P[Z > z]$	$P[-z < Z < z]$	$P[|Z| > z]$
3.4	.9997	.0003	.9993	.0007
3.5	.9998	.0002	.9995	.0005
3.6	.9998	.0002	.9997	.0003
3.719	.9999	.00010	.9998	.00020
3.8	.99992	.00007	.99986	.00014
3.9	.99995	.00005	.99990	.00010
4.0	.99997	.00003	.99994	.00006

Table B: Cumulative Terms of the Binomial Distribution

The table gives $P[X \geq x|n, p]$ for selected values of n, $p \leq \frac{1}{2}$, and $1 \leq x \leq n$. For values of $p > \frac{1}{2}$, use the relation

$$P[X \geq x|n, p] = 1 - P[X \geq n - x + 1|n, 1 - p].$$

For example, for $p = .90$ and $n = 10$, $P[X \geq 8] = 1 - P[X \geq 3]$ for $p = .10$, which equals $1 - .0702 = .9298$.

						p					
n	x	.05	.10	.15	.20	.25	.30	.35	.40	.45	.50
2	1	.0975	.1900	.2775	.3600	.4375	.5100	.5775	.6400	.6975	.7500
	2	.0025	.0100	.0225	.0400	.0625	.0900	.1225	.1600	.2025	.2500
3	1	.1426	.2710	.3859	.4880	.5781	.6570	.7254	.7840	.8336	.8750
	2	.0073	.0280	.0608	.1040	.1563	.2160	.2818	.3520	.4253	.5000
	3	.0001	.0010	.0034	.0080	.0156	.0270	.0429	.0640	.0911	.1250
4	1	.1855	.3439	.4780	.5904	.6836	.7599	.8215	.8704	.9085	.9375
	2	.0140	.0523	.1095	.1808	.2617	.3483	.4370	.5248	.6090	.6875
	3	.0005	.0037	.0120	.0272	.0508	.0837	.1265	.1792	.2415	.3125
	4	.0000	.0001	.0005	.0016	.0039	.0081	.0150	.0256	.0410	.0625
5	1	.2262	.4095	.5563	.6723	.7627	.8319	.8840	.9222	.9497	.9688
	2	.0226	.0815	.1648	.2627	.3672	.4718	.5716	.6630	.7438	.8125
	3	.0012	.0086	.0266	.0579	.1035	.1631	.2352	.3174	.4069	.5000
	4	.0000	.0005	.0022	.0067	.0156	.0308	.0540	.0870	.1312	.1875
	5	.0000	.0000	.0001	.0003	.0010	.0024	.0053	.0102	.0185	.0313
6	1	.2649	.4686	.6229	.7379	.8220	.8824	.9246	.9533	.9723	.9844
	2	.0328	.1143	.2235	.3446	.4661	.5798	.6809	.7667	.8364	.8906
	3	.0022	.0159	.0473	.0989	.1694	.2557	.3529	.4557	.5585	.6563
	4	.0001	.0013	.0059	.0170	.0376	.0705	.1174	.1792	.2553	.3438
	5	.0000	.0001	.0004	.0016	.0046	.0109	.0223	.0410	.0692	.1094
	6	.0000	.0000	.0000	.0001	.0002	.0007	.0018	.0041	.0083	.0156
7	1	.3017	.5217	.6794	.7903	.8665	.9176	.9510	.9720	.9848	.9922
	2	.0444	.1497	.2834	.4233	.5551	.6706	.7662	.8414	.8976	.9375
	3	.0038	.0257	.0738	.1480	.2436	.3529	.4677	.5801	.6836	.7734
	4	.0002	.0027	.0121	.0333	.0706	.1260	.1998	.2898	.3917	.5000
	5	.0000	.0002	.0012	.0047	.0129	.0288	.0556	.0963	.1529	.2266
	6	.0000	.0000	.0001	.0004	.0013	.0038	.0090	.0188	.0357	.0625
	7	.0000	.0000	.0000	.0000	.0001	.0002	.0006	.0016	.0037	.0078
8	1	.3366	.5695	.7275	.8322	.8999	.9424	.9681	.9832	.9916	.9961
	2	.0572	.1869	.3428	.4967	.6329	.7447	.8309	.8936	.9368	.9648
	3	.0058	.0381	.1052	.2031	.3215	.4482	.5722	.6846	.7799	.8555
	4	.0004	.0050	.0214	.0563	.1138	.1941	.2936	.4059	.5230	.6367
	5	.0000	.0004	.0029	.0104	.0273	.0580	.1061	.1737	.2604	.3633
	6	.0000	.0000	.0002	.0012	.0042	.0113	.0253	.0498	.0885	.1445
	7	.0000	.0000	.0000	.0001	.0004	.0013	.0036	.0085	.0181	.0352
	8	.0000	.0000	.0000	.0000	.0000	.0001	.0002	.0007	.0017	.0039
9	1	.3698	.6126	.7684	.8658	.9249	.9596	.9793	.9899	.9954	.9980
	2	.0712	.2252	.4005	.5638	.6997	.8040	.8789	.9295	.9615	.9805
	3	.0084	.0530	.1409	.2618	.3993	.5372	.6627	.7682	.8505	.9102
	4	.0006	.0083	.0339	.0856	.1657	.2703	.3911	.5174	.6386	.7461
	5	.0000	.0009	.0056	.0196	.0489	.0988	.1717	.2666	.3786	.5000

n	x	.05	.10	.15	.20	.25	.30	.35	.40	.45	.50
	6	.0000	.0001	.0006	.0031	.0100	.0253	.0536	.0994	.1658	.2539
	7	.0000	.0000	.0000	.0003	.0013	.0043	.0112	.0250	.0498	.0898
	8	.0000	.0000	.0000	.0000	.0001	.0004	.0014	.0038	.0091	.0195
	9	.0000	.0000	.0000	.0000	.0000	.0000	.0001	.0003	.0008	.0020
10	1	.4013	.6513	.8031	.8926	.9437	.9718	.9865	.9940	.9975	.9990
	2	.0861	.2639	.4557	.6242	.7560	.8507	.9140	.9536	.9767	.9893
	3	.0115	.0702	.1798	.3222	.4744	.6172	.7384	.8327	.9004	.9453
	4	.0010	.0128	.0500	.1209	.2241	.3504	.4862	.6177	.7340	.8281
	5	.0001	.0016	.0099	.0328	.0781	.1503	.2485	.3669	.4956	.6230
	6	.0000	.0001	.0014	.0064	.0197	.0473	.0949	.1662	.2616	.3770
	7	.0000	.0000	.0001	.0009	.0035	.0106	.0260	.0548	.1020	.1719
	8	.0000	.0000	.0000	.0001	.0004	.0016	.0048	.0123	.0274	.0547
	9	.0000	.0000	.0000	.0000	.0000	.0001	.0005	.0017	.0045	.0107
	10	.0000	.0000	.0000	.0000	.0000	.0000	.0000	.0001	.0003	.0010
11	1	.4312	.6862	.8327	.9141	.9578	.9802	.9912	.9964	.9986	.9995
	2	.1019	.3026	.5078	.6779	.8029	.8870	.9394	.9698	.9861	.9941
	3	.0152	.0896	.2212	.3826	.5448	.6873	.7999	.8811	.9348	.9673
	4	.0016	.0185	.0694	.1611	.2867	.4304	.5744	.7037	.8089	.8867
	5	.0001	.0028	.0159	.0504	.1146	.2103	.3317	.4672	.6029	.7256
	6	.0000	.0003	.0027	.0117	.0343	.0782	.1487	.2465	.3669	.5000
	7	.0000	.0000	.0003	.0020	.0076	.0216	.0501	.0994	.1738	.2744
	8	.0000	.0000	.0000	.0002	.0012	.0043	.0122	.0293	.0610	.1133
	9	.0000	.0000	.0000	.0000	.0001	.0006	.0020	.0059	.0148	.0327
	10	.0000	.0000	.0000	.0000	.0000	.0000	.0002	.0007	.0022	.0059
	11	.0000	.0000	.0000	.0000	.0000	.0000	.0000	.0000	.0002	.0005
12	1	.4596	.7176	.8578	.9313	.9683	.9862	.9943	.9978	.9992	.9998
	2	.1184	.3410	.5565	.7251	.8416	.9150	.9576	.9804	.9917	.9968
	3	.0196	.1109	.2642	.4417	.6093	.7472	.8487	.9166	.9579	.9807
	4	.0022	.0256	.0922	.2054	.3512	.5075	.6533	.7747	.8655	.9270
	5	.0002	.0043	.0239	.0726	.1576	.2763	.4167	.5618	.6956	.8062
	6	.0000	.0005	.0046	.0194	.0544	.1178	.2127	.3348	.4731	.6128
	7	.0000	.0001	.0007	.0039	.0143	.0386	.0846	.1582	.2607	.3872
	8	.0000	.0000	.0001	.0006	.0028	.0095	.0255	.0573	.1117	.1938
	9	.0000	.0000	.0000	.0001	.0004	.0017	.0056	.0153	.0356	.0730
	10	.0000	.0000	.0000	.0000	.0000	.0002	.0008	.0028	.0079	.0193
	11	.0000	.0000	.0000	.0000	.0000	.0000	.0001	.0003	.0011	.0032
	12	.0000	.0000	.0000	.0000	.0000	.0000	.0000	.0000	.0001	.0002
13	1	.4867	.7458	.8791	.9450	.9762	.9903	.9963	.9987	.9996	.9999
	2	.1354	.3787	.6017	.7664	.8733	.9363	.9704	.9874	.9951	.9983
	3	.0245	.1339	.3080	.4983	.6674	.7975	.8868	.9421	.9731	.9888
	4	.0031	.0342	.1180	.2527	.4157	.5794	.7217	.8314	.9071	.9539
	5	.0003	.0065	.0342	.0991	.2060	.3457	.4995	.6470	.7721	.8666
	6	.0000	.0009	.0075	.0300	.0802	.1654	.2841	.4256	.5732	.7095
	7	.0000	.0001	.0013	.0070	.0243	.0624	.1295	.2288	.3563	.5000
	8	.0000	.0000	.0002	.0012	.0056	.0182	.0462	.0977	.1788	.2905
	9	.0000	.0000	.0000	.0002	.0010	.0040	.0126	.0321	.0698	.1334
	10	.0000	.0000	.0000	.0000	.0001	.0007	.0025	.0078	.0203	.0461
	11	.0000	.0000	.0000	.0000	.0000	.0001	.0003	.0013	.0041	.0112
	12	.0000	.0000	.0000	.0000	.0000	.0000	.0000	.0001	.0005	.0017
	13	.0000	.0000	.0000	.0000	.0000	.0000	.0000	.0000	.0000	.0001
14	1	.5123	.7712	.8972	.9560	.9822	.9932	.9976	.9992	.9998	.9999
	2	.1530	.4154	.6433	.8021	.8990	.9525	.9795	.9919	.9971	.9991
	3	.0301	.1584	.3521	.5519	.7189	.8392	.9161	.9602	.9830	.9935
	4	.0042	.0441	.1465	.3018	.4787	.6448	.7795	.8757	.9368	.9713
	5	.0004	.0092	.0467	.1298	.2585	.4158	.5773	.7207	.8328	.9102

n	x	.05	.10	.15	.20	p .25	.30	.35	.40	.45	.50
	6	.0000	.0015	.0115	.0439	.1117	.2195	.3595	.5141	.6627	.7880
	7	.0000	.0002	.0022	.0116	.0383	.0933	.1836	.3075	.4539	.6047
	8	.0000	.0000	.0003	.0024	.0103	.0315	.0753	.1501	.2586	.3953
	9	.0000	.0000	.0000	.0004	.0022	.0083	.0243	.0583	.1189	.2120
	10	.0000	.0000	.0000	.0000	.0003	.0017	.0060	.0175	.0426	.0898
	11	.0000	.0000	.0000	.0000	.0000	.0002	.0011	.0039	.0114	.0287
	12	.0000	.0000	.0000	.0000	.0000	.0000	.0001	.0006	.0022	.0065
	13	.0000	.0000	.0000	.0000	.0000	.0000	.0000	.0001	.0003	.0009
	14	.0000	.0000	.0000	.0000	.0000	.0000	.0000	.0000	.0000	.0001
15	1	.5367	.7941	.9126	.9648	.9866	.9953	.9984	.9995	.9999	1.0000
	2	.1710	.4510	.6814	.8329	.9198	.9647	.9858	.9948	.9983	.9995
	3	.0362	.1841	.3958	.6020	.7639	.8732	.9383	.9729	.9893	.9963
	4	.0055	.0556	.1773	.3518	.5387	.7031	.8273	.9095	.9576	.9824
	5	.0006	.0127	.0617	.1642	.3135	.4845	.6481	.7827	.8796	.9408
	6	.0001	.0022	.0168	.0611	.1484	.2784	.4357	.5968	.7392	.8491
	7	.0000	.0003	.0036	.0181	.0566	.1311	.2452	.3902	.5478	.6964
	8	.0000	.0000	.0006	.0042	.0173	.0500	.1132	.2131	.3465	.5000
	9	.0000	.0000	.0001	.0008	.0042	.0152	.0422	.0950	.1818	.3036
	10	.0000	.0000	.0000	.0001	.0008	.0037	.0124	.0338	.0769	.1509
	11	.0000	.0000	.0000	.0000	.0001	.0007	.0028	.0093	.0255	.0592
	12	.0000	.0000	.0000	.0000	.0000	.0001	.0005	.0019	.0063	.0176
	13	.0000	.0000	.0000	.0000	.0000	.0000	.0001	.0003	.0011	.0037
	14	.0000	.0000	.0000	.0000	.0000	.0000	.0000	.0000	.0001	.0005
	15	.0000	.0000	.0000	.0000	.0000	.0000	.0000	.0000	.0000	.0000
16	1	.5599	.8147	.9257	.9719	.9900	.9967	.9990	.9997	.9999	1.0000
	2	.1892	.4853	.7161	.8593	.9365	.9739	.9902	.9967	.9990	.9997
	3	.0429	.2108	.4386	.6482	.8029	.9006	.9549	.9817	.9934	.9979
	4	.0070	.0684	.2101	.4019	.5950	.7541	.8661	.9349	.9719	.9894
	5	.0009	.0170	.0791	.2018	.3698	.5501	.7108	.8334	.9147	.9616
	6	.0001	.0033	.0235	.0817	.1897	.3402	.5100	.6712	.8024	.8949
	7	.0000	.0005	.0056	.0267	.0796	.1753	.3119	.4728	.6340	.7728
	8	.0000	.0001	.0011	.0070	.0271	.0744	.1594	.2839	.4371	.5982
	9	.0000	.0000	.0002	.0015	.0075	.0257	.0671	.1423	.2559	.4018
	10	.0000	.0000	.0000	.0002	.0016	.0071	.0229	.0583	.1241	.2272
	11	.0000	.0000	.0000	.0000	.0003	.0016	.0062	.0191	.0486	.1051
	12	.0000	.0000	.0000	.0000	.0000	.0003	.0013	.0049	.0149	.0384
	13	.0000	.0000	.0000	.0000	.0000	.0000	.0002	.0009	.0035	.0106
	14	.0000	.0000	.0000	.0000	.0000	.0000	.0000	.0001	.0006	.0021
	15	.0000	.0000	.0000	.0000	.0000	.0000	.0000	.0000	.0001	.0003
	16	.0000	.0000	.0000	.0000	.0000	.0000	.0000	.0000	.0000	.0000
17	1	.5819	.8332	.9369	.9775	.9925	.9977	.9993	.9998	1.0000	1.0000
	2	.2078	.5182	.7475	.8818	.9499	.9807	.9933	.9979	.9994	.9999
	3	.0503	.2382	.4802	.6904	.8363	.9226	.9673	.9877	.9959	.9988
	4	.0088	.0826	.2444	.4511	.6470	.7981	.8972	.9536	.9816	.9936
	5	.0012	.0221	.0987	.2418	.4261	.6113	.7652	.8740	.9404	.9755
	6	.0001	.0047	.0319	.1057	.2347	.4032	.5803	.7361	.8529	.9283
	7	.0000	.0008	.0083	.0377	.1071	.2248	.3812	.5522	.7098	.8338
	8	.0000	.0001	.0017	.0109	.0402	.1046	.2128	.3595	.5257	.6855
	9	.0000	.0000	.0003	.0026	.0124	.0403	.0994	.1989	.3374	.5000
	10	.0000	.0000	.0000	.0005	.0031	.0127	.0383	.0919	.1834	.3145
	11	.0000	.0000	.0000	.0001	.0006	.0032	.0120	.0348	.0826	.1662
	12	.0000	.0000	.0000	.0000	.0001	.0007	.0030	.0106	.0301	.0717
	13	.0000	.0000	.0000	.0000	.0000	.0001	.0006	.0025	.0086	.0245
	14	.0000	.0000	.0000	.0000	.0000	.0000	.0001	.0005	.0019	.0064
	15	.0000	.0000	.0000	.0000	.0000	.0000	.0000	.0001	.0003	.0012
	16	.0000	.0000	.0000	.0000	.0000	.0000	.0000	.0000	.0000	.0001
	17	.0000	.0000	.0000	.0000	.0000	.0000	.0000	.0000	.0000	.0000

n	r	.05	.10	.15	.20	p .25	.30	.35	.40	.45	.50
18	1	.6028	.8499	.9464	.9820	.9944	.9984	.9996	.9999	1.0000	1.0000
	2	.2265	.5497	.7759	.9009	.9605	.9858	.9954	.9987	.9997	.9999
	3	.0581	.2662	.5203	.7287	.8647	.9400	.9764	.9918	.9975	.9993
	4	.0109	.0982	.2798	.4990	.6943	.8354	.9217	.9672	.9880	.9962
	5	.0015	.0282	.1206	.2836	.4813	.6673	.8114	.9058	.9589	.9846
	6	.0002	.0064	.0419	.1329	.2825	.4656	.6450	.7912	.8923	.9519
	7	.0000	.0012	.0118	.0513	.1390	.2783	.4509	.6257	.7742	.8811
	8	.0000	.0002	.0027	.0163	.0569	.1407	.2717	.4366	.6085	.7597
	9	.0000	.0000	.0005	.0043	.0193	.0596	.1391	.2632	.4222	.5927
	10	.0000	.0000	.0001	.0009	.0054	.0210	.0597	.1347	.2527	.4073
	11	.0000	.0000	.0000	.0002	.0012	.0061	.0212	.0576	.1280	.2403
	12	.0000	.0000	.0000	.0000	.0002	.0014	.0062	.0203	.0537	.1189
	13	.0000	.0000	.0000	.0000	.0000	.0003	.0014	.0058	.0183	.0481
	14	.0000	.0000	.0000	.0000	.0000	.0000	.0003	.0013	.0049	.0154
	15	.0000	.0000	.0000	.0000	.0000	.0000	.0000	.0002	.0010	.0038
	16	.0000	.0000	.0000	.0000	.0000	.0000	.0000	.0000	.0001	.0007
	17	.0000	.0000	.0000	.0000	.0000	.0000	.0000	.0000	.0000	.0001
	18	.0000	.0000	.0000	.0000	.0000	.0000	.0000	.0000	.0000	.0000
19	1	.6226	.8649	.9544	.9856	.9958	.9989	.9997	.9999	1.0000	1.0000
	2	.2453	.5797	.8015	.9171	.9690	.9896	.9969	.9992	.9998	1.0000
	3	.0665	.2946	.5587	.7631	.8887	.9538	.9830	.9945	.9985	.9996
	4	.0132	.1150	.3159	.5449	.7369	.8668	.9409	.9770	.9923	.9978
	5	.0020	.0352	.1444	.3267	.5346	.7178	.8500	.9304	.9720	.9904
	6	.0002	.0086	.0537	.1631	.3322	.5261	.7032	.8371	.9223	.9682
	7	.0000	.0017	.0163	.0676	.1749	.3345	.5188	.6919	.8273	.9165
	8	.0000	.0003	.0041	.0233	.0775	.1820	.3344	.5122	.6831	.8204
	9	.0000	.0000	.0008	.0067	.0287	.0839	.1855	.3325	.5060	.6762
	10	.0000	.0000	.0001	.0016	.0089	.0326	.0875	.1861	.3290	.5000
	11	.0000	.0000	.0000	.0003	.0023	.0105	.0347	.0885	.1841	.3238
	12	.0000	.0000	.0000	.0000	.0005	.0028	.0114	.0352	.0871	.1796
	13	.0000	.0000	.0000	.0000	.0001	.0006	.0031	.0116	.0342	.0835
	14	.0000	.0000	.0000	.0000	.0000	.0001	.0007	.0031	.0109	.0318
	15	.0000	.0000	.0000	.0000	.0000	.0000	.0001	.0006	.0028	.0096
	16	.0000	.0000	.0000	.0000	.0000	.0000	.0000	.0001	.0005	.0022
	17	.0000	.0000	.0000	.0000	.0000	.0000	.0000	.0000	.0001	.0004
	18	.0000	.0000	.0000	.0000	.0000	.0000	.0000	.0000	.0000	.0000
	19	.0000	.0000	.0000	.0000	.0000	.0000	.0000	.0000	.0000	.0000
20	1	.6415	.8784	.9612	.9885	.9968	.9992	.9998	1.0000	1.0000	1.0000
	2	.2642	.6083	.8244	.9308	.9757	.9924	.9979	.9995	.9999	1.0000
	3	.0755	.3231	.5951	.7939	.9087	.9645	.9879	.9964	.9991	.9998
	4	.0159	.1330	.3523	.5886	.7748	.8929	.9556	.9840	.9951	.9987
	5	.0026	.0432	.1702	.3704	.5852	.7625	.8818	.9490	.9811	.9941
	6	.0003	.0113	.0673	.1958	.3828	.5836	.7546	.8744	.9447	.9793
	7	.0000	.0024	.0219	.0867	.2142	.3920	.5834	.7500	.8701	.9423
	8	.0000	.0004	.0059	.0321	.1018	.2277	.3990	.5841	.7480	.8684
	9	.0000	.0001	.0013	.0100	.0409	.1133	.2376	.4044	.5857	.7483
	10	.0000	.0000	.0002	.0026	.0139	.0480	.1218	.2447	.4086	.5881
	11	.0000	.0000	.0000	.0006	.0039	.0171	.0532	.1275	.2493	.4119
	12	.0000	.0000	.0000	.0001	.0009	.0051	.0196	.0565	.1308	.2517
	13	.0000	.0000	.0000	.0000	.0002	.0013	.0060	.0210	.0580	.1316
	14	.0000	.0000	.0000	.0000	.0000	.0003	.0015	.0065	.0214	.0577
	15	.0000	.0000	.0000	.0000	.0000	.0000	.0003	.0016	.0064	.0207
	16	.0000	.0000	.0000	.0000	.0000	.0000	.0000	.0003	.0015	.0059
	17	.0000	.0000	.0000	.0000	.0000	.0000	.0000	.0000	.0003	.0013
	18	.0000	.0000	.0000	.0000	.0000	.0000	.0000	.0000	.0000	.0002
	19	.0000	.0000	.0000	.0000	.0000	.0000	.0000	.0000	.0000	.0000
	20	.0000	.0000	.0000	.0000	.0000	.0000	.0000	.0000	.0000	.0000

					p					
n	*x*	.05	.10	.15	.20	.25	.30	.35	.40	.45	.50

n	*x*	.05	.10	.15	.20	.25	.30	.35	.40	.45	.50
21	1	.6594	.8906	.9671	.9908	.9976	.9994	.9999	1.0000	1.0000	1.0000
	2	.2830	.6353	.8450	.9424	.9810	.9944	.9986	.9997	.9999	1.0000
	3	.0849	.3516	.6295	.8213	.9255	.9729	.9914	.9976	.9994	.9999
	4	.0189	.1520	.3887	.6296	.8083	.9144	.9669	.9890	.9969	.9993
	5	.0032	.0522	.1975	.4140	.6326	.8016	.9076	.9630	.9874	.9964
	6	.0004	.0144	.0827	.2307	.4334	.6373	.7991	.9043	.9611	.9867
	7	.0000	.0033	.0287	.1085	.2564	.4495	.6433	.7998	.9036	.9608
	8	.0000	.0006	.0083	.0431	.1299	.2770	.4635	.6505	.8029	.9054
	9	.0000	.0001	.0020	.0144	.0561	.1477	.2941	.4763	.6587	.8083
	10	.0000	.0000	.0004	.0041	.0206	.0676	.1623	.3086	.4883	.6682
	11	.0000	.0000	.0001	.0010	.0064	.0264	.0772	.1744	.3210	.5000
	12	.0000	.0000	.0000	.0002	.0017	.0087	.0313	.0849	.1841	.3318
	13	.0000	.0000	.0000	.0000	.0004	.0024	.0108	.0352	.0908	.1917
	14	.0000	.0000	.0000	.0000	.0001	.0006	.0031	.0123	.0379	.0946
	15	.0000	.0000	.0000	.0000	.0000	.0001	.0007	.0036	.0132	.0392
	16	.0000	.0000	.0000	.0000	.0000	.0000	.0001	.0008	.0037	.0133
	17	.0000	.0000	.0000	.0000	.0000	.0000	.0000	.0002	.0008	.0036
	18	.0000	.0000	.0000	.0000	.0000	.0000	.0000	.0000	.0001	.0007
	19	.0000	.0000	.0000	.0000	.0000	.0000	.0000	.0000	.0000	.0001
	20	.0000	.0000	.0000	.0000	.0000	.0000	.0000	.0000	.0000	.0000
	21	.0000	.0000	.0000	.0000	.0000	.0000	.0000	.0000	.0000	.0000
22	1	.6765	.9015	.9720	.9926	.9982	.9996	.9999	1.0000	1.0000	1.0000
	2	.3018	.6608	.8633	.9520	.9851	.9959	.9990	.9998	1.0000	1.0000
	3	.0948	.3800	.6618	.8455	.9394	.9793	.9939	.9984	.9997	.9999
	4	.0222	.1719	.4248	.6680	.8376	.9319	.9755	.9924	.9980	.9996
	5	.0040	.0621	.2262	.4571	.6765	.8355	.9284	.9734	.9917	.9978
	6	.0006	.0182	.0999	.2674	.4832	.6866	.8371	.9278	.9729	.9915
	7	.0001	.0044	.0368	.1330	.3006	.5058	.6978	.8416	.9295	.9738
	8	.0000	.0009	.0114	.0561	.1615	.3287	.5264	.7102	.8482	.9331
	9	.0000	.0001	.0030	.0201	.0746	.1865	.3534	.5460	.7236	.8569
	10	.0000	.0000	.0007	.0061	.0295	.0916	.2084	.3756	.5650	.7383
	11	.0000	.0000	.0001	.0016	.0100	.0387	.1070	.2280	.3963	.5841
	12	.0000	.0000	.0000	.0003	.0029	.0140	.0474	.1207	.2457	.4159
	13	.0000	.0000	.0000	.0001	.0007	.0043	.0180	.0551	.1328	.2617
	14	.0000	.0000	.0000	.0000	.0001	.0011	.0058	.0215	.0617	.1431
	15	.0000	.0000	.0000	.0000	.0000	.0002	.0016	.0070	.0243	.0669
	16	.0000	.0000	.0000	.0000	.0000	.0000	.0003	.0019	.0080	.0262
	17	.0000	.0000	.0000	.0000	.0000	.0000	.0001	.0004	.0021	.0085
	18	.0000	.0000	.0000	.0000	.0000	.0000	.0000	.0001	.0005	.0022
	19	.0000	.0000	.0000	.0000	.0000	.0000	.0000	.0000	.0001	.0004
	20	.0000	.0000	.0000	.0000	.0000	.0000	.0000	.0000	.0000	.0001
	21	.0000	.0000	.0000	.0000	.0000	.0000	.0000	.0000	.0000	.0000
	22	.0000	.0000	.0000	.0000	.0000	.0000	.0000	.0000	.0000	.0000
23	1	.6926	.9114	.9762	.9941	.9987	.9997	1.0000	1.0000	1.0000	1.0000
	2	.3206	.6849	.8796	.9602	.9884	.9970	.9993	.9999	1.0000	1.0000
	3	.1052	.4080	.6920	.8668	.9508	.9843	.9957	.9990	.9998	1.0000
	4	.0258	.1927	.4604	.7035	.8630	.9462	.9819	.9948	.9988	.9998
	5	.0049	.0731	.2560	.4993	.7168	.8644	.9449	.9810	.9945	.9987
	6	.0008	.0226	.1189	.3053	.5315	.7312	.8691	.9460	.9814	.9947
	7	.0001	.0058	.0463	.1598	.3463	.5601	.7466	.8760	.9490	.9827
	8	.0000	.0012	.0152	.0715	.1963	.3819	.5864	.7627	.8848	.9534
	9	.0000	.0002	.0042	.0273	.0963	.2291	.4140	.6116	.7797	.8950
	10	.0000	.0000	.0010	.0089	.0408	.1201	.2592	.4438	.6364	.7976
	11	.0000	.0000	.0002	.0025	.0149	.0546	.1425	.2871	.4722	.6612
	12	.0000	.0000	.0000	.0006	.0046	.0214	.0682	.1636	.3135	.5000
	13	.0000	.0000	.0000	.0001	.0012	.0072	.0283	.0813	.1836	.3388
	14	.0000	.0000	.0000	.0000	.0003	.0021	.0100	.0349	.0937	.2024

n	x	.05	.10	.15	.20	p .25	.30	.35	.40	.45	.50
	15	.0000	.0000	.0000	.0000	.0001	.0005	.0030	.0128	.0411	.1050
	16	.0000	.0000	.0000	.0000	.0000	.0001	.0008	.0040	.0153	.0466
	17	.0000	.0000	.0000	.0000	.0000	.0002	.0010	.0048	.0173	
	18	.0000	.0000	.0000	.0000	.0000	.0000	.0000	.0002	.0012	.0053
	19	.0000	.0000	.0000	.0000	.0000	.0000	.0000	.0000	.0002	.0013
	20	.0000	.0000	.0000	.0000	.0000	.0000	.0000	.0000	.0000	.0002
	21	.0000	.0000	.0000	.0000	.0000	.0000	.0000	.0000	.0000	.0000
	22	.0000	.0000	.0000	.0000	.0000	.0000	.0000	.0000	.0000	.0000
	23	.0000	.0000	.0000	.0000	.0000	.0000	.0000	.0000	.0000	.0000
24	1	.7080	.9202	.9798	.9953	.9990	.9998	1.0000	1.0000	1.0000	1.0000
	2	.3392	.7075	.8941	.9669	.9910	.9978	.9995	.9999	1.0000	1.0000
	3	.1159	.4357	.7202	.8855	.9602	.9881	.9970	.9993	.9999	1.0000
	4	.0298	.2143	.4951	.7361	.8850	.9576	.9867	.9965	.9992	.9999
	5	.0060	.0851	.2866	.5401	.7534	.8889	.9578	.9866	.9964	.9992
	6	.0010	.0277	.1394	.3441	.5778	.7712	.8956	.9600	.9873	.9967
	7	.0001	.0075	.0572	.1889	.3926	.6114	.7894	.9040	.9636	.9887
	8	.0000	.0017	.0199	.0892	.2338	.4353	.6425	.8081	.9137	.9680
	9	.0000	.0003	.0059	.0362	.1213	.2750	.4743	.6721	.8270	.9242
	10	.0000	.0001	.0015	.0126	.0547	.1528	.3134	.5109	.7009	.8463
	11	.0000	.0000	.0003	.0038	.0213	.0742	.1833	.3498	.5461	.7294
	12	.0000	.0000	.0001	.0010	.0072	.0314	.0942	.2130	.3849	.5806
	13	.0000	.0000	.0000	.0002	.0021	.0115	.0423	.1143	.2420	.4194
	14	.0000	.0000	.0000	.0000	.0005	.0036	.0164	.0535	.1341	.2706
	15	.0000	.0000	.0000	.0000	.0001	.0010	.0055	.0217	.0648	.1537
	16	.0000	.0000	.0000	.0000	.0000	.0002	.0016	.0075	.0269	.0758
	17	.0000	.0000	.0000	.0000	.0000	.0000	.0004	.0022	.0095	.0320
	18	.0000	.0000	.0000	.0000	.0000	.0000	.0001	.0005	.0028	.0113
	19	.0000	.0000	.0000	.0000	.0000	.0000	.0000	.0001	.0007	.0033
	20	.0000	.0000	.0000	.0000	.0000	.0000	.0000	.0000	.0001	.0008
	21	.0000	.0000	.0000	.0000	.0000	.0000	.0000	.0000	.0000	.0001
	22	.0000	.0000	.0000	.0000	.0000	.0000	.0000	.0000	.0000	.0000
	23	.0000	.0000	.0000	.0000	.0000	.0000	.0000	.0000	.0000	.0000
	24	.0000	.0000	.0000	.0000	.0000	.0000	.0000	.0000	.0000	.0000
25	1	.7226	.9282	.9828	.9962	.9992	.9999	1.0000	1.0000	1.0000	1.0000
	2	.3576	.7288	.9069	.9726	.9930	.9984	.9997	.9999	1.0000	1.0000
	3	.1271	.4629	.7463	.9018	.9679	.9910	.9979	.9996	.9999	1.0000
	4	.0341	.2364	.5289	.7660	.9038	.9668	.9903	.9976	.9995	.9999
	5	.0072	.0980	.3179	.5793	.7863	.9095	.9680	.9905	.9977	.9995
	6	.0012	.0334	.1615	.3833	.6217	.8065	.9174	.9706	.9914	.9980
	7	.0002	.0095	.0695	.2200	.4389	.6593	.8266	.9264	.9742	.9927
	8	.0000	.0023	.0255	.1091	.2735	.4882	.6939	.8464	.9361	.9784
	9	.0000	.0005	.0080	.0468	.1494	.3231	.5332	.7265	.8660	.9461
	10	.0000	.0001	.0021	.0173	.0713	.1894	.3697	.5754	.7576	.8852
	11	.0000	.0000	.0005	.0056	.0297	.0978	.2288	.4142	.6157	.7878
	12	.0000	.0000	.0001	.0015	.0107	.0442	.1254	.2677	.4574	.6550
	13	.0000	.0000	.0000	.0004	.0034	.0175	.0604	.1538	.3063	.5000
	14	.0000	.0000	.0000	.0001	.0009	.0060	.0255	.0778	.1827	.3450
	15	.0000	.0000	.0000	.0000	.0002	.0018	.0093	.0344	.0960	.2122
	16	.0000	.0000	.0000	.0000	.0000	.0005	.0029	.0132	.0440	.1148
	17	.0000	.0000	.0000	.0000	.0000	.0001	.0008	.0043	.0174	.0539
	18	.0000	.0000	.0000	.0000	.0000	.0000	.0002	.0012	.0058	.0216
	19	.0000	.0000	.0000	.0000	.0000	.0000	.0000	.0003	.0016	.0073
	20	.0000	.0000	.0000	.0000	.0000	.0000	.0000	.0001	.0004	.0020
	21	.0000	.0000	.0000	.0000	.0000	.0000	.0000	.0000	.0001	.0005
	22	.0000	.0000	.0000	.0000	.0000	.0000	.0000	.0000	.0000	.0001
	23	.0000	.0000	.0000	.0000	.0000	.0000	.0000	.0000	.0000	.0000
	24	.0000	.0000	.0000	.0000	.0000	.0000	.0000	.0000	.0000	.0000
	25	.0000	.0000	.0000	.0000	.0000	.0000	.0000	.0000	.0000	.0000

n	x	.05	.10	.15	.20	p .25	.30	.35	.40	.45	.50
30	1	.7854	.9576	.9924	.9988	.9998	1.0000	1.0000	1.0000	1.0000	1.0000
	2	.4465	.8163	.9520	.9895	.9980	.9997	1.0000	1.0000	1.0000	1.0000
	3	.1878	.5886	.8486	.9558	.9894	.9979	.9997	1.0000	1.0000	1.0000
	4	.0608	.3526	.6783	.8773	.9626	.9907	.9981	.9997	1.0000	1.0000
	5	.0156	.1755	.4755	.7448	.9021	.9698	.9925	.9985	.9998	1.0000
	6	.0033	.0732	.2894	.5725	.7974	.9234	.9767	.9943	.9989	.9998
	7	.0006	.0258	.1526	.3930	.6519	.8405	.9414	.9828	.9960	.9993
	8	.0001	.0078	.0698	.2392	.4857	.7186	.8762	.9565	.9879	.9974
	9	.0000	.0020	.0278	.1287	.3264	.5685	.7753	.9060	.9688	.9919
	10	.0000	.0005	.0097	.0611	.1966	.4112	.6425	.8237	.9306	.9786
	11	.0000	.0001	.0029	.0256	.1057	.2696	.4922	.7085	.8650	.9506
	12	.0000	.0000	.0008	.0095	.0507	.1593	.3452	.5689	.7673	.8998
	13	.0000	.0000	.0002	.0031	.0216	.0845	.2198	.4215	.6408	.8192
	14	.0000	.0000	.0000	.0009	.0082	.0401	.1263	.2855	.4975	.7077
	15	.0000	.0000	.0000	.0002	.0027	.0169	.0652	.1754	.3552	.5722
	16	.0000	.0000	.0000	.0001	.0008	.0064	.0301	.0971	.2309	.4278
	17	.0000	.0000	.0000	.0000	.0002	.0021	.0124	.0481	.1356	.2923
	18	.0000	.0000	.0000	.0000	.0001	.0006	.0045	.0212	.0714	.1808
	19	.0000	.0000	.0000	.0000	.0000	.0002	.0014	.0083	.0334	.1002
	20	.0000	.0000	.0000	.0000	.0000	.0000	.0004	.0029	.0138	.0494
	21	.0000	.0000	.0000	.0000	.0000	.0000	.0001	.0009	.0050	.0214
	22	.0000	.0000	.0000	.0000	.0000	.0000	.0000	.0002	.0016	.0081
	23	.0000	.0000	.0000	.0000	.0000	.0000	.0000	.0000	.0004	.0026
	24	.0000	.0000	.0000	.0000	.0000	.0000	.0000	.0000	.0001	.0007
	25	.0000	.0000	.0000	.0000	.0000	.0000	.0000	.0000	.0000	.0002
	26	.0000	.0000	.0000	.0000	.0000	.0000	.0000	.0000	.0000	.0000
	27	.0000	.0000	.0000	.0000	.0000	.0000	.0000	.0000	.0000	.0000
	28	.0000	.0000	.0000	.0000	.0000	.0000	.0000	.0000	.0000	.0000
	29	.0000	.0000	.0000	.0000	.0000	.0000	.0000	.0000	.0000	.0000
	30	.0000	.0000	.0000	.0000	.0000	.0000	.0000	.0000	.0000	.0000
35	1	.8339	.9750	.9966	.9996	1.0000	1.0000	1.0000	1.0000	1.0000	1.0000
	2	.5280	.8776	.9757	.9960	.9995	.9999	1.0000	1.0000	1.0000	1.0000
	3	.2542	.6937	.9130	.9810	.9967	.9995	.9999	1.0000	1.0000	1.0000
	4	.0958	.4690	.7912	.9395	.9864	.9976	.9997	1.0000	1.0000	1.0000
	5	.0290	.2693	.6193	.8565	.9590	.9909	.9984	.9998	1.0000	1.0000
	6	.0073	.1316	.4311	.7279	.9024	.9731	.9942	.9990	.9999	1.0000
	7	.0015	.0552	.2652	.5672	.8089	.9350	.9830	.9966	.9995	.9999
	8	.0003	.0200	.1438	.4007	.6777	.8674	.9581	.9898	.9981	.9997
	9	.0000	.0063	.0689	.2550	.5257	.7659	.9110	.9740	.9943	.9991
	10	.0000	.0017	.0292	.1457	.3737	.6354	.8349	.9425	.9848	.9970
	11	.0000	.0004	.0110	.0747	.2419	.4900	.7284	.8877	.9646	.9917
	12	.0000	.0001	.0037	.0344	.1421	.3484	.5981	.8048	.9271	.9795
	13	.0000	.0000	.0011	.0142	.0756	.2271	.4577	.6943	.8656	.9552
	14	.0000	.0000	.0003	.0053	.0363	.1350	.3240	.5639	.7767	.9123
	15	.0000	.0000	.0001	.0018	.0158	.0731	.2109	.4272	.6624	.8447
	16	.0000	.0000	.0000	.0005	.0062	.0359	.1256	.2997	.5315	.7502
	17	.0000	.0000	.0000	.0001	.0022	.0160	.0682	.1935	.3976	.6321
	18	.0000	.0000	.0000	.0000	.0007	.0064	.0336	.1143	.2751	.5000
	19	.0000	.0000	.0000	.0000	.0002	.0023	.0150	.0615	.1749	.3679
	20	.0000	.0000	.0000	.0000	.0001	.0008	.0061	.0300	.1016	.2498
	21	.0000	.0000	.0000	.0000	.0000	.0002	.0022	.0133	.0536	.1553
	22	.0000	.0000	.0000	.0000	.0000	.0001	.0007	.0053	.0255	.0877
	23	.0000	.0000	.0000	.0000	.0000	.0000	.0002	.0019	.0109	.0448
	24	.0000	.0000	.0000	.0000	.0000	.0000	.0001	.0006	.0042	.0205
	25	.0000	.0000	.0000	.0000	.0000	.0000	.0000	.0002	.0014	.0083
	26	.0000	.0000	.0000	.0000	.0000	.0000	.0000	.0000	.0004	.0030
	27	.0000	.0000	.0000	.0000	.0000	.0000	.0000	.0000	.0001	.0009
	28	.0000	.0000	.0000	.0000	.0000	.0000	.0000	.0000	.0000	.0003

n	x	.05	.10	.15	.20	p .25	.30	.35	.40	.45	.50
	29	.0000	.0000	.0000	.0000	.0000	.0000	.0000	.0000	.0000	.0001
	30	.0000	.0000	.0000	.0000	.0000	.0000	.0000	.0000	.0000	.0000
	⋮										
			... remaining terms = .0000 ...								
	35										
40	1	.8715	.9852	.9985	.9999	1.0000	1.0000	1.0000	1.0000	1.0000	1.0000
	2	.6009	.9195	.9879	.9985	.9999	1.0000	1.0000	1.0000	1.0000	1.0000
	3	.3233	.7772	.9514	.9921	.9990	.9999	1.0000	1.0000	1.0000	1.0000
	4	.1381	.5769	.8698	.9715	.9953	.9994	.9999	1.0000	1.0000	1.0000
	5	.0480	.3710	.7367	.9241	.9840	.9974	.9997	1.0000	1.0000	1.0000
	6	.0139	.2063	.5675	.8387	.9567	.9914	.9987	.9999	1.0000	1.0000
	7	.0034	.0995	.3933	.7141	.9038	.9762	.9956	.9994	.9999	1.0000
	8	.0007	.0419	.2441	.5629	.8180	.9447	.9876	.9979	.9998	1.0000
	9	.0001	.0155	.1354	.4069	.7002	.8890	.9697	.9939	.9991	.9999
	10	.0000	.0051	.0672	.2682	.5605	.8041	.9356	.9844	.9973	.9997
	11	.0000	.0015	.0299	.1608	.4161	.6913	.8785	.9648	.9926	.9989
	12	.0000	.0004	.0120	.0875	.2849	.5594	.7947	.9291	.9821	.9968
	13	.0000	.0001	.0043	.0432	.1791	.4228	.6857	.8715	.9614	.9917
	14	.0000	.0000	.0014	.0194	.1032	.2968	.5592	.7888	.9249	.9808
	15	.0000	.0000	.0004	.0079	.0544	.1926	.4279	.6826	.8674	.9597
	16	.0000	.0000	.0001	.0029	.0262	.1151	.3054	.5598	.7858	.9231
	17	.0000	.0000	.0000	.0010	.0116	.0633	.2022	.4319	.6815	.8659
	18	.0000	.0000	.0000	.0003	.0047	.0320	.1239	.3115	.5609	.7852
	19	.0000	.0000	.0000	.0001	.0017	.0148	.0699	.2089	.4349	.6821
	20	.0000	.0000	.0000	.0000	.0006	.0063	.0363	.1298	.3156	.5627
	21	.0000	.0000	.0000	.0000	.0002	.0024	.0173	.0744	.2130	.4373
	22	.0000	.0000	.0000	.0000	.0000	.0009	.0075	.0392	.1331	.3179
	23	.0000	.0000	.0000	.0000	.0000	.0003	.0030	.0189	.0767	.2148
	24	.0000	.0000	.0000	.0000	.0000	.0001	.0011	.0083	.0405	.1341
	25	.0000	.0000	.0000	.0000	.0000	.0000	.0004	.0034	.0196	.0769
	26	.0000	.0000	.0000	.0000	.0000	.0000	.0001	.0012	.0086	.0403
	27	.0000	.0000	.0000	.0000	.0000	.0000	.0000	.0004	.0034	.0192
	28	.0000	.0000	.0000	.0000	.0000	.0000	.0000	.0001	.0012	.0083
	29	.0000	.0000	.0000	.0000	.0000	.0000	.0000	.0000	.0004	.0032
	30	.0000	.0000	.0000	.0000	.0000	.0000	.0000	.0000	.0001	.0011
	31	.0000	.0000	.0000	.0000	.0000	.0000	.0000	.0000	.0000	.0003
	32	.0000	.0000	.0000	.0000	.0000	.0000	.0000	.0000	.0000	.0001
	33	.0000	.0000	.0000	.0000	.0000	.0000	.0000	.0000	.0000	.0000
	⋮										
				... remaining terms = .0000 ...							
	40										
45	1	.9006	.9913	.9993	1.0000	1.0000	1.0000	1.0000	1.0000	1.0000	1.0000
	2	.6650	.9476	.9940	.9995	1.0000	1.0000	1.0000	1.0000	1.0000	1.0000
	3	.3923	.8410	.9735	.9968	.9997	1.0000	1.0000	1.0000	1.0000	1.0000
	4	.1866	.6711	.9215	.9871	.9984	.9999	1.0000	1.0000	1.0000	1.0000
	5	.0729	.4729	.8252	.9618	.9941	.9993	.9999	1.0000	1.0000	1.0000
	6	.0239	.2923	.6858	.9098	.9821	.9974	.9997	1.0000	1.0000	1.0000
	7	.0066	.1585	.5218	.8232	.9554	.9920	.9990	.9999	1.0000	1.0000
	8	.0016	.0757	.3606	.7025	.9059	.9791	.9967	.9996	1.0000	1.0000
	9	.0003	.0320	.2255	.5593	.8275	.9529	.9909	.9988	.9999	1.0000
	10	.0001	.0120	.1274	.4120	.7200	.9066	.9780	.9964	.9996	1.0000
	11	.0000	.0040	.0651	.2795	.5911	.8353	.9531	.9906	.9987	.9999
	12	.0000	.0012	.0302	.1741	.4543	.7380	.9104	.9784	.9964	.9996
	13	.0000	.0003	.0127	.0995	.3252	.6198	.8453	.9554	.9910	.9988
	14	.0000	.0001	.0048	.0521	.2159	.4912	.7563	.9164	.9799	.9967
	15	.0000	.0000	.0017	.0250	.1327	.3653	.6467	.8570	.9591	.9920
	16	.0000	.0000	.0005	.0110	.0753	.2538	.5248	.7751	.9238	.9822
	17	.0000	.0000	.0002	.0044	.0395	.1642	.4017	.6728	.8698	.9638
	18	.0000	.0000	.0000	.0017	.0191	.0986	.2887	.5564	.7944	.9324
	19	.0000	.0000	.0000	.0006	.0085	.0549	.1940	.4357	.6985	.8837

| | | | | | p | | | | | |
n	x	.05	.10	.15	.20	.25	.30	.35	.40	.45	.50
	20	.0000	.0000	.0000	.0002	.0035	.0283	.1215	.3214	.5869	.8144
	21	.0000	.0000	.0000	.0001	.0013	.0135	.0708	.2223	.4682	.7243
	22	.0000	.0000	.0000	.0000	.0005	.0060	.0382	.1436	.3526	.6170
	23	.0000	.0000	.0000	.0000	.0001	.0024	.0191	.0865	.2494	.5000
	24	.0000	.0000	.0000	.0000	.0000	.0009	.0089	.0483	.1650	.3830
	25	.0000	.0000	.0000	.0000	.0000	.0003	.0038	.0250	.1017	.2757
	26	.0000	.0000	.0000	.0000	.0000	.0001	.0015	.0120	.0582	.1856
	27	.0000	.0000	.0000	.0000	.0000	.0000	.0005	.0053	.0308	.1163
	28	.0000	.0000	.0000	.0000	.0000	.0000	.0002	.0021	.0150	.0676
	29	.0000	.0000	.0000	.0000	.0000	.0000	.0001	.0008	.0068	.0362
	30	.0000	.0000	.0000	.0000	.0000	.0000	.0000	.0003	.0028	.0178
	31	.0000	.0000	.0000	.0000	.0000	.0000	.0000	.0001	.0010	.0080
	32	.0000	.0000	.0000	.0000	.0000	.0000	.0000	.0000	.0004	.0033
	33	.0000	.0000	.0000	.0000	.0000	.0000	.0000	.0000	.0001	.0012
	34	.0000	.0000	.0000	.0000	.0000	.0000	.0000	.0000	.0000	.0004
	35	.0000	.0000	.0000	.0000	.0000	.0000	.0000	.0000	.0000	.0001
	36	.0000	.0000	.0000	.0000	.0000	.0000	.0000	.0000	.0000	.0000
	⋮										
		...remaining terms = .0000 ...									
	45										
50	1	.9231	.9948	.9997	1.0000	1.0000	1.0000	1.0000	1.0000	1.0000	1.0000
	2	.7206	.9662	.9971	.9998	1.0000	1.0000	1.0000	1.0000	1.0000	1.0000
	3	.4595	.8883	.9858	.9987	.9999	1.0000	1.0000	1.0000	1.0000	1.0000
	4	.2396	.7497	.9540	.9943	.9995	1.0000	1.0000	1.0000	1.0000	1.0000
	5	.1036	.5688	.8879	.9815	.9979	.9998	1.0000	1.0000	1.0000	1.0000
	6	.0378	.3839	.7806	.9520	.9930	.9993	.9999	1.0000	1.0000	1.0000
	7	.0118	.2298	.6387	.8966	.9806	.9975	.9998	1.0000	1.0000	1.0000
	8	.0032	.1221	.4812	.8096	.9547	.9927	.9992	.9999	1.0000	1.0000
	9	.0008	.0579	.3319	.6927	.9084	.9817	.9975	.9998	1.0000	1.0000
	10	.0002	.0245	.2089	.5563	.8363	.9598	.9933	.9992	.9999	1.0000
	11	.0000	.0094	.1199	.4164	.7378	.9211	.9840	.9978	.9998	1.0000
	12	.0000	.0032	.0628	.2893	.6184	.8610	.9658	.9943	.9994	1.0000
	13	.0000	.0010	.0301	.1861	.4890	.7771	.9339	.9867	.9982	.9998
	14	.0000	.0003	.0132	.1106	.3630	.6721	.8837	.9720	.9955	.9995
	15	.0000	.0001	.0053	.0607	.2519	.5532	.8122	.9460	.9896	.9987
	16	.0000	.0000	.0019	.0308	.1631	.4308	.7199	.9045	.9780	.9967
	17	.0000	.0000	.0007	.0144	.0983	.3161	.6111	.8439	.9573	.9923
	18	.0000	.0000	.0002	.0063	.0551	.2178	.4940	.7631	.9235	.9836
	19	.0000	.0000	.0001	.0025	.0287	.1406	.3784	.6644	.8727	.9675
	20	.0000	.0000	.0000	.0009	.0139	.0848	.2736	.5535	.8026	.9405
	21	.0000	.0000	.0000	.0003	.0063	.0478	.1861	.4390	.7138	.8987
	22	.0000	.0000	.0000	.0001	.0026	.0251	.1187	.3299	.6100	.8389
	23	.0000	.0000	.0000	.0000	.0010	.0123	.0710	.2340	.4981	.7601
	24	.0000	.0000	.0000	.0000	.0004	.0056	.0396	.1562	.3866	.6641
	25	.0000	.0000	.0000	.0000	.0001	.0024	.0207	.0978	.2840	.5561
	26	.0000	.0000	.0000	.0000	.0000	.0009	.0100	.0573	.1966	.4439
	27	.0000	.0000	.0000	.0000	.0000	.0003	.0045	.0314	.1279	.3359
	28	.0000	.0000	.0000	.0000	.0000	.0001	.0019	.0160	.0780	.2399
	29	.0000	.0000	.0000	.0000	.0000	.0000	.0007	.0076	.0444	.1611
	30	.0000	.0000	.0000	.0000	.0000	.0000	.0003	.0034	.0235	.1013
	31	.0000	.0000	.0000	.0000	.0000	.0000	.0001	.0014	.0116	.0595
	32	.0000	.0000	.0000	.0000	.0000	.0000	.0000	.0005	.0053	.0325
	33	.0000	.0000	.0000	.0000	.0000	.0000	.0000	.0002	.0022	.0164
	34	.0000	.0000	.0000	.0000	.0000	.0000	.0000	.0001	.0009	.0077
	35	.0000	.0000	.0000	.0000	.0000	.0000	.0000	.0000	.0003	.0033
	36	.0000	.0000	.0000	.0000	.0000	.0000	.0000	.0000	.0001	.0013
	37	.0000	.0000	.0000	.0000	.0000	.0000	.0000	.0000	.0000	.0005
	38	.0000	.0000	.0000	.0000	.0000	.0000	.0000	.0000	.0000	.0002
	39	.0000	.0000	.0000	.0000	.0000	.0000	.0000	.0000	.0000	.0000
	⋮										
		...remaining terms = .0000 ...									
	50										

Table C: Percentage Points of the χ^2 Distribution

Values of χ^2 in Terms of Upper Tail Probability and Degrees of Freedom ν

The table gives values $\chi^2_{\nu,p}$ such that a χ^2 variable with ν degrees of freedom exceeds $\chi^2_{\nu,p}$ with probability p, i.e., $P[\chi^2 > \chi^2_{\nu,p}] = p$.

Source: Reprinted by permission from *Biometrika Tables for Statisticians, Volume I* by E.S. Pearson and H.O. Hartley. Copyright 1954 Cambridge University Press, Cambridge, England.

$\nu \setminus p$	0.1	0.05	0.025	0.01	0.005	0.001	0.0005	0.0001
1	2.7055	3.8415	5.0239	6.6349	7.8794	10.828	12.116	15.137
2	4.6052	5.9915	7.3778	9.2103	10.5966	13.816	15.202	18.421
3	6.2514	7.8147	9.3484	11.3449	12.8381	16.266	17.730	21.108
4	7.7794	9.4877	11.1433	13.2767	14.8602	18.467	19.997	23.513
5	9.2364	11.0705	12.8325	15.0863	16.7496	20.515	22.105	25.745
6	10.6446	12.5916	14.4494	16.8119	18.5476	22.458	24.103	27.856
7	12.0170	14.0671	16.0128	18.4753	20.2777	24.322	26.018	29.877
8	13.3616	15.5073	17.5346	20.0902	21.9550	26.125	27.868	31.828
9	14.6837	16.9190	19.0228	21.6660	23.5893	27.877	29.666	33.720
10	15.9871	18.3070	20.4831	23.2093	25.1882	29.588	31.420	35.564
11	17.2750	19.6751	21.9200	24.7250	26.7569	31.264	33.137	37.367
12	18.5494	21.0261	23.3367	26.2170	28.2995	32.909	34.821	39.134
13	19.8119	22.3621	24.7356	27.6883	29.8194	34.528	36.478	40.871
14	21.0642	23.6848	26.1190	29.1413	31.3193	36.123	38.109	42.579
15	22.3072	24.9958	27.4884	30.5779	32.8013	37.697	39.719	44.263
16	23.5418	26.2962	28.8454	31.9999	34.2672	39.252	41.308	45.925
17	24.7690	27.5871	30.1910	33.4087	35.7185	40.790	42.879	47.566
18	25.9894	28.8693	31.5264	34.8053	37.1564	42.312	44.434	49.189
19	27.2036	30.1435	32.8523	36.1908	38.5822	43.820	45.973	50.796
20	28.4120	31.4104	34.1696	37.5662	39.9968	45.315	47.498	52.386
21	29.6151	32.6705	35.4789	38.9321	41.4010	46.797	49.011	53.962
22	30.8133	33.9244	36.7807	40.2894	42.7956	48.268	50.511	55.525
23	32.0069	35.1725	38.0757	41.6384	44.1813	49.728	52.000	57.075
24	33.1963	36.4151	39.3641	42.9798	45.5585	51.179	53.479	58.613
25	34.3816	37.6525	40.6465	44.3141	46.9278	52.620	54.947	60.140
26	35.5631	38.8852	41.9232	45.6417	48.2899	54.052	56.407	61.657
27	36.7412	40.1133	43.1944	46.9630	49.6449	55.476	57.858	63.164
28	37.9159	41.3372	44.4607	48.2782	50.9932	56.892	59.300	64.662
29	39.0875	42.5569	45.7222	49.5879	52.3356	58.302	60.735	66.152
30	40.2560	43.7729	46.9792	50.8922	53.6720	59.703	62.162	67.633
40	51.8050	55.7585	59.3417	63.6907	66.7659	73.402	76.095	82.062
50	63.1671	67.5048	71.4202	76.1539	79.4900	86.661	89.560	95.969
60	74.3970	79.0819	83.2976	88.3794	91.9517	99.607	102.695	109.503
70	85.5271	90.5312	95.0231	100.425	104.215	112.317	115.578	122.755
80	96.5782	101.879	106.629	112.329	116.321	124.839	128.261	135.783
90	107.565	113.145	118.136	124.116	128.299	137.208	140.782	148.627
100	118.498	124.342	129.561	135.807	140.169	149.449	153.167	161.319

Table D: Critical Values of the Durbin-Watson Statistic

Let r_i denote the residuals from a linear model with k explanatory factors (excluding the constant term). The Durbin-Watson statistic is

$$d = \sum_{i=2}^{n}(r_i - r_{i-1})^2 \Big/ \sum_{i=1}^{n} r_i^2.$$

To test the null hypothesis of zero serial correlation against an alternative of positive serial correlation, reject H_0 at one-sided level $\alpha = .05$ if $d < d_L$ given in the table. If $d > d_U$, conclude that d is not significant and do not reject H_0. If $d_L < d < d_U$, the test is deemed inconclusive.

To test H_0 against an alternative of negative serial correlation, use the statistic $(4 - d)$ in place of d.

For a two-tailed test of H_0 against serial correlation of any sign, reject H_0 at level $\alpha = .10$ if either $d < d_L$ or $4 - d < d_L$. If $d > d_U$ and $4 - d > d_U$, conclude d is not significant. Otherwise the test is deemed inconclusive.

Source:

J. Durbin and G.S. Watson, *Testing for serial correlation in least squares regression II*, 38 Biometrika 159–178 (1951).

Table D: Critical Values of the Durbin-Watson Statistic

n	$k=1$		$k=2$		$k=3$		$k=4$		$k=5$	
	d_L	d_U	d_L	d_U	d_L	d_U	d_L	d_U	d_L	d_U
15	1.08	1.36	0.95	1.54	0.82	1.75	0.69	1.97	0.56	2.21
16	1.10	1.37	0.98	1.54	0.86	1.73	0.74	1.93	0.62	2.15
17	1.13	1.38	1.02	1.54	0.90	1.71	0.78	1.90	0.67	2.10
18	1.16	1.39	1.05	1.53	0.93	1.69	0.82	1.87	0.71	2.06
19	1.18	1.40	1.08	1.53	0.97	1.68	0.86	1.85	0.75	2.02
20	1.20	1.41	1.10	1.54	1.00	1.68	0.90	1.83	0.79	1.99
21	1.22	1.42	1.13	1.54	1.03	1.67	0.93	1.81	0.83	1.96
22	1.24	1.43	1.15	1.54	1.05	1.66	0.96	1.80	0.86	1.94
23	1.26	1.44	1.17	1.54	1.08	1.66	0.99	1.79	0.90	1.92
24	1.27	1.45	1.19	1.55	1.10	1.66	1.01	1.78	0.93	1.90
25	1.29	1.45	1.21	1.55	1.12	1.66	1.04	1.77	0.95	1.89
26	1.30	1.46	1.22	1.55	1.14	1.65	1.06	1.76	0.98	1.88
27	1.32	1.47	1.24	1.56	1.16	1.65	1.08	1.76	1.01	1.86
28	1.33	1.48	1.26	1.56	1.18	1.65	1.10	1.75	1.03	1.85
29	1.34	1.48	1.27	1.56	1.20	1.65	1.12	1.74	1.05	1.84
30	1.35	1.49	1.28	1.57	1.21	1.65	1.14	1.74	1.07	1.83
31	1.36	1.50	1.30	1.57	1.23	1.65	1.16	1.74	1.09	1.83
32	1.37	1.50	1.31	1.57	1.24	1.65	1.18	1.73	1.11	1.82
33	1.38	1.51	1.32	1.58	1.26	1.65	1.19	1.73	1.13	1.81
34	1.39	1.51	1.33	1.58	1 27	1.65	1.21	1.73	1.15	1.81
35	1.40	1.52	1.34	1.58	1.28	1.65	1.22	1 73	1.16	1.80
36	1.41	1.52	1.35	1.59	1.29	1.65	1.24	1.73	1.18	1.80
37	1.42	1.53	1.36	1.59	1.31	1.66	1.25	1.72	1.19	1.80
38	1.43	1.54	1.37	1.59	1.32	1.66	1.26	1.72	1.21	1.79
39	1.43	1.54	1.38	1.60	1.33	1.66	1.27	1.72	1.22	1.79
40	1.44	1.54	1.39	1.60	1.34	1.66	1.29	1.72	1.23	1.79
45	1.48	1.57	1.43	1.62	1.38	1.67	1.34	1.72	1.29	1.78
50	1.50	1.59	1.46	1.63	1.42	1.67	1.38	1.72	1.34	1.77
55	1.53	1.60	1.49	1.64	1.45	1.68	1.41	1.72	1.38	1.77
60	1.55	1.62	1.51	1.65	1.48	1.69	1.44	1.73	1.41	1.77
65	1.57	1.63	1.54	1.66	1.50	1.70	1.47	1.73	1.44	1.77
70	1.58	1.64	1.55	1.67	1.52	1.70	1.49	1.74	1.46	1.77
75	1.60	1.65	1.57	1.68	1.54	1.71	1.51	1.74	1.49	1.77
80	1.61	1.66	1.59	1.69	1.56	1.72	1.53	1.74	1.51	1.77
85	1.62	1.67	1.60	1.70	1.57	1.72	1.55	1.75	1.52	1.77
90	1.63	1.68	1.61	1.70	1.59	1.73	1.57	1.75	1.54	1.78
95	1.64	1.69	1.62	1.71	1.60	1.73	1.58	1.75	1.56	1.78
100	1.65	1.69	1.63	1.72	1.61	1.74	1.59	1.76	1.57	1.78

Table E: Percentage Points of the t Distribution

The table gives values $t_{\nu,p}$ such that a Student t variable with ν degrees of freedom exceeds $t_{\nu,p}$ in absolute value with probability p, i.e., $P[|t| > t_{\nu,p}] = p$. The tail area probability in a single tail equals $p/2$.

Source:

Reprinted by permission from *Biometrika Tables for Statisticians, Volume I* by E.S. Pearson and H.O. Hartley. Copyright 1954 Cambridge University Press, Cambridge, England.

$\nu \setminus p$	0.800	0.500	0.200	0.100	0.050	0.020	0.010	0.005	0.002	0.001
1	0.325	1.000	3.078	6.314	12.706	31.821	63.657	127.321	318.309	636.619
2	0.289	0.816	1.886	2.920	4.303	6.965	9.925	14.089	22.327	31.598
3	0.277	0.765	1.638	2.353	3.182	4.541	5.841	7.453	10.214	12.924
4	0.271	0.741	1.533	2.132	2.776	3.747	4.604	5.598	7.173	8.610
5	0.267	0.727	1.476	2.015	2.571	3.365	4.032	4.773	5.893	6.869
6	0.265	0.718	1.440	1.943	2.447	3.143	3.707	4.317	5.208	5.959
7	0.263	0.711	1.415	1.895	2.365	2.998	3.499	4.029	4.785	5.408
8	0.262	0.706	1.397	1.860	2.306	2.896	3.355	3.833	4.501	5.041
9	0.261	0.703	1.383	1.833	2.262	2.821	3.250	3.690	4.297	4.781
10	0.260	0.700	1.372	1.812	2.228	2.764	3.169	3.581	4.144	4.587
11	0.260	0.697	1.363	1.796	2.201	2.718	3.106	3.497	4.025	4.437
12	0.259	0.695	1.356	1.782	2.179	2.681	3.055	3.428	3.930	4.318
13	0.259	0.694	1.350	1.771	2.160	2.650	3.012	3.372	3.852	4.221
14	0.258	0.692	1.345	1.761	2.145	2.624	2.977	3.326	3.787	4.140
15	0.258	0.691	1.341	1.753	2.131	2.602	2.947	3.286	3.733	4.073
16	0.258	0.690	1.337	1.746	2.120	2.583	2.921	3.252	3.686	4.015
17	0.257	0.689	1.333	1.740	2.110	2.567	2.898	3.223	3.646	3.965
18	0.257	0.688	1.330	1.734	2.101	2.552	2.878	3.197	3.610	3.922
19	0.257	0.688	1.328	1.729	2.093	2.539	2.861	3.174	3.579	3.883
20	0.257	0.687	1.325	1.725	2.086	2.528	2.845	3.153	3.552	3.850
21	0.257	0.686	1.323	1.721	2.080	2.518	2.831	3.135	3.527	3.819
22	0.256	0.686	1.321	1.717	2.074	2.508	2.819	3.119	3.505	3.792
23	0.256	0.685	1.319	1.714	2.069	2.500	2.807	3.104	3.485	3.768
24	0.256	0.685	1.318	1.711	2.064	2.492	2.797	3.090	3.467	3.745
25	0.256	0.684	1.316	1.708	2.060	2.485	2.787	3.078	3.450	3.725
26	0.256	0.684	1.315	1.706	2.056	2.479	2.779	3.067	3.435	3.707
27	0.256	0.684	1.314	1.703	2.052	2.473	2.771	3.057	3.421	3.690
28	0.256	0.683	1.313	1.701	2.048	2.467	2.763	3.047	3.408	3.674
29	0.256	0.683	1.311	1.699	2.045	2.462	2.756	3.038	3.396	3.659
30	0.256	0.683	1.310	1.697	2.042	2.457	2.750	3.030	3.385	3.646
40	0.255	0.681	1.303	1.684	2.021	2.423	2.704	2.971	3.307	3.551
60	0.254	0.679	1.296	1.671	2.000	2.390	2.660	2.915	3.232	3.460
120	0.254	0.677	1.289	1.658	1.980	2.358	2.617	2.860	3.160	3.373
∞	0.253	0.674	1.282	1.645	1.960	2.326	2.576	2.807	3.090	3.291

Table F: Percentage Points of the F Distribution—$P[F > F_{\nu_1,\nu_2;.05}] = .05$

The tables give values $F_{\nu_1,\nu_2;p}$ such that an F variable with ν_1 and ν_2 degrees of freedom exceeds $F_{\nu_1,\nu_2;p}$ with probability p, i.e., $P[F > F_{\nu_1,\nu_2;p}] = p$. The first table gives critical values for $p = 0.05$, the next table gives critical values for $p = 0.01$.

Source:

Reprinted by permission from *Biometrika Tables for Statisticians, Volume I* by E.S. Pearson and H.O. Hartley. Copyright 1954 Cambridge University Press, Cambridge, England.

$\nu_2 \setminus \nu_1$	1	2	3	4	5	6	8	12	15	20	30	60	∞
1	161.4	199.5	215.7	224.6	230.2	234.0	238.9	243.9	245.9	248.0	250.1	252.2	254.3
2	18.51	19.00	19.16	19.25	19.30	19.33	19.37	19.41	19.43	19.45	19.46	19.48	19.50
3	10.13	9.55	9.28	9.12	9.01	8.94	8.85	8.74	8.70	8.66	8.62	8.57	8.53
4	7.71	6.94	6.59	6.39	6.26	6.16	6.04	5.91	5.86	5.80	5.75	5.69	5.63
5	6.61	5.79	5.41	5.19	5.05	4.95	4.82	4.68	4.62	4.56	4.50	4.43	4.36
6	5.99	5.14	4.76	4.53	4.39	4.28	4.15	4.00	3.94	3.87	3.81	3.74	3.67
7	5.59	4.74	4.35	4.12	3.97	3.87	3.73	3.57	3.51	3.44	3.38	3.30	3.23
8	5.32	4.46	4.07	3.84	3.69	3.58	3.44	3.28	3.22	3.15	3.08	3.01	2.93
9	5.12	4.26	3.86	3.63	3.48	3.37	3.23	3.07	3.01	2.94	2.86	2.79	2.71
10	4.96	4.10	3.71	3.48	3.33	3.22	3.07	2.91	2.85	2.77	2.70	2.62	2.54
11	4.84	3.98	3.59	3.36	3.20	3.09	2.95	2.79	2.72	2.65	2.57	2.49	2.40
12	4.75	3.89	3.49	3.26	3.11	3.00	2.85	2.69	2.62	2.54	2.47	2.38	2.30
13	4.67	3.81	3.41	3.18	3.03	2.92	2.77	2.60	2.53	2.46	2.38	2.30	2.21
14	4.60	3.74	3.34	3.11	2.96	2.85	2.70	2.53	2.46	2.39	2.31	2.22	2.13
15	4.54	3.68	3.29	3.06	2.90	2.79	2.64	2.48	2.40	2.33	2.25	2.16	2.07
16	4.49	3.63	3.24	3.01	2.85	2.74	2.59	2.42	2.35	2.28	2.19	2.11	2.01
17	4.45	3.59	3.20	2.96	2.81	2.70	2.55	2.38	2.31	2.23	2.15	2.06	1.96
18	4.41	3.55	3.16	2.93	2.77	2.66	2.51	2.34	2.27	2.19	2.11	2.02	1.92
19	4.38	3.52	3.13	2.90	2.74	2.63	2.48	2.31	2.23	2.16	2.07	1.98	1.88
20	4.35	3.49	3.10	2.87	2.71	2.60	2.45	2.28	2.20	2.12	2.04	1.95	1.84
21	4.32	3.47	3.07	2.84	2.68	2.57	2.42	2.25	2.18	2.10	2.01	1.92	1.81
22	4.30	3.44	3.05	2.82	2.66	2.55	2.40	2.23	2.15	2.07	1.98	1.89	1.78
23	4.28	3.42	3.03	2.80	2.64	2.53	2.37	2.20	2.13	2.05	1.96	1.86	1.76
24	4.26	3.40	3.01	2.78	2.62	2.51	2.36	2.18	2.11	2.03	1.94	1.84	1.73
25	4.24	3.39	2.99	2.76	2.60	2.49	2.34	2.16	2.09	2.01	1.92	1.82	1.71
26	4.23	3.37	2.98	2.74	2.59	2.47	2.32	2.15	2.07	1.99	1.90	1.80	1.69
27	4.21	3.35	2.96	2.73	2.57	2.46	2.31	2.13	2.06	1.97	1.88	1.79	1.67
28	4.20	3.34	2.95	2.71	2.56	2.45	2.29	2.12	2.04	1.96	1.87	1.77	1.65
29	4.18	3.33	2.93	2.70	2.55	2.43	2.28	2.10	2.03	1.94	1.85	1.75	1.64
30	4.17	3.32	2.92	2.69	2.53	2.42	2.27	2.09	2.01	1.93	1.84	1.74	1.62
40	4.08	3.23	2.84	2.61	2.45	2.34	2.18	2.00	1.92	1.84	1.74	1.64	1.51
60	4.00	3.15	2.76	2.53	2.37	2.25	2.10	1.92	1.84	1.75	1.65	1.53	1.39
120	3.92	3.07	2.68	2.45	2.29	2.17	2.02	1.83	1.75	1.66	1.55	1.43	1.25
∞	3.84	3.00	2.60	2.37	2.21	2.10	1.94	1.75	1.67	1.57	1.46	1.32	1.00

numerator df

Table F: Percentage Points of the F Distribution—$P[F > F_{\nu_1, \nu_2; .01}] = .01$

<table>
<tr><td colspan="14" align="center">numerator df</td></tr>
<tr><th>$\nu_2 \setminus \nu_1$</th><th>1</th><th>2</th><th>3</th><th>4</th><th>5</th><th>6</th><th>8</th><th>12</th><th>15</th><th>20</th><th>30</th><th>60</th><th>∞</th></tr>
<tr><td>1</td><td>4052.</td><td>4999.5</td><td>5403.</td><td>5625.</td><td>5764.</td><td>5859.</td><td>5982.</td><td>6106.</td><td>6157.</td><td>6209.</td><td>6261.</td><td>6313.</td><td>6366.</td></tr>
<tr><td>2</td><td>98.50</td><td>99.00</td><td>99.17</td><td>99.25</td><td>99.30</td><td>99.33</td><td>99.37</td><td>99.42</td><td>99.43</td><td>99.45</td><td>99.47</td><td>99.48</td><td>99.50</td></tr>
<tr><td>3</td><td>34.12</td><td>30.82</td><td>29.46</td><td>28.71</td><td>28.24</td><td>27.91</td><td>27.49</td><td>27.05</td><td>26.87</td><td>26.69</td><td>26.50</td><td>26.32</td><td>26.13</td></tr>
<tr><td>4</td><td>21.20</td><td>18.00</td><td>16.69</td><td>15.98</td><td>15.52</td><td>15.21</td><td>14.80</td><td>14.37</td><td>14.20</td><td>14.02</td><td>13.84</td><td>13.65</td><td>13.46</td></tr>
<tr><td>5</td><td>16.26</td><td>13.27</td><td>12.06</td><td>11.39</td><td>10.97</td><td>10.67</td><td>10.29</td><td>9.89</td><td>9.72</td><td>9.55</td><td>9.38</td><td>9.20</td><td>9.02</td></tr>
<tr><td>6</td><td>13.75</td><td>10.92</td><td>9.78</td><td>9.15</td><td>8.75</td><td>8.47</td><td>8.10</td><td>7.72</td><td>7.56</td><td>7.40</td><td>7.23</td><td>7.06</td><td>6.88</td></tr>
<tr><td>7</td><td>12.25</td><td>9.55</td><td>8.45</td><td>7.85</td><td>7.46</td><td>7.19</td><td>6.84</td><td>6.47</td><td>6.31</td><td>6.16</td><td>5.99</td><td>5.82</td><td>5.65</td></tr>
<tr><td>8</td><td>11.26</td><td>8.65</td><td>7.59</td><td>7.01</td><td>6.63</td><td>6.37</td><td>6.03</td><td>5.67</td><td>5.52</td><td>5.36</td><td>5.20</td><td>5.03</td><td>4.86</td></tr>
<tr><td>9</td><td>10.56</td><td>8.02</td><td>6.99</td><td>6.42</td><td>6.06</td><td>5.80</td><td>5.47</td><td>5.11</td><td>4.96</td><td>4.81</td><td>4.65</td><td>4.48</td><td>4.31</td></tr>
<tr><td>10</td><td>10.04</td><td>7.56</td><td>6.55</td><td>5.99</td><td>5.64</td><td>5.39</td><td>5.06</td><td>4.71</td><td>4.56</td><td>4.41</td><td>4.25</td><td>4.08</td><td>3.91</td></tr>
<tr><td>11</td><td>9.65</td><td>7.21</td><td>6.22</td><td>5.67</td><td>5.32</td><td>5.07</td><td>4.74</td><td>4.40</td><td>4.25</td><td>4.10</td><td>3.94</td><td>3.78</td><td>3.60</td></tr>
<tr><td>12</td><td>9.33</td><td>6.93</td><td>5.95</td><td>5.41</td><td>5.06</td><td>4.82</td><td>4.50</td><td>4.16</td><td>4.01</td><td>3.86</td><td>3.70</td><td>3.54</td><td>3.36</td></tr>
<tr><td>13</td><td>9.07</td><td>6.70</td><td>5.74</td><td>5.21</td><td>4.86</td><td>4.62</td><td>4.30</td><td>3.96</td><td>3.82</td><td>3.66</td><td>3.51</td><td>3.34</td><td>3.17</td></tr>
<tr><td>14</td><td>8.86</td><td>6.51</td><td>5.56</td><td>5.04</td><td>4.69</td><td>4.46</td><td>4.14</td><td>3.80</td><td>3.66</td><td>3.51</td><td>3.35</td><td>3.18</td><td>3.00</td></tr>
<tr><td>15</td><td>8.68</td><td>6.36</td><td>5.42</td><td>4.89</td><td>4.56</td><td>4.32</td><td>4.00</td><td>3.67</td><td>3.52</td><td>3.37</td><td>3.21</td><td>3.05</td><td>2.87</td></tr>
<tr><td>16</td><td>8.53</td><td>6.23</td><td>5.29</td><td>4.77</td><td>4.44</td><td>4.20</td><td>3.89</td><td>3.55</td><td>3.41</td><td>3.26</td><td>3.10</td><td>2.93</td><td>2.75</td></tr>
<tr><td>17</td><td>8.40</td><td>6.11</td><td>5.18</td><td>4.67</td><td>4.34</td><td>4.10</td><td>3.79</td><td>3.46</td><td>3.31</td><td>3.16</td><td>3.00</td><td>2.83</td><td>2.65</td></tr>
<tr><td>18</td><td>8.29</td><td>6.01</td><td>5.09</td><td>4.58</td><td>4.25</td><td>4.01</td><td>3.71</td><td>3.37</td><td>3.23</td><td>3.08</td><td>2.92</td><td>2.75</td><td>2.57</td></tr>
<tr><td>19</td><td>8.18</td><td>5.93</td><td>5.01</td><td>4.50</td><td>4.17</td><td>3.94</td><td>3.63</td><td>3.30</td><td>3.15</td><td>3.00</td><td>2.84</td><td>2.67</td><td>2.49</td></tr>
<tr><td>20</td><td>8.10</td><td>5.85</td><td>4.94</td><td>4.43</td><td>4.10</td><td>3.87</td><td>3.56</td><td>3.23</td><td>3.09</td><td>2.94</td><td>2.78</td><td>2.61</td><td>2.42</td></tr>
<tr><td>21</td><td>8.02</td><td>5.78</td><td>4.87</td><td>4.37</td><td>4.04</td><td>3.81</td><td>3.51</td><td>3.17</td><td>3.03</td><td>2.88</td><td>2.72</td><td>2.55</td><td>2.36</td></tr>
<tr><td>22</td><td>7.95</td><td>5.72</td><td>4.82</td><td>4.31</td><td>3.99</td><td>3.76</td><td>3.45</td><td>3.12</td><td>2.98</td><td>2.83</td><td>2.67</td><td>2.50</td><td>2.31</td></tr>
<tr><td>23</td><td>7.88</td><td>5.66</td><td>4.76</td><td>4.26</td><td>3.94</td><td>3.71</td><td>3.41</td><td>3.07</td><td>2.93</td><td>2.78</td><td>2.62</td><td>2.45</td><td>2.26</td></tr>
<tr><td>24</td><td>7.82</td><td>5.61</td><td>4.72</td><td>4.22</td><td>3.90</td><td>3.67</td><td>3.36</td><td>3.03</td><td>2.89</td><td>2.74</td><td>2.58</td><td>2.40</td><td>2.21</td></tr>
<tr><td>25</td><td>7.77</td><td>5.57</td><td>4.68</td><td>4.18</td><td>3.85</td><td>3.63</td><td>3.32</td><td>2.99</td><td>2.85</td><td>2.70</td><td>2.54</td><td>2.36</td><td>2.17</td></tr>
<tr><td>26</td><td>7.72</td><td>5.53</td><td>4.64</td><td>4.14</td><td>3.82</td><td>3.59</td><td>3.29</td><td>2.96</td><td>2.81</td><td>2.66</td><td>2.50</td><td>2.33</td><td>2.13</td></tr>
<tr><td>27</td><td>7.68</td><td>5.49</td><td>4.60</td><td>4.11</td><td>3.78</td><td>3.56</td><td>3.26</td><td>2.93</td><td>2.78</td><td>2.63</td><td>2.47</td><td>2.29</td><td>2.10</td></tr>
<tr><td>28</td><td>7.64</td><td>5.45</td><td>4.57</td><td>4.07</td><td>3.75</td><td>3.53</td><td>3.23</td><td>2.90</td><td>2.75</td><td>2.60</td><td>2.44</td><td>2.26</td><td>2.06</td></tr>
<tr><td>29</td><td>7.60</td><td>5.42</td><td>4.54</td><td>4.04</td><td>3.73</td><td>3.50</td><td>3.20</td><td>2.87</td><td>2.73</td><td>2.57</td><td>2.41</td><td>2.23</td><td>2.03</td></tr>
<tr><td>30</td><td>7.56</td><td>5.39</td><td>4.51</td><td>4.02</td><td>3.70</td><td>3.47</td><td>3.17</td><td>2.84</td><td>2.70</td><td>2.55</td><td>2.39</td><td>2.21</td><td>2.01</td></tr>
<tr><td>40</td><td>7.31</td><td>5.18</td><td>4.31</td><td>3.83</td><td>3.51</td><td>3.29</td><td>2.99</td><td>2.66</td><td>2.52</td><td>2.37</td><td>2.20</td><td>2.02</td><td>1.80</td></tr>
<tr><td>60</td><td>7.08</td><td>4.98</td><td>4.13</td><td>3.65</td><td>3.34</td><td>3.12</td><td>2.82</td><td>2.50</td><td>2.35</td><td>2.20</td><td>2.03</td><td>1.84</td><td>1.60</td></tr>
<tr><td>120</td><td>6.85</td><td>4.79</td><td>3.95</td><td>3.48</td><td>3.17</td><td>2.96</td><td>2.66</td><td>2.34</td><td>2.19</td><td>2.03</td><td>1.86</td><td>1.66</td><td>1.38</td></tr>
<tr><td>∞</td><td>6.63</td><td>4.61</td><td>3.78</td><td>3.32</td><td>3.02</td><td>2.80</td><td>2.51</td><td>2.18</td><td>2.04</td><td>1.88</td><td>1.70</td><td>1.47</td><td>1.00</td></tr>
</table>

Table G1:   Critical Values of the Kolmogorov-Smirnov One-Sample Test

A sample of size n is drawn from a population with cumulative distribution function $F(x)$. Let $x_{(1)}, \ldots, x_{(n)}$ denote the sample values arranged in ascending order. Define the empirical distribution function $F_n(x)$ to be the step function $F_n(x) = k/n$ for $x_{(i)} \leq x < x_{(i+1)}$, where k is the number of observations not greater than x. Under the null hypothesis that the sample has been drawn from the specified distribution, $F_n(x)$ should be fairly close to $F(x)$. Define

$$D_n = \max |F_n(x) - F(x)|.$$

For a two-tailed test this table gives critical values of the sampling distribution of D_n under the null hypothesis. Reject the hypothetical distribution if D_n exceeds the tabulated value. If n is over 35, determine the critical values of D_n by the divisions indicated at the end of the table. For $\alpha = .01$ and $.05$ the asymptotic formulas give values which are slightly high—by 1.5% for $n = 80$.

A one-tailed test is provided by the statistic $D^+ = \max[F_n(x) - F(x)]$.

Source:

Adapted with permission from *Handbook of Tables for Probability and Statistics,* Table X.7 (Chemical Rubber Co., 2d ed., 1968).

Table G1:   Critical Values of the Kolmogorov-Smirnov One-Sample Test

Sample size	Significance Level				
n	.20	.15	.10	.05	.01
1	.900	.925	.950	.975	.995
2	.684	.726	.776	.842	.929
3	.565	.597	.642	.708	.829
4	.494	.525	.564	.624	.734
5	.446	.474	.510	.563	.669
6	.410	.436	.470	.521	.618
7	.381	.405	.438	.486	.577
8	.358	.381	.411	.457	.543
9	.339	.360	.388	.432	.514
10	.322	.342	.368	.409	.486
11	.307	.326	.352	.391	.468
12	.295	.313	.338	.375	.450
13	.284	.302	.325	.361	.433
14	.274	.292	.314	.349	.418
15	.266	.283	.304	.338	.404
16	.258	.274	.295	.328	.391
17	.250	.266	.286	.318	.380
18	.244	.259	.278	.309	.370
19	.237	.252	.272	.301	.361
20	.231	.246	.264	.294	.352
25	.21	.22	.24	.264	.32
30	.19	.20	.22	.242	.29
35	.18	.19	.21	.23	.27
40				.21	.25
50				.19	.23
60				.17	.21
70				.16	.19
80				.15	.18
90				.14	
100				.14	
Asymptotic Formula	$1.07/\sqrt{n}$	$1.14/\sqrt{n}$	$1.22/\sqrt{n}$	$1.36/\sqrt{n}$	$1.63/\sqrt{n}$

Table G2:   Critical Values of the Kolmogorov-Smirnov Two-Sample Test

A sample of size n_1 is drawn from a population with cumulative distribution function $F(x)$. Define the empirical distribution function $F_{n_1}(x)$ to be the step function $F_{n_1}(x) = k/n_1$, where k is the number of observations not greater than x. A second, independent sample of size n_2 is drawn with empirical distribution function $F_{n_2}(x)$. Under the null hypothesis that the samples have been drawn from the same distribution, the statistic

$$D_{n_1,n_2} = \max |F_{n_1}(x) - F_{n_2}(x)|$$

should not be too large. For a two-tailed test, this table gives critical values of the sampling distribution of D_{n_1,n_2} under the null hypothesis.

Reject the null hypothesis if D_{n_1,n_2} exceeds the tabulated value.

For large values of n_1 and n_2, determine the critical values by the approximate formulas indicated at the end of the table.

A one-tailed test is available using $D_{n_1,n_2}^+ = \max[F_{n_1}(x) - F_{n_2}(x)]$.

Source:

Adapted with permission from *Handbook of Tables for Probability and Statistics,* Table X.8 (Chemical Rubber Co., 2d ed., 1968).

Table G2:   Critical Values of the Kolmogrov-Smirnov Two-Sample Test

Sample size n_1

Sample size n_2	2	3	4	5	6	7	8	9	10	12	15
2	*	*	*	*	*	*	7/8	16/18	9/10		
	*	*	*	*	*	*	*	*	*		
3		*	*	12/15	5/6	18/21	18/24	7/9		9/12	
		*	*	*	*	*	*	8/9		11/12	
4			3/4	16/20	9/12	21/28	6/8	27/36	14/20	8/12	
			*	*	10/12	24/28	7/8	32/36	16/20	10/12	
5				4/5	20/30	25/35	27/40	31/45	7/10		10/15
				4/5	25/30	30/35	32/40	36/45	8/10		11/15
6					4/6	29/42	16/24	12/18	19/30	7/12	
					5/6	35/42	18/24	14/18	22/30	9/12	
7						5/7	35/56	40/63	43/70		
						5/7	42/56	47/63	53/70		
8							5/8	45/72	23/40	14/24	
							6/8	54/72	28/40	16/24	
9								5/9	52/90	20/36	
								6/9	62/90	24/36	
10									6/10		15/30
									7/10		19/30
12										6/12	30/60
										7/12	35/60
15											7/15
											8/15

Note 1: The upper value gives a significance level at most .05; the lower value gives a significance level at most .01.

Note 2: Where * appears, do not reject H_0 at the given level.

Note 3: For large values of n_1 and n_2, use the approximate critical value

$$c_p \cdot (n_1^{-1} + n_2^{-1})^{1/2},$$

where the multiplier c_p is given by

p	.10	.05	.025	.01	.005	.001
c_p	1.22	1.36	1.48	1.63	1.73	1.95

Table H1:   Critical Values of the Wilcoxon Signed Ranks Test

Let $d_i = x_i - y_i$ denote the difference between scores in the ith of n matched pairs of observations. Rank all the d_i without regard to sign, giving rank 1 to the smallest, rank 2 to the next smallest, etc. Then affix the sign of difference d_i to each corresponding rank. Let W denote the smaller sum of like-signed ranks. This table gives 1%, 2%, 5%, and 10% two-sided critical values W_α for selected values of n. Reject the null hypothesis of no difference between distributions of x and y for values of $W \leq W_\alpha$. For a one-sided test of level $\alpha/2$, reject H_0 only if $W \leq W_\alpha$ and W corresponds to a sum of ranks of proper sign.

Source:

Adapted with permission from *Handbook of Tables for Probability and Statistics,* Table X.2 (Chemical Rubber Co., 2d ed., 1968).

Table H1:   Critical Values of the Wilcoxon Signed Ranks Test

Two-sided α (One-sided α) n	.10 (.05)	.05 (.025)	.02 (.01)	.01 (.005)
5	1			
6	2	1		
7	4	2	0	
8	6	4	2	0
9	8	6	3	2
10	11	8	5	3
11	14	11	7	5
12	17	14	10	7
13	21	17	13	10
14	26	21	16	13
15	30	25	20	16
16	36	30	24	19
17	41	35	28	23
18	47	40	33	28
19	54	46	38	32
20	60	52	43	37
21	68	59	49	43
22	75	66	56	49
23	83	73	62	55
24	92	81	69	61
25	101	90	77	68
26	110	98	85	76
27	120	107	93	84
28	130	117	102	92
29	141	127	111	100
30	152	137	120	109
31	163	148	130	118
32	175	159	141	128
33	188	171	151	138
34	201	183	162	149
35	214	195	174	160
36	228	208	186	171
37	242	222	198	183
38	256	235	211	195
39	271	250	224	208
40	287	264	238	221
41	303	279	252	234
42	319	295	267	248
43	336	311	281	262
44	353	327	297	277
45	371	344	313	292
46	389	361	329	307
47	408	379	345	323
48	427	397	362	339
49	446	415	380	356
50	466	434	398	373

Table H2:   Critical Values of the Wilcoxon Rank-Sum Test

Given two samples of size m and n, with $m \leq n$, let S be the sum of the ranks from $1, \ldots, (m+n)$ corresponding to the smaller sample. Large or small values of S lead to rejection of the null hypothesis that assignment of ranks between groups is at random.

The table provides upper and lower critical values, S_U and S_L. For a two-sided level $\alpha = .05$ test, reject H_0 if $S > S_U$ or $S < S_L$. Do not reject H_0 if $S_L \leq S \leq S_U$. One-sided tests with these limits have level $\alpha = .025$. Enter the table with m and $n - m \geq 0$.

The relationship between the Mann-Whitney U-statistic and S is $U = mn + \frac{1}{2}m(m+1) - S$.

Source:

Adapted with permission from *Handbook of Tables for Probability and Statistics*, Table X.5 (Chemical Rubber Co., 2d ed, 1968).

Lower .05 Critical Values S_L

$n - m$	3	4	5	6	7	8	9	10	11	12	13	14	15	16	17	18	19	20	21	22	23	24	25
0	5	11	18	26	37	49	63	79	96	116	137	160	185	212	240	271	303	337	373	411	451	493	536
1	6	12	19	28	39	51	66	82	100	120	141	165	190	217	246	277	310	345	381	419	460	502	546
2	6	12	20	29	41	54	68	85	103	124	146	170	195	223	252	284	317	352	389	428	468	511	555
3	7	13	21	31	43	56	71	88	107	128	150	174	201	229	258	290	324	359	397	436	477	520	565
4	7	14	22	32	45	58	74	91	110	131	154	179	206	234	264	297	331	367	404	444	486	529	574
5	8	15	24	34	46	61	77	94	114	135	159	184	211	240	271	303	338	374	412	452	494	538	584
6	8	16	25	36	48	63	79	97	118	139	163	189	216	245	277	310	345	381	420	460	503	547	593
7	9	17	26	37	50	65	82	101	121	143	168	194	221	251	283	316	351	389	428	469	512	556	603
8	10	17	27	39	52	68	85	104	125	147	172	198	227	257	289	323	358	396	436	477	520	565	612
9	10	18	29	41	54	70	88	107	128	151	176	203	232	262	295	329	365	403	443	485	529	575	622
10	11	19	30	42	56	72	90	110	132	155	181	208	237	268	301	336	372	411	451	493	538	584	632
11	11	20	31	44	58	75	93	113	135	159	185	213	242	274	307	342	379	418	459	502	546	593	641
12	12	21	32	45	60	77	96	117	139	163	190	218	248	279	313	349	386	426	467	510	555	602	651
13	12	22	33	47	62	80	99	120	143	167	194	222	253	285	319	355	393	433	475	518	564	611	660
14	13	23	35	49	64	82	101	123	146	171	198	227	258	291	325	362	400	440	482	526	572	620	670
15	13	24	36	50	66	84	104	126	150	175	203	232	263	296	331	368	407	448	490	535	581	629	679
16	14	24	37	52	68	87	107	129	153	179	207	237	269	302	338	375	414	455	498	543	590	638	689
17	14	25	38	53	70	89	110	132	157	183	212	242	274	308	344	381	421	463	506	551	599	648	699
18	15	26	40	55	72	92	113	136	161	187	216	247	279	314	350	388	428	470	514	560	607	657	708
19	15	27	41	57	74	94	115	139	164	191	221	252	284	319	356	395	435	477	522	568	616	666	718
20	16	28	42	58	76	96	118	142	168	195	225	256	290	325	362	401	442	485	530	576	625	675	727
21	16	29	43	60	78	99	121	145	171	199	229	261	295	331	368	408	449	492	537	584	633	684	737
22	17	30	45	61	80	101	124	148	175	203	234	266	300	336	374	414	456	500	545	593	642	693	747
23	17	31	46	63	82	103	127	152	179	207	238	271	306	342	380	421	463	507	553	601	651	703	756
24	18	31	47	65	84	106	129	155	182	211	243	276	311	348	387	427	470	515	561	609	660	712	766
25	18	32	48	66	86	108	132	158	186	216	247	281	316	353	393	434	477	522	569	618	668	721	775

Table H2: Critical Values of the Wilcoxon Rank-Sum Test

Upper .05 Critical Values S_U

$n - m$	3	4	5	6	7	8	9	10	11	12	13	14	15	16	17	18	19	20	21	22	23	24	25
0	16	25	37	52	68	87	108	131	157	184	214	246	280	316	355	395	438	483	530	579	630	683	739
1	18	28	41	56	73	93	114	138	164	192	223	255	290	327	366	407	450	495	543	593	644	698	754
2	21	32	45	61	78	98	121	145	172	200	231	264	300	337	377	418	462	508	556	606	659	713	770
3	23	35	49	65	83	104	127	152	179	208	240	274	309	347	388	430	474	521	569	620	673	728	785
4	26	38	53	70	88	110	133	159	187	217	249	283	319	358	399	441	486	533	583	634	687	743	801
5	28	41	56	74	94	115	139	166	194	225	257	292	329	368	409	453	498	546	596	648	702	758	816
6	31	44	60	78	99	121	146	173	201	233	266	301	339	379	420	464	510	559	609	662	716	773	832
7	33	47	64	83	104	127	152	179	209	241	274	310	349	389	431	476	523	571	622	675	730	788	847
8	35	51	68	87	109	132	158	186	216	249	283	320	358	399	442	487	535	584	635	689	745	803	863
9	38	54	71	91	114	138	164	193	224	257	292	329	368	410	453	499	547	597	649	703	759	817	878
10	40	57	75	96	119	144	171	200	231	265	300	338	378	420	464	510	559	609	662	717	773	832	893
11	43	60	79	100	124	149	177	207	239	273	309	347	388	430	475	522	571	622	675	730	788	847	909
12	45	63	83	105	129	155	183	213	246	281	317	356	397	441	486	533	583	634	688	744	802	862	924
13	48	66	87	109	134	160	189	220	253	289	326	366	407	451	497	545	595	647	701	758	816	877	940
14	50	69	90	113	139	166	196	227	261	297	335	375	417	461	508	556	607	660	715	772	831	892	955
15	53	72	94	118	144	172	202	234	268	305	343	384	427	472	519	568	619	672	728	785	845	907	971
16	55	76	98	122	149	177	208	241	276	313	352	393	436	482	529	579	631	685	741	799	859	922	986
17	58	79	102	127	154	183	214	248	283	321	360	402	446	492	540	591	643	697	754	813	873	936	1001
18	60	82	105	131	159	188	220	254	290	329	369	411	456	502	551	602	655	710	767	826	888	951	1017
19	63	85	109	135	164	194	227	261	298	337	377	420	466	513	562	613	667	723	780	840	902	966	1032
20	65	88	113	140	169	200	233	268	305	345	386	430	475	523	573	625	679	735	793	854	916	981	1048
21	68	91	117	144	174	205	239	275	313	353	395	439	485	533	584	636	691	748	807	868	931	996	1063
22	70	94	120	149	179	211	245	282	320	361	403	448	495	544	595	648	703	760	820	881	945	1011	1078
23	73	97	124	153	184	217	251	288	327	369	412	457	504	554	606	659	715	773	833	895	959	1025	1094
24	75	101	128	157	189	222	258	295	335	377	420	466	514	564	616	671	727	785	846	909	973	1040	1109
25	78	104	132	162	194	228	264	302	342	384	429	475	524	575	627	682	739	798	859	922	988	1055	1125

Glossary of Symbols

Unless otherwise noted, $\log(x)$ refers to the natural logarithm of x, equivalent to the notation $\ln(x)$. Common logarithms (to base 10) are written $\log_{10}(x)$. References are to sections of first substantive use.

α	4.4	df	5.1	s.e.	5.3
β	4.4	e^x	2.2	SS_b	7.2
β_i	8.2	$\exp(\cdot)$	2.2	SS_w	7.2
Γ	11.1	EX	1.2	$S(t)$	11.1
δ	7.2	$F(\cdot)$	7.2	t	7.1
Δ	13.9	G^2	5.6	$t_{v,\alpha}$	7.1
$\Delta \bar{r}$	14.3	H_0, H_1	4.4	U	3.4
Θ	5.6	$H(t)$	11.1	Var	1.3
$\hat{\Theta}$	5.6	$\ln$	4.8	W	12.2
$\theta(t)$	11.1	$\log$	5.6	$\bar{X}$	1.1
μ	1.1	min	4.5	X^2	1.4
ν	7.1	max	4.5	z	4.3
π	2.1	MS_b	7.2	$\binom{n}{a}$	2.2
Π	4.2	MS_w	7.2	$_nC_r$	2.1
ρ	1.1	NPV	3.4	$\approx$	2.1
σ	1.1	OLS	13.2	$!$	2.1
Σ	12.2	PPV	3.4	$\cap$	3.1
τ	6.3	$P[\cdot]$	1.1	$\sim$	4.2
ϕ	4.3	$P[\cdot|\cdot]$	3.1	$\int$	11.1
Φ	12.1	r	1.1		
χ^2	5.6	R^2	13.3		
Ω	4.5	$R^2_{Z(X)}$	14.3		
$\bar{A}, \bar{B}$	3.3	r_S	12.4		
Bin	4.2	s^2	1.1		
cdf	4.1	S^2_p	7.1		
Cov	1.4	s.d.	13.6		

List of Cases

Bibliography

The books listed below span a broad range in subject matter and mathematical difficulty. Under each subject heading, we have included authoritative, standard works that should be consulted when issues arise involving expert opinion in that subject. The lawyer should be forewarned: many of the works—apart from the elementary texts—are quite technical and fully accessible only to specialists. In reading them, guidance should be sought from an expert. We also have yielded to the temptation to include some specialized books of particular topical or historical interest. Needless to say, the lists are not complete, and on some subjects are merely introductions to a vast literature.

Elementary and Intermediate Statistical Texts

Cochran, William G., and George W. Snedecor *Statistical Methods*, 7th ed., Iowa State University Press, 1980.

Fisher, Ronald A., *Statistical Methods for Research Workers*, Oliver & Boyd, 1925 (14th ed., revised and enlarged, Hafner, 1973).

Freedman, David, Robert Pisani, and Roger Purves, *Statistics*, 3d ed., W.W. Norton, 1998.

Freund, John E., *Modern Elementary Statistics*, 7th ed., Prentice Hall, 1988.

Hodges, J.L., Jr. and Erich L. Lehmann, *Basic Concepts of Probability and Statistics*, 2d ed., Holden-Day, 1970.

Hoel, Paul G., *Elementary Statistics*, 4th ed., John Wiley, 1976.

Mendenhall, William, *Introduction to Probability and Statistics*, 6th ed., Duxbury Press, 1983.

Mosteller, Frederick, Robert E.K. Rourke, and George B. Thomas, Jr., *Probability with Statistical Applications*, 2d ed., Addison-Wesley, 1970.

Rice, John A., *Mathematical Statistics and Data Analysis*, 2d ed., Duxbury Press, 1995.

Wonnacott, Ronald J., and Thomas H. Wonnacott, *Introductory Statistics*, 4th ed., John Wiley, 1985.

Advanced Statistical Texts

Cramér, Harald, *Mathematical Methods of Statistics*, Princeton University Press, 1946.

Hays, William L., *Statistics*, 3d ed., Holt, Rhinehart & Winston, 1981.

Hoel, Paul G., *Introduction to Mathematical Statistics*, 5th ed., John Wiley, 1984.

Hogg, Robert V. and Allen T. Craig, *Introduction to Mathematical Statistics*, 5th ed., MacMillan, 1994.

Lehmann, Erich L., *Testing Statistical Hypotheses: And Theory of Point Estimation*, 2d ed., John Wiley, 1986.

Mendenhall, William, Richard L. Scheaffer, and Dennis Wackerly, *Mathematical Statistics with Applications*, 3d ed., Duxbury Press, 1986.

Neter, John and William Wasserman, *Applied Linear Statistical Models*, 2d ed., Richard D. Irwin, 1985.

Rao, C. Radhakrishna, *Linear Statistical Inference and Its Applications*, 2d ed., John Wiley, 1973.

Wilks, Samuel S., *Mathematical Statistics*, John Wiley, 1962.

Winer, B.J., *Statistical Principles in Experimental Design*, 2d ed., McGraw-Hill, 1971.

Bayesian Methods

Box, George E.P., and George C. Tiao, *Bayesian Inference in Statistical Analysis*, Addison-Wesley, 1973.

DeGroot, Morris H., *Optimal Statistical Decisions*, McGraw-Hill, 1970.

Good, Irving J., *The Estimation of Probabilities: An Essay on Modern Bayesian Methods*, MIT Press, 1965.

Lindley, Dennis V., *Introduction to Probability and Statistics from a Bayesian Viewpoint*, Part 1: Probability, Part 2: Inference, Cambridge University Press, 1980.

Mosteller, Frederick, and David L. Wallace, *Applied Bayesian and Classical Inference: The Case of the Federalist Papers*, 2d ed., Springer-Verlag, 1984.

Savage, I. Richard, *Statistics = Uncertainty and Behavior*, Houghton-Mifflin, 1968.

Biostatistics and Epidemiology

Armitage, Peter, *Sequential Medical Trials*, 2d ed., Oxford University Press, 1975.

Blalock, H.M., Jr., ed., *Causal Models in the Social Sciences*, 2d ed., Aldine de Gruyter, 1971.

Breslow, Norman E., and N.E. Day, *Statistical Methods in Cancer Research*, Oxford University Press, Vol. 1, Design and Analysis of Case Control Studies, 1980, Vol. 2, The Design and Analysis of Cohort Studies, 1988.

Cox, David R., and David O. Oakes, *Analysis of Survival Data*, Rutledge Chapman and Hall, 1984.

Fleiss, Joseph L., *Statistical Methods for Rates and Proportions*, 2d ed., John Wiley, 1981.

Fleiss, Joseph L., *Design and Analysis of Clinical Experiments*, John Wiley, 1986.

Friedman, Lawrence M., Curt D. Furberg, and David L. DeMets, *Fundamentals of Clinical Trials*, 3d ed., Mosby-Year Book, 1996.

Gross, Alan J., and Virginia Clark, *Survival Distributions: Reliability Applications in the Biomedical Sciences*, John Wiley, 1975.

Kalbfleisch, John D., and Ross L. Prentice, *The Statistical Analysis of Failure Time Data*, John Wiley, 1980.

Kelsey, Jennifer L., Douglas Thompson, and Alfred S. Evans, *Methods in Observational Epidemiology*, Oxford University Press, 1986.

Lawless, Jerald F., *Statistical Models and Methods for Lifetime Data*, John Wiley, 1982.

McMahon, Brian and Thomas F. Pugh, *Epidemiology—Principles and Methods*, Little & Brown, 1970.

Miller, Rupert G., Jr., Gail Gong, and Alvaro Muñoz, *Survival Analysis*, John Wiley, 1981.

Rosner, Bernard, *Fundamentals of Biostatistics*, 4th ed., Duxbury Press, 1995.

Rothman, Kenneth J. and Sander Greenland, *Modern Epidemiology*, 2d ed., Lippincott–Raven Publishers, 1998.

Schlesselman, James J., and Paul D. Stolley, *Case-Control Studies: Design, Conduct, Analysis*, Oxford University Press, 1982.

Susser, Mervyn, *Causal Thinking in the Health Sciences: Concepts and Strategies of Epidemiology*, Oxford University Press, 1973.

Econometrics and Time Series

Anderson, Theodore W., *The Statistical Analysis of Time Series*, John Wiley, 1971.

Box, George E.P., and Gwilym M. Jenkins, *Time Series Analysis: Forecasting and Control.* rev. ed., Holden-Day, 1976.

Cox, David R., and P.A.W. Lewis, *The Statistical Theory of Series of Events*, Rutledge Chapman and Hall, 1966.

Hannan, E.J., *Time Series Analysis*, Rutledge Chapman and Hall, 1967.

Johnston, J., *Econometric Methods*, McGraw-Hill, 1963.

Theil, Henri, *Principles of Econometrics*, John Wiley, 1971.

Wonnacott, Ronald J., and Thomas H. Wonnacott, *Econometrics*, 2d ed., John Wiley, 1979.

Encyclopedias

Kendall, Maurice G., and William R. Buckland, *A Dictionary of Statistical Terms*, 4th ed., John Wiley, 1986.

Kotz, Samuel, Norman L. Johnson, and Campbell B. Read, *Encyclopedia of Statistical Sciences*, Vols. 1–9 plus Supplement, John Wiley, 1982–1989.

Kruskal, William H., and Judith M. Tanur, eds., *International Encyclopedia of Statistics*, The Free Press, Vols. 1 and 2, 1978.

Exploratory Data Analysis

Mosteller, Frederick, Stephen E. Fienberg, and Robert E.K. Rourke, *Beginning Statistics with Data Analysis*, Addison-Wesley, 1983.

Mosteller, Frederick, and John W. Tukey, *Data Analysis and Regression: A Second Course in Statistics*, Addison-Wesley, 1977.

Tukey, John W., *Exploratory Data Analysis*, Addison-Wesley, 1977.

Foundations and Statistical Theory

Cox, David R., and Dennis V. Hinkley, *Theoretical Statistics*, Rutledge Chapman and Hall, 1974.

Fisher, Ronald A., *Statistical Methods and Scientific Inference*, Oliver & Boyd, 1956 (3d ed., revised and enlarged, Hafner, 1973).

Fraser, D.A.S., *The Structure of Inference*, John Wiley, 1968.

Good, Irving J., *Probability and the Weighing of Evidence*, Hafner, 1950.

Kendall, Maurice G., and Allen Stuart, *The Advanced Theory of Statistics*, 4th ed., Vol. I, Distribution Theory, 1977; Vol. II, Inference and Relationship, 1979; Vol. III, Design and Analysis, and Time Series, Hafner, 1983.

Kolmogorov, A.N., *Foundations of the Theory of Probability*, 2d ed., Chelsea, 1956.

Savage, Leonard Jimmie, *The Foundations of Statistics*, Dover, 1972.

Historical Studies

Adams, William J., *The Life and Times of the Central Limit Theorem*, Kaedmon, 1974.

Hacking, Ian, *The Emergence of Probability: A Philosophical Study of Early Ideas About Probability, Induction, and Statistical Inference*, Cambridge University Press, 1984.

Porter, Theodore M., *The Rise of Statistical Thinking 1820–1900*, Princeton University Press, 1986.

Stigler, Stephen M., *The History of Statistics*, Belknap Press of Harvard University Press, 1986.

Todhunter, I., *A History of the Mathematical Theory of Probability*, Macmillan, 1865 (reprinted, Chelsea, 1965).

Legal and Public Policy

Baldus, David, and J. Cole, *Statistical Proof of Discrimination*, Shepards–McGraw-Hill, 1980.

Barnes, David W. and John M. Conley, *Statistical Evidence in Litigation, Methodology, Procedure, and Practice*, Little, Brown, 1986.

DeGroot, Morris H., Stephen E. Fienberg, and Joseph B. Kadane, *Statistics and the Law*, John Wiley, 1987.

Fairley, William B., and Frederick Mosteller, eds., *Statistics and Public Policy*, Addison-Wesley, 1977.

Fienberg, Stephen E., ed., *The Evolving Role of Statistical Assessments as Evidence in the Courts*, Springer-Verlag, 1989.

Finkelstein, Michael O., *Quantitative Methods in Law: Studies in the Application of Mathematical Probability and Statistics to Legal Problems*, The Free Press, 1978.

Gastwirth, Joseph L., *Statistical Reasoning in Law and Public Policy*, Vol. 1, Statistical Concepts and Issues of Fairness; Vol. 2, Tort Law, Evidence, and Health, Academic Press, 1988.

Gastwirth, Joseph L., *Statistical Science in the Courtroom.* Springer-Verlag New York, 2000.

Monahan, John, and Laurens Walker, *Social Science in Law: Cases and Materials*, The Foundation Press, 1985.

Reference Manual on Scientific Evidence, 2d ed. Federal Judicial Center, 2000.

Zeisel, Hans and David Kaye, *Prove it with Figures: Empirical methods in law and litigation*, Springer-Verlag New York, 1997.

Multiple Regression

Belsley, D.A., E. Kuh, and Roy E. Welsch, *Regression Diagnostics: Identifying Influential Data and Sources of Collinearity*, John Wiley, 1980.

Chatterjee, Samprit, and Bertram Price, *Regression Analysis by Example*, John Wiley, 1977.

Cook, Robert and Sanford Weisberg, *Residuals and Inference in Regression*, Rutledge Chapman and Hall, 1982.

Draper, Norman R., and Harry Smith, *Applied Regression Analysis*, 2d ed., John Wiley, 1981.

Hosmer, David W. and Stanley Lemeshow, *Applied Logistic Regression*, 2d ed., John Wiley, 2000.

Kleinbaum, David G., Lawrence L. Kupper, and Keith E. Muller, *Applied Regression Analysis and Other Multivariable Methods*, 2d ed., Duxbury Press, 1978.

Multivariate Statistics

Anderson, T.W., *An Introduction to Multivariate Statistical Analysis*, 2d ed., John Wiley, 1984.

Bishop, Yvonne M.M., Stephen E. Feinberg, and Paul W. Holland, *Discrete Multivariate Analysis*, MIT Press, 1974.

Bock, R. Darrell, *Multivariate Statistical Methods in Behavioral Research*, 2d ed., Scientific Software, 1975.

Dempster, Arthur P., *Elements of Continuous Multivariate Analysis*, Addison-Wesley, 1969.

Morrison, Donald F., *Multivariate Statistical Methods*, 2d ed., McGraw-Hill, 1976.

Nonparametric Methods

Gibbons, Jean D., *Nonparametric Methods for Quantitative Analysis*, 2d ed., American Sciences Press, 1985.

Kendall, Maurice G., *Rank Correlation Methods*, 4th ed., Griffin, 1970.

Lehmann, Erich L., *Nonparametrics: Statistical Methods Based on Ranks*, Holden-Day, 1975.

Mosteller, Frederick, and Robert E.K. Rourke, *Sturdy Statistics: Nonparametric and Order Statistics*, Addison-Wesley, 1973.

Probability

Bartlett, M.S., *An Introduction to Stochastic Processes*, 3d ed., Cambridge University Press, 1981.

Breiman, Leo, *Probability and Stochastic Processes: With a View Toward Applications*, 2d ed., Scientific Press, 1986.

Cramér, Harald, *The Elements of Probability Theory and Some of Its Applications*, 2d ed., Krieger, 1973.

Feller, William, *An Introduction to Probability Theory and Its Applications*, John Wiley, Vol. 1, 3d ed., 1968; Vol. 2, 2d ed., 1971.

Jeffreys, Harold, *Theory of Probability*, 3d ed., Oxford University Press, 1983.

Keynes, John Maynard, *A Treatise on Probability*, Macmillan, 1921 (reprinted, American Mathematical Society Press, 1975).

Loève, M., *Probability Theory*, Vol. 1, 4th ed., 1976; Vol. 2, 1978. Springer-Verlag.

Ross, Sheldon M., *Stochastic Processes*, John Wiley, 1983.

Uspensky, J.V., *Introduction to Mathematical Probability*, McGraw-Hill, 1937.

Von Mises, Richard, *Probability, Statistics and Truth*, Allen & Unwin, 1928 (2d ed., 1961).

Sampling

Cochran, William G., *Sampling techniques*, 3d ed., John Wiley, 1977.

Deming, W. Edwards, *Sample Design in Business Research*, John Wiley, 1960.

Deming, W. Edwards, *Some Theory of Sampling*, John Wiley, 1950.

Roberts, Donald M., *Statistical Auditing*, Amer. Inst. of Certified Public Accountants, 1978.

Yates, Frank, *Sampling Methods for Censuses and Surveys*, 4th ed., Oxford University Press, 1987.

Other Specialized Texts

Blalock, Hubert M., Jr., *Social Statistics*, 2d ed., McGraw-Hill, 1979.

Cohen, Jacob, *Statistical Power Analysis*, 2d ed., L. Erlbaum Assocs., 1988.

Cox, David R., *Analysis of Binary Data*, Rutledge Chapman and Hall, 1970.

Efron, Bradley and Rob Tibshirani, *An Introduction to the Bootstrap*, Chapman & Hall, 1994.

Gelman, Andrew, John B. Carlin, Hal S. Stem, and Donald Rubin, *Bayesian Data Analysis*, CRC Press, 1995.

Goodman, Leo A., and William H. Kruskal, *Measures of Association for Cross Classifications*, Springer-Verlag, 1979.

Johnson, Norman L. and Samuel Kotz, *Distributions in Statistics*, Vol. 1, Discrete Distributions, 1971; Vol. 2, Continuous Univariate Distributions, 1971; Vol. 3, Continuous Multivariate Distributions, 1972, John Wiley.

Li, Ching Chun, *Path Analysis—A Primer*, The Boxwood Press, 1975.

Miller, Rupert G., Jr., *Simultaneous Statistical Inference*, 2d ed., Springer-Verlag, 1981.

Scheffé, Henry, *The Analysis of Variance,* John Wiley, 1959.

Wald, Abraham, *Sequential Analysis*, Dover, 1947.

Tables

Abramowitz, Milton, and Irene A. Stegun, *Handbook of Mathematical Functions with Formulas, Graphs, and Mathematical Tables*, 10th ed., U.S. Govt. Printing Office, 1972.

Beyer, William H., ed., *Handbook of Tables for Probability and Statistics*, 2d ed., Chemical Rubber Co. Press, 1968.

Lieberman, Gerald L., and Donald B. Owen, *Tables of the Hypergeometric Probability Distribution*, Stanford University Press, 1961.

Institute of Mathematical Statistics, ed., *Selected Tables in Mathematical Statistics*, Vols. 1–6, American Mathematical Society Press, 1973.

Index